D1556265

International Series on Astronomy and Astrophysics

Series Editors

International Series on Astronomy and Astrophysics

1. E.N. Parker, *Spontaneous Current Sheets in Magnetic Fields with Applications to Stellar X-rays*
2. C.F. Kennel, *Convection and Substorms: Paradigms of Magnetospheric Phenomenology*
3. L.F. Burlaga, *Interplanetary Magnetohydrodynamics*

SPONTANEOUS CURRENT SHEETS IN MAGNETIC FIELDS

With Applications to Stellar X-rays

EUGENE N. PARKER

Laboratory for Astrophysics and Space Research
University of Chicago

New York Oxford
OXFORD UNIVERSITY PRESS
1994

Oxford University Press

Oxford New York Toronto
Delhi Bombay Calcutta Madras Karachi
Kuala Lumpur Singapore Hong Kong Tokyo
Nairobi Dar es Salaam Cape Town
Melbourne Auckland Madrid

and associated companies in
Berlin Ibadan

Published by Oxford University Press, Inc.,
200 Madison Avenue, New York, New York 10016

Library of Congress Cataloging-in-Publication Data
Parker, E.N. (Eugene Newman), 1927–
Spontaneous current sheets in magnetic fields:
with applications to stellar x-rays / Eugene N. Parker
p. cm. — (International series on astronomy and astrophysics ; 2)
Includes bibliographical references and index.
ISBN 0-19-507371-1
1. Sun—Corona. 2. Solar x-rays. 3. Magnetic fields (Cosmic physics)
4. Magnetostatics. I. Title. II. Series.
QB529.P36 1994
523.7′2—dc20 93-33780

1 3 5 7 9 8 6 4 2

Printed in the United States of America
on acid-free paper

Preface

The magnetic heating of stellar coronas and galactic halos has led to the realization over the years that the electric currents associated with the magnetic fields are universally partially concentrated into widely separated thin sheets. For otherwise there is insufficient dissipation of magnetic energy to provide the observed heating. The problem has been to understand why the currents should be concentrated, rather than spread entirely smoothly over the field. Various special circumstances, e.g., the collision of two distinct magnetic lobes, have been conceived and described (over the last four decades) with an eye to understanding the flare phenomenon as the most intense magnetic heating of all. However, the less intense and more continuous heating of the X-ray corona of a star has proved more difficult, because the heating appears where observation shows only a continuous field. The continuous form of magnetic fields is taken for granted unless discontinuous fluid motions are specified.

The need for this monograph arises, at least in part, from the widespread habit of thinking that fields are generally continuous. The classical linear Maxwell equations with continuous sources have continuous solutions (excluding such supraluminous phenomena as Čerenkov radiation) and we all learned field theory in that context. But only those fields described by fully elliptic equations, e.g., Laplace's equation or the wave equation $\nabla^2\phi + k^2\phi = 0$, have exclusively continuous solutions. The fact is that the field equations of magnetostatics in an electrically conducting medium have the field lines as a set of real characteristics in addition to the two sets of complex characteristics of the elliptic equation. So one should expect surfaces of tangential discontinuity extending along the field lines unless there is some special circumstance that would provide an entirely continuous field. That is to say, we should expect the electric currents to be concentrated into thin sheets unless conditions conspire to distribute the currents more smoothly.

The situation is summarized by the basic theorem of magnetostatics that, in relaxing to magnetostatic equilibrium in an infinitely conducting fluid, almost all field topologies form internal surfaces of tangential discontinuity (current sheets). The formation of the tangential discontinuities is caused by the balance of the Maxwell stresses, and if the formation of a true mathematical discontinuity is frustrated by the nonvanishing resistivity of real fluids, there can be no complete static equilibrium. The Maxwell stresses drive fluid motions in their constant pursuit of discontinuity. This is, of course, the phenomenon commonly called rapid reconnection, or neutral point reconnection, of the magnetic field across the site of the potential tangential discontinuity.

This monograph is the extension of the theory of the universal suprathermal activity of magnetic fields in both laboratory and astronomical settings, initiated in *Cosmical Magnetic Fields*. Chapter 14 of that writing dealt with the dynamical nonequilibrium of magnetic fields lacking invariance of one form or another. The nonequilibrium arises because, as already noted, complete magnetostatic equilibrium of a magnetic field in a conducting fluid requires either a simple symmetric or invariant field topology or, lacking the necessary symmetry, it requires the formation of surfaces of tangential discontinuity (current sheets) within the magnetic field. Any slight resistivity in the fluid prevents the full achievement of the necessary mathematical discontinuity, of course, so the absence of static equilibrium, viz. dynamical nonequilibrium, is the result. The present writing approaches the problem from another direction, beginning with the idealized case in which resistivity is identically zero and the field has time to relax into a final asymptotic magnetostatic state. Interest centers on fields with general topologies, lacking the special, and apriori unlikely, topologies that provide a static field that is everywhere continuous.

The length of the monograph arises from the need to understand the basis for the theorem and to understand the implications of the theorem. So the writing "begins at the beginning," with a brief development of the magnetohydrodynamic equations as the proper description of the large-scale properties of a magnetic field in a noninsulating fluid or plasma. There is a curious popular notion to the contrary that has arisen in the past decade.

The basic theorem of magnetostatics then follows from the magnetohydrodynamic equations for static equilibrium. To see the theorem in perspective a number of examples are presented to contrast the special character of the fully continuous field. In particular, it is shown by example that the specification of a magnetostatic field on the boundaries of a region provides a unique determination of a continuous field throughout the region, exercising the elliptic aspect of the field equations when there are no discontinuities along the real characteristics. On the other hand, considering that the field is frozen into the ponderable conducting fluid, the topology of the field in a region could have been manipulated into most any internal form, with no impact on the normal component of the field on the boundaries, etc. The discontinuity along the field lines is the means by which the mathematics accommodates the arbitrary topology. Without it the field equations would contradict the physics, and that would have far-reaching implications indeed!

Then the geometry and topology of the surfaces of discontinuity need to be studied, at least in a preliminary fashion. The magnetic field lines, as characteristics of the magnetostatic equilibrium equations, play a prominent role in the development so that the optical analogy is an appropriate device for understanding the form of the static field.

Finally, we come to the specific application of the basic theorem to the corona of the Sun, suggesting the origin of the X-ray corona of a solitary star like the Sun, so that the necessary observational tests can be described. For it is the observations now of the motion of the footpoints of the magnetic field of active regions and of the detailed space and time behavior of the X-ray emission that must carry on from the formal theoretical principles. For the basic theorem of magnetostatics asserts in effect that the magnetic heating of the solar atmosphere depends only on the rate at which the swirling and intermixing of the photospheric footpoints of the field introduces magnetic free energy into the field. The spontaneous tangential discontinuities automatically take care of the dissipation of that free energy into heat.

The observations have not yet established the necessary swirling and intermixing of the photospheric footpoints of the bipolar magnetic field on the Sun. It is that continuing quasi-static deformation of the footpoints to which the necessary magnetic free energy is attributed. Without such free energy there is nothing to dissipate into coronal heat. The most interesting discovery of all would be the absence of the assumed mixing of the footpoints. In that instance the only available theoretical possibility would seem to be the dissipation of intense high frequency Alfven waves (with periods of the general order of 1 sec or less). Their origin would be mysterious indeed, requiring a wholly new and arbitrary dynamical state beneath the visible surface of the Sun. And if in the Sun, what then in other stars? It follows that the necessary studies of the small-scale dynamics of the photosphere should be undertaken with a full appreciation of the implications of the results, whatever those results might prove to be.

This is perhaps the appropriate place to note that a semantic difficulty has arisen in the past year or so, based on the application of dynamical terminology to magnetostatic phenomena. Specifically, some authors have presented elaborate theories of magnetohydrodynamic "turbulence," with which they propose to describe the small-scale structure of quasi-static magnetic fields in the corona of the Sun, referring to the formation of current sheets as the "cascade of magnetic energy to large wave number k." But the asymptotic relaxation of a magnetic field to static equilibrium is neither "turbulent" nor "cascading," and the use of such terms is a disservice to both the authors and the readers. It is a fact that the formation of a tangential discontinuity represents an extension of a tail on the Fourier spectrum to large wave numbers. But the extension is not a dynamical cascade in any sense. The only dynamical aspect is the *inhibiting* effect of the inertia of the fluid being ejected in the process of forming the discontinuities. Rather the declining thickness of the magnetic shear layers and the associated ejection of fluid jets is in response to the requirement for ultimate *static* balance of the Maxwell stresses.

Acknowledgments

It is a pleasure to acknowledge the many contributions of colleagues to the development of this monograph. First of all, there is the intellectual stimulus, provided by rational discussion and a general stirring of ideas on the one hand, and enduring disbelief on the other. All of these press the author to provide a clearer exposition. Perhaps the most helpful of all have been the five published papers of which I am aware "proving" the impossibility of the spontaneous development of tangential discontinuities, or current sheets, by continuous deformation of an initially continuous field. For such papers have opened up new dimensions to the theory, showing solutions to the field equations to be added to the existing repertoire and disposing of special situations that are not part of the general picture.

No monograph is complete without figures, and I wish to express my appreciation to Dr. Gerard Van Hoven, Dr. D.D. Schnack, and Dr. Z. Mikic for permission to publish two figures from their important numerical simulation of the rapid formation of current sheets during the continuous deformation of an initially uniform field. Dr. W.H. Matthaeus and Dr. David Montgomery gave permission to publish figures created from their numerical simulation of the fields, flows, and current sheets that develop in 2D magnetohydrodynamic turbulence. Dr. B.C. Low generously agreed to my using several figures from his work, and from his work in collaboration with Dr. Y.Q. Hu, showing the properties of tangential discontinuities that form when a simple poloidal field is deformed. My thanks also to Dr. H.R. Strauss and Dr. N.F. Otani for permission to use their figures showing the formation of discontinuities in the numerical simulation of the ballooning mode. These graphic simulations show more vividly than words and theorems the nature of the spontaneous formation of tangential discontinuities as the continuous deformation of a magnetic field progresses away from the necessary simple symmetry of a continuous magnetostatic field.

Dr. Leon Golub furnished the spectacular high resolution X-ray picture of the Sun that is the frontispiece of this monograph. The detail is essential in evaluating the nature of the coronal heating and has become possible through years of technical development of the interference coating to make the mirror of the Normal Incidence X-Ray Telescope (NIXT). The intrinsic resolution is better than one second of arc. The picture was one of many taken during the few minutes of observing available from a sounding rocket.

Finally, this monograph was created by the nimble fingers, sharp eyes, and agile mind of Ms. Valerie Smith, working from my handwritten manuscript with its many erasures, insertions, deletions, rewritings, and general smudging and illegibility by the time I was through with it. Ms. Smith learned TeX from reading the manual. She honed her skill through endless hours of struggling with manuscripts such as this one. Much the same way as I learned physics, and I hope with the same satisfaction in the accomplishment.

Contents

1 INTRODUCTION, 3

1.1 The General Picture, 3
1.2 Activity of Stars and Galaxies, 6
1.3 The Nature of Active Magnetic Fields, 7
1.4 Rapid Dissipation, 12
1.5 The Magnetostatic Theorem, 16
1.6 Continuous Magnetostatic Fields, 19
1.7 Perturbations of a Continuous Field, 20
1.8 Tangential Discontinuities, 23
1.9 The Optical Analogy, 25
1.10 General Discussion, 26
1.11 Hydrodynamic Turbulence, 27

2 THE FIELD EQUATIONS, 29

2.1 Appropriate Field Concepts, 29
2.1.1 Momentum and Energy of the Particles, 30
2.1.2 Electromagnetic Momentum and Energy, 33
2.2 The MHD Limit, 35
2.2.1 Statistical Averages, 36
2.2.2 Electromagnetics of a Noninsulating Fluid, 37
2.2.3 Electromagnetics of a Weakly Insulating Fluid, 38
2.2.4 Alternative Formulations, 40
2.3 The MHD Equations in a Tenuous Plasma, 41
2.4 Deviations from MHD, 44
2.4.1 Parallel Electric Field, 47
2.4.2 Surfaces of Discontinuity, 49
2.4.3 Anomalous Resistivity, 50
2.4.4 Onset of Anomalous Resistivity and Electric Double Layers, 53

3 INVARIANCE, DEGENERACY, AND CONTINUOUS SOLUTIONS, 55

3.1 Magnetostatic Equilibrium Equations, 55
3.2 Field Configurations, 57
3.3 Basic Field Configuration, 58
3.4 Characteristics and Discontinuities, 61
3.5 Unlimited Winding, 66
3.6 Action at a Distance, 69

4 **FORMAL STRUCTURE OF THE MAGNETOSTATIC EQUATIONS, 71**
4.1 Introduction, 71
4.2 Continuous Solutions, 72
4.3 Invariant Fields, 73
4.4 Nearly Invariant Fields, 75
4.5 Infinitesimal Perturbation to a Force-Free Field, 77
4.6 Integral Constraints, 78
4.7 Further Integral Constraints, 81
4.8 Infinitesimal Perturbation Including Fluid Pressure, 84
4.9 Physical Basis for Tangential Discontinuities, 85
4.10 Discontinuities and Dynamical Nonequilibrium, 87
4.11 Hamiltonian Formulation of the Magnetostatic Field, 90
4.11.1 Reformulation of Hamiltonian Representation, 91
4.11.2 Destruction of Flux Surfaces, 94

5 **DIRECT INTEGRATION OF EQUILIBRIUM EQUATIONS, 95**
5.1 Description of the Field, 95
5.2 The Equilibrium Equations for a Force-Free Field, 99
5.3 Topology of Field Lines, 101
5.4 The Primitive Torsional Field, 104
5.4.1 Elementary Integration, 106
5.4.2 Asymptotic Solution, 111
5.5 Continuous Solutions, 114
5.5.1 Solution for Primitive Force-Free Field, 117
5.5.2 The Field Topology, 120
Appendix: Field Lines in a Canted Field, 123

6 **EXAMPLES OF FIELD DISCONTINUITIES, 125**
6.1 Introduction, 125
6.2 Compression of a Primitive Force-Free Field, 126
6.3 Tangential Discontinuities in Two Dimensions, 128
6.4 Free Energy of a Discontinuity, 131
6.5 The Y-Type Neutral Point in a Potential Field, 134
6.6 Free Energy Above a Plane Boundary, 137
6.7 Free Energy of an Extended Neutral Point, 141
6.8 The Creation of X-Type Neutral Points, 145
6.9 Direct Numerical Simulation, 154
6.10 Discontinuity Through Instability, 156
6.11 Discontinuities Between Twisted Flux Bundles, 161
6.12 Physical Construction of Continuous Fields, 166
6.13 Time-Dependent Fields, 170

7 **THE OPTICAL ANALOGY, 173**
7.1 The Basic Construction, 173
7.2 Special Cases, 178
7.3 The Field Rotation Effect, 182

7.4 Refraction Around a Maximum, 184
7.4.1 General Parabolic Maximum, 187
7.4.2 Gaussian Maximum, 190
7.5 Non-Euclidean Surfaces, 194
7.6 Refraction in a Slab and Field Line Topology, 198
7.7 Relative Motion of Index of Refraction, 202
7.8 Bifurcation of Fields, 205
7.9 Gaps and Various Discontinuities, 210
7.10 Topology Around a Field Maximum, 212
7.11 Displacement of Individual Flux Bundles, 219

8 TOPOLOGY OF TANGENTIAL DISCONTINUITIES, 225
8.1 Conservation of Discontinuities, 225
8.2 Fluid Pressure, 227
8.3 The Primitive Discontinuity, 229
8.4 Topology Around Neutral Points, 233
8.5 General Form of Discontinuities, 238
8.6 Field Structure Around Intersecting Discontinuities, 241
8.6.1 Magnetostatic Field in a Vertex, 244
8.6.2 Compatibility of Sectors, 248
8.7 Discontinuities Around Displaced Flux Bundles, 252

9 FLUID MOTIONS, 255
9.1 Collision of Separate Regions, 255
9.1.1 Inviscid Fluid, 255
9.1.2 Viscous Fluid, 261
9.1.3 Viscosity and Resistivity, 265
9.2 High Speed Sheets, 268
9.3 Dynamical Model, 272
9.4 Hydrodynamic Model of High Speed Sheet, 274
9.5 Displaced Flux Bundles, 281

10 EFFECTS OF RESISTIVITY, 286
10.1 Diffusion of a Current Sheet, 286
10.2 Dissipation with Continuing Shear, 289
10.3 Dynamical Dissipation, 291
10.3.1 Basic Concepts, 293
10.3.2 Basic Relations, 296
10.3.3 Limits of Reconnection, 299
10.3.4 Theoretical Developments, 300
10.3.5 Tearing Instability, 303
10.3.6 The Complementary Approach, 306
10.3.7 The Third Dimension, 308
10.4 Nonuniform Resistivity, 309
10.4.1 Quasi-static Plane Current Sheet, 312
10.4.2 Alfven Transit Time Effects, 315
10.4.3 Upward Extension of Current Sheet Thickness, 317
10.4.4 Nonuniform Current Sheets, 320
10.5 Changes in Field Topology, 321

11 SOLAR X-RAY EMISSION, 328

11.1 The Origin of X-Ray Astronomy, 328

11.1.1 Theoretical Considerations, 330

11.2 Conditions in the X-Ray Corona, 331

11.2.1 Variability in the X-Ray Corona, 336

11.3 General Considerations, 338

11.3.1 Detailed Modeling of Coronal Loops, 340

11.4 Toward a Theory of Coronal Heating, 341

11.4.1 Waves in the Corona, 342

11.4.2 Wave Dissipation, 344

11.4.3 The Wave Dilemma, 346

11.4.4 Quasi-static Fields, 348

11.4.5 Energy Input to Quasi-static Fields, 351

11.4.6 Dissipation of Quasi-static Fields, 353

11.4.7 Energy of the Basic Nanoflare, 356

11.4.8 Physical Properties of the Basic Nanoflare, 358

11.4.9 Variation Along a Coronal Loop, 361

11.5 Observational Tests, 363

12 UNIVERSAL NANOFLARES, 367

12.1 General Considerations, 367

12.2 The Solar Flare, 369

12.2.1 Basic Nature of the Flare Release, 371

12.2.2 Nanoflare Size, 373

12.2.3 Preflare Deformation, 375

12.2.4 Flare Size Distribution, 375

12.2.5 Internal Dynamics, 377

12.3 The Geomagnetic Field, 378

12.3.1 Aurorae, 379

12.4 Diverse Settings, 382

12.5 Cosmic Rays and Galactic Halos, 385

12.6 Observational and Experimental Studies of Nanoflaring, 388

References, 393

Index, 419

SPONTANEOUS CURRENT SHEETS IN MAGNETIC FIELDS

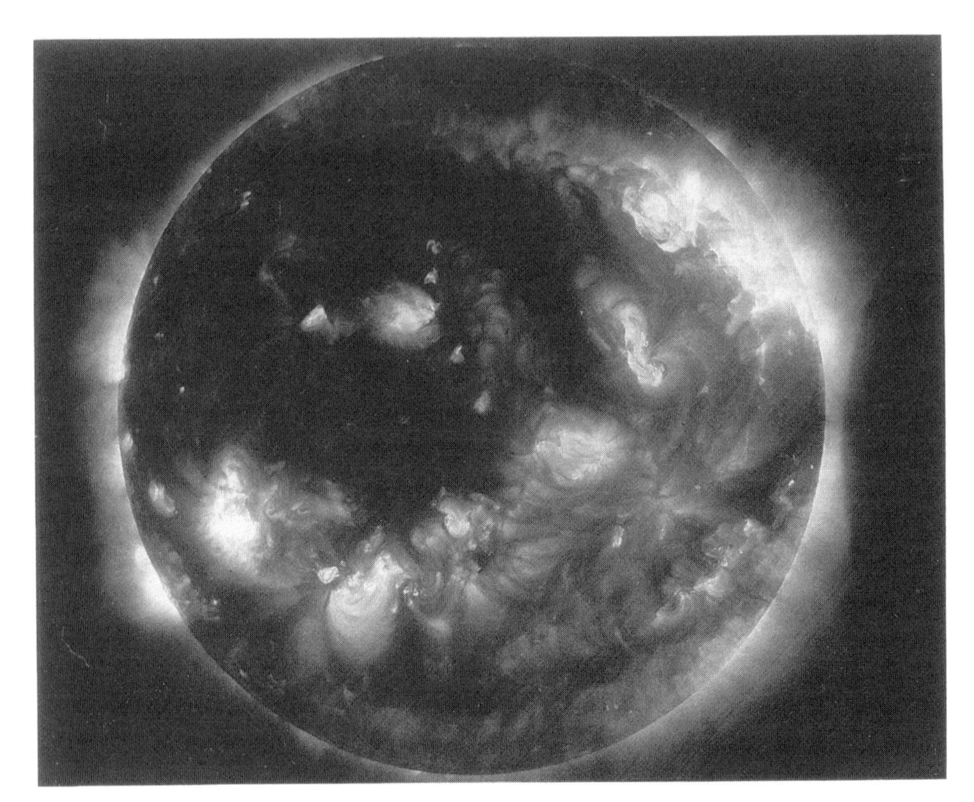

An X-ray photograph of the sun, seen in soft X-rays (17:03 UT, 11 July 1991) showing the denser coronal gas with temperature in the range 1–3 × 10^6 K. The photograph was made with the Normal Incidence X-ray Telescope (NIXT) above the atmosphere of earth (Golub, et al. 1990, Chapter 11) and was kindly furnished by Dr Leon Golub, Harvard Smithsonian Observatory.

1

Introduction

1.1 The General Picture

This monograph treats the basic theorem of magnetostatics, that the lowest available energy state of a magnetic field $\mathbf{B}(\mathbf{r})$ in an infinitely conducting fluid contains surfaces of tangential discontinuity (current sheets, across which the direction of the field changes discontinuously) for all but the most carefully tailored field topologies. That is to say, almost all continuous magnetic field configurations develop internal discontinuities as they relax to equilibrium. The theorem may be stated conversely to the effect that continuous fields are associated only with special topologies. The theorem is a consequence of the basic structure of the Maxwell stress tensor.

The magnetostatic theorem has broad application to the activity of the external magnetic fields of planets, stars, interstellar gas clouds, and galaxies, and to the magnetic fields in laboratory plasmas. In particular the theorem indicates that magnetic fields are highly dissipative, as a consequence of their internal current sheets, providing the principal heat source that creates the flares and X-ray coronas of stars and galaxies, and providing the aurora in the magnetic field of the Earth and other planets.

Observations show the remarkable fact that most stars emit X-rays[1] as thermal bremsstrahlung and line emission, indicating outer atmospheres (coronas) of 10^6–10^7 K. The Sun provides a laboratory to study the structure and the physics of the stellar X-ray corona, which are otherwise lost in the unresolved telescopic images of the more distant stars. Detailed observations of the Sun show that the X-ray emission arises from gas trapped in local bipolar regions of magnetic field and heated by some form of magnetic dissipation in the enclosing field. The theoretical dilemma has been that the very small electrical resistivity of the hot X-ray emitting gas is not conducive to dissipation of magnetic field. However, the ubiquitous tangential discontinuity is unique in that it causes the free energy of the field to dissipate by dynamical neutral-point reconnection at a rate determined more by the Alfven speed than by the slight resistivity of the medium. It appears that the X-ray luminosity of the Sun, and presumably, therefore, the X-ray luminosity of most solitary middle and late main sequence stars, is a consequence of a sea of small

[1] White dwarfs and most solitary red giants provide exceptions. In the opposite extreme, the extraordinary X–ray luminosities of certain special multiple star systems are attributed to the gravitational energy of matter from a giant star falling onto the surface of a compact star (white dwarf or neutron star), and, in somewhat less extreme cases, to the strong tidal churning of close companions.

reconnection events — nanoflares — in the local surfaces of tangential discontinuity throughout the bipolar magnetic fields of active regions. The degree of fluctuation, i.e., the duration and intensity of the individual nanoflare, is not quantitatively defined yet by the theory. The magnetic fields are continually deformed by the underlying convection, so that they continually develop new tangential discontinuities as the old discontinuities are dissipated, thereby providing an ongoing source of heat for the active X-ray corona. Thus the spontaneous discontinuity is the basis for much of X-ray astronomy.

The X-ray luminosity of solitary stars shows occasional transient increases as a result of concentrated outbursts, or flares, at the star. The individual flare can be studied at the Sun where it appears as an intense burst of dissipation of magnetic energy in the corona (Parker, 1957a) as the subphotospheric convection rams together two otherwise separate external magnetic lobes (usually bipoles) to produce a particularly strong magnetic discontinuity. Following the initial burst of dissipation at the discontinuity the flare continues with what appears to be a sea of nanoflares within the colliding bipoles, triggered by the initial burst and by the overall deformation of the colliding bipoles.

Much the same happens in the external magnetic fields of spiral galaxies, which are continually and rapidly (20–100 km/sec) inflated by the powerful relativistic cosmic ray gas generated within the disks of the galaxies. The current sheets produced in the geomagnetic field by the strong deformation of the field by the confining solar wind and by the dynamical reconnection with the field of the solar wind represent another facet of the same general situation, that deformation of magnetic field usually produces internal discontinuities.

In summary, wherever magnetic fields are deformed from the special geometrical form and internal topology of continuous fields, there arise internal surfaces of tangential discontinuity, providing strong dissipation of magnetic energy in an otherwise essentially dissipationless system. This process is manifest throughout the astronomical universe in the exotic phenomena of X-ray emission.

The spontaneous formation of tangential discontinuities in a magnetic field undergoing a simple (or complex) continuous deformation is a basic (but largely unfamiliar) physical phenomenon arising directly from the nonlinear character of the Maxwell stresses in the deformed magnetic field. In view of the unfamiliar character of the special properties of the magnetostatic equation giving rise to the discontinuities, the theoretical development progresses a step at a time, exploring in detail the properties of the field equations for magnetostatic equilibrium to show how the tangential discontinuity is a natural and necessary part of the equilibrium of almost all field topologies.

As we shall see, the equilibrium equations for a magnetic field in an infinitely conducting fluid are qualitatively different from the equilibrium equations for fields in vacuum. The equations for a vacuum field are fully elliptic, with two sets of imaginery characteristics. In a conducting medium the equations possess two sets of imaginery characteristics, but in addition the equations possess a set of real characteristics. The real characteristics are represented by the field lines, thereby providing for the surfaces of tangential discontinuity. As with all physical phenomena, the basic equations, with their stark economy of structure, possess precisely those features that are necessary to reconcile the diverse physical properties of the field. In the present case, it is the arbitrary topology of the field that must somehow be reconciled to the invariance of fluid pressure and/or the invariance of the torsion along the field lines. The tangential discontinuity is precisely the means by which the reconciliation is achieved. The essential point is that the convective motions in stars and galaxies, and sometimes in laboratory plasma devices, deform magnetic fields without regard for the special topological conditions necessary to avoid the formation of discontinuities. Hence the ubiquitous character of the tangential discontinuity in the astronomical universe with the exotic pyrotechnic consequences already mentioned.

Conventional mathematical methods are not particularly effective in dealing with the nonlinear magnetostatic field equations, so in Chapter 7 the optical analogy is introduced, which greatly facilitates the treatment the field line topology associated with deformation of a magnetic field. The optical analogy takes advantage of the fact that the lines of force of a static magnetic field $\mathbf{B}(\mathbf{r})$ in any isobaric surface follow the same pattern as the optical ray paths in an index of refraction $B(\mathbf{r}) = |\mathbf{B}(\mathbf{r})|$. Indeed, the optical analogy applies to the projection of any vector field $\mathbf{F}(\mathbf{r})$ onto the local flux surfaces of $\nabla \times \mathbf{F}$. Hence a sufficiently concentrated maximum in $B(\mathbf{r})$ causes a bifurcation in the field pattern, as the field lines pass around on either side, rather than over, the maximum. The bifurcation of the field pattern is the singular feature that creates the tangential discontinuity. The gap in the field pattern associated with the bifurcation is centered over the maximum and permits the otherwise separated regions of field on either side to come into contact through the gap. The separate fields create a tangential discontinuity at their contact surface in the gap.

The optical analogy applies to stationary flow of ideal inviscid incompressible fluid in the same special way that it applies to the magnetic field, because the stationary Euler equation and the magnetostatic equation are identical in form. In its general form the optical analogy applies to time-dependent turbulent hydrodynamic flows, showing the relation between local velocity maxima and vortex sheets. A brief discussion is provided in Chapter 7. The essential point is that the dynamical formation of vortex sheets in turbulent flows is a trend that is already conspicuous in the stationary flow, of which the vortex sheet is an intrinsic part.

The general phenomenon of spontaneous internal tangential discontinuities has received only limited attention in the literature, in most cases without appreciating its general occurrence and its importance for astrophysics. In particular, the optical analogy has not been previously recognized or exploited to describe the creation of a discontinuity by the adjoining regions of the field. Hence one of the goals of this monograph is to develop the optical analogy and then to exploit the analogy to extend the general theory of the spontaneous tangential discontinuities in magnetostatic fields. We apply the general theory to the universal magnetic activity of stars, planets, and galaxies and to the magnetic confinement of plasma in the laboratory.

Now the theoretical development is extensive, as is the range of applications. Hence this first chapter establishes a road map, a "cultural history," and a commentary on some of the major points of interest along the way, describing the general ideas involved in both the magnetostatic theorem and the astronomical settings in which the magnetostatic theorem is to be applied. The succeeding chapters provide detailed examination of the many individual aspects of the theory and its applications.

1.2 Activity of Stars and Galaxies

Consider the general nature of the activity of astronomical objects. Observations of the Sun show an active, rather than a placid, object. Observation leaves no alternative to the idea that the activity is a direct consequence of magnetic fields. Where there is magnetic field, there is activity, and vice versa.

Cowling (1958) gives a brief history of the study of magnetic fields in the Sun (see also Cowling, 1953; Kiepenheuer, 1953; Parker, 1979, pp. 739–746; Priest, 1982; Weiss, 1983; Foukal, 1990). The existence of magnetic fields was first inferred in 1889 by Bigelow from the filamentary appearance of the coronal streamers seen during total eclipse. Hale (1908a–d, 1913; see also Hale and Nicholson, 1938) was the first to observe the Zeeman effect, establishing that sunspots represent regions where the field is $2–3 \times 10^3$ gauss. Hale's instrumental noise was evidently about 50 gauss, because he thought (erroneously) that he detected a general dipole magnetic field of about that intensity at the North and South poles.

Detection and study of the magnetic fields outside sunspots had to wait for the development of electronics and the Babcock magnetograph (Babcock and Babcock, 1955; Babcock, 1959) to map the line-of-sight component of magnetic field over the solar photosphere (see also Howard, 1959; Leighton, 1959). The complex nature of the large-scale photospheric magnetic fields throughout three complete 11-year sunspot cycles is now a matter of record.

The outstanding aspects of the magnetic activity (besides the conspicuous sunspots) are the suprathermal effects and the violent mass motions, with the transient solar flare and the coronal mass ejection as the extreme examples, respectively. On a continuing basis there is coronal gas confined in the 100 gauss bipolar magnetic fields of active regions, and heated to $2–3 \times 10^6$ K with densities as large as 10^{10} H atoms/cm^3 so as to emit X-rays at a rate 10^7 ergs/cm^2 sec, to be compared to the photospheric radiation intensity of 6×10^{10} ergs/cm^2 sec (Withbroe and Noyes, 1977). The coronal gas in open field configurations (5–10 gauss) reaches $1.5–2 \times 10^6$ K and expands continually to produce the solar wind, requiring a heat input of about 5×10^5 ergs/cm^2 (Withbroe and Noyes, 1977; Withbroe, 1988). The relatively low density (10^8 atoms/cm^3) precludes significant emission of X-rays.

The coronal mass ejection represents a magnetic catapult that flings matter out into space (Illing and Hundhausen, 1986; Athay and Illing, 1986; Athay, Low, and Rompolt, 1987; Webb and Hundhausen, 1987) with individual ejections estimated to be sometimes as large as 10^{32} ergs (Hundhausen, 1990). The solar flare, which may also be as large as 10^{32} ergs, is an example of extreme intensity, emitting hard X-rays and gamma-rays, and accelerating ions and electrons, sometimes to relativistic energies (Svestka, 1976; Priest, 1981, 1982).

The more closely one looks at the Sun, the more activity there is to see on progressively smaller scales. There is continual microflaring in the small-scale network fields as small magnetic bipoles emerge in supergranule cells and are swept into the cell boundaries, where they accumulate to provide the network fields. It appears that this microflaring may be the principal source of heat in the regions of weak open field (Martin, 1984, 1988, 1990; Porter, et al. 1987; Porter and Moore, 1988; Parker, 1991a), as already noted. Dere, Bartoe, and Brueckner (1991), Brueckner and Bartoe (1983) and Brueckner, et al. (1986) find tiny jets and explosive events in the chromosphere–corona transition layer, evidently associated with the microflaring in the network fields (Porter and Dere, 1991).

The X-ray corona, even with the very high space and time resolution of the recent normal-incidence X-ray telescope (Walker, et al. 1988; Golub, et al. 1990), appears as a filamentary continuum, the individual nanoflares being unresolved. The existence of the nanoflares is indicated by the observed electromagnetic emission spectrum, showing excitation well above the mean temperature (Sturrock, et al. 1990; Laming and Feldman, 1992; Feldman, 1992; Feldman, et al. 1992), and implying that the temperature varies sharply and intermittently through a wide range.

A somewhat similar situation is inferred for the solar flare, where it appears that the principal emission arises from a sea of nanoflares in one or more of the magnetic bipoles whose collision creates the initial impulsive phase of hard radiation (Parker, 1987; Machado, et al. 1988).

The magnetic field of the Sun at the photosphere is composed of tiny, intense, and widely separated magnetic fibrils of $1\text{–}2 \times 10^3$ gauss across diameters as small as 10^7 cm (Beckers and Schröter, 1968; Livingston and Harvey, 1969, 1971; Simon and Noyes, 1971; Howard and Stenflo, 1972; Frazier and Stenflo, 1972; Stenflo, 1973; Chapman, 1973), with the fibrils expanding to fill the entire space in the chromosphere above (Kopp and Kuperus, 1968; Gabriel, 1976; Athay, 1981). The magnetic fibrils are unresolved for the most part (cf. Dunn and Zirker, 1973) and are carried with the photospheric convection (Title, et al. 1989).

Observations of stars and galaxies show universal X-ray emission, flaring, etc., suggesting that the active Sun is a paradigm rather than an anomaly. Evidently magnetic fields and magnetic activity are everywhere (Parker, 1979, p. 6). On the other hand, from the point of view of the physicist, the Sun is unique, being the only star for which the form of the activity can be seen. This is essential because, first, the nature of the magnetic activity is exotic, lying outside the realm of the terrestrial physics laboratory. Second, the basic equations of physics admit of so many different classes of solutions (for which general mathematical descriptions are not available) that the nature of the observed magnetic activity cannot be deduced from first principles. Hence a theoretical understanding can be developed only with quantitative and qualitative guidance from detailed observations.

1.3 The Nature of Active Magnetic Fields

There is an initial puzzle at the simplest theoretical level. For it must be remembered that the general occurrence of magnetic fields in astronomical objects can be understood only from the fact of their long life implied by the relative unimportance of resistive dissipation in the interior of the objects, whereas the observed continuing activity of the external fields of astronomical objects implies bursts of rapid dissipation, converting magnetic energy into heat, fast particles, etc. To elaborate, the existence of magnetic fields in planets, stars, gas clouds, and galaxies can be understood only from the fact of the relatively small effective resistive diffusion coefficient η and the relatively large scale ℓ, so that the characteristic resistive decay time ℓ^2/η is long compared to any convective turn over time ℓ/v in the internal fluid motion v. The appropriate dimensionless number is the magnetic Reynolds number $N_M = \ell v/\eta$, representing the ratio of these two characteristic times. It is sometimes convenient to define the Lundquist number N_L as $\ell C/\eta$, where C is the characteristic Alfven speed $C = B/(4\pi\rho)^{\frac{1}{2}}$ in the field. A typical value of N_L in the solar corona, where $\ell \cong 10^{10}$ cm, $\eta \cong 10^3$ cm^2/sec, and $C \cong 10^8$ cm/sec, is 10^{15}, indicating the relative smallness of resistive dissipation of the magnetic field.

Note, then, the limitations of the terrestrial plasma laboratory where ℓ may perhaps be as large as a meter, which is 10^{-7} or less of the gross scale of the fields in the solar corona. The resistive decay time ℓ^2/η in a cubic meter of laboratory plasma with a thermal energy of 10^2 eV may be 1 sec where the Lundquist number is generally 10^4 or less, making it impossible to study more than relatively transient dynamical effects. Needless to say, the laboratory experiments that have been performed on plasma confinement, on the formation and coalescence of islands, and on a variety of major instabilities, have been essential in guiding the theoretical development of the basic plasma phenomena. But the quasi-equilibrium magnetic configurations that ultimately ignite into extended sequences of flaring require the enormous Lundquist numbers of the astronomical setting.

Having established the essential resistive longevity of the magnetic fields in astronomical settings, how is it, then, that the fields observed in the Sun are in a state of perpetual internal dissipation (and sometimes explosive dissipation), diverting free magnetic energy to heat the X-ray corona and the solar wind, sometimes accelerating particles to cosmic ray energies, sometimes flinging mass out into space, etc.? These dissipative phenomena occur in seconds or minutes. They involve reconnection of field lines, which is intrinsically a resistive effect. For without resistivity the field lines are permanently connected and can do little or nothing to heat the ambient gases, nor can they cut loose from their moorings to depart into space. So on the one hand there is longevity because of the small resistivity and large scale, while on the other hand there is vigorous dissipation.

The resolution of the contradiction has gradually emerged over the years, beginning with studies of large solar flares, where it is well established that the explosive dissipation is a consequence of rapid neutral point reconnection of magnetic fields. Evidently this provides the conspicuous impulsive onset of a flare, as already noted, when separate lobes (topological regions) of field are rammed together (cf. Parker, 1957b; Sweet, 1958) squashing the X-type neutral point where the fields come into contact. Unless by chance the colliding fields are closely parallel, one field component meets its opposite number in the other bipole and dynamical annihilation occurs. The process is simply that the intervening gas is rapidly squeezed out from between the opposite components until the separation becomes so small that the opposite fields dissipate. No matter how small the effective resistivity the separation is soon sufficiently small, and the electric current density sufficiently large, as to dissipate the opposite components in a short time. The dissipation frees the gas from the field, but the gas thus liberated is continually squeezed out from between the opposite components. So the process of rapid reconnection continues as long as there is free energy available in the colliding fields (Parker, 1957b).

Flaring by rapid reconnection was not initially associated with the general heating of the active X-ray corona because the field within a magnetic bipole was assumed to be continuous throughout. A prescient paper by Gold (1964) noted that the photospheric convective turbulence deforms, wraps, and winds the bipolar magnetic fields above the surface of the Sun, continually increasing the magnetic free energy. He proposed that the convective turbulence twists the fields so tightly that some form of dissipation (dynamical instability, reconnection, etc.) must occur, continually converting magnetic energy into heat and causing the elevated temperature of the solar atmosphere. Syrovatskii (1971, 1978, 1981) was the first to recognize the universal vulnerability of the X-type neutral point in the projection of the magnetic field onto any plane perpendicular to **B**. He noted that any squeezing of the neutral point creates a current sheet, or pinch sheet as he sometimes referred to it. Parker (1972, 1973, 1979 pp. 511–519) pointed out that special invariance of the magnetic field is necessary to avoid current sheets or tangential discontinuities as an intrinsic part of the equilibrium of a magnetic field. He noted that the magnetic fields created in the convective fluid of a star generally do not have the necessary invariance or symmetry and so they may be expected to contain surfaces of tangential discontinuity, subject to rapid reconnection. He proposed that the otherwise runaway increase in the small-scale components (large wave number k) of a magnetic field in a turbulent fluid may be checked by such rapid reconnection. It was Glencross (1975, 1980) who first suggested explicitly that the general occurrence of current sheets pointed out by Parker provides the heat source for the X-ray corona of the Sun. Parker (1981, 1983a) sketched out some specific circumstances for heating the active corona in that way.

The basic idea behind the application of the magnetostatic theorem to magnetic activity and coronal heating can be expressed in the following way. A simple continuous magnetic field configuration $\mathbf{B}(\mathbf{r})$ is preserved by its large scale $L(N_L \gg 1)$ in the presence of small resistivity. But the large–scale field $\mathbf{B}(\mathbf{r})$ of a convective object, e.g., a star or galaxy, is internally wrapped and interwoven, producing strong local deformation $\Delta\mathbf{B}$ on intermediate scales, ℓ. These intermediate scales are sufficiently large that they too are preserved. However, the topology of $\mathbf{B} + \Delta\mathbf{B}$ is no longer the simple topology of the basic form $\mathbf{B}(\mathbf{r})$. The magnetostatic theorem asserts that the field $\mathbf{B} + \Delta\mathbf{B}$ develops internal discontinuities as it relaxes to equilibrium. The internal tangential discontinuities involve magnetic free energy, and, since η is small but not identically zero in the real physical world, the discontinuities provide rapid reconnection and quick dissipation of the free energy into heat. The dissipation consumes $\Delta\mathbf{B}$ but not **B**, of course, because the topology of $\mathbf{B}(\mathbf{r})$ is simple enough to permit a continuous equilibrium field, which is preserved by its large-scale L, as remarked in the beginning. Thus the dissipation is active so long as there is enough $\Delta\mathbf{B}(\mathbf{r})$ that the topology requires discontinuities for equilibrium.

This fundamental property of magnetostatic fields merits further elaboration, because there is at least some slight resistive dissipation everywhere in the gases in the astronomical universe or in the terrestrial laboratory. The resistivity converts each ideal surface of tangential discontinuity into a thin transition layer, or current sheet, in which magnetic energy is rapidly dissipated as a consequence of the high current density. In fact the characteristic thickness of the current sheet is prevented from reaching the zero of the true discontinuity in a ideal fluid with zero resistivity. The thickness falls only to a value sufficiently small that the dissipation balances the dynamical trend toward zero thickness. The rapid dissipation and the associated field line reconnection continues only so long as the field topology requires discontinuities for magnetostatic equilibrium. The thin current sheet is not in internal equilibrium, of course. It is the topology of the quasi-static equilibrium fields in the regions of continuous field, filling the volume between the current sheets, that drives the formation of the discontinuities. The dissipation continually reduces the topology of the continuous field to simpler forms through reconnection of field across the current sheet. So the rapid dissipation continues until the topology is so simple that it no longer requires the current sheets.

This point is often ignored and it deserves an illustration. Consider a tangential discontinuity in a static field in an infinite space, where the discontinuity is not required by the topology, i.e., a passive discontinuity. An example would be an initial equilibrium field extending uniformly in the z-direction ($B_x = B_y = 0$) from $z = -\infty$ to $z = +\infty$, with $B_z = +B_0$ in $y > 0$ and $B_z = -B_0$ in $y < 0$. The field is filled with an infinitely conducting incompressible fluid and is clearly in magnetostatic equilibrium with a surface of discontinuity (current sheet) at $y = 0$. Then suppose that at some time $t = 0$ a small uniform resistivity η is introduced throughout the fluid, so that the field remains in static equilibrium, evolving according to the familiar magnetohydrodynamic diffusion equation

$$\partial B_z/\partial t = \eta\, \partial^2 B_z/\partial y^2.$$

It is readily shown that

$$B_z = B_0 \operatorname{erf}[y/(4\eta t)^{\frac{1}{2}}]$$

for $t > 0$, where erf denotes the error function

$$\operatorname{erf}\chi = \frac{2}{\pi^{\frac{1}{2}}} \int_0^{\chi} ds \exp(-s^2).$$

The characteristic thickness of the current sheet increases from zero with the passage of time, in the form $(4\eta t)^{\frac{1}{2}}$. The dissipation rate per unit volume varies as $(B_0^2/4\pi t)\exp(-y^2/2\eta t)$. The total rate of dissipation of energy (throughout $-\infty < y < +\infty$) per unit area of the current sheet is $(\eta/2\pi t)^{\frac{1}{2}} B_0^2/8\pi$, declining as $t^{-1/2}$ with the passage of time.[2]

The essential point of this example of a passive discontinuity is that the initial rapid dissipation of magnetic energy into heat quickly declines to a low level because the current sheet rapidly thickens when it is not rejuvenated by the Maxwell stresses. It follows that the creation of tangential discontinuities or current sheets is not sufficient to guarantee continuing magnetic dissipation unless the discontinuities are required by the topology of the continuous portions of the field. In that case the Maxwell stresses continually extract the field and fluid from the thickening current sheet, thereby maintaining a thickness δ so small as to continue rapid dissipation and reconnection of field. This is, of course, the basis for rapid reconnection of fields (Sweet, 1958; Parker, 1957b, 1963a) across a scale ℓ in a time $\ell N_L^{1/2}/C$ or less, instead of the passive diffusion time $\ell N_L/C$, where C is the characteristic Alfven speed and N_L ($\gg 1$) is the Lundquist number $\ell C/\eta$.

One may wonder, then, if the resistive dissipation of magnetic energy can be maintained at a substantial level if some discontinuous motion (e.g., $v_x = +vkz$ in $y > 0$ and $v_y = -vkz$ in $y < 0$ in the uniform field $e_z B$) is introduced to regenerate the passive discontinuity at a fixed rate. It is a simple matter to show (in §10.2) that,

[2]The onset of the resistive tearing instability might enhance the dissipation somewhat, but that does not prevent the decline with increasing t.

even in that case, the dissipation is confined to a transition layer with the characteristic diffusion thickness $(4\eta t)^{\frac{1}{2}}$, which grows too slowly to consume a significant fraction of the available free magnetic energy.

In summary, rapid reconnection occurs only where the topology provides Maxwell stresses in a form to drive the current sheet toward vanishing thickness. Other surfaces of discontinuity are passive and do not drive the dynamical reconnection, so they are limited to the characteristic diffusion scale $(4\eta t)^{\frac{1}{2}}$.

If we view from a distance the theoretical problem of dissipation of magnetic free energy with small resistivity, there are two obvious approaches. One, adopted in this monograph, is to consider the ideal case of a magnetic field embedded in a fluid whose resistivity is identically zero, so that strong deformation of the field develops whatever tangential discontinuities are required for equilibrium by the field topology.

The other approach is to start with the nonvanishing resistivity and inquire into its effects in a field containing internal shear. This begins with the universal resistive magnetohydrodynamic instabilities shown by Spicer (1976, 1977, 1982) and Van Hoven (1976, 1979, 1981) many years ago to occur wherever the magnetic field is subject to shear or torsion. The resistive kink and tearing instabilities are the primary candidates. If h is the characteristic scale of the shear, of the order of $|\mathbf{B}|/|\nabla \times \mathbf{B}$, then the characteristic growth time τ for the resistive instability is given by (Furth, Killeen, and Rosenbluth, 1963; Parker, 1979; Van Hoven, 1981)

$$\begin{aligned} \tau &\sim h^2 (k^2/\eta^3 C^2)^{\frac{1}{5}} \\ &= (kh)^{\frac{2}{5}} \tau_R^{\frac{3}{5}} \tau_A^{\frac{2}{5}} \end{aligned}$$

for wavelength $2\pi/k$ along the current sheet, Alfven speed $C = |\mathbf{B}|/(4\pi\rho)^{\frac{1}{2}}$, and resistive diffusion coefficient η. The characteristic diffusion time τ_R is h^2/η and the characteristic Alfven transit time is h/C. Spicer and Van Hoven point out that τ is a relatively long time because of the magnitude of τ_R in the large dimension h of magnetic fields in the typical astronomical setting. The essential point, then, is that the trend toward a surface of tangential discontinuity provides a local shear scale h that tends rapidly toward zero. It follows that the resistive tearing instability arises primarily in the declining thickness of the current sheets when the sheets are well on their way to becoming tangential discontinuities. Thus the resistive approach leads to the same situation as the development that starts with a fluid of zero resistivity and notes that the thickness of the current sheet declines until resistive dissipation — presumably in the form of a resistive instability — prevents further decline (Parker, 1990d). In either case we end up in the same final state, with the dissipation arising at the site of the potential tangential discontinuities. Then since the theory of the spontaneous formation of tangential discontinuities does not depend upon the resistivity, it is easiest to develop the theory before bringing in the final limiting effect of resistivity. The conclusion is simply that any magnetic field in a fluid of small resistivity in a convective astronomical setting is subject to rapid internal dissipation of its magnetic free energy.

It follows that there is continual dissipation of magnetic energy into heat throughout the bipolar fields of magnetic active regions and throughout the strong small-scale bipoles of the network fields on the Sun. Quantitative estimates indicate that the rate of dissipation of magnetic energy is substantial, providing most of the heat that maintains the X-ray emitting gas trapped in the bipolar fields (Parker, 1983a). The individual reconnection fluctuations — nanoflares — are mostly below the limit of detection, so that the observer sees only the general glow that represents the sea of nanoflares. Similarly the magnetic bipoles that provide flares contain internal discontinuities that dissipate rapidly when bipoles collide to produce a flare. The observations of Machado, et al. (1988) indicate that well over half the energy of a flare comes from the nanoflares within the bipoles, as distinct from the initial explosive reconnection at the interface between two colliding bipoles.

With this general principle of spontaneous discontinuities in hand, it would appear that substantial internal generation of heat by magnetic dissipation in most active astronomical magnetic fields is inevitable. Planetary magnetospheres are subject to varying deformation in the fluctuating solar wind, while the footpoints of the planetary magnetic field move about in the ionosphere. The magnetic fields of galaxies are subject to the motion of the interstellar gas, and particularly to rapid inflation by the cosmic ray gas generated by supernovae, etc. We conjecture that the X-ray emission from the halos of many spiral galaxies is a consequence of the magnetic dissipation in the discontinuities associated with magnetostatic equilibrium of the deformed fields (Parker, 1990a, 1992). Needless to say, the X-ray emission from the Sun is the only case available for detailed telescopic observation, so it is the primary focus for the theoretical development.

1.4 Rapid Dissipation

It was realized from the outset that the explosive dissipation of magnetic energy to produce the flare must center around singular places in the magnetic field where dissipation can occur in spite of the small resistivity. Thus Giovanelli (1947) and Cowling (1953) considered the possibility of electrical discharges at neutral points in the field as the cause of a solar flare. Dungey (1953, 1958a,b) and Chapman and Kendall (1963) pursued the idea further, treating the stability of the (X-type) neutral point. Sweet (1958) considered the results of the neutral point created when, for instance, two bipolar regions on the Sun collide head to tail. Parker (1957b, 1963a, 1973) treated Sweet's scenario in the context of magnetohydrodynamics and showed that when two oppositely directed (antiparallel) magnetic fields $\pm B$ are pressed together over a width ℓ in the presence of an incompressible fluid of density ρ and resistive diffusion coefficient η, the fluid is squeezed out from between the two opposing fields, causing the field gradient to steepen until resistive dissipation creates a steady state. The configuration is sketched in Fig. 1.1. The steady state occurs when the characteristic thickness δ of the transition from $-B$ to $+B$ is reduced to the order of $\ell/N_L^{\frac{1}{2}}$, where again N_L is the Lundquist number $\ell C/\eta$, in terms of the characteristic Alfven speed $C = B/(4\pi\rho)^{\frac{1}{2}}$. With this steep field gradient (when $N_L \gg 1$) the two oppositely directed fields move into the transition layer from

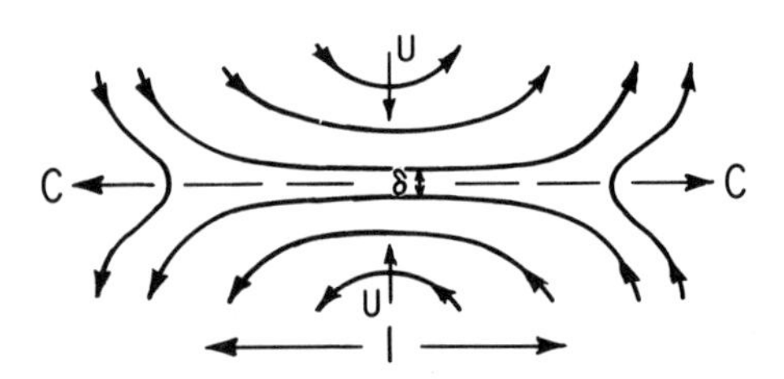

Fig. 1.1. A schematic drawing of the field lines undergoing rapid reconnection across the dashed center line.

either side with a speed u of the order of $\eta/\delta = C/N_L^{\frac{1}{2}}$. The fluid swept into the transition layer with the fields is expelled (by the magnetic tension and pressure) out the ends of the layer at a speed of the order of C. This expulsion of fluid is responsible for maintaining the small thickness δ so that the dissipation continues at a rapid pace $C/N_L^{\frac{1}{2}}$. In the absence of the expulsion of fluid, the characteristic scale is ℓ, rather than δ, and the rate of dissipation is C/N_L, smaller by the large factor $N_L^{\frac{1}{2}}$.

The essential point is that the magnetic stresses throughout the region of field are of such form as to push the thickness of the transition layer (from $-B$ to $+B$) continually toward zero, in an attempt to produce a tangential discontinuity. If the resistivity η were zero, then $N_L = \infty$ and δ would fall to zero, achieving a true discontinuity. Insofar as η is small but not zero, the enhanced field gradient, $O(B/\delta)$, remains large but finite, with the fields flowing into the region of dissipation from either side at a speed $u \cong C/N_L^{\frac{1}{2}}$. So, the discontinuities are really thin transition layers of small but nonvanishing thickness $O(\delta)$. The magnetic stresses throughout the continuous fields on either side of the transition layer drive the thickness toward zero while the resistivity tends to thicken the layer. A dynamical balance arises when δ is of the order of $\ell/N_L^{\frac{1}{2}}$. Thus the original theory of reconnection represents the essential feature of the spontaneous formation of the tangential discontinuity. What was lacking until recently was an understanding of the general occurrence of the scenario in all magnetic fields subject to continuous deformation.

Now the magnetic energy is dissipated by the reconnection speed u at a rate $uB^2/8\pi$ ergs/cm^2 sec. Some of the energy is converted directly into heat, with the rest into the kinetic energy of the ejected fluid depending upon detailed conditions. It is expected that the jet of ejected fluid is turbulent, so that it is quickly thermalized and converted to heat. The phenomenon is sketched in Fig. 1.1, and is referred to as rapid, or neutral point, reconnection, because the lines of force of the two initially separate and oppositely oriented fields $\pm B$ (parallel to the transition layer) are reconnected by the resistivity so as to lie across the transition layer of thickness δ. Magnetic energy is dissipated into heat across δ faster by a factor of the order of $N_L^{\frac{1}{2}}$ than by resistive diffusion across the scale ℓ.

The increased diffusion rate, by a factor $N_L^{\frac{1}{2}}$, is large and interesting, but entirely inadequate to account for the vigorous dissipation represented by a flare. The next step came from Petschek (1964) and Petschek and Thorne (1967) who suggested that two opposite fields $\pm B$ of scale ℓ may, as a consequence of the dynamics of the inflow and outflow, come into contact across only a narrow width $h(\ll \ell)$, rather than across the full width ℓ of the field as assumed by Parker and Sweet. Petschek argued from dynamical considerations that u might then be as large as $C/\ln N_L$, with h as small as $\ell(\ln N_L)^2/N_L$ and δ as small as $\ell \ln N_L/N_L$, in order of magnitude, in the limit of large N_L. The dynamics involve resistive diffusion in a small neighborhood of the neutral point, with oblique standing Alfven waves extending out from each of the four corners of the small central diffusion region (with dimensions $h \times \delta$) sketched in Fig. 1.2. It was clear that any intermediate merging rate $C/N_L^{\frac{1}{2}} < u < C/\ln N_L$ is possible, depending upon the boundary conditions (see discussion in Vasyliunas, 1975). This has been placed in a formal context recently by Priest and Forbes (1986) and particularly elegant analytical solutions have been provided by Hassam (1991) and Craig and Clymont (1991).

Sonnerup (1970, 1971) showed that with a somewhat different field profile, determined by the forces pushing the two opposite fields together (Fig. 1.3), the rate

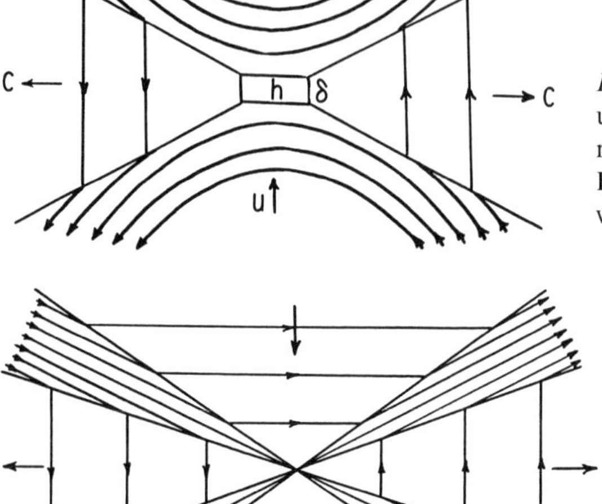

Fig. 1.2. A sketch of the field configuration for Petschek's model of rapid reconnection, with the configuration of Fig. 1.1 in the small central rectangle, where ℓ is replaced by h.

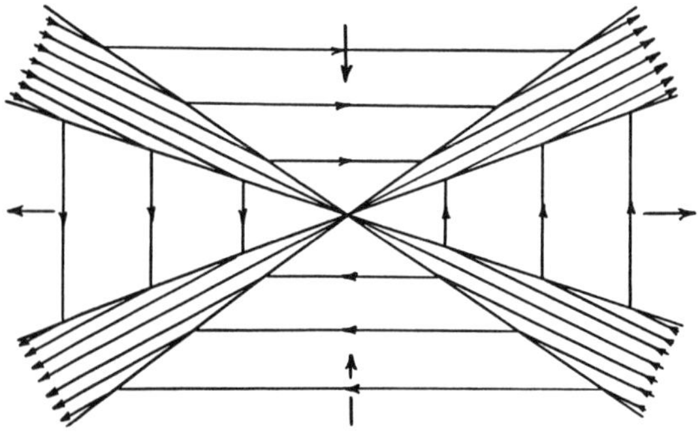

Fig. 1.3. A sketch of the field configuration in Sonnerup's model of rapid reconnection, with the central diffusion region shrunk to zero.

of reconnection u may be as large or as small as desired, depending upon how firmly the opposite fields are pushed together. In this case there are two slow waves in each quadrant extending obliquely into the origin from some (unexplained) point of creation at the periphery of the flow. The formal mathematical solution involves an infinitely sharp corner in the fluid velocity and in the magnetic field at the origin so that the resistivity of the fluid does not enter into the considerations.

Biskamp and Welter (1980) and Biskamp (1984, 1986) have constructed 2D-numerical simulations of reconnection and, so far, have found only slower reconnection rates, along the lines of Syrovatskii's (1971) current sheet model. Whether the failure to find the more rapid reconnection rates obtained by the analytical solutions can be attributed to the boundary conditions or to the modest Lundquist numbers to which numerical simulations are restricted, remains to be seen.

The theoretical developments provide reconnection velocities u that readily account for the initial intense phase of a solar flare as the rapid reconnection between two large-scale lobes of magnetic field pushed together by the motion of their footpoints in the photospheric convection. Priest (1981, pp. 139–216) Low (1987, 1989, 1991), Low and Wolfson (1987), and Jensen (1989), among others, provide explicit examples of the current sheets formed in this way.

To write down some specific numbers, note that with $\ell = 10^9$ cm, $\eta = 10^3\,\mathrm{cm}^2/$sec and $C = 10^8$ cm/sec the Lundquist number $N_L = \ell C/\eta$ may be as large as 10^{14} within a bipolar magnetic region on the Sun, so that the minimum reconnection rate $C/N_L^{\frac{1}{2}}$ is small, of the order of 10 cm/sec. But $\ln N_L$ is only about 30, so that the reconnection rate may be 10^6 times larger, or 10^2 km/sec, determined by the Alfven speed C. Thus the explosive burst of energy release at the onset of a flare (in a period of the order of 10^2 sec) is comprehensible in spite of the small resistivity.

The geomagnetic substorm, associated with geomagnetic reconnection with a southward component of the magnetic field in the solar wind and with the magnetic reconnection across the neutral sheet between the north and south lobes of the geomagnetic tail, is also understandable, if not precisely defined, by the theoretical rapid reconnection (cf. Hones, 1984).

In summary, the large-scale Fourier components of the magnetic fields in astronomical objects are preserved by the small resistivity. On the other hand, the magnetostatic theorem asserts that in the absence of resistivity the magnetic stresses cannot avoid creating small-scale Fourier components, in the form of tangential discontinuities, in almost all field topologies. The discontinuities arise spontaneously and asymptotically in time throughout the interior of each lobe of field simply as a consequence of an overall continuous deformation of the field. So small reconnection events are expected throughout any magnetic field subject to continuing deformation. The small and generally unresolved bursts of reconnection at the ubiquitous internal surfaces of discontinuity appear to be the principal heat source for the ongoing thermal X-ray emission from the solar corona (Parker, 1975, 1979, p. 359, 1981, 1983a; Glencross 1975, 1980) and appear (Parker, 1987; Machado, et al. 1988) to be responsible for the continuing X-ray emission of a solar flare, following the brief intense initial phase. One infers that the general X-ray emission from other stars and the intense flare phenomena of the *dM* dwarf stars are to be understood on a similar basis.

Now, it is apparent from this discussion that in the presence of a small resistivity the magnetostatic theorem leads to conditions that are anything but static at the surfaces of discontinuity. This does not invalidate the application of the theorem, however, because it is the quasi-static balance of magnetic pressure and tension throughout the continuous field filling the volume between surfaces of tangential discontinuity that drives the formation of the tangential discontinuities. The field throughout the volume moves only with the speed u that is small compared to the Alfven speed, and the rapid reconnection at the discontinuity does not significantly disturb the form of the quasi-static equilibrium throughout the volume. On the other hand the dynamical conditions within the actual transition layer δ, representing the potential surface of discontinuity, are not at all like the static form, and there is nothing that the magnetostatic theorem has to say about these dynamical conditions, except that they are created by the static conditions throughout the volume of the field. The situation is closely analogous to the strong shock in an otherwise continuous hydrodynamic flow. The hydrodynamic equations apply to the volume of continuous flow outside the thin shock transition and determine where the shock appears, without describing the structure of the shock transition.

In practice it appears that the reconnection proceeds somewhere in the neighborhood of the minimum rate $C/N_L^{\frac{1}{2}}$, except for sudden bursts of more rapid reconnection (Finn and Kaw, 1977; Van Hoven, 1981; Montgomery, 1982; Lichtenberg, 1984; Dahlburg, et al. 1986) perhaps triggered by passing magnetohydrodynamic waves (Sakai, 1983a,b; Matthaeus and Lamkin, 1985, 1986; Tajima and Sakai, 1986). We have referred to the individual (usually unresolved) small reconnection events as *nanoflares* (Parker, 1988) because it is estimated indirectly that in the solar X-ray corona they are 10^{-8}–10^{-9} of a large solar flare (at 10^{32} ergs/cm^2). We wish to distinguish them from microflares (typically 10^{-5}–10^{-6} of a large flare), which are small but individually observed where small magnetic bipoles collide in the converging flows at the boundaries of supergranule cells on the solar surface. The true nature of the nanoflare remains to be determined by direct means.

For the nanoflares, then, it appears that the internal winding and interweaving of the field lines slowly accumulates without much reconnection (perhaps at the slow rate $C/N_L^{\frac{1}{2}}$) until an individual discontinuity exceeds some critical strength, whereupon a local burst of reconnection (perhaps at a rate as large as $C/\ln N_L$) reduces the strength to where the rapid reconnection falls back toward $C/N_L^{\frac{1}{2}}$. Then the slow accumulation begins again, to be followed after a time by another burst of reconnection, etc.

1.5 The Magnetostatic Theorem

Consider the basic magnetostatic theorem on which the foregoing inferences are based. The theorem can be understood in a variety of ways. To begin, note the magnetostatic equation

$$4\pi \nabla p = (\nabla \times \mathbf{B}) \times \mathbf{B} \tag{1.1}$$

describing the balance between the gas pressure $p(\mathbf{r})$ and the Maxwell stresses in the magnetic field $\mathbf{B}(\mathbf{r})$. Arnold (1965, 1966, 1974) pointed out from formal mathematical considerations that almost all solutions to equations of the form of equation (1.1) contain discontinuities. He dealt explicitly with the Euler equation

$$-\nabla(p/\rho + \frac{1}{2} v^2) = (\nabla \times \mathbf{v}) \times \mathbf{v} \tag{1.2}$$

for the stationary flow of an ideal inviscid incompressible fluid, but, as emphasized by Moffatt (1985, 1990), the solutions of equations (1.1) and (1.2) have identical form. So there is a formal mathematical basis for the ubiquitous tangential discontinuity.

Moffatt (1986) showed that the solutions to equation (1.2) are dynamically unstable, in contrast to the general stability of many solutions of equation (1.1). The difference arises because the Maxwell stress provides a tension $B^2/4\pi$ along the field lines whereas the Reynolds stress provides a compressive force ρv^2 along the streamlines.

So far as we are aware, Syrovatskii (1971, 1978, 1981) was the first to argue that current sheets and rapid reconnection are a general and unavoidable consequence of the quasi-static deformation of a magnetic field. To understand his approach consider the projection of the field lines of $\mathbf{B}(\mathbf{r})$ onto the plane perpendicular to $\mathbf{B}(\mathbf{r})$ at some specified point $\mathbf{r}_0$ in the field. Projection of $\mathbf{B}(\mathbf{r})$ onto the plane perpendicular to $\mathbf{B}(\mathbf{r}_0)$ provides either an O-type or an X-type neutral point in the 2D field in that plane (see discussion in Parker, 1979, pp. 383–391). Consider a point $\mathbf{r}_0$ such that the neutral point is an X-type (Fig. 1.4a). Then apply external forces to the field so as to squash the whole region, and hence squash the X-type vertex, in one direction or another, splitting the X-type vertex into two Y-type vertices, as sketched in Fig. 1.4(b). There is then a tangential discontinuity extending from one Y-vertex to the other. There is also a weaker discontinuity extending away along the two arms of each Y. Syrovatskii pointed out that this effect occurs at one or more points in any quasi-static field subjected to anisotropic or inhomogeneous deformation.

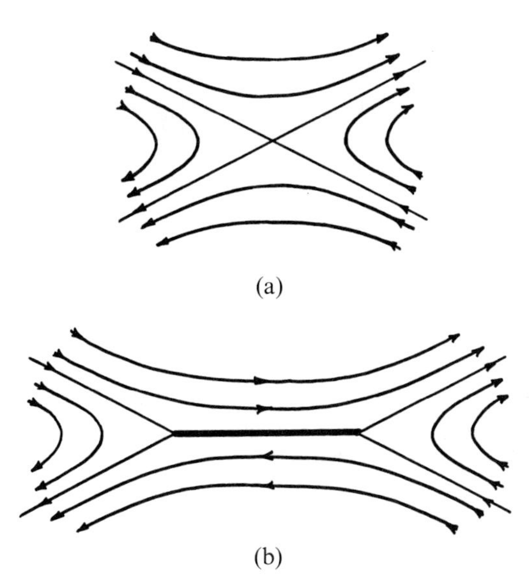

Fig. 1.4. (a) An X-type neutral point in the 2D field (on the plane perpendicular to $\mathbf{B}(\mathbf{r}_0)$), which is squeezed from above and below into two Y-type neutral points in (b). The heavy line indicates the location of the tangential discontinuity.

Subsequent development of the theory of tangential discontinuities shows that Syrovatskii's example illustrates the basic effect. The presentation of specific examples of the formation of discontinuities in Chapter 6 encounters the deformation of one X-type neutral point into two Y-type points in every case. The optical analogy, presented in Chapter 7, shows that the tangential discontinuity arises from gaps created by local field maxima in the flux surfaces, and Syrovatskii's example proves to be the elevation of the magnetic field whose plan view is the gap in the flux surface.

We begin, then, with a sketch of the standard setting for the construction of the magnetic field and the spontaneous appearance of the tangential discontinuities. The model is presented in detail in Chapter 3 and the sketch outlined here is intended only to define the physical context of the discussion. The essential physics is most simply described by starting with a uniform magnetic field B_0 extending in the z–direction through an infinitely conducting ($\eta = 0$) fluid between the boundary planes $z = 0$ and $z = L$ (Parker, 1972). Then at time $t = 0$ the fluid is set into the prescribed two dimensional incompressible transverse motion

$$v_x = +kz\partial\psi/\partial y,\ v_y = -kz\partial\psi/\partial x,\ v_z = 0, \tag{1.3}$$

determined by the arbitrary stream function $\psi(x, y, kzt)$. The function ψ is chosen to be a well-behaved, bounded, smooth, continuous, n–times differentiable function of its arguments. The deformation of the fluid and field is strong, and it is convenient to think of $\psi(x, y, kLt)$ as representing the introduction of a succession of unrelated mixing patterns of the footpoints of the field at the boundary $z = L$ while the footpoints are held fixed at $z = 0$.

After a time t the magnetic field, which is carried with the fluid, has the form

$$B_x = +B_0\,kt\partial\psi/\partial y,\ B_y = -B_0\,kt\partial\psi/\partial x,\ B_z = B_0 \tag{1.4}$$

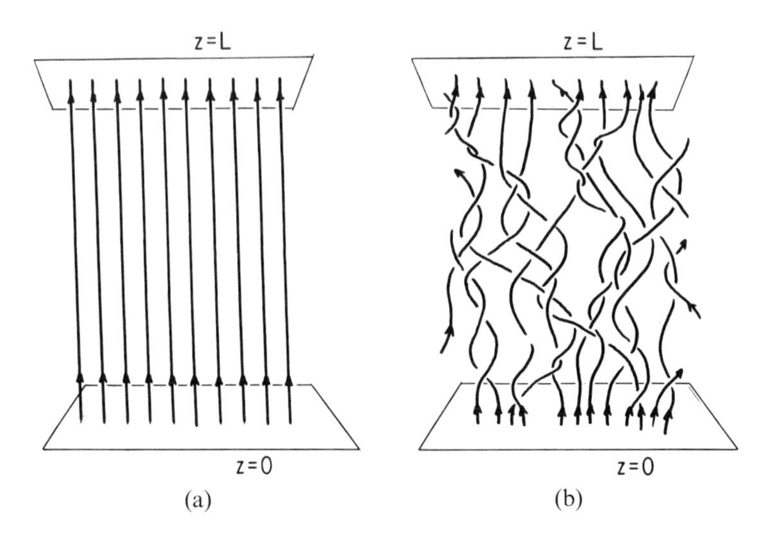

Fig. 1.5. A sketch of the arbitrary winding of the field lines of the continuous field described by equation (1.4), beginning with (a) the uniform field B_0 and become mixed and interlaced after a time t in (b).

The field lines wind and interweave among each other, following the same mixing in passing from $z = 0$ (where the footpoints remain fixed) to $z = L$ as the mapping $\psi(x, y, kLt)$ of the footpoints on $z = L$. Thus the field lines are strongly wrapped and randomly intermixed in passing from $z = 0$ to $z = L$. But note that the field is continuous, because of the prescribed smooth behavior of ψ. Fig. 1.5 is a sketch of the initial field and the interwoven state of that field after some time t.

The essential point is that any arbitrary, well-behaved function ψ is a physical possibility. To fix ideas suppose that $\psi(x, y, kLt)$ passes through n successive arbitrary, random swirling patterns at each value of (x, y) as t increases from 0 to τ. If the transverse swirling and mixing represented by ψ has a characteristic scale ℓ, then a footpoint of the field at $z = L$ moves a distance of the order of ℓ or more in each swirl, thereby accumulating a total path length $O(n\ell)$ or more on $z = L$. For purely random swirling the final distance of almost all footpoints from their initial positions is $O(n^{\frac{1}{2}}\ell)$. Note that L may be so large that $L \gg n\ell$ in the presence of strong winding, so that the inclination of the field to the z–direction may be, but need not be, small.

Stop the fluid motion at $t = \tau$, then, and hold the footpoints of the field lines fixed at $z = 0$ and at $z = L$. Release the fluid so that the continuous field is free to relax to the lowest available energy state, ultimately achieving static equilibrium after some sufficiently long period of time. The magnetostatic theorem states that almost all strongly deformed fields develop internal tangential discontinuities in the process of relaxing to static equilibrium (Parker, 1986a,b).

Consider some of the implications of this assertion. The first point is that the choice of the stream function ψ determines the topology — the winding and interweaving — of the field lines, and the topology is preserved by the infinite electrical conductivity of the fluid, i.e., by the vanishing of the electric field in the frame of reference moving with the fluid. The function ψ does not provide the final equilibrium field distribution $\mathbf{B}(\mathbf{r})$, which arises only in the relaxation of the field from the continuous form in equation (1.4) to static equilibrium.

1.6 Continuous Magnetostatic Fields

It comes to mind that there are extensive families of continuous solutions to the magnetostatic equilibrium equation (1.1). Infinitely many different continuous equilibrium fields! However, the known continuous equilibrium solutions involve either only weak deformation of the field from a uniform state (cf. the interesting numerical solutions by Van Ballegooijen, 1985, 1986; Mikic, Schnack, and Van Hoven, 1989 and the analytic solutions by Zweibel and Li, 1987) or a symmetry, degeneracy, or invariance (ignorable coordinate) of some form (see discussion in Grad, 1967, 1984; Parker, 1979, pp. 359–378; Tsinganos, 1981, 1982a,b). The requirement of invariance is not dissimilar to the well-known Taylor–Proudman theorem of hydrodynamics (Proudman, 1916; Taylor, 1917; see also discussion in Chandrasekhar, 1961) arising from the identical form of equations (1.1) and (1.2), to the effect that a stationary velocity field in a system rotating with angular velocity Ω must be invariant in the direction of Ω.

A noteworthy example of an equilibrium based on degeneracy was pointed out by Rosner and Knobloch (1982), involving force-free fields described by the solutions of the field equation (Lundquist, 1950)

$$\nabla \times \mathbf{B} = \alpha \mathbf{B} \tag{1.5}$$

for a single value of α, say $\alpha = q$. The equation is then linear and solutions may be superposed without limit, providing no end of complexity in the total field $\mathbf{B}(\mathbf{r})$ throughout the region.

Another class of continuous solution was found by Van Ballegooijen (1985) who showed that small perturbations $\delta\mathbf{B}$ about a uniform field $\mathbf{B}_0$ (which permits strong mixing over the length of the field in the limit of large L) provide a field described by an equation that is an exact analogy to the 2D time-dependent vorticity equation. The equation is nonlinear and few analytic solutions are known, but one expects that the equation has a variety of continuous solutions. This interesting class of solutions is possible because the zero-order uniform field has no scale of its own, on which more will be said in the sequel.

It is clear that strong winding of the field, wherein a substantial fraction of the field lines extend through at least two different wrapping patterns between $z = 0$ and $z = L$, permits neither an ignorable coordinate, nor a uniform α, nor an undetermined scale in any direction. It is readily seen that such a field, if it were to remain continuous in static equilibrium, would pose a contradiction. The nature of the contradiction is most readily exhibited in the case that the pressure externally applied to the fluid at the boundaries $z = 0, L$ is uniform ($p = p_0$). The scalar product of $\mathbf{B}$ with equation (1.1) yields the condition $\mathbf{B} \cdot \nabla p = 0$, that the fluid pressure extends uniformly along each field line. Then since all the field lines connect to the uniform pressure at $z = 0, L$, it follows that p is uniform throughout the entire region ($0 \leqslant z \leqslant L$). Equation (1.1) reduces to the force-free form (1.5). Equation (1.5) has some interesting mathematical properties. The coefficient $\alpha = \alpha(\mathbf{r})$ represents the torsion in the field, with

$$\alpha = \mathbf{B} \cdot \nabla \times \mathbf{B}/B^2 \tag{1.6}$$

The analogous quantity $\mathbf{v} \cdot \nabla \times \mathbf{v}$ in hydrodynamics is called the *helicity*. Since $\nabla \cdot \mathbf{B} = 0$, the divergence of equation (1.5) gives the well-known result

$$\mathbf{B} \cdot \nabla\alpha = 0, \tag{1.7}$$

stating that the torsion α is uniform along each field line. It is obvious that field lines extending through two or more unrelated winding patterns may be subjected to quite different torsions in the different patterns, because as each flux bundle wraps in arbitrary ways around the neighboring flux bundles, its own internal torsion must match the wrapping if it is to fit continuously against its neighbors. But the internal torsion is fixed at some uniform value which cannot accommodate different topological wrappings. The only reconciliation with equation (1.7) is the development of tangential discontinuities, across which there is a finite difference in field direction. The net torsion across the tangential discontinuity is not restricted by equation (1.7) because the surface of discontinuity is only the surface of contact between two adjacent regions of field. The surface contains no magnetic flux — no field lines — and the restriction posed by equation (1.7) is evaded.

The arbitrarily complicated continuous fields constructed by Rosner and Knobloch (1982) are not restricted by equation (1.7) because they are based on $\nabla\alpha = 0$, for which equation (1.7) is trivially satisfied. On the other hand, if any variation of α is present, the degeneracy is removed, superposition of solutions is not possible, and equation (1.7) becomes a nontrivial restriction, requiring tangential discontinuities if the field topology involves a change in the torsion along any flux bundle anywhere in $0 < z < L$.

Another way to state the problem is to note the two requirements (a) that the torsion α is uniform along each field line while (b) an arbitrary topology is imposed on the field through our choice of ψ in equation (1.4). These two conditions are not irreconcilable unless we also arbitrarily impose the condition that the field is continuous everywhere. To insist on continuity is to restrict the function $\psi(x, y, kzt)$ to special invariant or degenerate forms that are incompatible with the physical possibility of arbitrary weaving and wrapping of the field lines.

We are accustomed to continuous fields in classical field theory, enforced by the fully elliptical character of the basic field equations, and leading to a unique determination of the field throughout a volume by specification of the field on the boundary of the volume. The field equation and the boundary conditions determine the topology of the field. But with a magnetic field in an infinitely conducting fluid the topology is determined ahead of time by the choice of ψ, and the normal component of the field on the boundary can be manipulated at will, by local compression or expansion of the distribution of footpoints of the field, without affecting the topology. So there can be no unique determination of the field topology throughout the volume by the field at the boundaries. As already noted, the field equations reflect this different situation by possessing an extra set of characteristics, in addition to the two imaginery characteristics of the familiar elliptic field equations in a vacuum. The subject is taken up at some length in Chapter 3, but the essential point is simply that the field lines represent a family of real characteristics, across which the field may be discontinuous, thereby accommodating the physics.

1.7 Perturbations of a Continuous Field

An obvious mathematical approach to the problem is to consider magnetostatic fields in some functional neighborhood of a known invariant equilibrium solution. This subject is taken up in Chapter 4 in detail, but it is appropriate to make some general comments here to establish perspective. A convenient example (Parker, 1979, pp. 370–378) is the field with linear invariance ($\partial/\partial z = 0$), satisfying the familiar Grad–Shafranov equations

$$B_x = +\partial A/\partial y, B_y = -\partial A/\partial x, B_z = B_z(A), p = p(A), \tag{1.8}$$

and

$$\nabla^2 A + 4\pi p'(A) + B_z'(A)\, B_z(A) = 0 \tag{1.9}$$

(see Tsinganos, 1981, 1982a,b for the field equations and solutions in other invariant geometries). Treat the total field $\mathbf{B}(x, y) + \delta\mathbf{B}(x, y, z)$ where the perturbation does not share the invariance of $\mathbf{B}(x, y)$. The first order perturbation equations are

$$4\pi\delta p = (\nabla \times \mathbf{B}) \times \delta\mathbf{B} + (\nabla \times \delta\mathbf{B}) \times \mathbf{B}.$$

This equation can then be integrated along the real characteristics, which are the projection of the field lines of $\mathbf{B}(x, y)$ onto any plane $z = constant$, given by $A(x, y) = constant$. In the case of interest, most field lines of $\mathbf{B}$ do not extend to $x, y = \pm\infty$, and, hence, close locally on themselves. The equilibrium equation for $\delta\mathbf{B}$ can be integrated along any field line, and it is found that generally the $\delta\mathbf{B}$ computed after carrying the integration once around a closed field line is different from the initial $\delta\mathbf{B}$ from which the integration started. The special conditions necessary to produce the same final and initial $\delta\mathbf{B}$ can generally be met only on special field lines and not over a finite region of field. Hence there must be a discontinuity in $\delta\mathbf{B}$ somewhere along the contour of integration.

The only way to satisfy the equations and avoid a discontinuity is to require that $\partial\delta\mathbf{B}/\partial z = 0$, which provides a $\mathbf{B} + \delta\mathbf{B}$ that is a member of the same general class of invariant solutions, described by equations (1.8) and (1.9). In other words, there are discontinuous solutions $\delta\mathbf{B}(x, y, z)$ in the neighborhood of the continuous solutions $\mathbf{B}(x, y)$, but the only continuous solutions $\mathbf{B} + \delta\mathbf{B}$ necessarily possess the same invariance ($\partial/\partial z = 0$) as $\mathbf{B}$. So if $\delta\mathbf{B}$ were produced by winding the field lines (by moving the footpoints of the field at the end plates $z = 0, L$) in a pattern that varied along the field, the resulting equilibrium of $\mathbf{B} + \delta\mathbf{B}$ would possess internal discontinuities. The same general result has been demonstrated by Vainshtein and Parker (1986) for rotational invariance ($\partial/\partial\varphi = 0$) and more broadly by Tsinganos (1982c) in the presence of a stationary velocity field.

Tsinganos, Distler, and Rosner (1984) have taken a somewhat different approach, using a Hamiltonian formulation of the toroidal field, so that the field lines are precisely analogous to the dynamical trajectories of a mechanical system in phase space. They use the Kolmogoroff–Arnold–Moser theorem to show that if a symmetric continuous magnetostatic equilibrium is subjected to a perturbation of arbitrary symmetry, then for a finite range of pressure values, the 2D isobaric surfaces do not necessarily coincide with magnetic flux surfaces. But, as already noted, the scalar product of $\mathbf{B}$ with the magnetostatic equation (1.1) leads to $\mathbf{B} \cdot \nabla p = 0$, stating that the isobaric surfaces lie along the field. Hence magnetostatic equilibrium is no longer satisfied everywhere. They show that spatially symmetric magnetostatic equilibria are topologically unstable to finite amplitude perturbations which do not have the original symmetry properties. In other words, symmetric continuous magnetostatic equilibria are special (and hence unlikely) states in which to find a magnetic field.

This development makes Grad's (1967, 1984) earlier point, that magnetostatic equilibrium is unlikely in toroidal fields. More important, it extends the foregoing perturbation calculation to finite amplitudes, and it provides a quantitative demonstration of the very special form required for a field to avoid discontinuities while in a magnetostatic state. Invariance (an ignorable coordinate) is the simplest condition providing continuity in static equilibrium, although, as already noted, there are other special degenerate conditions permitting both static equilibrium and an absence of discontinuities throughout the field.

Coming back to the perturbation $\delta\mathbf{B}$ about a solution $\mathbf{B}$ to the Grad–Shafranov equations, the simplest example of $\mathbf{B}$ is a uniform field B_0 in the z–direction. As already noted this case is degenerate in the sense that there is no characteristic scale. In any field but a uniform field the winding of the field lines of $\mathbf{B}$ about each other provides a scale length in the z–direction that is related to the transverse scale, in the x- and y–directions. But the scale disappears for the uniform field and the scales of variation of $\delta\mathbf{B}$ along and across $\mathbf{B}$ are independent of each other. Thus, if the two scales are treated as being of the same order, the calculation is carried forward in the manner just described, integrating along the field lines of the unperturbed field. The necessary condition for equilibrium of a continuous field is again $\partial\delta\mathbf{B}/\partial z = 0$ (Parker, 1972; 1979 pp. 363–368).

More recently Van Ballegooijen (1985) pointed out that the scale of variation of $\delta\mathbf{B}$ along the uniform field $\mathbf{B}_0$ may be one order larger than across $\mathbf{B}_0$ providing a different ordering of the terms in the field equation (1.5) and an interesting variation in the requirement for solution (noted in §1.6) already noted in a different context in §1.6. In particular, the perturbation $\delta\mathbf{B}$ with a transverse scale ℓ and a magnitude εB_0 ($\varepsilon \ll 1$) has a winding pattern that extends a distance $O(\ell/\varepsilon)$ along $\mathbf{B}$ if the mutual wrapping of the field lines proceeds through one or more radians before the winding pattern changes. Then $\partial/\partial z$ is small $O(\varepsilon)$ compared to $\partial/\partial x$ and $\partial/\partial y$ so that $\partial\delta\mathbf{B}/\partial z$ is small $O(\varepsilon^2)$ and should be dropped from the first-order equations. Then if $\delta\alpha$ represents the torsion coefficient (the zero-order torsion coefficient being identically zero for the uniform field) equation (1.7) reduces to

$$\delta B_x \frac{\partial\delta\alpha}{\partial x} + \delta B_y \frac{\partial\delta\alpha}{\partial y} + B_0 \frac{\partial\delta\alpha}{\partial z} = 0, \tag{1.10}$$

each term being small to second order. Thus by making $\delta\mathbf{B}$ almost invariant $O(\varepsilon)$ with respect to z, the mathematics recovers a condition that is less stringent than the basic $\partial/\partial z = 0$. Equation (1.10) asserts that $\delta\alpha$ is invariant along the field lines of the perturbed field, rather than requiring that $\delta\alpha$ be invariant along the lines of the zero-order field. Van Ballegooijen goes on to point out the exact analogy between equation (1.10) and the equation for the time-dependent vorticity in the 2D motion of an ideal inviscid incompressible fluid. The variable t in the vorticity equation plays the role of z in equation (1.7), in this case, because the torsion $\delta\alpha$ is invariant along each field line and the vorticity is invariant along the world line of each element of fluid. This example is interesting because it establishes the existence of continuous fields which are not degenerate or symmetric or invariant in any simple way. Instead the solutions are characterized by a transverse scale small compared to the longitudinal scale.

There are no known continuous-time-dependent analytical solutions to the 2D vorticity equation, apart from the motion associated with one or two vortex lines with potential flow everywhere between, or the motion associated with many moving vortex lines arranged in special symmetric patterns that do not change with time. The equation describes the time behavior of a 2D fluid, which can be set in arbitrary continuous motion at time $t = 0$, and which continues to move for an indefinite period afterward. There is obviously a diversity of initially continuous-time-dependent flows. One may wonder whether an initially continuous flow of an ideal fluid may in a finite time form tangential discontinuities. The strict mathematical analogy with the magnetostatic equation suggests that there are no such solutions. As we shall see in Chapter 8, the structure of tangential discontinuities in magnetic fields in static equilibrium indicates that they do not begin or end somewhere in the volume between the rigid end plates, $z = 0, L$. The formal analogy between z in the magnetostatic field and t in the evolving hydrodynamic field implies that discontinuities do not begin or end in finite time t. Note, however, that this analogy does not exclude the asymptotic formation of vortex sheets in the limit of large t. On the other hand, in the presence of a small but nonvanishing viscosity, the final asymptotic approach of a vortex layer to zero thickness is irrelevant, and it is well known that a slightly viscous fluid in inhomogeneous motion becomes turbulent in a finite time, with the vortex sheet thickness limited by the viscosity. In a similar way, a magnetostatic field may have an arbitrary winding pattern at $z = 0$, but if the field is continuous, the pattern at $z = 0$ determines the continuous pattern for all finite $z > 0$, producing whatever topology is compatible with the invariance of α along each field line, just as the vorticity ω is transported bodily along the world line of each element of fluid so that $d\omega/dt = 0$.

1.8 Tangential Discontinuities

In nature the winding of the field lines (carried out by random convective transport of the footpoints of the field) is arbitrary so that the field is not expected to have the symmetry, invariance, or degeneracy necessary for continuous solutions. To repeat the point made earlier (following equation (1.7) in §1.6) for arbitary ψ a flux bundle threads around first one way and then the other through the spaghetti of neighboring interwoven flux bundles. The torsion within each elemental flux bundle must vary along the bundle in precise coordination with the winding around the neighboring flux bundles if the bundle is to fit continuously against each flux bundle that it encounters. But $\mathbf{B} \cdot \nabla\alpha = 0$ requires that there is no variation in the torsion α along the flux bundle. So the condition for a continuous field cannot generally be met. There are unavoidable tangential discontinuities where the flux bundle winds around neighboring flux bundles. There is no way that an arbitary winding and interweaving of the field lines can relax to equilibrium and preserve continuity (Parker, 1986a,b,c, 1989a).

The braiding of three flux bundles provides an elementary illustration of the principles. A single wavelength of braiding is sketched in Fig. 1.6, in which each flux bundle wraps first one way and then the other around each of the other two bundles. Hence the net torsion along each flux bundle is zero, so α, being constant along each line of force, can be set equal to zero. Then since $\nabla \times \mathbf{B} = \alpha\mathbf{B}$, it follows that $\nabla \times \mathbf{B} = 0$ so that $\mathbf{B} = -\nabla\phi$ and $\nabla^2\phi = 0$. Within each flux bundle the field is a potential field without torsion. There are no continuous solutions to Laplace's

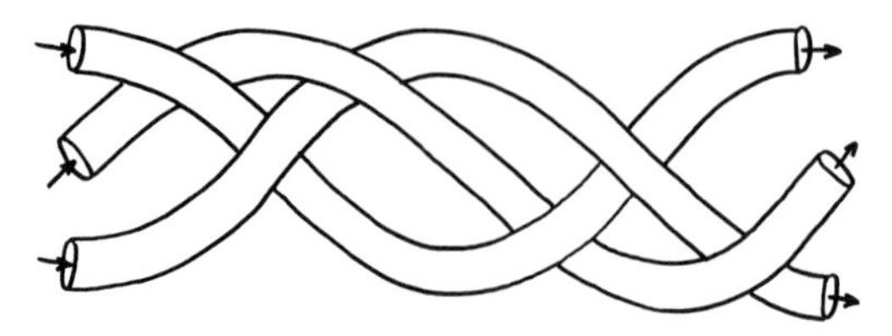

Fig. 1.6. A sketch of one wavelength of the braiding of three flux bundles.

equation representing a braided field. It follows that the fields are not aligned where each flux bundle passes obliquely across a neighbor, so that there is necessarily a tangential discontinuity. Note again, then, that the net torsion (i.e., the discontinuity in field direction) at each mathematical surface of tangential discontinuity does not violate $\mathbf{B} \cdot \nabla\alpha = 0$ because there is no magnetic flux in the mathematical surface. The surface of discontinuity is merely the surface of contact between two regions of nonparallel field.

The tangential discontinuity makes its appearance in the theory of magnetostatics in other ways than through considerations on $\mathbf{B} \cdot \nabla\alpha = 0$. For instance, one may construct any continuous magnetostatic field (e.g., a solution of equation (1.9)) containing more than one winding pattern, i.e., with two or more distinct topological regions, and show from elementary considerations that almost any overall deformation of that field creates tangential discontinuities, by upsetting the X-type vertices of the topological separatrices (Parker, 1982, 1983b, 1990b). The process can be seen in a variety of formal mathematical examples of tangential discontinuities from simple compression or expansion of a continuous field (Priest, 1981, pp. 144–171; Kulsrud and Hahm, 1982; Hahm and Kulsrud, 1985; Low, 1987, 1989; Low and Wolfson, 1987; Jensen, 1989). The discontinuities form along the topological separatrices of the original field. Chapter 6 treats several examples in detail.

A variation of this is to consider the construction of a continuous field by the actual physical displacement of the fluid and field in a hypothetical laboratory apparatus. It is then easy to show that the slightest random error in the mechanical manipulation misses the desired mathematical continuity, producing tangential discontinuities with two or more Y-type neutral points where one X-type is necessary for continuity (Parker, 1982, 1990b).

As already noted, the magnetostatic equation (1.1) has two imaginery characteristics (Parker, 1979, pp. 361–363). This fact in itself would make equation (1.1) a fully elliptic equation, so that the internal field would be uniquely determined by the arbitrary distribution of field at the boundaries. This is physically absurd, of course, because the internal winding and intertwining of the field, i.e., the internal topology, is arbitary, determined separately by the sequence of random swirling of the footpoints of the field (at $z = 0, L$). Following the random intertwining of the field lines the footpoints can be pushed together so that the normal component of the field on $z = 0, L$ has an arbitary distribution, without affecting the internal topology. We remark again that the escape from this contradiction lies in the additional set of characteristics, represented by the field lines. Discontinuities along these real characteristics are permitted. The discontinuities avoid the uniqueness implied by the imaginery characteristics alone, because the uniqueness depends on the assumption that the field is continuous throughout. Thus, in the foregoing example of the braided field, in which the field reduces to a solution of Laplace's equation,

$\nabla^2\phi = 0$, the well known uniqueness of $\nabla^2\phi = 0$ is avoided by the surfaces of tangential discontinuity (the topological separatrices) between the three individual flux bundles.

A direct approach to the formal problem is through integration of the nonlinear magnetostatic equation (1.1). As already noted, there are many families of known continuous solutions. Standard mathematical techniques automatically generate continuous solutions, so they provide the special topologies required to achieve continuity everywhere, but at the price of an ignorable coordinate or degeneracy. Fortunately, there are some simple cases where the integration of equation(1.1) can be carried through in the absence of the symmetry required for continuity (Parker, 1990c). One finds that, if the solution has the specified topology and fits the boundary conditions, then it contains surfaces of tangential discontinuity. If, on the other hand, the solution is required to be continuous, then there appears a unique solution which satisfies the essential boundary condition on the normal component of the field, but which necessarily has an internal topology different from the specified topology. Yet the specified topology is physically well posed, because it can be produced by simple precise hypothetical physical manipulation of a magnetic field in a ponderable fluid under ideal circumstances of zero resistivity. So complete continuity of a field simply does not conform to equation (1.1). An extended example is explored in Chapter 5.

A straightforward integration of $\mathbf{B} \cdot \nabla\alpha = 0$ provides a general expression for α in terms of $\mathbf{B}$, in the form of a contour integral along the field lines and along the orthogonal family of lines. It is shown (Parker, 1986a,b) that α is generally discontinuous, which comes about through the discontinuity of $\mathbf{B}$.

1.9 The Optical Analogy

Finally, we construct the optical analogy (Parker, 1981, 1989b,c, 1991b) to show how the tangential discontinuity arises from the local properties of the magnetic field in the neighborhood of a maximum in the field magnitude $\mathbf{B}(\mathbf{r})$. The optical analogy applies to the projection $\mathbf{F}_s$ of any vector field $\mathbf{F}(\mathbf{r})$ onto the local flux surfaces of $\nabla \times \mathbf{F}$, because $\mathbf{F}_s$ can be described as the gradient of a scalar potential ϕ in that surface. For the force-free field $\nabla \times \mathbf{B} = \alpha\mathbf{B}$, the local flux surfaces of $\nabla \times \mathbf{B}$ coincide with the local flux surfaces of $\mathbf{B}$. So in any flux surface of $\mathbf{B}(\mathbf{r})$ the field can be written as $\mathbf{B} = -\nabla\phi$, (noting that $\nabla \cdot \mathbf{B} = -\nabla^2\phi \neq 0$ in the 2D flux surface). The equations for the field lines in the flux surface are the same as the equations for the optical ray paths in a medium with index of refraction $B(\mathbf{r}) = |\mathbf{B}(\mathbf{r})|$. This is the optical analogy in its form for magnetostatic fields. Its importance lies in the fact that it shows the direct connection between the field pattern and the local variation of $B(\mathbf{r})$. In particular, the field lines are refracted around a local maximum in $B(\mathbf{r})$.

If the maximum is sufficiently concentrated, the optical pathlength around the maximum is shorter than across the maximum. So Fermat's principle specifies a bifurcation in the field pattern where the field lines pass around the maximum rather than across the maximum. This bifurcation or gap in the field pattern is the singular property that leads to the formation of the tangential discontinuity. The gap permits the fields on either side of the flux surface to come in contact, as already remarked. These otherwise separate fields are generally not parallel, with the result that the contact surface, or separatrix, between the regions becomes a surface of tangential discontinuity throughout the gap.

There is a theorem by Yu (1973) that demonstrates the association of a bifurcation in the field lines with a tangential discontinuity. Yu considered field lines in the isobaric surfaces ($p = constant$) in a magnetic field in static equilibrium, described by equation (1.1). Since $\mathbf{B} \cdot \nabla p = \nabla \times \mathbf{B} \cdot \nabla p = 0$, the isobaric surfaces are flux surfaces of both $\mathbf{B}$ and $\nabla \times \mathbf{B}$. Treating two near neighboring field lines he showed that the current density $j_\perp$ perpendicular to the field varies in direct proportion to the separation of the lines. A bifurcation of the field pattern in an isobaric surface represents an increase in the separation of adjacent field lines from infinitesimal to finite distances in a finite distance along the field. Hence the bifurcation is automatically associated with an infinite current density, i.e., a current sheet or tangential discontinuity. It follows from the optical analogy that any field with localized internal maxima produces internal tangential discontinuities, unless there are rigid boundaries so close at to prevent any bifurcation in the field pattern.

1.10 General Discussion

The magnetostatic theorem is a statement about the relation between the topology and the continuity of a magnetostatic field, and as such it touches upon several other properties of the magnetic fields to be found in nature. Consider first the general stochastic nature of the field lines of any magnetic field in the physical universe. The field lines are defined as the integrals of

$$\frac{dx}{B_x} = \frac{dy}{B_y} = \frac{dz}{B_z}$$

at any instant in time. Pick any two neighboring points $\mathbf{r}_0$ and $\mathbf{r}_0 + \delta\mathbf{r}_0$ with $\delta\mathbf{r}_0$ perpendicular to $\mathbf{B}(\mathbf{r}_0)$. These two points define two field lines to the specified field $\mathbf{B}(\mathbf{r})$ with characteristic scale ℓ. Assuming that $|\delta\mathbf{r}| \ll \ell$, the separation $\delta\mathbf{r}(s)$ of the two field lines increases more or less exponentially with distance s measured along either line from $\mathbf{r}_0$, as a consequence of the fluctuating gradients in $\mathbf{B}(\mathbf{r})$. The exponential separation continues until $|\delta\mathbf{r}(s)|$ becomes comparable to ℓ, beyond which the two field lines random walk relative to each other and their separation increases more like $s^{\frac{1}{2}}$ (Jokipii and Parker, 1968; Parker, 1979, pp. 274–297, 1992).

A Hamiltonian formulation of the field, in which the individual field line corresponds to the trajectory of a particle in phase space, is particularly useful to treat the topology when the field closes on itself, i.e., toroidal geometry. In any such closed configuration the field lines are ergodic, as may be seen from the Hamiltonian formulation (Kerst, 1962; Parker, 1969; Boozer, 1983; Cary and Littlejohn, 1983). Since $\mathbf{B} \cdot \nabla p = 0$ it follows that $\nabla p = 0$ throughout an ergodic toroidal field in static equilibrium, from which it follows that there can be no true static equilibrium if the plasma is confined by the field or vice versa (Moser, 1962; Grad, 1967, 1984). The construction of global flux surfaces and isobaric surfaces, which project along the field, is problematical because any such surface fills the entire toroidal volume. Needless to say a *local* flux surface is generated by the projection of any transverse curve along the field lines.

The presence of a suprathermal plasma component (e.g., the multimillion degree gas of the solar corona, the 10^8 K plasma generated in a solar flare, or the cosmic rays that fill the gaseous disk of the galaxy) provides another source of nonequilibrium in any field rooted in the dense thermal gas of a self-gravitating body such as a star or galaxy. For the free flow of the suprathermal gas (without significant gravitational confinement) along the field lines inflates without limit the outer lobes of the field, where the field falls asymptotically to zero. This nonequilibrium is the means by which the open magnetic field regions are formed on the Sun to provide the coronal holes and the solar wind (Parker, 1963b). It appears to be, at least in part, the basis for the extended galactic magnetic halos (Parker, 1965, 1968, 1969, 1979, pp. 274–297, 1992) and for the resulting tangential discontinuities, magnetic dissipation, and X-ray emission from those halos (Parker, 1990a, 1992).

In summary, the stochastic field lines extending between $z = 0$ and $z = L$ (equivalent to the mirror geometry employed in the plasma laboratory) become the ergodic field of toroidal geometry (in which the field circles incommensurably through itself).

The present writing is concerned primarily with the bounded "mirror geometry" of a magnetic field extending in the z–direction between "end plates" $z = 0$ and $z = L$ (Parker, 1972, 1979, p. 364), thereby avoiding questions of the existence of global flux surfaces, global isobaric surfaces, and static equilibrium and avoiding the absence of equilibrium when a suprathermal gas component is present. The flux surfaces are simply and uniquely defined (extending only between the end plates $z = 0, L$) and play a key role in the development of the theory.

1.11 Hydrodynamic Turbulence

Consider the fact discussed in §1.5 that the magnetostatic equation (1.1) and the stationary Euler equation (1.2) are analogous. Thus the surfaces of tangential discontinuity (current sheets) in magnetostatic fields have an exact counterpart in the surfaces of tangential discontinuity (vortex sheets) in stationary flows. Arnold pointed out that almost all solutions of the Euler equation, and hence of the magnetostatic equation, involve tangential discontinuities. It can now be stated that there are no continuous solutions for almost all field topologies.

Now, we are familiar with the idea that the dynamical instability of a stationary flow at high Reynolds number leads to turbulence and the formation of vortex sheets (Batchelor, 1947). The interesting point is that the formation of vortex sheets is already an intrinsic part of the initial stationary flow, arising from the nature of the static Reynolds stresses (Parker, 1989b, 1991b). Turbulence is, then, the unstable time-dependent form of the ubiquitous vortex sheet formation, driven by the Reynolds stress in both the time-dependent and time-independent flow. Specifically, the vorticity ω is transported bodily with the fluid velocity $\mathbf{v}$, with

$$\partial\omega/\partial t = \nabla \times (\mathbf{v} \times \omega),$$

and regions of different vorticity, i.e., separate eddies, coming into contact without their velocities being compelled to match smoothly, creating a vortex sheet at the surface of contact, i.e., at the topological separatrix.

The formation of vortex sheets in a turbulent flow is conventionally referred to as a "cascade" to large wave number $\mathbf{k}$. The large wave number arises mainly from the abrupt change of $\mathbf{v}$ across the vortex sheet, remembering that the sheet itself is relatively broad, involving small wave numbers. The broad vortex sheet is continually complicated by the Kelvin–Helmholtz instability which produces corrugations at increasingly large wave number, of course. The cascade to large wave numbers in statistically stationary turbulence may be thought of as the time-dependent unstable form of the tangential discontinuities of the stationary flow with any but the simplest topology.

The intrinsic role of tangential discontinuities in stationary velocity fields and in static magnetic fields could also be referred to as a cascade to large wave numbers, but in practice the term *cascade* suggests the concept of dynamical effects in turbulence, which is absent in the quasi-static formation of tangential discontinuities. On the other hand, the optical analogy, which is not restricted to quasi-static fields provides an appropriate mathematical tool for treating the dynamical tendency toward tangential discontinuities in the time-dependent turbulent flow, showing that any sufficiently localized maximum in v_s presses toward a gap in the pattern of $\mathbf{v}_s$, where $v_s = |\mathbf{v}_s|$ and $\mathbf{v}_s$ represents the instantaneous projection of the fluid velocity $\mathbf{v}$ onto a flux surface of the vorticity $\nabla \times \mathbf{v}$. This is developed at length in Chapter 7.

2

The Field Equations

2.1 Appropriate Field Concepts

Maxwell's equations can be written in terms of various field vectors **E**, **D**, **H**, **B**, **A**, **j**, etc. The first step is to establish the most convenient vector quantities for formulating the basic equations of magnetohydrodynamics (MHD). We find it simplest to deal directly with the electric field **E** and the magnetic field **B**, in a system of units (e.g., cgs) that is compatible with the basic symmetry of Maxwell's equations. In these units Maxwell's equations are

$$\nabla \cdot \mathbf{E} = 4\pi s, \tag{2.1}$$

$$\nabla \cdot \mathbf{B} = 0, \tag{2.2}$$

$$4\pi \mathbf{j} + \partial \mathbf{E}/\partial t = +c\nabla \times \mathbf{B}, \tag{2.3}$$

$$\partial \mathbf{B}/\partial t = -c\nabla \times \mathbf{E}, \tag{2.4}$$

where s is the electric charge density and **j** is the electric current density. The electric field **E** is defined in terms of the force on a static charge, and **B** can be defined by the Lorentz force $\mathbf{j} \times \mathbf{B}/c$ or by the equivalent Lorentz transformation in the presence of an electric field, there being no experimental evidence of magnetic charges in the universe. The field pressures $E^2/8\pi$ and $B^2/8\pi$ and the field tensions $E^2/4\pi$ and $B^2/4\pi$ are measured in dynes/cm^2 and the electric and magnetic fields E and B are of equal magnitude (gm$^{\frac{1}{2}}$ cm$^{-\frac{1}{2}}$ sec^{-1}) when their stress densities are equal. The effect of matter on the electromagnetic fields **E** and **B** appears through s and **j**.

The molecular electric and magnetic polarization of ionized gases is negligible in most circumstances, in contrast to the dominant effect of the inertia of the bulk motion **v** of the plasma. Thus, while the dielectric coefficient ε and the magnetic permeability μ of gaseous materials are close to one, the inertia of the plasma coupled to the magnetic field causes the phase velocity of a plane transverse MHD (low frequency) wave to be of the order of the Alfven speed $C = B/(4\pi\rho)^{\frac{1}{2}}$, which is small compared to the speed of light in any case with which we shall be concerned. So the formulation of the MHD equations is carried out in the nonrelativistic regime, neglecting the atomic polarization. That is to say, the development neglects $\varepsilon - 1$, $\mu - 1$, and v^2/c^2 compared to one. The effect of the inertia of the plasma is to produce an effective plasma index of refraction kc/ω large compared to one, even if $\varepsilon, \mu \cong 1$.

Now the formulation of the physical concepts involved in the MHD equations is most simply understood in the cgs system of units, as already noted. It is possible to accomplish the development in the mks system, with the accessory coefficients ε_0 and μ_0, of course, but the procedure with **E** and **B** measured in different units too often serves to obscure the simple basic concepts. Once the MHD equations are formulated, of course, it makes relatively little difference what system of units is used because the electric field is suppressed in the MHD field equations in favor of **B** and $\nabla \times \mathbf{B}$.

In view of current popular misunderstandings of the applicability of MHD theory, the development of MHD is presented here to illustrate the minimum assumptions necessary for the validity of MHD. The exposition begins with the concepts of electromagnetic momentum and energy (Poynting's theorem), treating an aggregate of moving electrically charged particles as the primitive physical system.

2.1.1 *Momentum and Energy of the Particles*

The equation of motion for an individual particle of mass M, charge q, position x_i, and velocity $w_i = dx_i/dt$ is

$$M dw_i/dt = q(E_i + \varepsilon_{ijk} w_j B_k/c) \tag{2.5}$$

$$\equiv F_i, \tag{2.6}$$

in an electric field E_i and magnetic field B_i, where ε_{ijk} is the usual permutation tensor used in forming vector products. The quantity F_i represents the total force on the particle.

Describe the velocity distribution of a large number of particles in some neighborhood of the coordinate position **r** by $f(\mathbf{r}, \mathbf{w}, t)$, normalized so that the number density of the particles, denoted by $N(\mathbf{r}, t)$, is described by

$$N(\mathbf{r}, t) = \int d^3\mathbf{w} \quad f(\mathbf{r}, \mathbf{w}, t). \tag{2.7}$$

The integration is over velocity space and $f(\mathbf{r}, \mathbf{w}, t)$ includes all particle species. It is assumed that $f(\mathbf{r}, \mathbf{w}, t)$ is nonvanishing only for $w \equiv |\mathbf{w}| \ll c$. The bulk velocity $v_i(\mathbf{r}, t)$, is given by the first moment of f,

$$N(\mathbf{r}, \mathbf{t}) v_i(\mathbf{r}, \mathbf{t}) = \int d^3\mathbf{w} w_i f(\mathbf{r}, \mathbf{w}, t). \tag{2.8}$$

The thermal velocity u_i of a particle is the difference between the individual particle velocity w_i and the bulk velocity $v_i(\mathbf{r}, t)$, so that

$$w_i = v_i(\mathbf{r}, t) + u_i, \tag{2.9}$$

noting that $\int d^3\mathbf{w} u_i f$ vanishes.

The distribution function f satisfies the Boltzmann equation

$$\frac{\partial f}{\partial t} + w_j \frac{\partial f}{\partial x_j} + \frac{F_j}{M} \frac{\partial f}{\partial w_j} = \left(\frac{\partial f}{\partial t} \right)_c, \tag{2.10}$$

where $(\partial f/\partial t)_c$ represents the transfer of particles from one velocity to another through collisions, conserving particles, particle momentum, and particle energy in the present case. It is assumed here that the collisions excite no internal degrees of freedom in the particles.

Proceeding in the conventional manner, consider the integral of the Boltzmann equation over velocity space. The collision term gives nothing, because the collisions conserve particles. The integral of $F_j \partial f/\partial w_j$ gives nothing because the part of F_j that depends upon w_j is perpendicular to w_j, so that it follows from equation (2.5) that $\partial F_j/\partial w_j = 0$. Hence integration by parts causes that term to vanish. The first two terms, then, give the familiar result,

$$\partial N/\partial t + \partial N v_j/\partial x_j = 0, \tag{2.11}$$

for conservation of particles.

Next multiply the Boltzmann equation by Mw_k and integrate over velocity space. The first two terms yield,

$$\begin{aligned} M\int d^3\mathbf{w}\left(w_k\frac{\partial f}{\partial t}+w_kw_j\frac{\partial f}{\partial x_j}\right) &= \frac{\partial}{\partial t}M\int d^3\mathbf{w}w_kf+\frac{\partial}{\partial x_j}M\int d^3\mathbf{w}w_jw_kf \\ &= \frac{\partial}{\partial t}MNv_k+\frac{\partial}{\partial x_j}M\int d^3\mathbf{u}(v_j+u_j)(v_k+u_k)f \\ &= \frac{\partial}{\partial t}MNv_k+\frac{\partial}{\partial x_j}MNv_jv_k \\ &\quad +\frac{\partial}{\partial x_j}M\int d^3\mathbf{u}u_ju_kf, \end{aligned} \tag{2.12}$$

remembering that the mean value of u_i is zero.

Note that $M\int d^3\mathbf{w}u_ju_kf$ represents the rate of transport of momentum Mu_k in the j-direction by the thermal motion u_j or the transport of momentum Mu_j in the k-direction by the thermal motion u_k. It is exactly analogous to the Reynolds stress tensor $R_{ij} = -\rho v_j v_k$, representing the bulk transport of momentum ρv_k by v_j or the bulk transport of momentum ρv_j by v_k. Hence, it may be called the thermal stress tensor p_{ij} (pressure tensor), writing

$$\begin{aligned} p_{ij} &= M\int d^3\mathbf{u}\,u_j\,u_kf \\ &= M\int d^3\mathbf{u}(w_j-v_j)(w_k-v_k)f. \end{aligned}$$

The thermal transport of momentum represented by p_{ij} does not depend upon collisions, which conserve momentum and contribute nothing to its transport. However, collisions enter into the computation of p_{ij} (cf. Chapman and Cowling, 1958; Sommerfeld, 1964) through the viscosity, which is neglected in the present calculation.

The third term on the left-hand side of the Boltzmann equation can also be evaluated formally, which shows that it is the net force per unit volume, easily computed by direct summation of the right-hand side of equation (2.5). We have

$$\begin{aligned} \int d^3\mathbf{w}w_k\,\mathbf{F}_j\frac{\partial f}{\partial w_j} &= \int d^3\mathbf{w}\left[\frac{\partial}{\partial wi}w_k\,F_jf-F_kf-w_k\frac{\partial F_j}{\partial w_j}f\right] \\ &= -\int d^3\mathbf{w}F_kf, \end{aligned}$$

since f vanishes at $w_k = \pm\infty$ and it is obvious from equations (2.5) and (2.6) that $\partial F_j/\partial w_j = 0$. Working directly with F_k, then, the sum of the charges q over all particles in a unit volume gives the total charge density s, while summing qw_j over all particles gives the net current density j_j. The final result from the integration of the Boltzmann equation is, then,

$$\frac{\partial}{\partial t} NMv_i = \frac{\partial R_{ij}}{\partial x_j} - \frac{\partial p_{ij}}{\partial x_j} + sE_i + \varepsilon_{ijk} j_j B_k / c \tag{2.13}$$

for the time rate of deposition of momentum at any given point in the plasma. The last term is just $\mathbf{j} \times \mathbf{B}/c$, of course.

The energy equation can be deduced along the same lines. Multiply the Boltzmann equation by $\frac{1}{2} M w_k w_k$ and integrate over velocity space again. The first two terms give

$$\frac{1}{2} M \int d^3\mathbf{w} w^2 \left(\frac{\partial f}{\partial t} + w_j \frac{\partial f}{\partial x_j} \right)$$

$$= \frac{\partial}{\partial t} \int d^3\mathbf{w} \frac{1}{2} M w^2 f + \frac{1}{2} M \frac{\partial}{\partial x_j} \int d^3\mathbf{u} f w^2 (v_j + u_j)$$

$$= \frac{\partial}{\partial t} \left(\frac{1}{2} NMv^2 + \frac{1}{2} M \int d^3\mathbf{u}\, u^2 f \right)$$

$$+ \frac{\partial}{\partial x_j} \left(\frac{1}{2} NMv^2 v_j + \frac{1}{2} v_j p_{kk} + \int d^3\mathbf{u} \frac{1}{2} M w^2 u_j f \right).$$

The quantity differentiated with respect to time on the right-hand side is obviously the total kinetic energy density of the particles, including the bulk motion and the thermal motion, for which the thermal energy density is

$$U = \frac{1}{2} M \int d^3\mathbf{u}\, u^2 f. \tag{2.14}$$

Note, then, that $p_{kk} = 2U$ and the term differentiated with respect to x_j represents the divergence of the energy transport flux. The first term is the bulk transport of the kinetic energy of bulk motion. The second term is

$$\frac{1}{2} v_j p_{kk} = v_j U \tag{2.15}$$

and represents the bulk transport of the thermal energy. The third term represents the net transport of total kinetic energy (bulk plus thermal) by the thermal motion. It can be expressed as the sum of two terms,

$$\int d^3\mathbf{u} M v_k u_k u_j f + \int d^3\mathbf{u} \frac{1}{2} M u_j u^2 f.$$

The first is the transport of the cross product energy $Mv_k u_k$ by the thermal velocity u_j, and can be written $v_k p_{kj}$, if preferred. The second represents the transport of thermal energy by the thermal motion u_j, and is the heat conduction term. In sum, then, the terms represent conservation of kinetic energy.

The remaining term on the left-hand side of the Boltzmann equation (2.10) represents the rate at which the force F_j does work on the particles, which is readily demonstrated by multiplying the term by $\frac{1}{2}Mw_k w_k$ and integrating by parts,

$$\begin{aligned}\frac{1}{2}M\int d^3\mathbf{w}w_k w_k F_j\frac{\partial f}{\partial w_j} &= \frac{1}{2}\int d^3\mathbf{w}\left[\frac{\partial}{\partial w_j}(w_k w_k F_j f) - 2w_k F_k f - w_k w_k\frac{\partial F_j}{\partial w_j}f\right]\\ &= -\int d^3\mathbf{w}w_k F_k f \\ &= -qw_k E_k N \\ &= -j_k E_k.\end{aligned} \tag{2.16}$$

Finally the right-hand side of the Boltzmann equation represents collisions, which are taken to be elastic so that there is no loss or gain of energy. The complete energy equation is, then,

$$\frac{\partial}{\partial t}\left(\frac{1}{2}NMv^2 + U\right) + \frac{\partial}{\partial x_j}\left(\frac{1}{2}Nmv^2 v_j + v_j U + v_k p_{kj} + \int d^3\mathbf{u}\frac{1}{2}Mu^2 u_j f\right) = j_k E_k. \tag{2.17}$$

Multiplying the Boltzman equation by $Mw_k w_\ell$ and integrating over velocity space yields an equation for $\partial p_{k\ell}/\partial t$ in terms of $p_{k\ell}$ and the general heat flow tensor $\int d^3\mathbf{u}u_k u_\ell u_m f$. Note again that the integral over the collision term generally does not vanish, because the finite mean free path allows particles from one location to propagate to another, taking their mean momentum with them to provide the momentum transfer referred to as viscosity. Heat transport and viscosity play important roles in astrophysical plasmas, but they are not central to the spontaneous occurrence of tangential discontinuities, so we pursue p_{ij} no further.

The next step is to rewrite the terms $sE_i + \varepsilon_{ijk}j_j B_k/c$ and $j_k E_k$ on the right-hand sides of equations (2.13) and (2.17), respectively, in a form comparable to the momentum and energy transport terms for the particles. The procedure is well known and can be found in any textbook on electrodynamics. A brief outline is provided below.

2.1.2 Electromagnetic Momentum and Energy

Consider the momentum equation (2.13). Use equation (2.1) to eliminate s from the right-hand side. Add $\mathbf{B}\nabla\cdot\mathbf{B}/4\pi$ to the right-hand side and then use equation (2.3) to eliminate $\mathbf{j}$. Rewrite the term $+(\mathbf{B}/4\pi c)\times\partial\mathbf{E}/\partial t$ as

$$\frac{\mathbf{B}}{4\pi c}\times\frac{\partial\mathbf{E}}{\partial t} = \frac{1}{c^2}\frac{\partial\mathbf{P}}{\partial t} + \frac{(\nabla\times\mathbf{E})\times\mathbf{E}}{4\pi}$$

by using equation (2.4), where $\mathbf{P} = c\mathbf{E}\times\mathbf{B}/4\pi$ is the Poynting vector. With the vector identity

$$(\nabla\times\mathbf{E})\times\mathbf{E} = -\nabla E^2/2 + (\mathbf{E}\cdot\nabla)\mathbf{E},$$

the result can be written as

$$\frac{\partial}{\partial t}(NMv_i + P_i/c^2) = \frac{\partial}{\partial x_j}(R_{ij} - p_{ij} + M_{ij}), \tag{2.18}$$

where M_{ij} is the Maxwell stress tensor,

$$M_{ij} = -\delta_{ij}(E^2 + B^2)/8\pi + (E_i E_j + B_i B_j)/4\pi, \tag{2.19}$$

and P_i is just the Poynting vector $\varepsilon_{ijk} c E_j B_k / 4\pi$.

The physical interpretation of equation (2.18) is unique. The term P_i/c^2 appears on an equal footing with NMv_i inside the time derivative, and therefore can be interpreted only as the momentum density of the electromagnetic field. It follows that M_{ij} represents the electromagnetic stress. That is to say, one region transfers momentum electromagnetically to another region with the stress density M_{ij}. The electromagnetic momentum density P_i/c^2 and the Maxwell stresses are just as real (if less familiar) as the momentum density NMv_i, the pressure tensor p_{ij}, and the Reynolds stress R_{ij} of the particles. The Maxwell stress is transferred to the particles through the usual force term $s\mathbf{E} + \mathbf{j} \times \mathbf{B}/c$, representing the divergence $(\partial M_{ij}/\partial x_j)$ of M_{ij}, which extracts the momentum from the momentum density P_i/c^2 of the electromagnetic field. Inspection of the form of M_{ij} shows the electromagnetic stress to be made up of an isotropic pressure $(E^2 + B^2)/8\pi$ and a tension $E^2/4\pi$ along the electric field lines and $B^2/4\pi$ along the magnetic field lines. In the MHD limit (neglecting terms second order in v/c compared to one) the electric stress drops out, so that the total pressure reduces to $B^2/8\pi$ and the tension along the field lines reduces to $B^2/4\pi$.

The energy equation (2.17) can be rewritten in a similar fashion using equation (2.3) again to eliminate $\mathbf{j}$, and writing

$$\begin{aligned}\mathbf{E} \cdot \nabla \times \mathbf{B} &= \mathbf{B} \cdot \nabla \times \mathbf{E} - \nabla \cdot (\mathbf{E} \times \mathbf{B}) \\ &= -\frac{1}{c}\mathbf{B} \cdot \partial \mathbf{B}/\partial t - \nabla \cdot (\mathbf{E} \times \mathbf{B})\end{aligned}$$

with the aid of equation (2.4). The final result is

$$\begin{aligned}&\frac{\partial}{\partial t}\left(\frac{1}{2}NMv^2 + U + \frac{E^2 + B^2}{8\pi}\right) \\ &+ \frac{\partial}{\partial x_j}\left(\frac{1}{2}NMv^2 v_j + v_j U + v_k p_{kj} + \int d^3w \frac{1}{2} Mu^2 u_j f + P_j\right) = 0.\end{aligned} \tag{2.20}$$

Thus, the energy density is $(E^2 + B^2)/8\pi$, and the electromagnetic energy flux is represented by the Poynting vector P_i. The electromagnetic energy density and the energy flux P_i are just as real as the kinetic energy $\frac{1}{2}NMv^2$ and energy flux $\frac{1}{2}NMv^2 v_j$ of the bulk motion of the particles. The energy of the field is transferred to the particles through the work term $\mathbf{j} \cdot \mathbf{E}$, representing the negative divergence of the Poynting vector.

The essential point of this well-known formulation of the momentum and energy equations is that the electromagnetic force (momentum transport) is transmitted through space by the Maxwell stresses M_{ij} and the electromagnetic energy is transmitted through space by the Poynting vector P_i. The divergence of M_{ij} represents the transfer of electromagnetic momentum to the particles at each point in space, and the negative divergence of P_i represents the transfer of electromagnetic energy to the particles at each point. Most important for MHD is the fact that the energy of the system is made up entirely of the kinetic energy of the bulk motion and the thermal motion, and the energy $B^2/8\pi$ in the field. The energy represented by the electric current is $\frac{1}{2}Mmv_c^2$ (where v_c is the mean electron conduction velocity) and is negligible in any but the most extreme small-scale current densities (see §2.4.3). To be precise the kinetic energy density $\frac{1}{2}Nmv_c^2$ of the electric current is smaller than the kinetic energy of the bulk motion v by the factor $(m/M)(v_c/v)^2$, and smaller than the ion thermal energy by $\frac{1}{3}(m/M)(v_c/u_I)^2$, and smaller than the magnetic energy by $(m/M)(v_c/C)^2$, where u_I is the ion thermal velocity $(kT_I/M)^{\frac{1}{2}}$ and C is the Alfven speed $B/(4\pi NM)^{\frac{1}{2}}$. It is pointed out in §2.4.3 that anomalous resistivity is created by plasma turbulence when v_c becomes as large as u_I, and very strong anomalous resistivity and electric double layers are expected when v_c becomes comparable to $(M/m)^{\frac{1}{2}}C$ in tenuous plasmas.

2.2 The MHD Limit

Consider the reduction of the equations of motion and Maxwell's equations to a form convenient for the treatment of the bulk motion and broad magnetic fields in astrophysical settings and in large laboratory plasma apparatus. The MHD limit is the appropriate form of the equation, providing an accurate description of the dynamics on any scale sufficiently large that the particles can be represented in the fluid approximation. The MHD induction equation then follows from the magnetic induction equation (2.4) and a Lorentz transformation, with only *the assumption that the fluid cannot sustain an electric field in its own frame of reference* without the presence of an enormous electric current density. That is to say, the only assumption is that the fluid is not a significant electrical insulator. Most gases in astronomical settings have enough free ions and electrons that they cannot be regarded as insulators.

There are exceptions, of course. The atmosphere in which this monograph was written, and is at this moment being read, is an obvious case in point. Perhaps the interiors of some very dense cold molecular clouds are another. The interior of an intense current sheet may sustain an electric field as a consequence of the large electric current density and the anomalous resistivity caused by that current, in which case MHD might not describe the interior of that current sheet.

To proceed, then, the development is carried out in the MHD limit, in which the particle velocities are small compared to the speed of light c, and the characteristic scale ℓ and characteristic time τ of variation of the plasma are large compared to the ion cyclotron radius R_{cI} and the ion cyclotron period $\tau_{cI} = 2\pi/\omega_{cI}$ of the particles that collectively make up the fluid ($\omega_{cI} = qB/Mc$), respectively. Except in the most extreme cases, there are so many ions and electrons that a smooth bulk velocity $\mathbf{v}(\mathbf{r}, t)$ can be defined in terms of the local mean velocity of the ions. The mean number density of the ions $N_I(\mathbf{r}, t)$ and the number density of the electrons $N_E(\mathbf{r}, t)$ are also well defined. The plasma may, for simplicity, be assumed to be fully and singly ionized (e.g., hydrogen above 2×10^4 K). Electrostatic forces guarantee that the electron number density $N_E(\mathbf{r}, t)$ and the ion number density $N_I(\mathbf{r}, t)$ are closely equal, provided that the number of particles is averaged over dimensions somewhat larger than the Debye radius $R_D = (kT_E/4\pi N_E e^2)^{\frac{1}{2}}$, where T_E is the electron temperature. The electric charge density $s(\mathbf{r}, t)$ is defined by the small difference

$$s(\mathbf{r}, t) = e[N_I(\mathbf{r}, t) - N_E(\mathbf{r}, t)]. \tag{2.21}$$

Since $\nabla \cdot \mathbf{E} = 4\pi s$ and (as will be shown) $\mathbf{E} = -\mathbf{v} \times \mathbf{B}/c$, it follows that s is of the order of $(B/\ell)(v/c)$, where ℓ is again the characteristic scale of the field. The electrostatic force on s is $s\mathbf{E}$, which is of the order of $(B^2/\ell)v^2/c^2$. It is smaller than the Lorentz force $O(B^2/\ell)$ by the factor v^2/c^2 and therefore is neglected in MHD theory.

There is also a small charge density associated with the electrostatic field that maintains approximate charge neutrality in the presence of an electron pressure gradient$\nabla(NkT_E)$. That electric field is of the order of $m\langle u_E^2\rangle/\ell$, where $\langle u_E^2\rangle$ represents the mean square electron thermal velocity. The associated charge density is of the order of $m\langle u_E^2\rangle/e\ell^2$, and the force per unit volume exerted on this charge density is $m^2\langle u_E^2\rangle^2/e^2\ell^3$. It is smaller than the Lorentz factor $O(B^2/\ell)$ by the factor $(R_{cE}^2/\ell^2)\langle u_E^2\rangle/c^2$, where $R_{cE} = m\langle u_E^2\rangle^{\frac{1}{2}} c/eB$ is the characteristic electron cyclotron radius. Assuming that the electron and ion temperatures are comparable in magnitude, $m\langle u_E^2\rangle$ can be replaced by $M\langle u_I^2\rangle$. The factor can then be written as $(R_{cI}^2/\ell^2)\langle u_I^2\rangle/c^2$, which can be neglected on the basis of either of the two factors alone.

The electric current density is

$$\mathbf{j}(\mathbf{r}, t) = e\mathbf{v}_c(\mathbf{r}, t)N(\mathbf{r}, t), \tag{2.22}$$

where the mean velocity $\mathbf{v}_E(\mathbf{r}, t)$ is written as $\mathbf{v}(\mathbf{r}, t) + \mathbf{v}_c(\mathbf{r}, t)$, so that $\mathbf{v}_c(\mathbf{r}, t)$ represents the (usually small) electron conduction velocity relative to the ions. The number density N can be set equal to either N_I or N_E here. As we shall see, the current density is determined by the curl of the magnetic field through Ampere's law, and vice versa.

The matter density $\rho(\mathbf{r}, t)$ is equal to $(M + m)N(\mathbf{r}, t)$, where M and m are the ion and electron masses, respectively ($m \ll M$). The total momentum density of the plasma is $\rho\mathbf{v}(\mathbf{r}, t)$ neglecting the inertia of the electron conduction velocity $\mathbf{v}_c$ and recognizing that the energy density $B^2/8\pi$ of the magnetic field is small $O(v^2/c^2)$ compared to the rest energy density ρc^2 of the plasma.

In this connection, note that the ion thermal velocity $u_I = (kT_I/M)^{\frac{1}{2}}$ is associated with the ion temperature T_I, the electron thermal velocity $u_E = (kT_E/m)^{\frac{1}{2}}$ is associated with the electron temperature T_E (there is no general reason to expect T_E and T_I to differ in order of magnitude), and the Alfven speed $C = B/(4\pi NM)^{\frac{1}{2}}$ is associated with the magnetic field B. The theoretical development is based on the restriction that the largest of the quantities u, u_I, u_E, and C is small compared to the speed of light c. Hence the Debye radius

$$R_D = (kT_E/4\pi Ne^2)^{\frac{1}{2}}, \tag{2.23}$$

and the characteristic ion cyclotron radius

$$R_{cI} = Mu_I c/eB, \tag{2.24}$$

are related by

$$R_D = R_{cI}(C/c)(T_E/T_I)^{\frac{1}{2}}.$$

That is to say, the Debye radius is small $O(C/c)$ compared to the ion cyclotron radius.

2.2.1 *Statistical Averages*

This is perhaps the appropriate place to consider the number of electrons or ions N_D in a Debye sphere,

$$N_D = 4\pi R_D^3 N/3. \tag{2.25}$$

This number is large in laboratory plasmas and in astrophysical settings outside stellar interiors. It is of the order of one within stellar interiors. Numerically, $R_D \cong 7(T/N)^{\frac{1}{2}}$ cm and

$$N_D \cong (10^2 T/N^{\frac{1}{3}})^{\frac{3}{2}}. \tag{2.26}$$

At the center of a star like the sun, $T \cong 1.5 \times 10^7$ K and $N \cong 2 \times 10^{26}/\text{cm}^3$, with the result that $R_D \cong 3 \times 10^{-9}$ cm and $N_D \cong 5$. At a depth of 2×10^3 km below the surface of a star like the Sun, where the temperature T is 2×10^4 K and the hydrogen is almost fully ionized, the density is $N \cong 10^{19}/\text{cm}^3$ so that $R_D \cong 3 \times 10^{-7}$ cm and $N_D \cong 1$. It follows, then, that the number of electrons or ions

$$N_{cI} = 4\pi N R_{cI}^3/3$$

within an ion cyclotron sphere is large, of the order of $N_D(c/C)^3$, because $C \ll c$ and $N_D \geqslant O(1)$. Hence in treating magnetic fields in stellar interiors it is sufficient for most purposes to average over a cyclotron sphere in computing the mean values of $N, \mathbf{v}$ etc.

Outside stars, in their outer atmospheres, in the heliosphere, and in interstellar and intergalactic space, and in laboratory plasmas the densities are so small that $N_D \gg 1$ and it is sufficient for most purposes to average over the Debye sphere.

The essential point is that the large-scale properties of the plasma in an electromagnetic field **E** and **B** under the conditions assumed in the foregoing paragraphs ($\ell \gg R_c \gg R_D; v, u_E, C \ll c$) can be approximated by a fluid. Conservation of fluid requires the familiar continuity equation

$$\partial\rho/\partial t + \nabla \cdot (\rho\mathbf{v}) = 0. \tag{2.27}$$

which has already been written as equation (2.11). The momentum equation (2.13) can be written as

$$\partial\rho\mathbf{v}/\partial t = -\nabla \cdot (\rho\mathbf{vv}) - \nabla p + \mathbf{j} \times \mathbf{B}/c, \tag{2.28}$$

where the electrostatic force $s\mathbf{E}$ has been neglected, it being small $O(v^2/c^2)$ and where p is the thermal pressure of the plasma (conveniently assumed to be isotropic for the moment) and equal to $Nk(T_E + T_I)$. The dyadic $-\rho\mathbf{vv} = -\rho v_i v_j = R_{ij}$ is the Reynolds stress (the momentum density $\rho\mathbf{v}$ transported by the bulk motion $\mathbf{v}$).

2.2.2 *Electromagnetics of a Noninsulating Fluid*

The electric field $\mathbf{E}'$ in the frame of reference of the fluid, moving with velocity $\mathbf{v}$, is related to **E** and **B** in the fixed coordinate frame by the familiar Lorentz transformation

$$\mathbf{E}' = \mathbf{E} + \mathbf{v} \times \mathbf{B}/c \tag{2.29}$$

at each point, neglecting terms $O(v^2/c^2)$ compared to one. Then if the plasma cannot sustain $\mathbf{E}'$, it follows that $\mathbf{E} = -\mathbf{v} \times \mathbf{B}/c$ and the induction equation (2.4) becomes the familiar MHD induction equation,

$$\partial\mathbf{B}/\partial t = \nabla \times (\mathbf{v} \times \mathbf{B}). \tag{2.30}$$

This equation states that the magnetic field is transported bodily by the bulk velocity $\mathbf{v}$ of the plasma, or fluid.

With this picture in mind, then, note that the electric field is small $O(v/c)$ compared to the magnetic field. Hence the energy density of the electric field is small $O(v^2/c^2)$ and can be neglected. The Maxwell stress tensor equation (2.19) reduces to

$$M_{ij} = -\delta_{ij} B^2/8\pi + B_i B_j/4\pi, \tag{2.31}$$

representing an isotropic pressure $B^2/8\pi$ and a tension $B^2/4\pi$ along the field.

The characteristic time of variation of **B** is $O(\ell/v)$. This is therefore also the time of variation of **E**. On the other hand, the term $c\nabla \times \mathbf{B}$ on the right-hand side of equation (2.3) is cB/ℓ, from which it follows that $\partial\mathbf{E}/\partial t$ is small $O(v^2/c^2)$ compared to $c\nabla \times \mathbf{B}$. Hence, in the MHD limit, equation (2.3) reduces to Ampere's law

$$4\pi\mathbf{j} = c\nabla \times \mathbf{B}. \tag{2.32}$$

This equation plays a basic role in MHD, permitting immediate calculation of the induced current density for any given field configuration $\mathbf{B}(\mathbf{r}, t)$. Unfortunately, the converse relationship for computing **B** from a specified **j** is more complicated, taking the integral form

$$\mathbf{B}(\mathbf{r}, t) = \frac{1}{c}\int \frac{d^3\mathbf{r}'\mathbf{j}(\mathbf{r}', t) \times (\mathbf{r} - \mathbf{r}')}{|\mathbf{r} - \mathbf{r}'|^3} \tag{2.33}$$

of the Biot–Savart law. Thus, to compute **B** at any point, the current density **j** must be specified over the entire volume. Any omission leads to error in **B**. Formulated in terms of **j** MHD is a nonlocal theory. Recall that the energy density and the stress (momentum flux) are both directly related to **B** rather than **j**. Hence, **j** is one step removed from the basic physical properties of the field. To get from **j** to **B** requires the global integral of the Biot–Savart law.

The essential point is that the magnetic field interacts with electrical conductors and must be considered a tangible elastic medium. It contains elastic energy $B^2/8\pi$ and exerts elastic forces (momentum flux) described by equation (2.31). In MHD the magnetic field is tangible, and ponderable as well, because the conducting fluid is locked into the field. The composite medium exerts both magnetic stresses and Reynolds stresses. The important scientific feature is the Maxwell stress introduced by the magnetic field, with tension rather than compression along the field, in contrast with the Reynolds stress in hydrodynamics which has compression $-\rho v_i v_j$. It is the tension along the field that causes the phenomenon of rapid reconnection of magnetic fields, providing the mystique of MHD in the astronomical universe and in the terrestrial plasma laboratory. In summary, then, magnetic fields are physical entities that poke and shove their way across the stage, providing the exotic suprathermal activity that is so conspicuous in plasmas on all scales. Magnetic fields are accompanied by an entourage of electric current and electric field, which play only subsidiary roles in most astrophysical cases.

2.2.3 Electromagnetics of a Weakly Insulating Fluid

Suppose as the next step in the development that $\mathbf{E}'$ is not precisely zero, although relatively small, and is related to the current density by the simple scalar Ohm's law,

$$\mathbf{j} = \sigma\mathbf{E}', \tag{2.34}$$

in terms of the electrical conductivity ($\sigma \cong 2 \times 10^7 T^{\frac{3}{2}}$/sec in fully ionized hydrogen, Cowling, 1953). The current density is fixed in terms of **B** by Ampere's law (equation 2.32). Hence, from equation (2.29) it follows that

$$\mathbf{E} = -\mathbf{v} \times \mathbf{B}/c + (c/4\pi\sigma)\nabla \times \mathbf{B}.$$

Write η for the resistive diffusion coefficient $c^2/4\pi\sigma$ and substitute this expression for **E** into equation (2.4), obtaining the familiar form,

$$\partial \mathbf{B}/\partial t = \nabla \times (\mathbf{v} \times \mathbf{B}) - \nabla \times (\eta \nabla \times \mathbf{B}), \tag{2.35}$$

for the MHD induction equation for the magnetic field in a fluid that is something less than a perfect conductor of electricity. This can be rewritten as

$$\partial \mathbf{B}/\partial t = \nabla \times (\mathbf{v} \times \mathbf{B}) + \eta \nabla^2 \mathbf{B} + (\nabla \times \mathbf{B}) \times \nabla \eta. \tag{2.36}$$

In most cases the spatial variation of η is neglected, so that

$$\partial \mathbf{B}/\partial t = \nabla \times (\mathbf{v} \times \mathbf{B}) + \eta \nabla^2 \mathbf{B}. \tag{2.37}$$

In the limit of large conductivity the resistive diffusion can be neglected, providing the ideal form (2.30), in which field and fluid are bound to move together. It is a statement of Lenz's law, that any motion of the field relative to the fluid induces an electric field $\mathbf{E}'$ providing electric currents that oppose motion of the field relative to the fluid. Formal mathematical demonstrations of this "frozen in" field condition can be found in standard works on MHD (Lundquist, 1952; Elsasser, 1954; Spitzer, 1956; Cowling, 1957; Dungey, 1958; Jeffrey, 1966; Ferraro and Plumpton, 1966; Roberts, 1967; Moffatt, 1978; Parker, 1979; Priest, 1982).

Note that the Poynting vector is now

$$\begin{aligned} \mathbf{P} &= -(\mathbf{v} \times \mathbf{B}) \times \mathbf{B}/4\pi, \\ &= [\mathbf{v}B^2 - \mathbf{B}(\mathbf{v} \cdot \mathbf{B})]/4\pi, \\ &= \mathbf{v}_\perp B^2/4\pi, \end{aligned} \tag{2.38}$$

where $\mathbf{v}_\perp$ is the component of **v** perpendicular to **B**. Note, too, that with Ampere's law (equation 2.32) the Lorentz force $\mathbf{j} \times \mathbf{B}/c$ (exerted by the field on the fluid) on the right-hand sides of the momentum equations (2.13) and (2.28) becomes

$$(\nabla \times \mathbf{B}) \times \mathbf{B}/4\pi = -\nabla(B^2/8\pi) + (\mathbf{B} \cdot \nabla)\mathbf{B}/4\pi \tag{2.39}$$

with the first term on the right-hand side representing the gradient of the isotropic magnetic pressure $B^2/8\pi$ and the second term representing the force exerted by the tension $B^2/4\pi$ as a consequence of the local curvature of the field. The right-hand side of this equation is precisely $\partial M_{ij}/\partial x_j$, of course.

The role of the resistive diffusion η is to permit a drift of the field relative to the fluid, the diffusion also reconnecting and altering the topology of the field lines. If the fluid is held motionless ($\mathbf{v} = 0$), then equation (2.37) can be written as

$$\frac{\partial B_i}{\partial t} = \eta \nabla^2 B_i. \tag{2.40}$$

The individual Cartesian components of the field each satisfy the classical diffusion equation. The inhomogeneities in each component spread out with the passage of time, so that in the long time limit ($t \gg \ell^2/\eta$) each component everywhere approaches the uniform mean value throughout the space.

The relative importance of η is determined by the magnetic Reynolds number or Lundquist number N_L (Lundquist, 1952), which represents the ratio of the magnitude of $\nabla \times (\mathbf{v} \times \mathbf{B})$ to $\eta\nabla^2\mathbf{B}$ on the right-hand side of equation (2.37). For a characteristic scale of variation ℓ, these magnitudes are vB/ℓ and $\eta B/\ell^2$, respectively, giving

$$N_L = v\ell/\eta = C\ell/\eta. \tag{2.41}$$

for v with the characteristic value C. The Lundquist number is large compared to one in most astrophysical settings, because of the large ℓ. If v is characterized by the Alfven speed $C = B/(4\pi NM)^{\frac{1}{2}}$, N_L has values of 10^{10} or more in stellar coronas and interiors.

2.2.4 Alternative Formulations

There are other possible formulations of the problem, of course. For instance, in electric circuits, e.g., electromagnets, transformers, and laboratory plasma devices, the electric current paths are largely determined by the topology of metal conductors threading an otherwise insulating volume, and the whole system is activated and controlled by applied electromotive forces. The convenient choice of basic quantities then may be the electric current and applied voltage. The energy input is the current times voltage, and the formalism of the Poynting vector, while correct, is relatively clumsy. There is a well-defined electrical potential so that the energy carried by the input current is expressible in terms of the potential energy of the conduction electrons. The field is deduced from the current.

Unfortunately, this point of view has led some to the idea that the electric current is the "fundamental" quantity, and the magnetic field associated with the current is a "dependent" quantity, overlooking the fact that the current is of primary interest only when the activity is controlled primarily by an applied emf, as in the familiar terrestrial laboratory. In astronomical settings the fields are manipulated by fluid velocities, rather than an applied emf, and it is generally simplest and safest to work with the basic fields **E** and **B**. The electric current follows as $\mathbf{j} = c\nabla \times \mathbf{B}/4\pi$ from Ampere's law as a consequence of the field. The electromagnetic energy flow is through the fields **E** and **B**, and is described by the Poynting vector $\mathbf{P} = c\mathbf{E} \times \mathbf{B}/4\pi$. The motion of the plasma continually deforms **B** and induces the electric current to flow, so as to satisfy Ampere's law in spite of the electrical resistivity of the fluid medium. Note again that the kinetic energy of the conduction electrons is negligible except in the most extreme cases. The local rate of dissipation of field energy is $j^2/\sigma = (\eta/4\pi)(\nabla \times \mathbf{B})^2$ in terms of the resistive diffusion coefficient $\eta = c^2/4\pi\sigma$, if a simple scalar conductivity σ can be properly defined (see below). The dissipated energy is automatically supplied by $\nabla \cdot \mathbf{P}$ through the electric field $\mathbf{E}(\mathbf{j} = \sigma\mathbf{E})$, of course, to drive the conduction electrons through the resistive fluid medium. The concepts of *cause* and *effect* are conventionally based on the direction of energy flow, from the *cause* to the *effect*. Hence, in astrophysical settings, where the field and fluid are the mutual contenders, the electric current (required by Ampere's law, see equation (2.31)) is *caused* by the magnetic field. The current flows as a result of $\nabla \cdot \mathbf{P} \neq 0$, at the expense of the energy in **B** and in **v**. The field **B** is the "fundamental" quantity and **j** is the "dependent" quantity.

On the other hand, if the field **B** is known and the Joule heating is required, then the convenient quantity is the current density, of course, computed from **B** through Ampere's law. But it is generally simplest to carry through the initial dynamical calculation in terms of **B**, with considerations on **j**, and such complications as anomalous resistivity, double layers, and Joule heating taken up after the fact.

Now one is not forced to adopt the point of view expressed here. It is possible to formulate the MHD equations (2.3), (2.4) and (2.11) in terms of the electric current density **j**, if one prefers to do so, with **B** given by equation (2.33) where the integral is taken over the entire system. With the MHD relation $\mathbf{E} = \mathbf{j}/\sigma - \mathbf{v} \times \mathbf{B}/c$, it is possible to replace **E** by the volume integral of **j**. The MHD equations

$$\frac{\partial \mathbf{B}}{\partial t} = \nabla \times (\mathbf{v} \times \mathbf{B}) - c\nabla \times (\mathbf{j}/\sigma),$$

$$\rho d\mathbf{v}/dt = -\nabla p + \mathbf{j} \times \mathbf{B}/c,$$

can then be written in terms of the Biot–Savart integral (equation 2.33). The resulting integrodifferential equations are essentially intractable, with the elementary physical concepts of local energy density, stress, etc. obscured. In view of this complexity, the procedure usually adopted in a current-based formulation is to declare the form of the electric currents (which is possible in some simple symmetrical cases) and then use the Biot–Savart law (equation 2.33) to deduce **B**. The literature is graced with a variety of "exotic" results obtained in this way through overlooking the currents induced elsewhere in the region by the initiation of **B**.

Experience shows that the direct formulation of the MHD equations in terms of the fields **E** and **B** is complex enough for most of us, working with **B** and the associated Maxwell stresses $M_{ij} = -\delta_{ij}B^2/8\pi + B_iB_j/4\pi$ as the electromagnetic quantities.

2.3 The MHD Equations in a Tenuous Plasma

The MHD induction equation (2.30) is well known. It relates the magnetic field $\mathbf{B}(\mathbf{r}, t)$ to the nonrelativistic velocity $\mathbf{v}(\mathbf{r}, t)$ of a highly conducting fluid. The derivation of this equation is provided in textbooks and articles on MHD (cf. Lundquist, 1952; Elsasser, 1954), and is often based on assuming the scalar form of Ohm's law $\mathbf{j} = \sigma\mathbf{E}'$, as in the foregoing development leading to equation (2.35). Unfortunately, there is sometimes confusion regarding the general validity of equation (2.30) when the scalar Ohm's law $\mathbf{j} = \sigma\mathbf{E}$ is not applicable, as in a tenuous and nearly collisionless plasma (in which the electron collision time greatly exceeds the electron cyclotron period in **B**). It is not always fully appreciated that equation (2.30) does not depend upon the particular relation between **j** and $\mathbf{E}'$. As already emphasized, equation (2.30) applies in any fluid that is not an electrical insulator, i.e., any fluid that is unable to sustain a significant static electric field $\mathbf{E}'$ in it own frame of reference. The form of Ohm's law can at most influence the form of the dissipation term, but in gases so tenuous that the scalar Ohm's law does not apply the dissipation is usually negligible, except in extreme cases (discussed in §2.4). This is readily demonstrated with the following example.

Consider the derivation of equation (2.30) using a generalized form of Ohm's law (cf. Schlüter, 1950) appropriate for a tenuous plasma. The essential point is that

Ampere's law (equation 2.32) determines **j** in terms of the field **B**, which contains the energy, transmits the stress, and is the prime mover of **j**. Whenever **j** strays from the value $c\nabla \times \mathbf{B}/4\pi$, there is a rapid and sustained growth of **E**, according to

$$\partial \mathbf{E}/\partial t = c\nabla \times \mathbf{B} - 4\pi\,\mathbf{j}, \tag{2.42}$$

which creates whatever **j** is demanded by $\nabla \times \mathbf{B}$.

Consider, then, how **j** might respond to **E**. An appropriate example is a fully ionized hydrogen plasma, composed of $N(\mathbf{r}, t)$ electrons and $N(\mathbf{r}, t)$ protons, each of mass m and M, respectively. Express the motion of the plasma in terms of the local average ion velocity $\mathbf{v}(\mathbf{r}, t)$, which is then the local bulk motion of the plasma. The electrons have a small average conduction velocity $\mathbf{v}_c$ relative to the ions so that the current density is $\mathbf{j} = -Ne\mathbf{v}_c$. The nimble electron gas is obliged by **E** to move with the massive proton gas. The speed of sound in the electron gas is so much larger than the ion speed of sound that the electron gas is in quasi-static equilibrium except in strong shock fronts. The equation of motion for the electron gas velocity $\mathbf{v}_E$ is

$$\begin{aligned} Nm\frac{d\mathbf{v}_E}{dt} &= -\nabla NkT_E - Ne(\mathbf{E} + \mathbf{v}_E \times \mathbf{B}/c) - Nm\nabla\Phi \\ &\quad - 2Nm\Omega \times \mathbf{v}_E - Nm\Omega \times (\Omega \times \mathbf{r}) - Nm\mathbf{v}_c/\tau \end{aligned}$$

in a coordinate system with angular velicity Ω. The last term on the right represents the transfer of momentum by collisions with the ions, where τ is the mean electron–ion collision time. The gravitational potential is represented by Φ.

Write $\mathbf{v}_E = \mathbf{v} + \mathbf{v}_c$, replacing $\mathbf{v}_c$ by $-\mathbf{j}/Ne$ and **j** by $c\nabla \times \mathbf{B}/4\pi$ except in the Coriolis term. The result can be solved for $\mathbf{E}'$, defined by equation (2.29), to give

$$\begin{aligned} \mathbf{E}' &= [(\nabla \times \mathbf{B}) \times \mathbf{B}/4\pi - \nabla NkT_E]/eN + (\eta/c)\nabla \times \mathbf{B} \\ &\quad - (m/e)\nabla(\Phi + \Psi) - 2(m/e)\Omega \times \mathbf{v}_E - Nmd\mathbf{v}_E/dt, \end{aligned} \tag{2.43}$$

where

$$\Psi = -\frac{1}{2}\left[r^2\Omega^2 - (\Omega \cdot \mathbf{r})^2\right] \tag{2.44}$$

is the centrifugal potential. It is convenient to replace $1/\tau$ by

$$\eta = mc^2/4\pi Ne^2\tau = c^2/\omega_{pE}^2\tau \tag{2.45}$$

for the resistive diffusion coefficient $c^2/4\pi\sigma$ for the Ohmic conductivity $\sigma = Ne^2\tau/m$, where $\omega_{pE} = (4\pi Ne^2/m)^{\frac{1}{2}}$ is the electron plasma frequency. Note, then, that $\mathbf{E}'$ is the electric field in the frame of reference of the fluid, moving with velocity $\mathbf{v}(\mathbf{r}, t)$.

Solve equation (2.29) for **E**, using $\mathbf{E}'$ from equation (2.43), and substitute into the induction equation (2.4). The result reduces to

$$\frac{\partial \mathbf{B}}{\partial t} = \nabla \times (\mathbf{v} \times \mathbf{B}) - \nabla \times (\eta\nabla \times \mathbf{B}) - (c/e)\nabla \times F, \tag{2.46}$$

where

$$F \equiv [(\nabla \times \mathbf{B}) \times \mathbf{B}/4\pi - \nabla NkT_E]/N - 2m\Omega \times \mathbf{v}_E - md\mathbf{v}_E/dt. \tag{2.47}$$

The first two terms on the right-hand side of equation (2.46) represent the familiar result of equation (2.35). The third term $\nabla \times F$ represents the Hall effect.

To evaluate the MHD consequences of the Hall effect, add the momentum equations for the electrons and ions, to obtain

$$\begin{aligned} NMd\mathbf{v}/dt + Nmd\mathbf{v}_E/dt = &-\nabla(NkT_E + NkT_I) + (\nabla \times \mathbf{B}) \times \mathbf{B}/4\pi \\ &- 2NM\Omega \times \mathbf{v} - 2Nm\Omega \times \mathbf{v}_E - N(M+m)\nabla(\Phi + \Psi) \\ &+ O(Ne\mathbf{E}'\varepsilon^2), \end{aligned} \tag{2.48}$$

where the last term on the right-hand side represents the electrostatic force arising from the slight difference in electron and ion density, already mentioned, where $\varepsilon = R_D/\ell$. Note that the magnitude of this term can be estimated from Ampere's law and the scalar Ohm's law to be of the order of $\varepsilon^2(Ne/c)\eta\nabla \times \mathbf{B}$.

Solving equation (2.48) for the terms that make up $\mathbf{F}$, it is readily shown that

$$\nabla \times \mathbf{F} = k\nabla T_I \times \nabla \ell n\, N + M\nabla \times (d\mathbf{v}/dt + 2\Omega \times \mathbf{v}). \tag{2.49}$$

The first term on the right-hand side of equation (2.49) provides some thermal effects that might be of interest in the absence of a magnetic field, or over long periods of time, in a spinning star. For it should be noted that in a rotating star the equipotentials of $\Phi + \Psi$ are oblate rather than spherical and generally do not coincide with the isotherms, determined by the outflow of heat. Hence $\nabla T_I \times \nabla N$ is not zero and there is both a Biermann battery effect (Biermann, 1950; Mestel and Roxburgh, 1961; Cattani and Sacchi, 1966; Roxburgh, 1966; Cattani, 1967; Thorne, 1967 and references therein) and a meridional circulation from the Eddington–Sweet effect (Von Zeipel, 1924; Eddington, 1929, 1959; Sweet, 1950; Mestel and Moss, 1977; Moss, 1977a,b and references therein). There is also a small battery effect driven by the viscosity (i.e., finite mean free path) in the presence of nonuniform rotation (Browne, 1968, 1985, 1988). However, these effects are small and can be overlooked in the context of the present development (see discussion in Parker, 1979, p. 467, 768).

Finally, the term in parentheses on the right-hand side of (2.49) vanishes for static equilibrium ($\mathbf{v} = 0$), so that apart from the small thermal effects of the first term, the Hall effect contributes nothing to the MHD equations for static equilibrium ($\mathbf{v} = 0$). In that case equation (2.25) reduces to the familiar MHD diffusion equation

$$\begin{aligned} \partial\mathbf{B}/\partial t &= -\nabla \times (\eta\nabla \times \mathbf{B}) \\ &= \eta\nabla^2\mathbf{B} - \nabla\eta \times \nabla \times \mathbf{B}, \end{aligned}$$

and there is only the very slow resistive decay of the field. If $\mathbf{v} \neq 0$ the first term in the parentheses on the right-hand side of equation (2.49) is M times the curl of the plasma acceleration in a nonrotating frame of reference (recalling that the centrifugal potential contributes nothing to $\nabla \times \mathbf{F}$). Hence the linearized middle term can be written as

$$(c/e)\nabla \times \mathbf{F} \cong (Mc/e)\partial(\nabla \times \mathbf{v})/\partial t.$$

from which it is evident that $\nabla \times \mathbf{F}$ is nonvanishing in the presence of a time-dependent vorticity, as in an Alfven wave or in a turbulent flow. To estimate the contribution to $\nabla \times \mathbf{F}$, consider a plane shear Alfven wave in a nonrotating uniform fluid containing a uniform magnetic field $\mathbf{B}_0$. Denote by $\delta\mathbf{B}$ some small perturbation of the field with an associated fluid velocity $\mathbf{v}$ in the form of an Alfven wave propagating along $\mathbf{B}_0$ with a phase velocity $C = B_0/(4\pi NM)^{\frac{1}{2}}$. In the absence of any

Hall effect, the field perturbation $\delta\mathbf{B}$ and the plasma velocity are related by the familiar expression

$$\delta\mathbf{B} = \pm(4\pi NM)^{\frac{1}{2}}\,\mathbf{v}.$$

Denote the frequency of the wave by ω and the wave number by k, so that $\omega = kC$. Then

$$\partial\mathbf{B}/\partial t = i\omega\delta\mathbf{B} = \pm i\omega(4\pi NM)^{\frac{1}{2}}\,\mathbf{v}.$$

The middle term on the right-hand side of equation (2.49) becomes

$$\begin{aligned}(c/e)|\nabla\times\mathbf{F}| &= (Mc/e)|k\times\partial\mathbf{v}/\partial t|,\\ &\cong M\omega ck|\mathbf{v}|.\end{aligned}$$

The ratio α of this term to $\partial\mathbf{B}/\partial t$ is

$$\begin{aligned}\alpha &= kc/\omega_{pI},\\ &= \omega/\omega_{cI},\end{aligned}$$

where ω_{pI} is the ion plasma frequency $(4\pi Ne^2/M)^{\frac{1}{2}}$ and ω_{cI} is the ion cyclotron frequency, eB_0/Mc. Thus the relative contribution α of the Hall effect is small for wave frequencies that are small compared to the ion cyclotron frequency (which is 10^4 radians/sec in a field of 1 gauss). In fact it is just this Hall effect that converts the plane shear Alfven wave of MHD into the whistler wave mode as the frequency ω increases to the ion cyclotron frequency ω_{cI}. In macroscopic MDH one deals with wave frequencies ω that are very small compared to ω_{cI}. The Hall effect causes a Faraday rotation of the plane of polarization of a wave, that becomes significant only as ω increases toward ω_{cI}. For $\omega \gg \omega_{cI}$, the electron inertia becomes important and the electron plasma frequency $\omega_p = (4\pi Ne^2/m)^{\frac{1}{2}}$ enters into the theory.

The net result is that for magnetostatics and for macroscopic bulk plasma motions the Hall effect makes no discernible contribution, with equation (2.46) reducing to equation (2.35). So the magnetic field is carried bodily with the fluid velocity $\mathbf{v}(\mathbf{r}, t)$, except insofar as the resistivity η permits the field to diffuse and reconnect relative to the plasma. There are no other significant effects.

2.4 Deviations from MHD

It is important to understand where the MHD equations do not apply. As already noted the MHD equations apply to any fluid that cannot sustain an electric field $\mathbf{E}'$ in its own frame of reference. This can be stated more generally, viz. the MHD induction equation (2.35) applies in any moving frame of reference, with velocity $\mathbf{U}(\mathbf{r}, t)$, in which there is no electric field $\mathbf{E}'$, whatever may be the motion of the fluid. As will become clear in the sequel, the MHD induction equation for a collisionless plasma applies in the frame of reference of the electric drift velocity $\mathbf{U} = c\mathbf{E}\times\mathbf{B}/B^2$, relative to which a collisionless plasma moves with the gradient and curvature drifts. The relative drift is of the order of the ion thermal velocity multiplied by the ratio of the characteristic ion cyclotron radius R_{cI} divided by the characteristic scale of variation ℓ of the magnetic field, so that in large-scale fields the plasma moves essentially with the electric drift velocity. Thus the MHD induction equation (2.30) and a form of the momentum equation (2.28) for motion perpendicular to **B** are recovered with $\mathbf{v}_\perp$ replaced by **U**. The difference between $\mathbf{v}_\perp$ and **U** declines to zero in the limit of large ℓ. The real possibility for deviation from MHD appears in the motion parallel to **B**.

Specifically, the bulk motion $\mathbf{v}(\mathbf{r}, t)$ of the plasma perpendicular to **B** is given by the electric drift velocity **U** plus the gradient and curvature drifts, so that the mean ion velocity, representing the bulk velocity of the plasma, is

$$\mathbf{v} = \mathbf{U} + (\mathbf{B}c/qB^4) \times [kT_{I\perp}\nabla B^2/2 + kT_{I\parallel}(\mathbf{B}\cdot\nabla)\mathbf{B}], \tag{2.50}$$

where $q = Ze$ is the charge on each ion, and where $T_{I\perp}$ and $T_{I\parallel}$ represent the ion temperature for thermal motions perpendicular and parallel to the field, respectively (Watson, 1956). In terms of the characteristic lengths $\ell_\perp$ and $\ell_\parallel$ perpendicular and parallel to **B**, respectively,

$$\ell_\perp^{-1} = |\nabla_\perp \ln B|,\ \ell_\parallel^{-1} = |\nabla_\parallel \ln B|, \tag{2.51}$$

and the perpendicular and parallel ion thermal velocities $u_{I\perp}$ and $u_{I\parallel}$, respectively ($kT_{I\perp} = \frac{1}{2}Mu_{I\perp}^2$, $kT_{I\parallel} = Mu_{I\parallel}^2$), and the respective characteristic cyclotron radii $R_{I\perp} = Mu_{I\perp}\ c/qB$, $R_{I\parallel} = Mu_{I\parallel}c/qB$, the gradient and curvature drift velocities are $u_{I\perp}R_{I\perp}/2\ell_\perp$ and $u_{I\parallel}R_{I\parallel}/\ell_\parallel$, respectively. Thus in the large-scale magnetic fields that appear in the astronomical universe, these two drift effects are small. They may also be small in relatively large-scale fields in the laboratory.

Brueckner and Watson (1956) solved the linearized collisionless Boltzman equation (the Vlasov equation) for small perturbations about an equilibrium field, using the guiding center description of the particle motions. The essential result of their calculation is that the dynamical evolution of any small perturbation is described by the hydrodynamic equation including the Lorentz force $(\nabla \times \mathbf{B}) \times \mathbf{B}/4\pi$. We subsequently took a more direct approach, starting again with the guiding center motions of equation (2.50) and summing over the cycloidal–spiral trajectories of the ions and electrons. The result, irrespective of the velocity distribution of the particles, is the total current density

$$\begin{aligned}\mathbf{j} = \frac{1}{2}(c\mathbf{B}/p_m) \times \{&-N(M+m)d\mathbf{U}/dt - \nabla(p_\perp + p_m) \\ &+ [(\mathbf{B}\cdot\nabla)\mathbf{B}][1 + (p_\perp - p_\parallel)/2p_m]\},\end{aligned} \tag{2.52}$$

where p_m is the magnetic pressure $B^2/8\pi$. The total plasma pressures perpendicular and parallel to the field are

$$p_\perp = \frac{1}{2}N(M\langle u_{I\perp}^2\rangle + m\langle u_{E\perp}^2\rangle), p_\parallel = N(M\langle u_{I\parallel}^2 + m\langle u_{E\parallel}^2\rangle), \tag{2.53}$$

where the subscripts I and E and the masses M and m refer to the ions and electrons, respectively both with the same number density N (singly ionized). Substituting this expression for the current density into Maxwell's equation (2.42), the result is (Parker, 1957)

$$\begin{aligned}\partial\mathbf{E}_\perp/\partial t = (c\mathbf{B}/2p_m) \times \{&-N(M+m)d\mathbf{U}/dt - \nabla_\perp(p_\perp + p_m) \\ &+ [(\mathbf{B}\cdot\nabla)\mathbf{B}]_\perp[1 + (p_\perp - p_\parallel)/2p_m]/4\pi\}.\end{aligned}$$

Equating the right-hand side to zero yields Ampere's law, which takes the form of the equation of motion

$$N(M+m)d\mathbf{U}/dt = -\nabla_\perp(p_\perp + p_m) + [(\mathbf{B}\cdot\nabla)\mathbf{B}]_\perp[1 + (p_\perp - p_\parallel)/2p_m^2] \qquad (2.54)$$

for the electric drift velocity $\mathbf{U}$. This is, of course, the familiar MHD momentum equation with the Lorentz force $(\nabla\times\mathbf{B})\times\mathbf{B}$, but with the additional term $(p_\perp - p_\parallel)/2p_m^2$ to take account of the centrifugal force of any net anisotropy of the thermal motions along curved lines of force. This effect vanishes for an isotropic plasma, of course.

The basic fact arising from these calculations is that any deviation of the bulk plasma motion $\mathbf{U}$ from the hydrodynamic equation (2.54) invokes a huge nonvanishing $\partial\mathbf{E}_\perp/\partial t$, quickly bringing $\mathbf{U}$ back into line with the hydrodynamic equation (2.54). The guiding center motions of the ions and electrons automatically supply the electric currents required by Ampere's law, and the hydrodynamic equation (2.54) is nothing more than the statement that the Maxwell stress and thermal particle momentum flux $p_\parallel$ and $p_\perp$ provide the net momentum change of the bulk motion $\mathbf{U}$. Indeed, if the particle motions failed to accomplish these things, it would indicate a fundamental incompatibility between Newton's and Maxwell's equations. The automatic fulfillment of Ampere's law means that the current density is taken care of, and we need concern ourselves only with $p_\parallel$ and $p_\perp$, and, of course, with the evolution of $\mathbf{B}$, as in any magnetohydrodynamic system.

Now if we form the vector product of $\mathbf{U} = c\mathbf{E}\times\mathbf{B}/B^2$ with $\mathbf{B}$, the result is

$$\mathbf{E} = -\mathbf{U}\times\mathbf{B}/c + \mathbf{E}_\parallel. \qquad (2.55)$$

Then it follows from the induction equation (2.4) that

$$\partial\mathbf{B}/\partial t = \nabla\times(\mathbf{U}\times\mathbf{B}) - c\nabla\times\mathbf{E}_\parallel. \qquad (2.56)$$

When $\nabla\times\mathbf{E}_\parallel = 0$ (see §2.4.1), this reduces to the familiar MHD induction equation (2.30). In this way the MHD momentum and induction equations are recovered for the collisionless plasma, written in terms of the electric drift velocity $\mathbf{U}$ instead of the perpendicular bulk velocity $\mathbf{v}_\perp$ of the plasma. These two velocities ($\mathbf{U}$ and $\mathbf{v}_\perp$) become equal in the limit that the characteristic field scale ℓ is large compared to the cyclotron radii of the ions and electrons. It is sufficient to work in that limit in most astrophysical circumstances, so that the collisionless plasma is effectively an ideal MHD fluid, as argued originally by Schlüter (1950, 1952, 1958).

On the other hand, it must be kept in mind that this monograph deals with the spontaneous tangential discontinuities that arise when a magnetic field is subject to continuous deformation, and it is these surfaces of discontinuity at which the collisionless plasma does *not* behave according to the MHD equations. In the real physical world the ideal surfaces of mathematical discontinuity in an ideal infinitely conducting fluid become shear layers of finite thickness, of the order of the ion cyclotron radius or more so that in principle the gradient drift may be non-negligible. The particle behavior may, or may not, be accurately represented by the guiding center approximation (equation 2.50). Then $\mathbf{v}$ and $\mathbf{U}$ are distinctly different quantities and the plasma equations do not reduce to the simple MHD forms of equations (2.54) and (2.56). On the other hand, in most instances resistive dissipation and related effects thicken the layers to scales large compared to the ion cyclotron radius, so that the MHD equations are applicable all the way through, with $\mathbf{U}\cong\mathbf{v}$. But whatever the thickness and the applicability of the MHD equations to the shear layer, the important point for the magnetostatic theorem is that the MHD equations apply to the broad expanse of field in the volume between the surfaces of tangential discontinuity. It is the balance of magnetic tension against magnetic pressure throughout the volume between surfaces that causes the formation of the surfaces of tangential discontinuity (shear layer). Within the shear layer, however thin it may be, the fluid and field are in the dynamical nonequilibrium state of the familiar neutral point rapid reconnection. The pressure and tension continually extract the dissipated, reconnected field from the region of the neutral point so as to maintain the steep field gradient across the layer. Which is to say that the field stresses in the MHD field continue to push toward a tangential discontinuity, the ultimate achievement of which is eluded only by the rapid dissipation. The intrinsic drive to produce discontinuities steepens the field gradient to the point that the dissipation is important, no matter how small the resistivity of the fluid. As we shall see, even in so highly conducting a fluid as the X-ray corona of a star, the resistive dissipation takes over when the thickness of the shear layer is still large compared to the ion cyclotron radius.

2.4.1 Parallel Electric Field

Consider, then, the additional term $\nabla \times \mathbf{E}_{\parallel}$ on the right-hand side of equation (2.56). In the presence of collisions and a weak $j_{\parallel}$, there is a free flow of electric current along the field, as discussed below, and $\mathbf{E}_{\parallel}$ is short circuited and can be neglected. However, in sufficiently tenuous plasmas this need not be the case, because an anisotropic thermal velocity of either the ions or electrons can produce a $\nabla \times \mathbf{E}_{\parallel}$. Such configurations are highly unstable and cannot endure for long because of the large $\partial B_{\parallel}/\partial t$ associated with them, but they may occur as a transient condition in strong shock fronts (cf. Tidman and Krall, 1971) and one should be aware of their nature.

In the absence of collisions the equation for the individual particle velocity $u_{\parallel}$ along the field lines can be written as

$$M du_{\parallel}/dt = -\mu dB/ds + qE_{\parallel}, \tag{2.57}$$

where μ is the conserved diamagnetic moment $\mu = \frac{1}{2} M u_{\perp}^2 / B$ of the thermal cyclotron motion of the particle and ds is an element of arc length along the field. This motion is the familiar mirroring effect, whereby particles are reflected from regions of strong field. The plasma density is diminished in regions of strong B if $p_{\perp} > p_{\parallel}$, and enhanced if $p_{\perp} < p_{\parallel}$ by the combined convergence of the field lines and the mirroring effect. For an isotropic plasma with $E_{\parallel} = 0$ the density is uniform along the field, so that there is no net force between the field and plasma in the direction parallel to $\mathbf{B}$, just as in conventional MHD, where the Lorentz force $(\nabla \times \mathbf{B}) \times \mathbf{B}/4\pi$ has no component parallel to $\mathbf{B}$. On the other hand, in an anisotropic plasma $(p_{\perp} \neq p_{\parallel})$ there is a pressure gradient along the field, balanced by the mirroring forces. If $F(u, s, \theta) \sin\theta d\theta du$ denotes the number of particles per unit volume with velocities in the range $(u, u + du)$ and pitch angles in the range $(\theta, \theta + d\theta)$ $(u_{\perp} = u \sin\theta$, $u_{\parallel} = u \cos\theta)$ at a distance s along the magnetic field $B(s)$, then under stationary conditions in the absence of $E_{\parallel}$ (Schlüter, 1950, 1952; Spitzer, 1952),

$$\frac{\partial F}{\partial s} \cos\theta + \frac{\partial F}{\partial \theta} \frac{\sin\theta}{2B} \frac{dB}{ds} = 0,$$

and $F = F(\mu)$ where again μ is the diamagnetic moment $Mu^2 \sin^2 \theta/2B$. Then if, for instance, $F \propto \mu^\alpha$ it is readily shown that the plasma density, as well as $p_\parallel$ and $p_\perp$, vary as $B(s)^{-\alpha}$, with $\alpha = 0$ for an isotropic distribution. Note, then, that a difference in the pitch angle distribution of electrons and ions produces a charge separation and an $\mathbf{E}_\parallel$, which provides effects not included in MHD. However, the difference in pitch angle distribution is conducive to plasma instabilities and plasma turbulence, which scatters the pitch angles toward an isotropic distribution. So the effect must be maintained by some external prime mover.

Chew, Goldberger, and Low (1956) give a simplified and useful treatment of the variation of $p_\parallel$ and $p_\perp$ in a spatially uniform but time varying plasma. The essential point here is that it is possible to construct situations in which there is an electric field $\mathbf{E}_\parallel$ as a consequence of different electron and ion thermal anisotropies. However, as already noted, the condition is usually transient because of the plasma instabilities produced by the anisotropies. A strong electric current parallel to $\mathbf{B}$ (such as arises in a tangential discontinuity) may sustain an anisotropy, but the large conduction velocity u of the electrons generates strong plasma turbulence, which scatters the particles and pushes their distribution toward isotropy. This is discussed in the next subsections.

Foukal and Hinata (1991) provide a detailed review of the electric fields of plasma waves and current sheets (tangential discontinuities) in magnetic activity on the Sun, together with the possible means for their observation. Present state-of-the-art instrumentation can detect fields of 5–10 volts/cm at the Sun. They point out that the electric fields of limb flares and post-flare loops may be detectable, and the successful observation of electric fields would be an important step in refining the theory of flares and related activity.

Schindler, Hesse, and Birn (1991) work out the relation between $E_\parallel$ and the deviation of the field form the pure MHD induction equation (2.30), without considering the physical conditions in the plasma that would permit the $E_\parallel$. The net effect is to allow slippage of the field line connections across the region of $E_\parallel$, equivalent to an effective parallel resistivity $\eta_\parallel$.

Finally, it should be noted that the mirroring of electrons and ions in converging fields does not generally block the free flow of electric currents. The current flows for any potential difference in excess of the kinetic energy of the lowest energy electrons. Hence so long as there are enough low-energy electrons that the electron conduction velocity remains well below the ion thermal velocity, there is no serious impediment to the free flow of electric current.

A similar question arises with regard to the plasma sheath in the abrupt transition from stellar coronal temperatures $T_c \gtrsim 10^6$ K to stellar chromospheric and photospheric temperatures of $T_p \lesssim 10^4$ K or less. The thermal electrons and ions of the hotter gas are largely absorbed upon entering the chromosphere — the wall — with the more mobile electrons held back by an effective negative wall potential ϕ_w of the order of 10^2 volts (kT_c/e) so that the net flow of charge is zero. A slight change $\Delta\phi \ll \phi_w$ in the wall potential causes an electric current to flow in one direction or another, of course, and conversely the passage of a current, to satisfy Ampere's law, changes the potential, but the problem is complicated (cf. Tanenbaum, 1967; Bittencourt, 1986). It is not obvious that there are noticeable effects because even if the maximum power level $j_\parallel \phi_w$ is assumed for a current $j_\parallel$ parallel to the magnetic field, the rate is negligible in the cases with which we are concerned. A variation of T_p over the lower boundary of a corona provides a varying ϕ_w which maps upward along the field lines and provides an electric field across the magnetic field. But again the resulting plasma drift $c\mathbf{E} \times \mathbf{B}/B^2$ proves to be negligible in the cases of interest. So the plasma sheath is an interesting phenomenon to be kept in mind for special situations, but it does not appear to provide interesting deviations from conventional MHD.

2.4.2 Surfaces of Discontinuity

The most important deviations from MHD occur in the extreme field gradients in shock fronts and in spontaneous tangential discontinuities.

In shock waves there can be an electric field $\mathbf{E}'$ in the frame of reference of the fluid, associated with the sudden acceleration of the plasma across the shock front. Indeed, the formation of a shock front in a tenuous plasma depends on non-MHD (high-frequency) oscillations of the plasma replacing the coulomb collisions that dominate the shock front in a dense plasma (cf. Tidman and Krall, 1971). The velocity gradient across a collisionless shock front may be so strong that some of the particles are accelerated rapidly to high energies. The situation is that the MHD equations apply throughout the space in which the shock occurs, with the exception of the shock front itself. Hence MHD determines the fields on each side of the shock surface, but not within the thin shock front itself. Conservation of matter, momentum, energy, and magnetic flux (the Rankine–Hugoniot relations) in passage between the MHD regions on either side serve to match the MHD fields across the small thickness of the front.

The surfaces of tangential discontinuity are to be viewed in much the same way as shock fronts in hydrodymamics, with the MHD equations describing the magnetic field throughout the volume between the surfaces as already noted. As a matter of fact, even the internal structure of the tangential discontinuity (with a finite thickness in the real physical world) can often be approximated by the MHD equations (except in extreme cases) employing an anomalous resistivity $\eta = \eta*$. The essential point is that plasma turbulence and the associated anomalous resistivity arise when the thickness of the transition layer — the discontinuity — is still large compared to the ion cyclotron radius. The large value of anomalous resistivity prevents the thickness from becoming smaller (to where MHD might not apply).

This point merits closer scrutiny. Consider an electric current density $\mathbf{j} = c\nabla \times \mathbf{B}/4\pi$ in a sheared magnetic field. The electron conduction velocity v_c is equal to $|\mathbf{j}|/Ne$ in a plasma with electron density N. If again we write $|\nabla \times \mathbf{B}| = |\mathbf{B}|/\ell = B/\ell$, then $v_c \cong cB/4\pi Ne\ell$. In the limit of large ℓ (weak shear) the electron conduction velocity is small compared to the ion thermal velocity $u_I = (kT_I/M)^{\frac{1}{2}}$. In this weak gradient limit the electrical conductivity and the thermal conductivity are provided by the Chapman–Enskog method (cf. Chapman, 1954; Spitzer, 1956). However, the mean free path for coulomb collisions depends upon the fourth power of the thermal velocity, so that in the tenuous plasma of a stellar corona the electrons moving several times the mean electron thermal velocity have mean free paths that may exceed the characteristic length ℓ of the field and temperature gradients. Scudder (1992a,b) finds that the standard methods begin to show error when the mean free path for the mean thermal velocity is only a small fraction of ℓ. He goes on to treat thermal conductivity in the more complicated circumstances of a strong flow of heat.

With this limitation on the conventional expression for electrical conductivity (which does not suffer quite as quickly as thermal conductivity), it is important to note that coulomb collisions become unimportant in the face of plasma turbulence and anomalous resistivity that may arise in the tangential discontinuities in a tenuous plasma. The problem takes a different turn.

2.4.3 Anomalous Resistivity

Linear perturbation theory indicates that plasma turbulence arises when the electron conduction velocity v_c exceeds $(m/M)^{\frac{1}{2}} u_I \cong 0.02 u_I$ (Buneman, 1958, 1959; Stringer, 1964; Kadomtsev, 1965; Sagdeev, 1967; Kalinin, et al. 1970; Liewer and Krall, 1973; Goedbloed, 1973; Gekelman, Stenzel, and Wild, 1982; Spicer, 1982), provided that the electron temperature T_e is much larger than the ion temperature T_I so as to avoid Landau damping. The possibility of avoiding the Landau damping with the special Bernstein–Greene–Kruskal (*BGK*) waves is taken up later. The present discussion is directed to $T_I \ll T_e$. The electron drift instability is the first to appear (Haerendel, 1977), but it produces only weak turbulence. Strong turbulence arises from the lower hybrid drift instability, when v_c exceeds $(m/M)^{\frac{1}{4}} u_I \cong 0.15 u_I$ (Coroniti and Eviatar, 1977; Huba, Gladd, and Papadopoulus, 1977; Tanaka and Sato, 1981). The stronger ion acoustic turbulence appears when v_c exceeds u_I (Hamburger and Friedman, 1968; Friedman and Hamburger, 1969; Diamond et al, 1984; Haerendel, 1990). Strong ion acoustic turbulence reduces the characteristic electron scattering (collision) time τ to about the ion plasma period $2\pi/\omega_{pI}$, where the ion plasma frequency is $\omega_{pI} = (4\pi N e^2/M)^{\frac{1}{2}}$. The electrical conductivity $\sigma = Ne^2\tau/m$ falls from the molecular value $\sigma \cong 2 \times 10^7 T^{\frac{3}{2}}$ esu to $(M/4m)^{\frac{1}{2}} \omega_{pE}$, in order of magnitude, where ω_{pE} is the electron plasma frequency $(4\pi N e^2/m)^{\frac{1}{2}} = (M/m)^{\frac{1}{2}} \omega_{pI}$. The resistive diffusion coefficient $\eta = c^2/4\pi\sigma$ becomes $(m/M)^{\frac{1}{2}} c^2/2\pi\omega_{pE}$. It is not clear where these effects appear in solar activity, where usually $T_e \cong T_i$ so that Landau damping suppresses the plasma waves. The intense heating in the current sheet of an ideal solar flare may provide the most favorable conditions if $T_e \gg T_i$ there (cf. Coppi and Friedland, 1971).

Now, if the electron conduction velocity v_c becomes comparable to the electron thermal velocity, $u_E = (M/m)^{\frac{1}{2}} u_I$, then Langmuir plasma oscillations, at the electron plasma frequency ω_{pE}, are strongly excited without regard for the ratio T_e/T_i. The criterion is $v_c \gtrsim 0.3 u_E$, approximately, when $T_E \cong T_I$. The electron scattering time τ becomes comparable to the electron plasma period $2\pi/\omega_{pE}$, so that the effective conductivity is reduced to $\frac{1}{2}\omega_{pE}$, with $\eta* \cong c^2/2\pi\omega_{pE}$. In the active solar corona this implies the value $\sigma \cong 3 \times 10^9$ esu and $\eta* \cong 3 \times 10^{10}$ cm^2/sec, comparable to the ordinary photospheric values. The effective resistivity is enhanced by a factor of the order of 10^7.

Now, high current density electrical discharges in MHD fields have been a subject of theoretical speculation for decades (cf. Giovanelli, 1947; Dungey, 1953, 1958), although lacking in certain essentials until an example of the electric double layer was first pointed out by Alfven (cf. Alfven and Carlquist, 1967; Block, 1972, 1978; Hayvaerts, 1981; Gekelman, Stenzel, and Wild, 1982; Stenzel, Gekelman, and Wild, 1982, 1983). The electric double layer is essentially a plasma diode, with a space charge limited current. The double layer may play a role in the same circumstances (of large electron conduction velocity v_c) in which there is anomalous resistivity, with much the same effect of causing an electric field parallel to **B**. The electric double layer may be defined as any *local* region across which there is an anomalous electrostatic potential drop, that may be small, comparable, or large compared to the electron thermal energy. A weak double layer may arise at the interface between two regions of plasma at different temperatures, serving to block the electric current caused by electron diffusion. Williams (1986) points out that a double layer may be created by intersecting beams of particles, in the absence of any net current. More conventionally, a double layer may arise through electron drift excitation of various plasma waves, creating plasma turbulence and the anomalous resistivity in the manner described above. That is to say, a double layer may form wherever a plasma oscillation (e.g., an ion acoustic wave or a Langmuir wave) becomes so strong that its internal electric field begins to control the electron conduction drift through the wave, providing a quasi-static potential well with a space charge limited electric current across it. The theoretical work of Berman, Tetreault, and Dupree (1985) and Tetreault (1988) provides some insights into how this may come about with only modest electron conduction velocities, for which the plasma may be linearly stable. The theory begins with the nonlinear BGK plasma waves. Bernstein, Greene, and Kruskal (1957) showed that there exist traveling electrostatic undamped wave solutions of arbitrary amplitude and form. Trapped particles are an essential part of the waves. The point is that the waves are not subject to Landau damping. Therefore, their creation does not depend upon the competition between the driving and the damping mechanisms, as it does in the usual linear stability analysis for plane sinusoidal waves without a special choice of trapped particles. So any suitably tailored energy input, no matter how small, may cause the BGK waves to grow.

Berman, Tetreault, and Dupree (1985) consider a BGK wave in the form of an isolated region of depleted ion density (Dupree, 1972, 1982). They show that an electron conduction velocity $v_c \sim u_I$ causes the isolated ion hole to grow in both depth and in velocity width in phase space, as the hole accelerates towards regions of higher average phase space density (Berman, Tetreault, and Dupree, 1983; Dupree, 1983). The hole moves at speeds comparable to the ion sound velocity. The picture is one of randomly moving ion holes, restricted to the dimension along the magnetic field. They suggest that a single isolated hole may grow for arbitrarily small v_c, although at a correspondingly diminished rate, of course.

The dissipation of the wandering ion holes arises from collisions between holes, upsetting the proper nonlinear relation between the wave form and the trapped particle distribution (Tetreault, 1988, 1989, 1990). Berman, Tetreault, and Dupree (1983) and Tetreault (1983) provide numerical simulations of interacting holes, showing that the threshold for the growth of ion holes is $v_c \sim 1.5u_I$ when the holes occupy about one-half of the volume. It is evident, then, that with diminishing electron conduction velocity v_c the holes grow more slowly and are more sparsely distributed. So there is no specific threshold, but rather a quantitative question of the effect of small or large numbers of individual holes on the mean value of the parallel component of the electric field.

The point emphasized by Tetreault (1991) is that the ion hole, or clump, instability may be the origin of the electric double layers associated with the terrestrial aurora. Observations show that electric double layers appear in regions where the electron conduction velocity is below the threshold for the known linear instabilities, suggesting that their origin may lie in the ion hole instability for which there is no hard and fast threshold. Detailed studies show that the idea is consistent with the observations. So we may tentatively assume that electron conduction velocities of the order of the ion thermal velocity u_I, or a little less, produce a significant number of double layers in the auroral acceleration region of the terrestrial magnetosphere, and, in all probability, in the magnetospheres of other planets as well. It is generally believed that the electric double layers accelerate the downward moving electrons that produce the aurora. We may conjecture that electric double layers appear quite generally in the stronger tangential discontinuities in the magnetic fields in active regions in the Sun. The MHD equations would not apply in the interior of any current sheet in which there were enough double layers to produce a significant electric field parallel to the magnetic field. The criterion may be an electron conduction velocity comparable to the ion thermal velocity, on which more will be said below.

Laboratory studies show strong steady double layers, with potential drops large compared to the electron thermal energy, when the electron conduction velocity v_c exceeds the electron thermal velocity u_E (Block, 1972; Quon and Wong, 1976; Hershkowitz, 1985). Hollenstein, Guyot, and Weibel (1980) report strong double layers with $v_c \cong 0.2u_E$. But for the most part, laboratory experience finds only relatively weak transient ion acoustic double layers when $v_c < u_E$. Weak double layers exist with v_c as small as the ion thermal velocity u_I. Chan, et al. (1986) provide a useful summary of experimental results along with their own laboratory identification of slow ion acoustic double layers, predicted by Perkins and Sun (1981), Stern (1981), and Schamel (1983). The formation of double layers, and particularly weak double layers, in the laboratory is not clearly dissociated from the effective cathode potential well where the electrons are injected. Hence it is not immediately obvious how to apply the laboratory results to current sheets in planetary magnetospheres (Westcott, et al, 1976; Block, 1978; Smith and Goetz, 1978; Temerin, et al. 1982; Stenzel, Gekelman, and Wild, 1982) solar flares (Heyvaerts, 1981), accretion onto neutron stars (Williams, et al. 1986), and the tangential discontinuities in the external magnetic fields of convecting stars, where there is no well-defined cathode surface. One may ask whether the weak laboratory double layers with $v_c < u_E$ are able to form at all. The initiation of strong double layers with large potential drops, that may accelerate particles to high energy, would seem to require $v_c > u_E$ to accomplish some of the feats attributed to them. There is much to be done yet in developing the theoretical application of the electric double layer to the many astronomical circumstances in which it may play an important role, providing a mechanism in some cases for direct, efficient conversion of magnetic energy into the kinetic energy of a small number of fast particles. A fundamental problem is to distinguish between the consequences of strong anomalous resistivity and one, or several, double layers, contributing to a parallel electric field.

With regard to MHD, the essential point is that the electric double layer, like the shock front and the tangential discontinuity, is a local non-MHD process that may arise at special locations within a general MHD field. That is to say, the double layer involves thin layers, idealized as surfaces, within an MHD field across which the electric field and magnetic field may be discontinuous. In this respect, it resembles the quasi-static tangential discontinuity in the magnetic field, across which the direction of the magnetic field is discontinuous.

The next section takes up the requirements on a magnetic field **B** in order to produce anomalous resistivity and electric double layers. We shall find that the conditions are sometimes met, but perhaps not as universally as some of the more ardent boosters of double layers have urged.

2.4.4 Onset of Anomalous Resistivity and Electric Double Layers

Consider the necessary magnetic conditions under which anomalous resistivity and electric double layers may arise. Specifically, what are the conditions under which the electron conduction velocity v_c exceeds the ion thermal velocity u_I or the electron thermal velocity u_E? The electron conduction velocity is determined from Ampere's law (equation 2.32), so that

$$v_c = cB_\perp / 4\pi Ne\ell,$$

where now $B_\perp$ represents the transverse component of the field in which $\nabla \times \mathbf{B}$ is formed, and ℓ represents the transverse scale so that $|\nabla \times \mathbf{B}| = B_\perp/\ell$. It is convenient in the calculations that follow to express $B_\perp$ as the fraction γ of the total field B. The scale ℓ may be expressed as νR_{cI} in terms of the numerical factor ν and the characteristic ion cyclotron radius, $R_{cI} = Mu_Ic/eB$. Then if $v_c \geqslant u_I$, it follows that $\gamma C^2 > \nu u_I^2$, where C is the local Alfven speed $B/(4\pi NM)^{\frac{1}{2}}$. If $v_c \geqslant v_E$, then $\gamma C^2 > \nu u_I u_E$.

Now the characteristic scale ℓ cannot be less than an ion cyclotron radius, and in fact we expect that $\nu \gg 1$ except in the most extreme cases. As an example, in the active X-ray corona of the sun ($T = 2-3 \times 10^6$ K, $N = 10^{10}$ H atoms/cm^3, $B \cong 10^2$ gauss) we have $C = 2 \times 10^8$ cm/sec, $u_I \cong 2 \times 10^7$ cm/sec, $u_E \cong 10^9$ cm/sec and $\gamma \cong \frac{1}{4}$. It follows that $R_{cI} \cong 20$ cm, and $\gamma C^2 = \nu u_I^2$ for $\nu \cong 25$. That is to say, the conduction velocity v_c becomes as large as u_I, producing strong plasma turbulence, when ℓ declines to $25R_{cI} = 5 \times 10^2$ cm. Note, however, that v_c reaches u_E only if ℓ is as small as $0.5R_{cI}$ ($\nu = 0.5$). We expect that ℓ may be prevented by the anomalous resistivity from declining below the value of about $30R_{cI}$, where anomalous resistivity appears, blocking further enhancement of the electric current density. And in any case, there is no way of which we are aware that ℓ can be smaller than the ion cyclotron radius R_{cI}. So we do not expect strong electric double layers to appear under ordinary circumstances in the active X-ray corona.

As already noted, a flare provides more extreme conditions, which might favor larger values of v_c, through larger ΔB. But note that an enhanced temperature is counterproductive here, so that it appears that the electric double layer does not play a central role (see discussion in Heyvaerts, 1981).

Low densities in planetary magnetospheres (say, 1 atom/cm in a magnetic field of 10^{-2} gauss or more) yield $C \geqslant 2 \times 10^9$ cm/sec, which is conducive to the production of electric double layers if the ambient plasma temperature is 10^7 K or less (see Block, 1978; Smith and Goertz, 1978, Temerin, et al. 1982; Stenzel, Gekelman, and Wild, 1982).

It appears, then, that the MHD equations are applicable almost everywhere in space, with the exceptions confined to the interior of thin sheets (shock fronts, tangential discontinuities, e.g., auroral current sheets and the many small current sheets in bipolar magnetic fields on the Sun, in the Galaxy, etc., and perhaps the extreme conditions in the central current sheet producing the intense impulsive phase of flare). The internal structure of the exceptional sheets has to be determined from more detailed considerations on the kinetic theory of plasmas. However, the large-scale dynamical behavior of the magnetic fields and plasma throughout the volume of space (excluding the current sheets) is properly determined from the induction and momentum equations of MHD, thereby providing the boundary conditions for determining the internal dynamical structure of the sheets.

3

Invariance, Degeneracy, and Continuous Solutions

3.1 Magnetostatic Equilibrium Equations

The formation of tangential discontinuities, as an intrinsic part of the static equilibrium of a magnetic field **B** embedded in an infinitely conducting fluid, follows directly from the mathematical properties of the equilibrium equations. The basic condition is that the field is solenoidal

$$\nabla \cdot \mathbf{B} = 0$$

(equation 2.2), which asserts that field lines have no ends. The field lines, and the tension $B^2/4\pi$ along the field lines, extend all the way through the field, from the rigid boundary at $z = 0$ to the rigid boundary at $z = L$. The static balance of Maxwell stress against fluid pressure p and fluid weight $-\rho\nabla\Phi$ in a gravitational potential Φ can be written

$$(\nabla \times \mathbf{B}) \times \mathbf{B}/4\pi = +\nabla p + \rho\nabla\Phi.$$

The scalar product with **B** yields

$$\mathbf{B} \cdot (\nabla p + \rho\nabla\Phi) = 0,$$

which asserts that the barometric law,

$$\frac{\partial p}{\partial s} = -\rho\frac{\partial \Phi}{\partial s},$$

applies to the variation of p and ρ with distance s along the field. In the simple case that the level surfaces of ρ coincide with those of Φ, ρ can be written as $\Pi'(\Phi)$ and it follows from the barometric relation that $p = p(\Phi)$. Then

$$(\nabla \times \mathbf{B}) \times \mathbf{B} = 4\pi\nabla P, \tag{3.1}$$

where

$$\begin{aligned} P &= P(\Phi), \\ &= p(\Phi) + \Pi(\Phi). \end{aligned}$$

We are not concerned with the origins of Φ, so P in equation (3.1) is considered to be merely a function of position **r**. It is the basic mathematical form of equation

(3.1), usually in the simple force-free case that P is uniform, with which we are concerned here.

Note that the scalar product of $\mathbf{B}$ with equation (3.1) yields the well-known result,

$$\mathbf{B} \cdot \nabla P = 0, \tag{3.2}$$

that the pressure P is rigorously constant along each field line when the field is in static equilibrium. If P is uniform throughout some finite region, then $\nabla P = 0$ and equation (3.1) reduces to

$$(\nabla \times \mathbf{B}) \times \mathbf{B} = 0.$$

That is to say, on the basis of Newton's third law, the field can exert no force on the fluid if the fluid does not push back on the field. In this case, we have the familiar force-free condition that

$$\nabla \times \mathbf{B} = \alpha(\mathbf{r})\mathbf{B}, \tag{3.3}$$

where the torsion coefficient α is given by

$$\alpha(\mathbf{r}) = \mathbf{B} \cdot \nabla \times \mathbf{B}/B^2. \tag{3.4}$$

The divergence of the force-free equation yields the familiar result,

$$\mathbf{B} \cdot \nabla\alpha = 0, \tag{3.5}$$

that α is rigorously constant along each field line. The curl of equation (3.3) leads to

$$\mathbf{B} \times \nabla\alpha = \nabla^2\mathbf{B} + \alpha^2\mathbf{B} \tag{3.6}$$

upon using equation (3.3) to eliminate $\nabla \times \mathbf{B}$. This relation prescribes the gradient of α across the field in terms of the elliptical differential form $\nabla^2\mathbf{B} + \alpha^2\mathbf{B}$.

The torsion coefficient α can be understood in terms of the magnetic circulation Γ around an elemental flux bundle of infinitesimal cross-sectional area S and total infinitesimal flux Φ. Denote an element of length around the periphery of S by $\mathbf{w}$. Then, using Stokes theorem, the force-free equilibrium equation (3.3), and the assumption that α and $\mathbf{B}$ vary smoothly across the flux bundle, it follows for infinitesimal S that

$$\begin{aligned}
\Gamma &\equiv \oint d\mathbf{w} \cdot \mathbf{B} \\
&= \int_S d\mathbf{S} \cdot \nabla \times \mathbf{B} \\
&= \alpha \int_S d\mathbf{S} \cdot \mathbf{B} \\
&= \alpha\Phi
\end{aligned}$$

where $d\mathbf{S}$ is an element of area in S. Since both α and Φ are constant along the flux bundle, it follows that the magnetic circulation is also constant along the bundle. The torsion coefficient α, then, represents the magnetic circulation per unit magnetic flux. When it comes to applying this fact to an actual flux bundle, note that in a force-free field with nonvanishing torsion α a flux bundle with a more or less circular cross-section S at one location spreads out into a fan with a width of the order of $\alpha S^{\frac{1}{2}}s$ at a distance s elsewhere along the bundle.

3.2 Field Configurations

In the discussion of magnetostatic equilibrium it is useful to distinguish two classes of field topology, based on the anchoring of the field lines in nearby, or only in distant or absent, rigid boundaries. The essential point is that the stresses in a region of line-tied field (near a rigid boundary in which the field lines are anchored) are transmitted across the field by the rigid boundary, forming a vital part of the equilibrium stress balance. Tangential discontinuities form in these line-tied fields when a continuous displacement (mapping) of the footpoints of the field lines at the rigid boundaries causes one lobe of field to be compressed against another. Examples can be found in Sweet (1969), Syrovatskii (1969), Low (1972), Priest and Raadu (1975), Hu and Low (1982), Low and Hu (1983), Low (1987, 1989), Low and Wolfson (1988), and Jensen (1989) showing how neutral points and current sheets are formed by the continuous displacement of the footpoints at the boundaries.

On the other hand, a wound and interwoven field whose field lines connect only into a distant rigid boundary is locally in equilibrium with its own stresses. The stresses exerted on the field by the distant boundary have no direct effect on the local equilibrium. For instance, it does not matter to the local equilibrium field whether the field lines pass through many different random winding patterns before connecting into the distant rigid boundaries or whether the field lines connect into no distant boundaries at all but instead circle around and re-enter at the opposite end of the field region, forming an overall toroidal geometry.[1] The essential point is that the local self-contained balance of magnetic stresses — the isotropic pressure $B^2/8\pi$ and the field aligned tension $B^2/4\pi$ — requires the presence of surfaces of tangential discontinuity in all but the most carefully tailored field topologies.

It is not surprising, then, to find a general mathematical distinction between the analytic solutions to the equilibrium equations (3.1) and (3.3) for nearby and distant boundaries. The distinction is that there are a variety of 3D analytic solutions for the anchored field (cf. Dicke, 1970; Raadu and Nakagawa, 1971; Nakagawa, et al. 1971; Nakagawa and Raadu, 1972; Nakagawa, 1973; Low, 1975a,b, 1987, 1989; Bogdan and Low, 1986; Lou and Hu, 1990) whereas there are no known solutions for the local self-equilibrium of a magnetic field far from any boundary without some symmetry, invariance, or degeneracy of one form or another. These solutions are discussed in §4.1. If there are analytic solutions under nondegenerate and fully 3D conditions far from any boundaries, they remain to be discovered. This "experimental" fact has led to the conjecture that without a degeneracy or suitable symmetry there are no solutions to equation (3.1) that are free of tangential discontinuities. Arnold (1965, 1966, 1972, 1974; see also Moffatt, 1985) pointed out that almost all solutions to the equilibrium equations involve discontinuities.

In the context of plasma confinement in toroidal magnetic fields Grad (1967) conjectured that there are no continuous equilibrium solutions in toroidal field geometries. He suggested that magnetostatic equilibria arising in the presence of suitable symmetry, invariance, or degeneracy are topologically unstable to any small perturbation that breaks the basic symmetry. That is to say, the introduction of any small nonsymmetric perturbation causes the field topology to diverge along the direction of the field from the initial symmetry, never to return. In fact in a re-entrant

[1]The toroidal topology is important in the case that $\partial/\partial\varphi = 0$, so that there may be resonance effects around the torus.

(toroidal) field topology the gradual divergence along the field feeds back through the system each time around, providing an ergodic topology throughout some finite region of the field. It follows from equation (3.2) in that case that the pressure P is uniform throughout the region. Hence equation (3.3) is applicable and it follows from equation (3.5) that the torsion α is uniform throughout the ergodic region. But the torsion α involves the actual winding of the field lines around each other, prescribed in the symmetric unperturbed solution. For if α is nonuniform before the introduction of the perturbation, then there is a redistribution of torsion (in the form of torsional Alfven waves) along the field, readjusting α to a uniform value throughout the ergodic region. Taylor's maxim (1974, 1986) that toroidal magnetic fields relax in the presence of a small resistivity to the minimum energy state, while conserving helicity, $\int d^3\mathbf{r}\mathbf{A}\cdot\mathbf{B}$, so that $\nabla\alpha = 0$, is not unrelated to this situation. The essential point is that the readjustment of α across the ergodic region generally changes α at the surface of the ergodic region, with the result that the surface field no longer matches smoothly to the field immediately outside the region. The result is a surface of tangential discontinuity.

The same remarks apply to a magnetic field extending between boundary planes $z = 0$ and $z = L$, which is the primary scenario for the theoretical development here. In that case there is no ergodic region because each field line passes only once across $0 \leqslant z \leqslant L$, but the tangential discontinuities appear for essentially the same reasons (Parker, 1972, 1979). The introduction of a small perturbation $O(\varepsilon)$ that breaks the symmetry or degeneracy of the zero-order field equilibrium causes the flux surfaces (and the associated isobaric surfaces, or surfaces of uniform torsion α) with the original symmetry to diverge along the field, so that the surfaces lose any memory of their initial form in a distance $O(\ell/\varepsilon)$, where ℓ is the characteristic scale of variation of both the zero-order field and the perturbation. But the initial form of the flux surfaces was essential to the local equilibrium everywhere along the field. So equilibrium in some continuous form is rendered impossible by the small perturbation. The only alternative is for the field to develop surfaces of discontinuity as the regions between these surfaces readjust to achieve static equilibrium. For it must be appreciated that in an infinitely conducting fluid the field sooner or later relaxes to the lowest available energy state while preserving the initial topology of the field lines. If the infinitely conducting fluid, in which the field is embedded, is given a small viscosity, it is clear that the field cannot remain in a nonequilibrium state in the limit of large time. Hence it eventually relaxes into a state containing surfaces of tangential discontinuity.

3.3 Basic Field Configuration

As already stated, the field configuration with which this theoretical development is primarily concerned consists of field lines extending through an infinitely conducting fluid between the rigid boundaries $z = 0$ and $z = L$. The development is carried out in the limit of large L/ℓ, where ℓ is the local scale of variation of the field. Hence, throughout the interior of the region (say in $m\ell < z < L - m\ell$, where m is a bounded integer, the rigid boundaries at $z = 0, L$ are far away and have no sensible effect on the local balance of stresses. In the general case there are no symmetries and invariances, there are no known analytical methods for solving equation (3.1) or (3.3), and the general presence of tangential discontinuities suggests that there are no general methods of solution along conventional lines

(Parker, 1972; Tsinganos, 1982). The tangential discontinuities make conventional numerical codes ineffective.

This is perhaps an appropriate place to be reminded again of the precise analogy between equation (3.1) for static equilibrium of a magnetic field **B** and the Euler equation

$$(\nabla \times \mathbf{v}) \times \mathbf{v} = -\nabla\Pi,$$

where $\Pi = p/\rho + \frac{1}{2}v^2$ for the stationary $(\partial/\partial t = 0)$ flow of an ideal inviscid incompressible fluid (Moffatt, 1985). Hence, the foregoing remarks on the formation of tangential discontinuities (current sheets) as an intrinsic part of the static equilibrium of a magnetic field **B** apply equally well to the formation of tangential discontinuities (vortex sheets) in a stationary flow **v**. There is one fundamental difference, and that is in the general instability of the stationary **v** as opposed to the stability of the lowest available static energy state of **B** (cf. Moffatt, 1986). Hence, the application of theory to actual hydrodynamical flows is a dynamical problem and much more difficult that the simple relaxation to the discontinuous static equilibrium of the lowest energy state of a magnetic field of specified topology. We will need the optical analogy of Chapter 7 to treat the problem.

Now, to define the physical system and the topology of the field, consider an initially uniform field B_0 in the z-direction, in which all the lines of force connect across an infinitely conducting fluid from $z = 0$ to $z = L$, with the region of field and fluid extending to $x, y = \pm\infty$ in both transverse directions. The field is uniform at time $t = 0$, sketched in Fig. 1.5(a), and at that time the 2D incompressible fluid motion, described earlier in §1.5,

$$v_x = +kz\partial\psi/\partial y,\ v_y = -kz\partial\psi/\partial x,\ v_z = 0 \tag{3.7}$$

is switched on, where $\psi = \psi(x, y, kzt)$ is a bounded, smoothly varying, continuous, n-times differentiable function of all three of its arguments. The fluid motion leaves the footpoints of the field fixed at $z = 0$, while swirling and mixing the fluid faster and faster with increasing z, the field lines at $z = L$ being transported fastest of all. Consider the form into which the initially uniform field is carried by the eddying, swirling fluid.

Write $B_x = +B_0\partial A/\partial y$ and $B_y = -B_0\partial A/\partial x$. The x and y components of the magnetohydrodynamic induction equation (2.30) can be written out, and it is readily seen that the x-component is given by the expression

$$\frac{\partial A}{\partial t} = kz\left(\frac{\partial\psi}{\partial x}\frac{\partial A}{\partial y} - \frac{\partial\psi}{\partial y}\frac{\partial A}{\partial x}\right) + \frac{\partial}{\partial z}kz\psi, \tag{3.8}$$

operated on by $+B_0\partial/\partial y$ while the y-component is obtained by operating with $-B_0\partial/\partial x$. This is, then, a suitable equation for A. In terms of the Langrangian, or total, derivative,

$$\frac{d}{dt} = \frac{\partial}{\partial t} + v_j\frac{\partial}{\partial x_j},$$

the equation can be written

$$\frac{d}{dt}(A - kt\psi) = 0. \tag{3.9}$$

Since A is zero at time $t = 0$, it follows that $A - kt\psi$ vanishes for all t, so that

$$A = kt\psi(x, y, kzt). \tag{3.10}$$

The field components are, then, given by equation (1.4), viz.

$$B_x = +B_0 kt\partial\psi/\partial y,\ B_y = -B_0 kt\partial\psi/\partial x,\ B_z = B_0, \tag{3.11}$$

as a result of the mechanical manipulation of the field (Parker, 1986).

Given that $\psi(x, y, kzt)$ is a smooth bounded continuous, n-times differentiable, and generally well-behaved function of its three arguments, it follows that the resulting **B** is a smooth, bounded, continuous and generally well-behaved function of position. With arbitrary choice of the function $\psi(x, y, kzt)$, it is evident that the field lines of **B** are subject to winding and interweaving in arbitrary patterns along the length of the field from $z = 0$ to $z = L$. Figure 1.5(a) is a sketch of the initial uniform field $e_z B_0$ at time $t = 0$ and Fig. 1.5(b) is a sketch of an arbitrarily interwoven field at a time $t > 0$. Note that while the wrapping and winding of field lines about their neighbors is arbitrary, depending only on the choice of the arbitrary function ψ, there can be no knots in the field lines, because the connectivity of each field line from $z = 0$ to $z = L$ is preserved throughout the operation.

To be precise, note again that the characteristic scale of transverse variation of ψ is ℓ, with $\ell \ll L$. Writing $\zeta = kzt$, the characteristic scale ℓ_z of variation of B_x and B_y in the z-direction is given by

$$\frac{1}{\ell_z} = ktO\left[\left[\frac{\partial}{\partial\zeta}\left\{\ln\left[\left(\frac{\partial\psi}{\partial x}\right)^2 + \left(\frac{\partial\psi}{\partial y}\right)^2\right]\right\}\right]\right], \tag{3.12}$$

which is assumed to be of the same general order of magnitude as the transverse scale ℓ in a strongly wound field. Hence the boundaries do not interfere significantly with the equilibrium throughout the deep interior $m\ell < z < L - m\ell$ where m is some integer larger than one.

Note that the projection of the stream lines of the fluid motion onto any plane $z = z_0$ $(0 < z_0 < L)$ at any instant in time is given by $\psi(x, y, kz_0 t) =$ *constant*. The field lines have the same pattern. We suppose that almost all of the streamlines of the 2D incompressible flow ψ, and hence almost all of the field lines, form localized eddies of scale ℓ, so that there is no net flow, or magnetic flux, across dimensions large compared to ℓ. Then since $\partial v_x/\partial x + \partial v_y/\partial y = 0$, it follows that the streamlines all close on themselves, except for an occasional line that might extend to infinity. It follows that the projection of almost all field lines onto each plane $z =$ *constant* form closed curves.

The next step in the formation of the final magnetostatic field is to shut off the fluid motion at time $t = \tau$. Then fix the footpoints of the field at $z = 0$ and $z = L$, whereupon the fluid is released so that the system is free to relax to the lowest available energy state, with the field line topology (Fig. 1.5b) preserved by the infinite electrical conductivity of the fluid. It is immaterial whether the complete relaxatin is only asymptotic in time or occurs in a finite time. It is assumed for the present development that the fluid pressure P is maintained uniform at the boundaries $z = 0, L$, so that in view of equation (3.2) P is uniform throughout $O < z < L$. Hence the final static equilibrium is described by equations (3.3) and (3.5).

Before going any further with the discussion it is essential to understand the general analytical properties of the magnetostatic equation (3.1) or (3.3). These properties are expressed in terms of the characteristics of the equations, which, as will be seen, permit the spontaneous development of tangential discontinuities. Following the development of the theory of characteristics in §3.4, the narrative returns in §3.5 to the physics of the discontinuities that develop in the static equilibrium state of the interwoven field described by equation (3.11).

3.4 Characteristics and Discontinuities

The general appearance of tangential discontinuities in the static equilibrium of almost all field topologies is a phenomenon that does not arise in the linear field equations of classical physics. Thus, for instance, a field ϕ described by Laplace's equation $\nabla^2\phi = 0$, or by a wave equation $\nabla^2\phi + k^2\phi = 0$, does not have internal discontinuities because a discontinuity would violate the basic equation. The question that arises, then, is what is the special property of the magnetostatic equilibrium equation (3.1), and the special case in equation (3.3) for the force-free fields, which permits discontinuities? The answer is that the magnetostatic equations have the same imaginery characteristics as $\nabla^2\phi$, but they have in addition a set of real characteristics (Parker, 1979). The magnetostatic equations are mixed hyperbolic–elliptic equations, whereas $\nabla^2\phi = 0$ and $\nabla^2\phi + k^2\phi = 0$ are fully elliptic, having two families of imaginery characteristics and no others.

To understand the implications of the characteristics, i.e., the characteristic curves, of a system of linear (or quasi-linear) partial differential equations, consider the simple case of a function $\phi(x_k)$ in an n-dimensional space restricted by v linear (or quasi-linear) first order partial differential equations. Presumably $v \leqslant n - 1$. Specify some arbitrary curve C, defined by $x_i = x_i(s)$ in the space, where s designates distance measured along C from some arbitrary point on C. Denote the direction cosines of C at each point by

$$\gamma_i(s) = dx_i/ds.$$

Suppose, then, that the function $\phi(s)$ is specified on C. The question is to what extent does this boundary condition determine $\phi(x_k)$ elsewhere in the space. In particular, given $\phi(s)$ on $x_i(s)$, under what circumstances is it possible to compute ϕ on neighboring parallel curves $x_i = x_i(s) + \delta x_i(s)$?

The computation of $\phi(x_k + \delta x_k)$ given $\phi(s) = \phi[x_k(s)]$ is equivalent to computing the spatial derivatives of ϕ on $x_i(s)$ in the $n - 1$ directions normal to $x_i(s)$. Given that the derivative parallel to $x_i(s)$ is known from $\phi(s)$, this is equivalent to computing all n derivatives $\partial\phi/\partial x_i$. Once the derivatives are known, the function ϕ on $x_i(s) + \delta x_i(s)$ follows from the Taylor expansion

$$\phi(x_k + \delta x_k) = \phi(x_k) + \delta x_j(s)\partial\phi/\partial x_j + \dots.$$

Now prescribing $\phi(s)$ determines $d\phi/ds$, of course, so that the relation

$$\frac{d\phi}{ds} = \gamma_j \frac{\partial\phi}{\partial x_j}$$

provides an inhomogeneous linear equation restricting the n components of $\partial\phi/\partial x_i$. In addition to this equation there are the v partial differential equations describing

ϕ, for a total of $v+1$ linear equations for the n unknown quantities $\partial\phi/\partial x_i$. If $v+1=n$, the equations provide the n components of $\partial\phi/\partial x_i$. It follows that, in this case, the specification of $\phi(s)$ on $x_i(s)$ determines ϕ on all neighboring curves $x_i(s)+\delta x_i(s)$. The calculation can be repeated on any one, or more, of the neighboring curves, and the solution $\phi(x_k)$ can be constructed in a stepwise fashion throughout any finite region. That is to say, the boundary condition that $\phi=\phi(s)$ on some curve C provides a unique solution throughout the region.

On the other hand, suppose that $v+1<n$. Then only $v+1$ of the n derivatives $\partial\phi/\partial x_i$ can be computed from $\phi(s)$ and a unique solution is provided if and only if the remaining $(n-v-1)$ derivatives are specified on C in addition to the function $\phi(s)$ itself.

But now suppose that the determinant of the coefficients of the $\partial\phi/\partial x_j$ in the $v+1$ partial differential equations should vanish on the curve C. There is generally one or more directions γ_i at each point for which this happens, and for such γ_i the $\partial\phi/\partial x_j$ are not determined by specifying $\phi(s)$ and $(n-v-1)$ of the $\partial\phi/\partial x_j$ on C. On the other hand, all other directions γ_i permit a unique determination of $v+1$ of the n derivatives $\partial\phi/\partial x_i$. Each set of direction cosines that causes the determinant to vanish defines a unique curve through the space, and, therefore, defines a family of curves that fills the space. These are called the characteristics of the system of partial differential equations. There are one or several families of characteristic curves for a set of partial differential equations, because the determinant of the coefficients vanishes for one or several directions γ_i.

Each individual characteristic has the unique property that specification of $\phi(s)$ and $n-v-1$ of the derivatives $\partial\phi/\partial x_i$ on that curve does not determine all n of the $\partial\phi/\partial x_i$. Hence, ϕ on neighboring characteristic curves is not related by the system of partial differential equations to ϕ on $x_i(s)$. That is to say ϕ need not vary continuously from one characteristic curve to a neighboring curve. It is stated (cf. Courant and Hilbert, 1962; Abbott, 1966), that the function ϕ is propagated independently along each characteristic. It follows that if the initial conditions, i.e., the boundary conditions, at some surface S extending across the family of characteristics provide a continuous ϕ, then that continuous form propagates away from S along the characteristics. In that case ϕ is a continuous function of position everywhere throughout the space penetrated by the characteristics. But note that ϕ is continuous because of the continuous boundary conditions. It is not constrained to be continuous by the system of partial differential equations. For if ϕ is specified as being discontinuous in some way on the surface S, that discontinuity extends away from S throughout the region along the characteristics. That phenomenon is obvious in the magnetostatic field extending from $z=0$ to $z=L$, wherein the field lines form the characteristics and a discontinuous fluid motion produces a discontinuous field. A vortex sheet produces a current sheet. The simple case $\psi=\pm vy$ for $y\gtrless 0$, respectively, yields $v_x=\pm vkz$ and $B_x=\pm B_0 vk\tau$ for $y\gtrless 0$, respectively, after a time τ. The field is subject to infinite shear across the flux surface $y=0$. In this case $\partial B_x/\partial y=2B_0 vk\tau\delta(y)$. But it is obvious from the physics of magnetic fields that, with $\nabla\cdot\mathbf{B}=0$ and the tension $B^2/4\pi$ along the field that a discontinuity cannot cut across the field.

This elementary discontinuity is not generally important in the astronomical world, however. The more interesting case arises where the discontinuity is not imposed at the boundary S but arises from the topological bifurcation of the characteristic curves into two or more separate channels as a consequence of the complicated topology of the characteristic, i.e., the field lines. It will be shown in Chapter 7 through the optical analogy that such bifurcations appear in the presence of a local maximum in the field strength. In that case the characteristic curves from one region of S come into contact with characteristic curves from a separate region of S, each with their own local values of ϕ. The result is automatically a discontinuity in ϕ where the distinct topological domains of the characteristics come into contact. That is to say, the discontinuities appear at the topological separatrices. Note, then, that the discontinuities created by the topology of the characteristics extend along the characteristics to the boundary so the field becomes discontinuous at the boundary S when there is a bifurcation of the characteristics somewhere in the region.

It is the tangential discontinuities produced by the topology of the field that are important in the physics of magnetostatic fields, because they are required by the topology and cannot go away so long as the topology is demands them. This is in contrast to the field discontinuity formed only because of a discontinuity in the velocity, which disappears as soon as a small amount of resistivity is introduced. This is taken up in detail in §10.2.

Consider the familiar Laplace's equation $\nabla^2\phi = 0$. It has only imaginery characteristics, e.g., $x = \pm i(y + y_0)$, which lie outside the real domain of the physics. So there can be no discontinuities for real x and y. As is well known, a solution ϕ to Laplace's equation is uniquely determined throughout a volume V by specification of ϕ (or the normal derivative $\partial\phi/\partial n$, or a linear combination of the two) on the enclosing boundary, or by specification of both ϕ and $\partial\phi/\partial n$ on any segment of finite length. An equation that has imaginery characteristics and no others is said to be elliptic. There are no characteristics of a purely elliptic equation in real coordinate space. Hence specification of the function ϕ and $\partial\phi/\partial n$ on a real curve provides a unique designation throughout the region.

In contrast with the fully elliptic equation, the magnetostatic equilibrium equations possess real characteristics in addition to the two families of imaginery characteristics. The fields on opposite sides of a real characteristic are not uniquely related. It is the real characteristics that allow the tangential discontinuities, providing an escape from an otherwise unique determination of the field throughout the interior of a volume V by specification of the field at the surface. For it is essential to remember the basic physics of the magnetic field embedded in an infinitely conducting fluid. The magnetic field can be deformed by arbitrary winding and interweaving of the field lines, providing an arbitrarily complicated internal field topology throughout $0 < z < L$. At the same time the field at the boundaries $z = 0, L$ reflects only the local winding and the local compression or expansion of the field, within some characteristic distance ℓ_z of $z = 0, L$. Note in particular that the topology of the interweaving field lines is established by the fluid motion (equation 3.11), but it is possible to vary the field components, and their derivatives with respect to z, at the boundaries by introducing *arbitrary* inhomogeneous compression and expansion of the distribution of footpoints on $z = 0$ and $z = L$. This local compressible mapping of the footpoints has no affect on the topology of the field throughout the interior of $0 < z < L$ because it involves no winding or unwinding of the already interwoven field lines. It serves only to massage the boundary conditions while the internal topology remains fixed. This demonstrates that there can be no unique physical connection between the boundary conditions and the topology. Hence the internal topological state of the field cannot be uniquely determined by specification of the field on $z = 0, L$. But if the field were continuous throughout $0 < z < L$, there would, in fact, be a unique relation between the internal field and the field at the boundaries. So a continuous field is generally excluded by the physics of magnetostatic equilibrium.

The known analytic solutions, based on an invariance or degeneracy of one form or another, all have an internal topology that is uniquely related to the field at the boundary, as required by the continuity of the field. The appearance of discontinuities destroys the unique relation between the internal topology and the field at the boundary, thereby accommodating the true physics of the arbitrarily interwoven magnetostatic field. That is to say, the discontinuities are an essential part of the physics of magnetostatic fields, for without them the mathematics would contradict the physics. Section 3.5 provides an example illustrating the problem for a field that extends through a large number n of different and unrelated winding patterns between $z = 0$ and $z = L$.

Consider, then, the characteristics of the magnetostatic equation (3.1). Write equation (3.1) in the form

$$4\pi \frac{\partial P}{\partial x_i} + B_j \frac{\partial B_j}{\partial x_i} = B_j \frac{\partial B_i}{\partial x_j}, \tag{3.13}$$

together with the divergence condition

$$\partial B_j / \partial x_j = 0. \tag{3.14}$$

These equations provide four independent linear relations between the 12 spatial derivatives of P and B_i. Denote by γ_i the direction cosines of some curve on which P and B_i are specified. Then dP/ds and dB_i/ds are known functions of distance s along the curve, providing the four relations

$$\frac{dP}{ds} = \gamma_j \frac{\partial P}{\partial x_j}, \tag{3.15}$$

$$\frac{dB_i}{ds} = \gamma_j \frac{\partial B_i}{\partial x_j}. \tag{3.16}$$

There are, then, a total of eight linear relations between the spatial derivatives of P and B_i, allowing computation of any eight of the 12, given P and B_i on the curve.

Suppose, then, that $\partial P/\partial z$ and $\partial B_i/\partial z$ as well as P and B_i are specified on the curve with direction cosines γ_i. All eight of the remaining derivatives $\partial P/\partial x$, $\partial P/\partial y$, $\partial B_i/\partial x$, $\partial B_i/\partial y$ can be computed. The exception arises for the direction γ_i at which the determinant of the coefficients of the eight equations vanishes. It is that direction that defines the characteristic curves, along which the derivatives cannot be computed, and across which tangential discontinuities are compatible with the equilibrium equation (3.1).

It is an elementary exercise to write out the 8×8 determinant D of the equations, taking the equations in the order in which they are written and placing the coefficients in the order of $\partial P/\partial x$, $\partial P/\partial y$, $\partial B_x/\partial x$, $\partial B_x/\partial y$, $\partial B_y/\partial x$, etc. The result is

$$D = \begin{vmatrix} +1 & 0 & 0 & -B_y & +B_y & 0 & B_z & 0 \\ 0 & +1 & 0 & +B_x & -B_x & 0 & 0 & B_z \\ 0 & 0 & 0 & 0 & 0 & 0 & -B_x & -B_y \\ 0 & 0 & 1 & 0 & 0 & 1 & 0 & 0 \\ \gamma_x & \gamma_y & 0 & 0 & 0 & 0 & 0 & 0 \\ 0 & 0 & \gamma_x & \gamma_y & 0 & 0 & 0 & 0 \\ 0 & 0 & 0 & 0 & \gamma_x & \gamma_y & 0 & 0 \\ 0 & 0 & 0 & 0 & 0 & 0 & \gamma_x & \gamma_y \end{vmatrix}.$$

Setting the determinant equal to zero leads to (Parker, 1979)

$$(\gamma_x^2 + \gamma_y^2)(\gamma_x B_y - \gamma_y B_x)^2 = 0. \tag{3.17}$$

We see immediately the two imaginery characteristics of the elliptic equation, $\gamma_x = \pm i\gamma_y$. The real characteristics are apparent as $\gamma_x B_y = \gamma_y B_x$, appearing to the second order in D. Noting that the equations for the field lines are

$$\gamma_i = \frac{dx_i}{ds} = \frac{B_i}{B},$$

where $B = (B_j B_j)^{\frac{1}{2}}$ is the field magnitude, it is evident that $\gamma_x/\gamma_y = B_x/B_y$ represents the projection of the field lines on the local plane $z = constant$. It is immediately obvious that if, instead of specifying $\partial P/\partial z, \partial B_i/\partial \mathbf{z}$, we had specified $\mathbf{e} \cdot \nabla P$ and $\mathbf{e} \cdot \nabla B_i$, where $\mathbf{e}$ is a given unit vector of arbitrary direction, the real characteristics would be the projection of the field lines on a plane perpendicular to $\mathbf{e}$. The essential point is that the field lines represent the real characteristics of the magnetostatic equation (3.1).

In the event that $\nabla P = 0$, the pressure P disappears from the equations and there are only the nine field derivatives $\partial B_i/\partial x_j$ to compute. Specification of B_i and $\partial B_i/\partial z$ on any curve γ_i determines the remaining six spatial derivatives of B_i. There are six linear equations, and the determinant of the coefficients is sixth order instead of eighth order. The first two columns in the eighth-order determinant D are missing, as is the fifth row, representing the coefficients in dP/ds. The third row, representing the Lorentz force in the z-direction, is also dropped, in view of the specification of $\partial B_i/\partial z$. The remaining 6×6 determinant yields

$$(\gamma_x^2 + \gamma_y^2)(\gamma_x B_y - \gamma_y B_x) = 0, \tag{3.18}$$

when set equal to zero. The characteristics are the same as before, except the field lines $\gamma_x B_y - \gamma_y B_x = 0$ now appear only to first order.

Other specifications of the field derivatives provide parallel results. For instance if we retain the vanishing of the z-component of the Lorentz force as one of the equations and then specify $\partial B_x/\partial z$ and $\partial B_y/\partial z$, but not $\partial B_z/\partial z$, the result is just the single real characteristic $\gamma_x B_y - \gamma_y B_x = 0$, etc.

A simpler way to derive the characteristics of the force-free field is to examine equations (3.5) and (3.6). The Laplacian operator in equation (3.6) indicates the imaginery characteristics $\gamma_x = \pm i\gamma_y$. Equation (3.5) is a first-order linear differential equation for α. The characteristics are obviously the field lines of $\mathbf{B}$. Hence equation (3.3), from which these results are derived, has the imaginery characteristics $\gamma_x = \pm i\gamma_y$ and the real characteristics represented by the field lines.

The essential point is that the magnetostatic equations (3.1) and (3.3) are qualitatively different from the familiar purely elliptic or purely hyperbolic equations of classical physics. They have the same imaginery characteristics as the elliptic equations, but in addition they have a set of real characteristics represented by the field lines. The field line characteristics provide an entirely different structure to the solutions, permitting surfaces of tangential discontinuity to accommodate the arbitrary winding and interweaving of the field lines, which is not possible with the purely elliptic equation.

With these principles in mind, consider the essential nature of the wound and interwoven field described by equation (3.11) when that field relaxes to the lowest available energy state.

3.5 Unlimited Winding

A simple but striking example of the limitations of continuous solutions of the force-free equilibrium equation (3.3) arises in the limit that the magnetic field described by equation (3.11) passes through a large number n of random statistically independent winding patterns (Parker, 1989). The essential point is that the field described by equation (3.11) is a mechanical construction, for which the smooth continuous stream function $\psi(x, y, kzt)$ can be specified at will. Hence the winding and interweaving of the field lines of the field of equation (3.11) can be constructed to any recipe that is desired.

Suppose then that the fluid motion (see equation 3.7) has a transverse scale ℓ and suppose that L is extremely large, essentially infinite. The fluid motion $\psi(x, y, kzt)$ is switched on for a long period of time τ such that ψ passes through a sequence of n statistically unrelated, strong, complex, close-packed swirls and eddies as kzt increases from 0 to $kL\tau$. The continuous function $\psi(x, y, kzt)$ is statistically homogeneous in space and time with a correlation length $\lambda = O(\ell)$ in the transverse (x, y) directions and a correlation length $\Lambda = O(\lambda)$ in the z-direction. Each individual winding pattern traversed by any given line of force has a length $m\Lambda$ where m is a finite number larger than one, say $m = 10$. Almost all the field lines extend through n such patterns, with $L = nm\Lambda$. Another way to state the situation is to note that the field lines are fixed at $z = 0$ while the footpoints of the field lines at $z = L$ are swirled by $\psi(x, y, kLt)$ through n successive unrelated mixing patterns, tracing out on $z = L$ the winding pattern of the field throughout $0 < z < L$. The simple sketch of interwoven fields shown in Fig. 1.5(b) may be considered to represent a single one of the n strong random patterns through which an individual field line passes on its way from $z = 0$ to $z = L$. Berger (1986) formulates the topological invariants for such a field.

Following the winding and interweaving episode ($0 < t < \tau$), the footpoints of the field are held fixed at $z = 0, L$ and the fluid is released so that the fluid and field are free to move in response to the unbalanced Lorentz forces $\partial M_{ij}/\partial x_j$ exerted by the field. With the help of a small viscosity the system eventually settles into the lowest available energy state and achieves a static equilibrium described by the force-free equation (3.3). The slightly viscous fluid maintains its infinite electrical conductivity, so that connectivity of the field lines between their footpoints at $z = 0$ and $z = L$ is preserved, along with the topology of their internal winding and interweaving.

Consider the mean values of the transverse field components B_x and B_y, and their transverse gradients $\partial B_x/\partial x$, $\partial B_x/\partial y$, $\partial B_y/\partial x$, $\partial B_y/\partial y$ along a typical field line. The overall mean value of the longitudinal field B_z is just B_0, of course, conserving the initial magnetic flux in the z-direction. The strong winding and wrapping produces transverse components with zero mean, but with rms values of the same general order as B_0, essentially $B_0 O(\lambda/\Lambda)$, over the full length L of the field. But over a single winding pattern, in which a given field line loops and circles a few times around, the mean values of B_x and B_y are generally less than B_0, but not zero. Say

$$\langle B_x \rangle, \langle B_y \rangle = \beta B_0, \tag{3.19}$$

where β is a number less than one because of partial cancellation in the averaging process. The magnitudes of the mean values of the transverse gradients, then, are of the order of $\beta B_0/\lambda$,

$$\left|\left\langle \frac{\partial B_x}{\partial x} \right\rangle\right|, \quad \left|\left\langle \frac{\partial B_x}{\partial y} \right\rangle\right|, \quad \left|\left\langle \frac{\partial B_y}{\partial x} \right\rangle\right|, \quad \left|\left\langle \frac{\partial B_y}{\partial y} \right\rangle\right| = O\left(\frac{\beta B_0}{\lambda}\right). \tag{3.20}$$

With these mean values along each individual field line extending through a single winding pattern, consider the mean along the entire length of the given field line from $z = 0$ to $z = L$, extending through n independent patterns. The essential point is that the signs of the mean transverse gradients, given by equation (3.20), vary at random along a field line as the line extends through one pattern after another. Therefore, the average along the entire length of the line, through n independent patterns is of the order of $(\beta B_0/\lambda)n^{-\frac{1}{2}}$ along almost all field lines. Using double angular brackets to denote the mean along the entire line from $z = 0$ to $z = L$,

$$|\langle\langle \partial B_x/\partial x \rangle\rangle|, \ |\langle\langle \partial B_x/\partial y \rangle\rangle|, \ |\langle\langle \partial B_y/\partial x \rangle\rangle|, \ |\langle\langle \partial B_y/\partial y \rangle\rangle| = O(\beta B_0/\lambda n^{\frac{1}{2}}) \tag{3.21}$$

in order of magnitude on almost all lines. In the limit of large n, this is zero. It should be added that in the same limit, the relative measure of the field lines on which the mean values do not approach zero is itself zero.

Consider, then, the mean of the z-component of the equilibrium equation (3.3) along a typical field line. With

$$\frac{\partial B_y}{\partial x} - \frac{\partial B_x}{\partial y} = \alpha B_z,$$

the average is

$$\left\langle\left\langle \frac{\partial B_y}{\partial x} \right\rangle\right\rangle - \left\langle\left\langle \frac{\partial B_x}{\partial y} \right\rangle\right\rangle = \alpha \langle\langle B_z \rangle\rangle \tag{3.22}$$

with the torsion coefficient α outside the averaging brackets because α is rigorously constant along the field line, according to equation (3.5). Note, then, that $\langle\langle B_z \rangle\rangle = O(B_0)$ and, therefore, is nonvanishing. On the other hand, the left hand side of the equation vanishes in the limit of large n. With the result in equation (3.21), it follows that the torsion coefficient is

$$\alpha = O(\beta/\lambda n^{\frac{1}{2}}) \tag{3.23}$$

on almost all field lines. In fact,

$$\lim_{n\to\infty} \alpha = 0 \tag{3.24}$$

on all but a vanishing fraction of the field lines. But if α vanishes, equation (3.3) becomes

$$\nabla \times \mathbf{B} = 0.$$

Hence $\mathbf{B}$ can be written as $-\nabla\phi$, and in view of the vanishing divergence of $\mathbf{B}$,

$$\nabla^2\phi = 0, \tag{3.25}$$

almost everywhere throughout the field. This is Laplace's equation. It is a fully elliptic equation, and its properties are well known. There are no continuous solutions to Laplace's equation that describe the extended interweaving of the field lines.

To provide a simple picture of the difficulty, imagine that the footpoints of the field at $z = 0, L$ are subjected to an irrotational motion with vanishing mean that serves to expand the regions where $B_z > B_0$ and contract the regions where $B_z < B_0$ so that $B_z = B_0$ over both boundary planes $z = 0, L$. Such a motion always exists, and it has no effect on the interwoven field extending the infinite distance L through the infinitely many unrelated winding patterns. But the solution to Laplace's equation for uniform B_z at $z = 0, L$ is just $\phi = -B_0 z$, so that $B_z = B_0$ throughout $0 < z < L$. This was the initial field before it was wound and interwoven by the fluid motion (equation 3.7) to the final form (equation 3.11). It does not match the topology of the final infinitely interwoven field (equation 3.11).

Perhaps the most obvious contradiction is the simple fact of equation (3.24), that there is no torsion in the field, in spite of the topological wrapping and mixing of the field lines between $z = 0$ and $z = L$. In fact, the real characteristics of equation (3.3) permit surfaces of tangential discontinuity along the flux surfaces of $\mathbf{B}$. The field between adjacent surfaces is irrotational and is a solution to Laplace's equation within the bounding surfaces of the tangential discontinuities. The torsion, that cannot appear in the irrotational field, is confined to the surfaces of discontinuity. There is no restriction imposed by equations (3.24)–(3.25) on the form of the surfaces of discontinuity. There is always a solution to Laplace's equation with the boundary condition that the normal component of the field vanishes on the surfaces of discontinuity. The surfaces intersect (see Chapter 8), forming ducts, with a local segment of one such duct sketched in Fig. 3.1. The field within each duct is irrotational and the existence of a suitable solution to Laplace's equation is readily seen

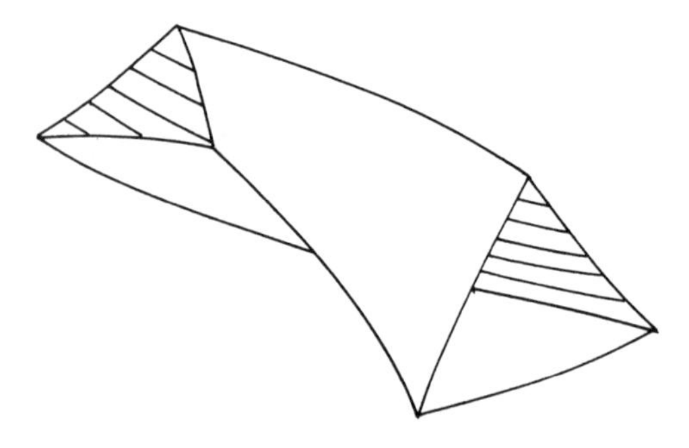

Fig. 3.1. A sketch of a segment of a duct defined by three intersecting surfaces of discontinuity, bounding the irrotational field flux extending along the interior of the duct.

from the simple fact that a potential field, such as arises under static conditions in a vacuum, can be trapped within a honeycomb of superconducting surfaces and manipulated by deformation of the enclosing surfaces in any way that does not cut off the magnetic flux and violate the divergence condition.

3.6 Action at a Distance

As a final exercise consider the evolution of the equilibrium magnetic field in the first winding pattern ($0 < z \lesssim m\Lambda$) as the subsequent winding patterns $n = 2, 3, 4, \ldots$ are added one by one. We start with the continuous interwoven field of equation (3.11) held firmly in the fluid of infinite electrical conductivity. Instead of releasing the fluid all at once at time τ, with the footpoints of the field held fixed at $z = 0$ and $z = L$ so that the field relaxes to the lowest available energy state simultaneously throughout $0 < z < L$, suppose instead that the lfuid and field are released from the fixed configuration of equation (3.11) by a slowly moving front $z = (t - \tau)V$ starting at $z = 0$ at time τ and progressing with the very small velocity $V \ll C = B_0/(4\pi\rho_0)^{\frac{1}{2}}$ toward $z = L$. The front drifts slowly across the first winding pattern in the long time $m\Lambda/V$, leaving behind that region in static equilibrium. In view of the arbitrary complex form of that pattern, the relaxation to equilibrium involves the formation of surfaces of tangential discontinuity with characteristic transverse scales of the order of ℓ or λ. The continuous field filling the regions between the surfaces of discontinuity is described by the force-free equation (3.3) with $|\alpha|$ of the order of $1/\lambda$ in the strong winding.

The slow moving liberation front crawls on across the second winding pattern ($m\Lambda < z < 2m\Lambda$), and the second pattern relaxes to equilibrium, forming its own internal surfaces of tangential discontinuity with essentially the same overall statistical properties as the first pattern. The readjustments of the field to static equilibrium in the second pattern lead to minor readjustments in the first pattern, of course, and the surfaces of discontinuity created in the second pattern extend into the first pattern, where they meet and merge with the surfaces of discontinuity of the first pattern.

The slow moving front creeps gradually across the third pattern ($2m\Lambda < z < 3m\Lambda$) which is then left in static equilibrium with its own surfaces of tangential discontinuity. These third pattern discontinuities extend into the second pattern, causing minor readjustments. However, with $m\Lambda$ large compared to the scale (ℓ, λ, Λ) of variation of the field through the patterns, the first pattern is largely buffered from the changes by the intervening second pattern.

The effects on the first pattern of the relaxation to equilibrium in the subsequent patterns ($n = 4, 5, \ldots$) are progressively less. As the slow moving relaxation front recedes into the distance ($z \gg m\Lambda$) further effects on the first pattern decline to zero, presumably exponentially with the passage of time. In Chapter 8 it is shown that the strength of an isolated discontinuity diminishes more or less inversely with distance from the location of the winding pattern of its creation. With the random signs of the tangential discontinuities created in successive patterns, and the tendency for weak discontinuities of distant origin to be caught up by the stronger discontinuities of nearby origin, the net effect declines at least as fast as the inverse of the square of the distance. So the slow quasi-equilibrium changes in the first pattern caused by the receding front decline at least as fast as $(\Lambda/Vt)^2$ and can eventually be ignored as the relaxation front recedes in the distance.

The interesting feature is that there is another effect that lingers on much longer in the first pattern, indicating the continuing recession of the distant relaxation front. That is the slow decline of the torsion coefficient, given by equation (3.23), which declines only as $n^{-\frac{1}{2}}$, so that in order of magnitude

$$\alpha = O(\beta/\lambda)(\Lambda/Vt)^{\frac{1}{2}}.$$

It is the slow asymptotic decline of α to zero, long after the other local effects of the distant receding relaxation front have disappeared, that indicates the continuing existence of the front. The magnetic field in the first pattern gradually and slowly approaches a potential form only long after the front is "out of sight." This shows the powerful consequences of the property of equation (3.5) that the torsion coefficient is rigorously constant along each field line. In this way the conditions at one point in a field are directly affected by distant deformation projecting along the set of real characteristics of the force-free equation (3.7). Thus, for instance, a section $(0 < z < L_1 < L)$ of continuous invariant $(\partial/\partial z = 0)$ equilibrium field, described by the Grad–Shafranov equation (1.9), would develop surfaces of tangential discontinuity as the relaxation front progresses beyond $z = L$, if the topology of the field is not invariant somewhere in $L_1 < z < L$.

4

Formal Structure of the Magnetostatic Equations

4.1 Introduction

The previous chapter has described the role of the real characteristics of the magnetostatic equations (3.1) and (3.3) in providing tangential discontinuities, enabling the mathematics to provide the arbitrary winding and interweaving of the field lines extending from $z = 0$ to $z = L$. The discussion is §3.5 illustrates the dilemma that arises with unlimited field line interweaving if the discontinuities are not allowed.

The purpose of the present chapter is to initiate a study of the formal mathematical solutions to the magnetostatic equations in the near vicinity of continuous invariant solutions. In Chapter 5 the investigation is extended to the general exact solutions of the magnetostatic equations, in the absence of any invariance. We begin with a cursory exposition of some of the known analytic solutions for continuous fields in three dimensions. The known continuous solutions all involve a degeneracy or invariance of one form or another, i.e., in some sense they are all 2D. Therefore they exhibit the familiar property that the fields on the boundaries of a region are uniquely and simply related to the winding of the fields throughout the interior of the region. The development proceeds, then, to construct solutions by formal integration under circumstances with no invariance or degeneracy, showing the explicit appearance of discontinuities as an essential part of the mathematical properties of the solutions. Specifically, we find that there are integral constraints on the field components and their derivatives if it is assumed that the field is continuous throughout the region. The integral constraints cannot be satisfied in a continuous field of any significant topological diversity, but the constraints disappear in the presence of the invariance that permits the continuous solutions. In the topologically complex field the integral constraints provide quantitative relations between the total strength of the discontinuities and the topology of the field.

Recalling that continuity is an essential part of the formal mathematical proof of the uniqueness of the solution of an elliptic partial differential equation for a given set of boundary conditions, it follows that the discontinuous solutions cannot be uniquely determined throughout $0 < z < L$ by specification of the fields on $z = 0, L$. Hence it is the discontinuous solutions, and not the continuous solutions, that have the mathematical freedom demanded by the physics, that the field at the boundaries $z = 0, L$ is generally not uniquely determined by the arbitrary winding and interweaving of the field throughout the interior. Thus, for instance, the footpoints of the field can be arbitrarily expanded and contracted in different locations to provide quite an arbitrary field strength at the boundary without changing the internal topology of a field (extending through a diversity of independent winding patterns throughout the interior $0 < z < L$).

The calculations show too that the intrinsic tangential discontinuities are not to be considered as the limiting case of a continuous change in the field over a small distance $\delta\ell$ as $\delta\ell \to 0$. That limit does not satisfy the mathematical requirements of the magnetostatic equations. The mathematics shows that the discontinuities are intrinsic, representing the surface of contact between two regions of different field topology.

In the real physical world, where the electrical resistivity of the fluid is not identically zero, the tangential discontinuity is, of course, replaced by a continuous but rapid change in field direction over a small distance $\delta\ell$. This cannot satisfy the magnetostatic equations within $\delta\ell$, so that thin sheet $\delta\ell$ is in dynamical none-quilibrium, usually involving dynamical reconnection of the field.

4.2 Continuous Solutions

Consider some examples of continuous analytic solutions of equations (3.1) and (3.3) made possible by a suitable degeneracy or invariance. As a first example, consider the solution of equation (3.3) based on the degeneracy $\nabla\alpha = 0$. The interesting features of this simple case were first emphasized by Rosner and Knobloch (1982). Setting $\alpha = q$, where q is a constant, they pointed out the unlimited topological complexity represented by the solutions. The essential point is that the equilibrium equation

$$\nabla \times \mathbf{B} = q\mathbf{B}, \tag{4.1}$$

is linear, and an arbitrary superposition of all solutions with the same value of q is permitted. The familiar solutions $B_\varphi = B_0 J_1(q\varpi)$, $B_\chi = B_0 J_0(q\varpi)$ can be employed, where the χ-direction is arbitrary and ϖ represents distance measured from any straight line extending in the χ-direction. Azimuth measured around the line is indicated by φ. The field magnitude B_0 for each χ-axis is arbitrary. It is left to the reader to imagine the unlimited complexity of the topology of the field lines that can be constructed by superposition of the solutions for an arbitrary array of arbitrarily differently oriented line segments χ. The solution is possible when $\nabla\alpha = 0$ because the restriction $\mathbf{B} \cdot \nabla\alpha = 0$ is satisfied for any $\mathbf{B}$.

Another class of solution, based on a small perturbation $\varepsilon\mathbf{b}(\mathbf{r})$ of the field from a uniform field $\mathbf{e}_z B_0$, with the degeneracy $\nabla B_0 = 0$, was pointed out by Van Ballegooijen (1985), assuming that the correlation scale ℓ_z along the field is of the order of ℓ/ε so that over a distance ℓ_z the field is twisted by one or more radians. In that case $\partial/\partial z$ is small $O(\varepsilon)$ compared to $\partial/\partial x$ and $\partial/\partial y$ rather than of the same order. The torsion coefficient is small $O(\varepsilon)$, and can be written εf. In lowest order, the divergence condition $\nabla \cdot \mathbf{b} = 0$ reduces to the 2D divergence in x and y. Therefore, b_x and b_y can be expressed in terms of a vector potential $a(x, y, z)$ with

$$b_x = +B_0\partial a/\partial y, = -B_0\partial a/\partial y \tag{4.2}$$

upon neglecting terms $O(\varepsilon)$ compared to one. The x and y components of $\nabla \times \mathbf{B} = \alpha\mathbf{B}$ yield b_z small $O(\varepsilon^2)$, while the z-component gives

$$f = -(\partial^2 a/\partial x^2 + \partial^2 a/\partial y^2). \tag{4.3}$$

Then all the terms in $\mathbf{B} \cdot \nabla\alpha = 0$ are second order in ε, yielding the relation

$$\frac{\partial f}{\partial z} = \frac{\partial a}{\partial x}\frac{\partial f}{\partial y} - \frac{\partial a}{\partial y}\frac{\partial f}{\partial x}. \tag{4.4}$$

Van Ballegooijen points out that equations (4.3) and (4.4) are the precise analog of the equations for the time-dependent 2D vorticity equation for an inviscid incompressible fluid

$$\frac{\partial \omega}{\partial t} = \frac{\partial S}{\partial x}\frac{\partial \omega}{\partial y} - \frac{\partial S}{\partial y}\frac{\partial \omega}{\partial x}, \tag{4.5}$$

where the vorticity ω is related to the stream function S by

$$\omega = -(\partial^2 S/\partial x^2 + \partial^2 S/\partial y^2). \tag{4.6}$$

and

$$v_x = +\partial S/\partial x, \; v_y = -\partial S/\partial y. \tag{4.7}$$

If z is replaced by t, equation (4.4) reduces to (4.5). Thus the magnetic field varies in the z-direction in the same manner that the vorticity varies with time. That is to say, the field variation mimics 2D turbulence in an inviscid fluid, providing complicated winding and interweaving of the field lines as they extend in the z-direction. Unfortunately there are only a few known analytic solutions to the 2D vorticity equation, but it is well known that there is an infinite family of states of 2D turbulence. Tangential discontinuities (vortex sheets) are an intrinsic part of most of the solutions (Arnold, 1965, 1966, 1978; Arnold and Avez, 1968). The essential point here is that the vorticity equation (4.5) can also be written

$$d\omega/dt = (\partial/\partial t + v_x \partial/\partial x + v_y \partial/\partial y)\omega = 0, \tag{4.8}$$

which is the restriction that the vorticity moves with the fluid and is conserved in the process just as α is constant along each field line in 3D space. Hence the continuous solutions cannot accommodate an arbitrary topological winding pattern of the vortex lines (introduced in the magnetic analogy by the arbitrary stream function ψ in equation (3.8)). In such cases, then, where the choice of ψ introduces winding of the field lines whose topological variation along z contradicts equations (4.4) and (4.8), there can be no entirely continuous solutions. The field adjusts throughout the volume so that the torsion is uniform along each field line, leaving discontinuities between finite regions within which the field is continuous.

4.3 Invariant Fields

Most continuous analytic solutions of equations of the form

$$4\pi\nabla P = (\nabla \times \mathbf{B}) \times \mathbf{B}, \tag{4.9}$$

where P is scalar function (equal to the fluid pressure plus gravitational potential energy per unit volume) are based on an invariance in the form of an ignorable coordinate. Axisymmetry and linear invariance ($\partial/\partial z = 0$) were treated by Lüst and Schlüter (1954), Gjellestad (1954), Prendergast (1956, 1957), Chandrasekhar (1956, 1958), Chandrasekhar and Kendall (1957), and Chandrasekhar and Woltjer (1958).

For linear invariance ($\partial/\partial z = 0$) the equation reduces to

$$B_x = +\partial A/\partial y,\ B_y = -\partial A/\partial x,\ B_z = B_z(A),\ p = p(A), \tag{4.10}$$

where A satisfies the quasi-linear elliptic Grad–Shafranov equation,

$$\nabla^2 A + 4\pi F'(A) = 0, \tag{4.11}$$

with

$$F(A) = p(A) + B_z(A)^2/8\pi. \tag{4.12}$$

If $\nabla p = 0$, the field is force free, with $\alpha = B_z'(A)$ and

$$\nabla \times \mathbf{B} = B_z'(A)\mathbf{B}. \tag{4.13}$$

The mathematical forms are similar for the other symmetries, e.g., linear, rotational, and helical (Edenstrasser, 1980; Tsinganos, 1981a,b; 1982a,b,c).

It is convenient now to continue the discussion in terms of general orthogonal curvilinear coordinates x^i for which the line elements are written (h_1, h_2, h_3).

The magnetostatic equation (4.9) in general orthogonal curvilinear coordinates (x^i) has been shown by Tsinganos (1981a,b; 1982a; Tsinganos, Distler, and Rosner, 1984) to reduce to a quasi-linear elliptic partial differential equation of the same general form as the Grad–Shafranov equation (4.11) when one of the coordinates is ignorable. The Grad-Shafranov equation is the special case that arises when z is the ignorable coordinate. Helical coordinates can be similarly reduced (Morozov and Solovev, 1966; Edenstrasser, 1980; Tsinganos, 1982b). The essential point is that in the presence of an ignorable coordinate, there is always one integral of the equations, allowing reduction of the three first-order components of equation (4.9) to one equation of second order. The field lines lie on cylindrical surfaces $A(x^1, x^2) = C$ (when x^3 is designated the ignorable coordinate) in x^i coordinate space. The field components can be written in terms of the line elements $h_1(x^1, x^2)$, $h_2(x^1, x^2)$, $h_3(x^1, x^2)$ as

$$B_1 = +\frac{1}{h_2 h_3}\frac{\partial A}{\partial x^2},\ B_2 = -\frac{1}{h_3 h_1}\frac{\partial A}{\partial x^1},\ h_3 B_3 = f(A). \tag{4.14}$$

The third component of equation (4.9) is then automatically satisfied. With $P = P(A)$, the first and second components of equation (4.9) are individually satisfied if

$$\frac{1}{h_1 h_2 h_3}\left[\frac{\partial}{\partial x^1}\left(\frac{h_2}{h_1 h_3}\frac{\partial A}{\partial x^1}\right) + \frac{\partial}{\partial x^2}\left(\frac{h_1}{h_2 h_3}\frac{\partial A}{\partial x^2}\right)\right] + \frac{B_3}{h_3}\frac{df}{dA} + 4\pi\frac{dP}{dA} = 0. \tag{4.15}$$

Since the flux surfaces $A(x^1, x^2) = \mathit{constant}$ in an infinite space have no edges, it follows that the surfaces either extend to infinity or form nested tori (Kruskal and Kulsrud, 1958). That is to say, the projection of the field onto any surface $x^3 = \mathit{constant}$ provides a 2D field $(B_1, B_2, 0)$ that is divergence free. Hence the projections of the field lines onto $x^3 = \mathit{constant}$ either extend to infinity or form closed loops. In the context of the present writing, treating the fields formed from an initially uniform field by finite close-packed localized swirls in ψ with scale ℓ, it follows that the projection of almost all field lines onto $z = constant$ form closed curves, as noted in the previous chapter.

4.4 Nearly Invariant Fields

The next step is to consider the consequences of introducing a weak dependence of the field on x^3, to see the effect of breaking the symmetry on the flux surfaces, and ultimately on the continuity of the field. Before delving into the mathematics, it is instructive to consider from simple geometrical considerations the qualitative effects to be expected. Denote the invariant field described above by $B_i(x^1, x^2)$. Add to this a small arbitrary perturbation εb_i that depends on x^3 as well as on x^1 and x^2, so that the total field is $B_i(x^1, x^2) + \varepsilon b_i(x^1, x^2, x^3)$ where $\varepsilon \ll 1$. It is obvious that the field lines, described by

$$\frac{dx^1}{B_1 + \varepsilon b_1} = \frac{dx^2}{B_2 + \varepsilon b^2} = \frac{dx^3}{B_3 + \varepsilon b_3},$$

gradually wander from the unperturbed paths of $\mathbf{B}$ for almost all choices of b_i. It follows that the cylindrical flux surfaces of $B_i(x^1, x^2)$ become increasingly corrugated with distance x^3 along the field. In fact, for most choices of b_i the wandering of the field lines is incommensurable, i.e., stochastic. Field lines that are near neighbors at $x^3 = 0$ wander apart, either diverging without bound in the limit of large x^3 or filling some portion of (x^1, x^2) ergodically. The cylindrical flux surfaces of B_i at $x^3 = z = 0$ are mapped along x^3 by the stochastic field lines, becoming increasingly contorted and rolled up within themselves with increasing x^3, sketched in Fig. 4.1. In the limit of large L the flux surface becomes infinitely broad and infinitely folded, spreading sinuously in ever-increasing folds across some finite portion of the cross-section of the field.

In a toroidal field geometry the field lines circle around the toroidal space without limit. There are no unique flux surfaces, since any one surface projects around and around the torus until it fills some finite volume. Hence it is said that the

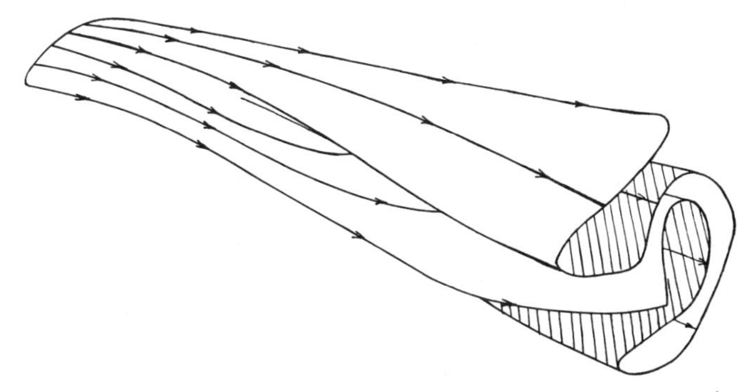

Fig. 4.1. A schematic drawing of a portion of a flux surface defined by $A(x^1, x^2) = constant$ at $x^3 = 0$, and projected toward positive x^3 along the field lines of the perturbed field $B_i + \varepsilon b_i$ to show the increasing convolution.

introduction of an unsymmetric perturbation $\varepsilon\mathbf{b}(\mathbf{r})$ into a toroidal field geometry destroys the flux surfaces of the initially symmetric toroidal field $\mathbf{B}$ (in which the flux surfaces are well defined) (Henon and Heiles, 1964; Rosenbluth, et al. 1966; Morozov and Solovev, 1966; Filonenko, Sagdeev, and Zaslavsky, 1967; Hamada, 1972).

To work out the consequences of the growing corrugations of a flux surface with increasing z recall that in most cases the fluid pressure $P(A)$, or the torsion coefficient $B_z'(A)$, of the zero-order field varies significantly over x^1 and x^2 at $z = x^3 = 0$. Now the perturbed pressure $P(A) + \varepsilon p$, or torsion coefficient $\alpha(A) + \varepsilon\lambda$, is rigorously constant along each field line of the perturbed field and yet is supposed to differ only by $O(\varepsilon)$ from the zero-order value as the field line wanders (with ever increasing x^3) into regions where $P(A)$ or $\alpha(A)$ is radically different. These two requirements are incompatible. Formally, a field line at $A = A_0$ at $x^3 = 0$, where $P = P(A_0)$, eventually finds its way to where $A = A_1$ differs strongly from A_0, and $P = P(A_1)$ differs strongly from $P = P(A_0)$, at $z = 0$. Yet the perturbation scheme requires that P vary along the perturbed field line by no more than $O(\varepsilon)$. It follows that a small perturbation εb_i of the field destroys the possibility of a static equilibrium by virtue of the unlimited wandering of the lines of the perturbed field. The wandering, however slow it may be, breaks the symmetry that is essential for static equilibrium.

The same contradiction arises in a force-free field where it is the torsion coefficient α, rather than P, that is rigorously constant along each field line.

This dilemma is avoided, of course, if b_i has the same invariance as B_i (i.e., $\partial/\partial z = \partial/\partial x^3 = 0$). Obviously when α is equal to some constant q in a force-free field, satisfying equation (4.1), there is no difficulty with the wander field lines of the perturbed field $B_i + \varepsilon b_i$, because the stochastic field lines encounter no value of α other than q no matter how far they wander away from their initial location (at $z = 0$). There are special forms of b_i that vary with x^3 in such a way as to leave the flux surfaces only slightly offset and perhaps slightly eccentric, but otherwise intact, thereby avoiding the stochastic wandering of the field lines and the ultimate dilemma. But such forms are a set of measure zero, as indicated by Floquet's theory of periodic solutions of differential equations with periodic coefficients.

Floquet's theory (cf. Ince, 1926) provides the special restrictions on b_i that are necessary for the perturbed field lines to circle periodically rather than unstably wandering ever farther away. The transverse displacement Δx^i of the perturbed field lines from the unperturbed lines $A(x^1, x^2) = A_0$ satisfies

$$\frac{d\Delta x^1}{dx^3} = \varepsilon\frac{b_1}{B_3}, \quad \frac{d\Delta x^2}{dx^3} = \varepsilon\frac{b_2}{B_3}. \tag{4.16}$$

Starting with $\Delta x^1 = \Delta x^2 = 0$ at $x^3 = 0$, the integration is along the unperturbed field line $A = A_0$, around which b_i and B_i vary periodically. For most choices of b_i the mean value of $B_i b_i$ along an unperturbed field line does not average to zero. The integration is unstable, with Δx^1 and Δx^2 increasing without bound. Floquet's theory shows that only special choices of b_i provide closed field lines.

We encounter Floquet's theory again in subsequent sections when we construct formal solutions to the field equations.

Consider, then, how the dilemma described above turns up in the formal mathematical solution of the force-free equilibrium field equation (3.3) As we might guess, the mathematics escapes the contradiction by developing surfaces of tangential discontinuity which take up the discrepant torsion α, leaving a consistent continuous solution throughout the regions between the surfaces.

4.5 Infinitesimal Perturbation to a Force-Free Field

As a first step in examining the analytic solutions to the magnetostatic equation (4.9), consider the small perturbation $\varepsilon b_i(x,y,z)$ to a force-free field $B_i(x,y)$ created by some arbitrary continuous mapping of the footpoints of the field at the boundaries $z = 0, L$. The equilibrium of the zero-order field $B_i(x,y)$ is described by equation (3.3), or by equations (4.10)–(4.13) with $p(A) = 0$. It follows that the vector potential $A(x,y)$ satisfies equation (4.11) or (4.15),

$$\frac{\partial^2 A}{\partial x^2} + \frac{\partial^2 A}{\partial y^2} + B_z'(A)B_z(A) = 0, \tag{4.17}$$

which is just the z-component of equation (4.9). The torsion coefficient α is equal to $B_z'(A)$, and the x and y components of $\nabla \times \mathbf{B} = \alpha\mathbf{B}$ are satisfied identically. The magnitude B of the field is

$$B = +[B_z^2(A) + |\nabla A|^2]^{\frac{1}{2}}.$$

With the field obtained through the deformation of equation (3.7) of an initial uniform field B_0 in the z-direction, it follows that $B_z(A)$ is everywhere positive, and presumably of the general order of B_0.

Suppose, as in the previous section, that an additional deformation of the field introduces the small perturbation $\varepsilon\mathbf{b}(x,y,z)$, where ε is the infinitesimal expansion parameter and $\mathbf{b}$ is comparable in magnitude to $\mathbf{B}$. Now the torsion coefficient is written as

$$\alpha = B_z'(A) + \varepsilon\lambda(x,y,z) \tag{4.18}$$

in the perturbed field. The first-order terms in the force-free equilibrium equations are

$$\nabla \times \mathbf{b} = \lambda\mathbf{B} + B_z'(A)\mathbf{b}. \tag{4.19}$$

The scalar product of this equation with $\mathbf{B}$ can be solved for λ, yielding

$$\begin{aligned}\lambda &= \mathbf{B}\cdot\left[\nabla \times \mathbf{b} - B_z'(A)\mathbf{b}\right]/B^2,\\ &= \left[(\nabla \times \mathbf{b})_{\parallel} - B_z'(A)b_{\parallel}\right]/B,\end{aligned} \tag{4.20}$$

where the subscript $\parallel$ denotes the component in the direction $\mathbf{B}/B$. Substituting this expression for λ into the equilibrium equation (4.19) results in

$$(\nabla \times \mathbf{b})_{\perp} = B_z'(A)\mathbf{b}_{\perp}, \tag{4.21}$$

where the subscript $\perp$ denotes the component perpendicular to $\mathbf{B}$, with

$$\mathbf{b}_{\perp} = \mathbf{b} - b_{\parallel}\mathbf{B}/B = \mathbf{b} - \mathbf{b}_{\parallel}, \tag{4.22}$$

$$(\nabla \times \mathbf{b})_{\perp} = \nabla \times \mathbf{b} - (\nabla \times \mathbf{b})_{\parallel}. \tag{4.23}$$

This same result follows from setting to zero the perturbation $(\nabla \times \mathbf{B}) \times \mathbf{b} + (\nabla \times \mathbf{b}) \times \mathbf{B}/4\pi$ of the Lorentz force and noting that $\nabla \times \mathbf{B} = B_z'(A)\mathbf{B}$. The essential point is that equilibrium requires that the torsion coefficient for the perpendicular component of $\mathbf{b}$ and $\nabla \times \mathbf{b}$ is unchanged from the value $B_z'(A)$ of the zero-order field. On the other hand there is no reason why we cannot wind in a different torsion in $\mathbf{b}$ through our choice of the winding function ψ in equation (3.7). So the simple perturbation expansion finds itself in contradiction with the physical freedom to introduce a perturbation εb_i with arbitrary torsion.

To continue with the calculation, note that the coefficients of $\mathbf{b}$ and λ in equation (4.19) are independent of z. Hence z is a separable coordinate and the solutions have a z-dependence of the form $\exp \kappa z$ where κ is an arbitrary complex constant. The equations are linear, so that solutions for different κ may be superposed. Write

$$b_i(x, y, z) = \sum_{\kappa} \beta_i(\kappa, x, y) \exp \kappa z, \tag{4.24}$$

$$\lambda(x, y, z) = \sum_{\kappa} \sigma(\kappa, x, y,) \exp \kappa z. \tag{4.25}$$

The requirement that $\nabla \cdot \mathbf{b} = 0$ reduces to

$$\frac{\partial \beta_x}{\partial x} + \frac{\partial \beta_y}{\partial y} + \kappa \beta_z = 0 \tag{4.26}$$

for each value of κ, while the three components of equation (4.19) can be written

$$\frac{\partial \beta_z}{\partial y} = \sigma \frac{\partial A}{\partial y} + B_z'(A)\beta_x + \kappa \beta_y, \tag{4.27}$$

$$\frac{\partial \beta_z}{\partial x} = \sigma \frac{\partial A}{\partial x} - B_z'(A)\beta_y + \kappa \beta_x, \tag{4.28}$$

$$\frac{\partial \beta_y}{\partial x} - \frac{\partial \beta_x}{\partial y} = \sigma B_z(A) + B_z'(A)\beta_z. \tag{4.29}$$

The divergence of equations (4.27)-(4.29), i.e., the first-order terms of $(\mathbf{B} + \varepsilon \mathbf{b}) \cdot \nabla \alpha = 0$, yields

$$\frac{\partial \sigma}{\partial x}\frac{\partial A}{\partial y} - \frac{\partial \sigma}{\partial y}\frac{\partial A}{\partial x} + \kappa B_z(A)\sigma + B_z''(A)\left(\beta_x \frac{\partial A}{\partial x} + \beta_y \frac{\partial A}{\partial y}\right) = 0. \tag{4.30}$$

4.6 Integral Constraints

A number of integral constraints can be deduced from these equations. The most important example follows from multiplying equation (4.27) by $-(\partial A)/\partial x)/|\nabla A|$ and equation (4.28) by $+(\partial A/\partial y)/|\nabla A|$. Note that these two factors are the y and x components, respectively, of the unit vector $\mathbf{e}_p$ tangent to the projection $A(x, y) = A_0$ of the field line onto any plane $z = constant$. The sum of the two expressions then gives

$$\mathbf{e}_p \cdot \nabla \beta_z = -B_z'(A)(\mathbf{e}_p \times \beta)_z + \kappa \mathbf{e}_p \cdot \beta. \tag{4.31}$$

Now, as described in §4.3, the projection $A(x, y) = A_0$ of the field lines onto $z = constant$ yields families of closed curves except for the occasional line forming a separatrix between closed regions. Denote by ds the element of arc length on any closed curve $A(x, y) = A_0$, measured in the counterclockwise direction. Consider the case that (B_x, B_y) points in the counterclockwise direction around that closed curve. The outward normal component of (b_x, b_y) is

$$\beta_n = \mathbf{e}_n \cdot \beta = -(\mathbf{e}_p \times \beta)_z, \tag{4.32}$$

where $\mathbf{e}_n$ is the outward unit normal vector

$$\mathbf{e}_n = \nabla A / |\nabla A|,$$

while the component of β_i tangent to $A(x, y) = A_0$ is

$$\beta_p = \mathbf{e}_p \cdot \beta,$$

of course. Equation (4.31) can be rewritten as

$$\partial \beta_z / \partial s = B_z'(A)\beta_n + \kappa \beta_p. \tag{4.33}$$

Integrate this relation around the closed contour $A(x, y) = A_0$ so that

$$\oint ds \partial \beta_z / \partial s = B_z'(A_0) \oint ds \beta_n + \kappa \oint ds \beta_p. \tag{4.34}$$

It follows from Gauss's theorem and equation (4.26) that the line integral of the normal component of β can be written as

$$\oint ds \beta_n = \int dx \int_{A_0} dy (\partial \beta x / \partial x + \partial \beta y / \partial y), \tag{4.35}$$

$$= -\kappa \int dx \int_{A_0} dy \beta_z, \tag{4.36}$$

where the integration on the right-hand side is over the area enclosed by the contour. It follows from Stokes' theorem and equation (4.29) that the integral of the tangential component can be written

$$\oint ds \beta_p = \int dx \int_{A_0} dy (\partial \beta y / \partial x - \partial \beta_x / \partial y), \tag{4.37}$$

$$= \int dx \int_{A_0} dy [\sigma B_z(A) + B_z'(A)\beta_z], \tag{4.38}$$

where again the integration is over the interior of $A = A_0$. Substituting equations (4.36) and (4.38) into (4.34) yields

$$\oint ds \partial \beta_z / \partial s = \kappa B_z(A_0) \int dx \int_{A_0} dy \sigma \tag{4.39}$$

for every closed field line contour $A(x, y) = A_0$.

Now suppose that β_z is everywhere continuous. Then obviously

$$\oint ds \partial \beta_z / \partial s = \oint d\beta_z = 0, \tag{4.40}$$

and equation (4.39) reduces to

$$\kappa B_z(A_0) \int dx \int_{A_0} dy\sigma = 0.$$

This requirement is automatically satisfied if the perturbation field has the same invariance ($\partial/\partial z = 0$ so that $\kappa = 0$) as the zero-order field. On the other hand, the arbitrary perturbed field generally lacks this invariance, so it is necessary that

$$\int dx \int_{A_0} dy\sigma = 0. \tag{4.41}$$

This result applies to the interior of all closed field lines, $A(x, y) = A_0$. Hence it applies to every annulus $(A, A + \delta A)$ bounded by closed field lines, so that

$$\oint ds\sigma/|\nabla A| = 0, \tag{4.42}$$

where the width of the annulus δA is proportional to $|\nabla A|^{-1}$. The requirement is that the mean value of σ around each annulus $(A, A + \delta A)$ in $z = constant$ must vanish if β_z is to be continuous. On the other hand, the arbitrary perturbation may involve introduction of a torsion that is nonvanishing and predominantly of one sign, so that the mean value around an annulus does not vanish. So again the formal expansion cannot accommodate the physical realities of arbitrary winding.

The calculation can be carried one step further, noting from equation (4.20) that

$$\sigma = [(\nabla \times \beta)]_{\|} - B_z'(A)\beta_{\|}]/B. \tag{4.43}$$

Then equation (4.42) becomes

$$\oint ds[(\nabla \times \beta)_{\|} - B_z'(A)\beta_{\|}]/B|\nabla A| = 0.$$

Denote the weighted average value around the contour $A = A_0$ by angular brackets, so that

$$\langle(\nabla \times \beta)_{\|}\rangle = \left(\oint ds(\nabla \times \beta)_{\|}/B|\nabla A|\right) \Big/ \oint ds/B|\nabla A|, \tag{4.44}$$

$$\langle\beta_{\|}\rangle = \left(\oint ds\beta_{\|}/B|\nabla A|\right) \Big/ \oint ds/B|\nabla A|. \tag{4.45}$$

The result is

$$\langle(\nabla \times \beta)_{\|}\rangle = B_z'(A_0)\langle\beta_{\|}\rangle. \tag{4.46}$$

The mean value of the torsion coefficient for the parallel field component $\beta_{\|}$ must be $B_z'(A)$, the same as in the (topological different) zero-order field **B**. Equation (4.21) requires the torsion coefficient for $\beta_{\perp}$ to have the unperturbed value $B_z'(A)$ at every point, and not merely in the weighted average around the contour. This is the result of the condition (4.42), that there can be no net change σ in the torsion coefficient of the perturbed field if the field is to be continuous everywhere.

The problem is that the arbitrary topology of the perturbation field bears little if any relation to the topology of **B**, so that the change in the torsion coefficient cannot be expected to average to zero everywhere. Hence, somewhere in the region the necessary conditions for continuity are not fulfilled and a discontinuity is the result.

One may wonder if there is not some trick by which the perturbed field may contrive to satisfy the requirement of equation (4.42) in spite of the arbitrary form of β relative to $\mathbf{B}$. There are two obvious reasons why the answer is negative in most cases. The first is that equation (4.42) is not the only restriction imposed on the equilibrium field perturbation. Two more restrictions are developed in the next section §4.7, and the simultaneous conformity to all three restrictions is a necessary condition for the absence of discontinuities. It is shown in §5.2 that there is insufficient freedom in the functional form of the perturbation field to satisfy all the necessary conditions with an arbitrary winding pattern in the perturbation field.

The second reason is a simple example in which the perturbation involves a net increase in the torsion of a well-defined flux bundle of the zero-order field $\mathbf{B}$. Consider the twisting of a broad flux bundle of the field $\mathbf{B}$, so that the torsion is increased from the zero-order $B_z'(A)$ throughout the interior of the flux bundle. The flux bundle itself may be a whole topological region of $\mathbf{B}$ bounded by a separatrix, or it may be a somewhat smaller flux bundle spiralling within the whole topological region, or it may involve more than one complete topological region, so that there are internal separatrices, as well as the separatrix that defines its boundary. When the perturbed field is released so that it can relax to equilibrium, the flux bundle contracts or expands, and may even develop a slightly $(O(\varepsilon))$ corkscrew form.

The essential point is that the field finds itself in a minimum energy state, i.e., static equilibrium. The equilibrium requires the condition in equation (4.42), that the net torsion is unchanged in a region where the torsion has been increased. We may guess that the field readjusts itself (without altering its topology) so as to reduce the mean torsion to the original $B_z'(A)$ on almost every contour $A = A_0$, thereby relaxing to a continuous form almost everywhere. But if there is a net overall increase in the torsion, then somewhere in the region the mean torsion must increase, in violation of equation (4.42), and that is where the discontinuities appear.

4.7 Further Integral Constraints

Consider the additional integral requirements for a continuous equilibrium field. Note that equation (4.30) can be written

$$\mathbf{e}_p \cdot \nabla\sigma + \kappa B_z(A_0)\sigma/|\nabla A| + B_z''(A_0), (\mathbf{e}_p \times \beta)_z = 0, \tag{4.47}$$

after dividing by $|\nabla A|$, where again $\mathbf{e}_p$ is the unit vector tangent to the curve $A = A_0$. This can be rewritten

$$\frac{\partial\sigma}{\partial s} + \kappa B_z(A)\sigma/|\nabla A| - B_z''(A)\beta_n = 0. \tag{4.48}$$

Integrating around the closed contour $A = A_0$ yields

$$\oint ds\partial\sigma/\partial s + \kappa B_z(A) \oint ds\sigma/|\nabla A| - B_z''(A) \oint ds\beta_n = 0. \tag{4.49}$$

If σ is continuous, the first integral vanishes and equation (4.42) requires that the second integral vanish, leaving

$$\oint ds\beta_n = 0 \tag{4.50}$$

for a continuous field. It follows from equation (4.36) that

$$\int dx \int dy\beta_z = 0, \tag{4.51}$$

where the integration is over the interior of $A = A_0$. Since the mean value of β_z vanishes over the interior of a finite range of A_0, it follows that the mean value over each annulus $(A_0, A_0 + \delta A_0)$ must vanish, so that the condition

$$\oint ds\beta_z/|\nabla A| = 0 \tag{4.52}$$

is necessary on all contours $A = A_0$ if the field is everywhere continuous. Note again that the perturbation $\beta(x, y)$ is arbitrary and generally exhibits a pattern unrelated to the winding pattern of the zero-order field A. Therefore β_z is generally nonvanishing, and, with a largely unrelated pattern, the mean value around any annulus $A = A_0$ cannot be expected to vanish everywhere. Where equation (4.52) is not satisfied there is a discontinuity in the torsion coefficient σ. The torsion coefficient would be continuous everywhere if the field were continuous everywhere. Hence the discontinuity in σ, or λ, implies a discontinuity in β or $\mathbf{b}$.

Another requirement for continuity can be obtained from formal integration of equation (4.48) for σ, yielding

$$\sigma(A_0, s) = \sigma(A_0, s_0)\exp\left[-\kappa B_z(A_0)\int_{s_0}^{s} ds'/|\nabla A|\right]$$
$$+ B_z''(A_0)\int_{s_0}^{s} ds'\beta_n(A_0, s')\exp\left[-\kappa B_z(A_0)\int_{s'}^{s} ds''/|\nabla A|\right],$$

where the integration is over arc length s along the closed contour $A = A_0$ from some arbitrary starting point $s = s_0$. If σ is continuous, the integration once around the contour, of length $\mathcal{L}(A_0)$, leads to

$$\sigma(A_0, s_0) + \mathcal{L} = \sigma(A_0, s_0)$$

for all s_0. Hence

$$\sigma(A_0, s_0)\left\{1 - \exp\left[-\kappa B_z(A_0)\oint ds/|\nabla A|\right]\right\}$$
$$= B_z''(A_0)\int_{s_0}^{s_0+\mathcal{L}} ds'\beta_n(A_0, s')\exp\left[-\kappa B_z(A_0)\int_{s'}^{s_0+\mathcal{L}} ds''/|\nabla A|\right] \tag{4.53}$$

for all values of s_0. This is another restriction on β_n, in addition to equation (4.34) or (4.35).

The restriction of equation (4.42) is just a special case of Floquet's theorem (Ince, 1926), which states that the solution to equation (4.30), in which the coefficients are periodic functions of s, is periodic only for special values of $\kappa B_z(A)$ or for special choices of $\beta_\perp(A, s)$ (see examples in Parker, 1979, pp. 370–378).

A perturbation **b** which fails to have the special form required by equation (4.36) or (4.42) must nonetheless satisfy the equilibrium field equations (4.17)–(4.21), of course. If the winding pattern of **b** is invariant along the field, then $\kappa = 0$ and the restriction of equation (4.36) disappears; equation (4.34) is automatically satisfied by the invariance. It is only insofar as the winding pattern of **b** varies with z that equations (4.42), (4.45), (4.46) and (4.53) are required.

Note that in the special case $\kappa = 0$ the divergence condition of equation (4.31) implies that b_x and b_y can be written in terms of a vector potential $a(x, y)$,

$$b_x = +\partial a/\partial y,\ b_y = -\partial a/\partial x. \tag{4.54}$$

The equations (4.19)–(4.21) are automatically satisfied by $\lambda = \lambda(A)$, $a = a(A)$, and $b_z = b_z(A)$ with

$$b_z'(A) = \lambda(A) + B_z'(A)a'(A), \tag{4.55}$$

and

$$(\nabla A)^2 a''(A) - a'(A)\frac{d}{dA}B_z^2(A) + \frac{d}{dA}b_z(A)B_z(A) = 0. \tag{4.56}$$

These perturbation solutions are, of course, a member of the general class of solutions for $\partial/\partial z = 0$, satisfying equation (4.17) for some suitable other choice of $B_z(A)$.

The more likely situation is that the condition of equations (4.42) and/or (4.45) and (4.46) are not satisfied and $\kappa \neq 0$, which brings us back to equation (4.39), from which it follows that the quantity

$$\sum \equiv \oint ds\partial\beta_z/\partial s$$

does not vanish. This is possible only if β_z is discontinuous at one or more points on the contour $A = A_0$. The value of $\sum$, given by

$$\sum = \kappa B_z(A_0)\int ds \int dy\sigma, \tag{4.57}$$

represents the algebraic sum of the discontinuous jumps in β_z around the contour.

This result applies to each contour $A = A_0$ over a finite range of A_0 and a finite range of z. It implies that the discontinuities lie on one or more surfaces intersecting the zero-order contours $A = A_0$.

The restrictions imposed by these various integral relations in §§4.6–4.8 become tractable for a continuous field in the special case that the field has rotational symmetry about some axis extending in the z-direction, as already noted. It is interesting that the same conclusion follows if the problem is approached from the point of view of constructing an appropriate solution of the Grad–Shafranov equations. Vainshtein and Parker (1986) set about the task of describing the equilibrium of a cluster of long straight twisted flux bundles extending uniformly ($\partial/\partial z = 0$) in the z-direction and confined to a finite region of xy-space by a uniform external pressure P. They showed that no continuous field solution can satisfy the boundary conditions unless the field has rotational symmetry about the z-direction. That is to say, there is a smooth continuous solution to the Grad–Shafranov equations only if the field consists of a single twisted flux bundle, in which the field lines all lie on concentric circular cylinders.

4.8 Infinitesimal Perturbation Including Fluid Pressure

Consider the equilibrium of a magnetic field in the presence of a fluid pressure p, so that the full magnetostatic equation (3.1) or (4.9) is employed. We have equation (3.2), $\mathbf{B} \cdot \nabla p = 0$, in place of the condition of equation (3.5) $\mathbf{B} \cdot \nabla \alpha = 0$ for $\nabla p = 0$. Then equation (4.52) applies with σ or α replaced by δp, if the field is to be continuous throughout the region (Parker, 1979, pp. 374–376).

The expression equivalent to equation (4.34) is also readily deduced. The first-order terms in equation (3.1) are

$$4\pi \nabla \xi = (\nabla \times \mathbf{B}) \times \beta + (\nabla \times \beta) \times \mathbf{B}, \tag{4.58}$$

upon writing the fluid pressure as $p + \varepsilon\xi(x, y) \exp \kappa z$, where $p = p(A)$ and A satisfies equation (4.11). Multiply the x-component of this equation by $(\partial A/\partial y)/|\nabla A|$ and the y-component by $-(\partial A/\partial x)/|\nabla A|$ and add. The result can be written

$$4\pi \frac{\partial}{\partial s}[\xi + B_z(A)\beta_z/4\pi] = \beta_n \nabla^2 A + \kappa B_z(A)\beta_p,$$

with β_p representing the component parallel to the contour $A = constant$, $z = constant$ and β_n the outward normal component $\mathbf{e}_n \cdot \beta = \beta \cdot \nabla A/|\nabla A|$ (see equation (4.32)). The Laplacian $\nabla^2 A$ can be eliminated with the aid of equations (4.11) and (4.12), so that

$$\frac{\partial}{\partial s}[\xi + B_z(A)\beta_z/4\pi] = -F'(A)\beta_n + (\kappa B_z(A)/4\pi)\beta_p. \tag{4.59}$$

The similarity of this equation to equation (4.33) for a force-free field is obvious. Integrating around a closed contour $A = A_0$ and requiring that the pressure perturbation $\xi + B_z(A)\beta_z/4\pi$ encounters no discontinuity, it follows that the left-hand side vanishes, leaving

$$4\pi F'(A_0) \oint ds \beta_n = \kappa B_z(A_0) \oint ds \beta_p \tag{4.60}$$

as the necessary condition on the perturbation β for the absence of discontinuity. The line integrals of β_n and β_p around every contour $A = A_0$ are related by a factor $4\pi F'(A_0)/B_z(A_0)$ that is prescribed by the zero-order field. No other combination of β_n and β_p is free of discontinuity.

This restriction for continuity can be cast in other forms. For instance, equations (4.36) and (4.37) can be used to rewrite the condition as

$$\kappa \int dx \int_{A_0} dy \{(\nabla \times \beta)_z - [4\pi F'(A_0)/B_z(A_0)]\beta_z\} = 0, \tag{4.61}$$

where the integration is over the interior of $A = A_0$. We see again that if the perturbation field has the same invariance $(\partial/\partial z = 0)$ as the zero-order field, then $\kappa = 0$ and the requirement is automatically satisfied. For the general perturbation, however, $\kappa \neq 0$ and the condition can be written

$$\langle\langle(\nabla \times \beta_z)\rangle\rangle = [4\pi F'(A_0)/B_z(A_0)]\langle\langle\beta_z\rangle\rangle, \tag{4.62}$$

where the double brackets represent the mean value over the interior of $A = A_0$. This condition asserts that the mean values of $(\nabla \times \beta)_z$ and β_z are related by a force-free field condition in which the torsion coefficient $4\pi F'(A_0)/B_z(A_0)$ is uniquely determined by the properties of the zero-order field at $A = A_0$, regardless of the form of the supposedly arbitrary topology of the perturbation. The force-free form of this condition is surprising because the fields are not force-free. It shows again the special form of the winding if the final β is to be free of discontinuities.

Now the condition of equation (4.62) for continuity of β applies for all value of A_0, and hence it applies to the annulus $(A_0, A_0 + \delta A_0)$. Recalling that the width of the annulus is $\delta A_0/|\nabla A|$, it is readily shown that

$$\oint_{A_0} \frac{ds(\nabla \times \beta)z}{|\nabla A|} = \frac{4\pi F_0'(A_0)}{B_z(A_0)} \oint_{A_0} \frac{ds\beta_z}{|\nabla A|} + 4\pi \frac{d}{dA_0}\left[\frac{F_0'(A_0)}{B_z(A_0)}\right]\langle\langle\beta_z\rangle\rangle. \tag{4.63}$$

The mean values of $(\nabla \times \beta)_z$ and β_z on the annulus δA_0 are related by the same torsion coefficient $4\pi F_0'(A_0)/B_z(A_0)$ plus a term involving the mean β_z throughout the interior of $A = A_0$. The arbitrary winding that produces β generally does not conform to this restriction, so β is generally not free of discontinuities.

An additional constraint on the field follows from the condition of equation (3.2), with

$$\mathbf{B} \cdot \nabla\xi + \beta \cdot \nabla p = 0.$$

With the fact that $p = p(A)$ this can be rewritten as

$$\partial\xi/\partial s + B_z(A)\kappa\xi + p'(A)\beta_n = 0.$$

Integrate around any closed contour $A = A_0$ and assume that ξ is continuous so that the line integral of $\partial\xi/\partial s$ vanishes. The result is

$$\kappa B_z(A_0) \oint ds\xi + p'(A) \oint ds\beta_n = 0.$$

With equation (4.36), this can be written

$$\kappa\left[B_z(A_0) \oint \frac{ds\xi}{|\nabla A|} - p'(A_0) \int dx \int_{A_0} dy\beta_z\right] = 0,$$

where the double integration is over the interior of $A = A_0$. Thus if $\kappa \neq 0$, it is necessary that the pressure perturbation satisfies

$$\oint \frac{ds\xi}{|\nabla A|} = \frac{P'(A_0)}{B_z(A_0)} \int dx \int dy\beta_z. \tag{4.64}$$

But the integration on the right-hand side is constrained by equation (4.61) or (4.62), while the pressure perturbation, introduced along with β at the boundaries, is arbitrary. So we encounter the same fundamental difficulty here with the pressure perturbation that was encountered in the force-free field with the torsion. Each is physically arbitrary, subject only to projecting uniformly along the field lines, so that in almost all cases the special requirements are not satisfied by β and the result is the appearance of tangential discontinuities.

4.9 Physical Basis for Tangential Discontinuities

The physical basis for the foregoing restrictions (e.g., equations (4.42), (4.45), (4.50), (4.52), (4.53), (4.62)–(4.64), etc. for continuity of the 3D perturbation $\varepsilon\mathbf{b}$ (of a field $\mathbf{B}$

with an ignorable coordinate) can be established in other ways from the basic equations $\mathbf{B} \cdot \nabla p = 0$ or $\mathbf{B} \cdot \nabla \alpha = 0$ (Parker, 1979, pp. 373, 377). First of all, note the physical requirement that the perturbation **b**, described by equation (4.24), must remain bounded in the limit of large L/ℓ. Hence, the real part of κ can represent nothing more than the surface perturbations $\exp[-(R\ell\kappa)z]$ or $\exp[-(R\ell\kappa)(L-z)]$ which die out rapidly with distance from the boundaries. On the other hand, we are interested in the solutions that are nonvanishing throughout the interior of $0 < z < L$, which arise through a purely imaginery κ. Write $\kappa = ik$, so that the solutions have the bounded periodic form $\exp ikz$. Note that k is nonvanishing if the perturbation **b** is to vary at all in the z-direction, and assume that L is so large that $kL \gg 1$. Then the field $\mathbf{B} + \varepsilon\mathbf{b}$ is periodic in z. That is to say, the field and fluid pressure or torsion at $z = z_0$ are reproduced again at $z_0 \pm 2\pi/k$, etc. But since p and α project along the field lines, the variations of $\delta p = \varepsilon\xi(x, y) \exp ikz$ and $\delta\alpha = \varepsilon\sigma(x, y) \exp ikz$ around the contour $A = A_0$ at $z = z_0$ are duplicated at $z = z_0 + 2\pi/k$ only if the field lines of the zero-order field **B** *all* make an integral number n turns[1] in the distance $2\pi/k$. So in general the field lines of **B** do not have the necessary property of an integral number of turns around $A = A_0$ in passing from z_0 to $z_0 + 2\pi/k$. There may, of course, be some special value of A_0 for which the field lines make an integral number of turns. However, it is obvious that if the field lines of **B** intersecting the closed contour $A = A_0$ at $z = z_0$ should happen to make an integral number of turns around $A = A_0$ in passing to $z_0 + 2\pi/k$, then in almost all field configurations **B** the lines do not make an integral number of turns for any neighboring values of A.

To go one step farther, note that the perturbed pressure p or torsion α projects along the field lines of the perturbed field, and the perturbed field lines intersecting $A = A_0$ at $z = z_0$ generally do not intersect $A = A_0$ at $z_0 + 2\pi/k$. Formally the field lines $x(z), y(z)$ of the zero-order field are given by the integrals of

$$B_z(A)dx/dz = +\partial A/\partial y, \quad B_z(A)dy/dz = -\partial A/\partial x. \tag{4.65}$$

The lines of the perturbed field are $x(z) + \varepsilon\delta x(z)$, $y(z) + \varepsilon\delta y(z)$, with

$$\frac{d\delta x}{dz} = \frac{b_x}{B_z(A)} \quad \frac{d\delta y}{dz} = \frac{b_y}{B_z(A)}, \tag{4.66}$$

to first order. These equations for the field line perturbation are integrated along the field lines of the zero-order field, so that

$$\delta x(z) = \frac{1}{B_z(A_0)} \int_{z_0}^{z} d\zeta \beta_z[x(\zeta), y(\zeta)] \exp ik\zeta, \tag{4.67}$$

$$\delta y(z) = \frac{1}{B_z(A_0)} \int_{z_0}^{z} d\zeta \beta_y[x(\zeta), y(\zeta)] \exp ik\zeta. \tag{4.68}$$

These integrals generally do not vanish at $z = z_0 + 2\pi/k$, so the contour $A = A_0$ at $z = z_0$ maps into a different contour,

$$A(x, y) + \varepsilon a(x, y) = A_0,$$

[1]In the unlikely case that the perturbation $\varepsilon\xi$ to p, or $\varepsilon\sigma$ to α, varies through m identical cycles around $A = A_0$, it is necessary that the field lines pass through n/m turns.

at $z_0 + 2\pi/k$. This contour at $z_0 + 2\pi/k$ encloses the same total magnetic flux as $A = A_0$ at z_0, because it represents a direct map of the closed curve $A = A_0$ along the field lines. The field **B** is the same at z_0 and $z_0 + 2\pi/k$. Hence the perturbed contour $A + \varepsilon a = A_0$ at $z_0 + 2\pi/k$ lies partly inside and partly outside $A = A_0$, crossing $A = A_0$ at least twice. If the perturbed contour crosses $A = A_0$ at the two points (x_1, y_1) and (x_2, y_2), then at those two locations the quantity ξ or σ must be identical at z_0 and $z_0 + 2\pi/k$. If ξ varies around the contour $A = A_0$ in some prescribed manner, this requires an integral number of turns of the field lines of **B** around $A = A_0$ for these lines through (x_1, y_1) and (x_2, y_2). Thus, again we are confronted with the fact that, for any given choice of k, there may be some discrete set of values of A_0 for which this requirement is satisfied. But for all values of A_0 except these special ones, the requirement is generally not fulfilled. This presents the contradiction that the magnetostatic solutions are mathematically restricted to the periodic z-dependence $\exp ikz$, but topologically unable to project p or α along the field lines in such a way as to achieve this periodicity. The escape from the contradiction is the introduction of tangential discontinuities.

Another way to say this is that the original flux surfaces $A(x, y) = A_0$ are broken up by all nonsymmetric perturbations, i.e., they are broken by almost all perturbations that do not share the same invariance $(\partial\partial z = 0)$ as the zero-order field. This was the basis for Grad's (1967) assertion that there are no continuous equilibrium solutions in a toroidal geometry, mentioned in §3.2. The same result applies here in linear geometry in the limit of large L and a finite rate of variation k of the perturbation **b** in the z-direction. The formal mathematical restrictions (e.g., equations (4.45), (4.50), (4.52), (4.63), and (4.64)) are necessary restrictions on the field perturbation so that the flux surfaces (along which π and σ are projected) are preserved. Almost all topological perturbations are excluded by these restrictions.

In summary, the topology of the perturbation **b**, wound into the zero-order field **B** by arbitrary prescribed continuous fluid motions, is largely unrestricted and unrelated in essential respects to the topology of **B**. In particular, the topology of **b** varies with z. Hence the foregoing restrictions on **b** and $\delta\alpha$ or δp for a continuous field are generally not satisfied. But the field relaxes into the lowest available energy state regardless of whether the conditions for the absence of discontinuity are satisfied. In most cases, then, there is discontinuity somewhere in the magnetostatic equilibrium of the perturbed field.

4.10 Discontinuities and Dynamical Nonequilibrium

It is shown by the formal relations in equations (4.39) and (4.42) and from equations (4.49) and (4.52) $\oint ds \partial\beta_z/\partial s$ and $\oint ds \partial\sigma/\partial s$ generally do not vanish on all contours $A = A_0$, except for special cases of the topology of the perturbation field. But it is obvious that these integrals vanish identically if β_z and σ are continuous functions of s, no matter how rapidly varying the continuous functions may be. Therefore, the jumps in β_z and σ necessary to cause the nonvanishing of these integrals must be of the nature of a true discontinuity. That is to say, $\sigma(A_0, s) = \sigma_1$ for $s < s_1$ and $\sigma(A_0, s) = \sigma_2 (\neq \sigma_1)$ for $s > s_1$, with $\lim \sigma(A_0, s)$ equal to σ_1 as s approaches s_1 from below and equal to σ_2 as s approaches from above. The discontinuity represents the surface of contact between the regions of continuous field, pointed out in §4.1.

Now in the real world, where there is some slight electrical resistivity, it is clear that a true discontinuity does not exist. A field quantity, e.g., β_z or σ, may change rapidly over s small distance $\delta\ell$ may be small, but it provides a continuous, rather than a discontinuous change, in the field. In the presence of a simple resistive diffusion coefficient η an a characteristic time t the scale $\delta\ell$ may be as small as $O[(\eta\, t)^{\frac{1}{2}}]$. But this does not satisfy the necessary discontinuity condition for static equilibrium, described in the foregoing section. The field outside the thin transition layer $\delta\ell$ is not affected by the conditions inside $\delta\ell$, so it is in equilibrium just as if there were a true discontinuity. But within $\delta\ell$ the conditions for static equilibrium are not fulfilled. Only the external field is in equilibrium, whereas in $\delta\ell$ the field is in a state of dynamical nonequilibrium.

The nonequilibrium takes the form of neutral point rapid reconnection described in §1.4 and taken up at some length in Chapter 10. Resistive diffusion across $\delta\ell$ is important, being the effect that keeps $\delta\ell$ from going to zero. There is resistive reconnection of the field, and the associated redistribution of the tension along the reconnected field lines retracts the reconnected flux from the region at the characteristic Alfven speed C (cf. Parker, 1957, 1963, 1979, pp. 392–395; Priest, 1982). Thus, if the characteristic breadth of the transition layer is ℓ, the extraction occurs in a time of the order of $t = \ell/C$. This is a dynamical process, driven by the Maxwell stresses in their quest for the minimum energy state of complete magnetostatic equilibrium. The stages of resistive reconnection and dynamical retraction of each successive flux layer are sketched in Fig. 4.2. The basic facts of the dynamics are repeated briefly to emphasize that the process is driven and continues to be driven so long as the global field topology requires a tangential discontinuity for the minimum energy state. The extraction of thin layer $\delta\ell$ of field does not alter the local absence of magnetostatic equilibrium, because the finite thickness $\delta\ell$ of the transition is maintained by η. So the process continues, reconnecting and extracting layers of field of thickness $\delta\ell$ in times $t = \ell/C$. It follows that $\delta\ell = O[(\eta\, t)^{\frac{1}{2}}] = O[(\eta\ell/C)^{\frac{1}{2}}]$. The speed at which the process reconnects and removes the equilibrium fields pressing against $\delta\ell$ on either side is of the order of

$$\begin{aligned} u &= \delta\ell/t, \\ &= (\eta C/\ell)^{\frac{1}{2}}, \\ &= C/N_L^{\frac{1}{2}}, \end{aligned} \tag{4.69}$$

where $N_L = \ell C/\eta$ is the characteristic Lundquist number (magnetic Reynolds number) (Parker, 1957; Sweet, 1958a,b; 1969). The thickness $\delta\ell$ is then $O(\ell/N_L^{\frac{1}{2}})$.

This result is a lower limit on the rate at which resistivity cuts across the field at a surface of discontinuity, because as first pointed out by Petschek (1964) and later by Petschek and Thorne (1967), Parker (1979, pp. 417–423), Priest (1982) and Priest and Forbes (1986) the scale ℓ across which the two nonparallel fields from each side come in contact may, in special situations, be smaller than the actual characteristic transverse scale ℓ of the field itself. Petschek showed that the effective width of the region of resistive diffusion may be smaller than the characteristic scale ℓ of the fields by a factor as large as the order of $N_L/(\ell n N_L)$, so that u may be as large as $O(C/\ell n N_L)$, with $\delta\ell$ as small as $\ell\ \ell n N_L/N_L$.

Sonnerup (1970, 1971), Sonnerup and Priest (1975), Parker (1979, pp. 423–428), Priest (1982) and Priest and Forbes (1986) have shown that the rate u of reconnection

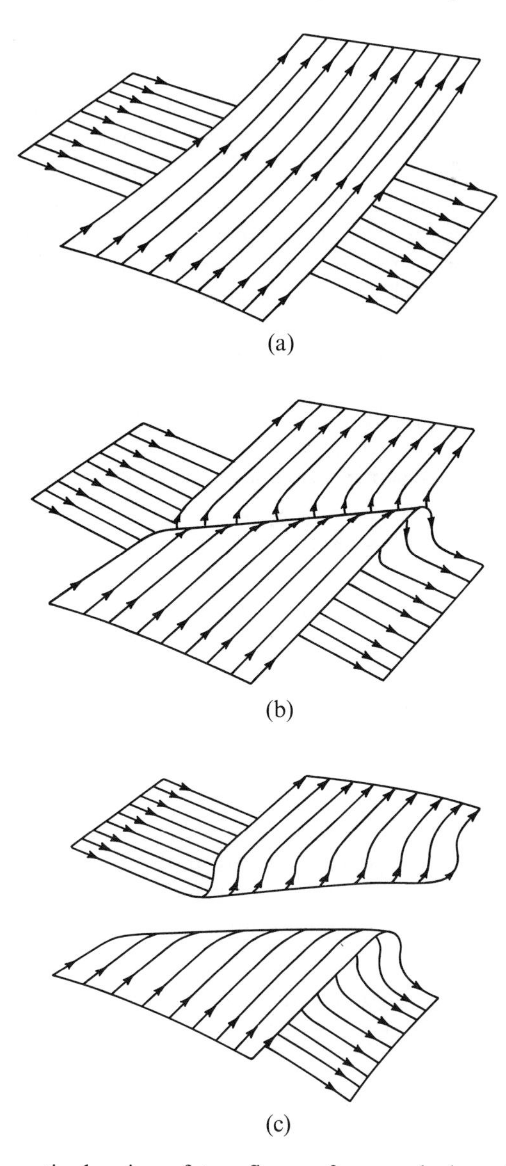

Fig. 4.2. (a) A schematic drawing of two flux surfaces pushed so close together that their opposite fields reconnect by resistive diffusion to produce the configuration (b). The tension along the field lines rapidly retracts the surfaces from the region, shown in (c), so that the next flux surfaces come close together.

can be even higher, as large as C, in the limit of large N_L, if the system is driven by a suitable pressure distribution over the surrounding fields (Parker, 1979, pp. 423–428). The compressibility of the fluid in which the field is embedded also affects the speed u at which the resistivity cuts across the field, generally enhancing u by a factor of the order of the ratio of the Alfven speed C to the sound speed v_s when $v_s < C$ (Parker, 1963) (A general development of these processes can be found in Vasyliunas (1975), Parker (1979, pp. 392–439), Priest (1982), and Priest and Forbes (1986).

It follows that in the highly (but not infinitely) conducting fluids in the real world, the field stresses over the broad regions of continuous field between the surfaces of tangential discontinuity are essentially the same as described here in an infinitely conducting fluid. On the other hand, conditions within the transition layers of thickness $\delta\ell$ (replacing the true surface of discontinuity) are not at all the same as required for static equilibrium, and it is in $\delta\ell$ that dynamical fast reconnection of the field arises in place of the ideal static true tangential discontinuity.

These facts provide an answer to the long-standing question (cf. Parker, 1979, pp. 383–391) of just where the rapid neutral point reconnection may arise in a deformed field. We expect the reconnection at the discontinuities.

4.11 Hamiltonian Formulation of the Magnetostatic Field

The symmetry breaking described in §4.8 and providing the underlying cause of the formation of discontinuities, can be treated effectively by employing the elegant formal mathematical analogy between the field lines of a magnetostatic field and the trajectory of a Hamiltonian system in phase space. The Hamiltonian formulation has been developed and exploited in connection with the toroidal fields employed in laboratory plasma confinement (cf. Kerst, 1962; Boozer, 1983; Cary and Littlejohn, 1983). Tsinganos, Distler, and Rosner (1984) have used the Hamiltonian formulation of fields to exploit the Kolmogoroff–Arnold–Moser theorem of mechanics to extend the perturbation results for infinitesimal ε in the foregoing sections to finite ε. This result extends the topological domain (from infinitesimal ε to finite ε) in which it can be shown that invariance is an essential property of continuous magnetostatic fields, demonstrating that the absence of invariance in the interwoven fields of stars and galaxies implies internal tangential discontinuities.

Consider the 1D Hamiltonian $H^{(1)}(p,q)$, in which $q = x^1$ and $p = x^2$, designed to describe the magnetostatic field of equation (4.15). The canonical equations of motion are

$$dq/dt = +\partial H^{(1)}/\partial p, \; dp/dt = -\partial H^{(1)}/\partial q, \tag{4.70}$$

wherein the time analog t is defined as

$$t = \frac{1}{f(A)} \int dx^3 h_3/h_1 h_2. \tag{4.71}$$

Here $f(A) \equiv h_3 B_3$ and h_1, h_2, h_3 represent the metric coefficients of the orthogonal curvilinear coordinate system x^i. Set $H^{(1)}(p,q)$ equal to the vector potential A, satisfying equation (4.15). Then since $dt = dx^3/h_1 h_2 B_3$, it is readily seen that equation (4.70) becomes

$$\begin{aligned} B_3 \frac{h_1 dx^1}{h_3 dx^3} &= +\frac{1}{h_2 h_3} \frac{\partial A}{\partial x^2}, \\ &= B_1, \end{aligned} \tag{4.72}$$

and

$$\begin{aligned} B_3 \frac{h_2 dx^2}{h_3 dx^3} &= -\frac{1}{h_3 h_1} \frac{\partial A}{\partial x^1}, \\ &= B_2. \end{aligned} \tag{4.73}$$

These equations are immediately recognizable as the differential equations for the field lines. Thus, the canonical equations automatically represent the field lines provided that we are able to supply the proper Hamiltonian function $A(x^1, x^2) = A(q, p)$ from the solution of equation (4.15). The Hamiltonian is "time" independent so the energy integral $H = E$ exists, which is just

$$A(x^1, x^2) = E,$$

recognizable as the cylindrical flux surfaces of the invariant field. This is the central feature of the Hamiltonian formulation of the field, that the flux surfaces $A = constant$ correspond to the integral (energy) surfaces of the Hamiltonian system, with the field lines corresponding to the dynamical trajectory of $q(t), p(t)$ in phase space. The addition of an unsymmetric field perturbation $\varepsilon b_i(x^1, x^2, x^3)$ to the zero-order field is equivalent to the addition of an unsymmetric perturbation $\delta H(x^1, x^2, x^3)$ to the Hamiltonian $H^{(1)}(x^1, x^2)$, causing the integral surfaces $H^{(1)} + \delta H = E + \delta E$ to become corrugated, duplicating the corrugation of the flux surfaces.

4.11.1 Reformulation of Hamiltonian Representation

To take advantage of the Kolmogoroff–Arnold–Moser theorem, it is necessary (following Tsinganos, Distler, and Rosner, 1984) to construct a Hamiltonian in which the trajectory of the system in phase space corresponds to a flux surface, rather than an individual field line, in the magnetic field. To reformulate the problem, then, recall that the n-dimensional Lagrangian $L^{(n)}(\mathbf{q}, \dot{\mathbf{q}})$ of a system described by the Hamiltonian $H^{(n)}(\mathbf{q}, \mathbf{p})$ is just

$$L^{(n)}(\mathbf{q}, \dot{\mathbf{q}}) = \mathbf{p} \cdot \dot{\mathbf{q}} - H^{(n)}(\mathbf{q}, \mathbf{p}), \tag{4.74}$$

where

$$\mathbf{p} = \partial L^{(n)}(\mathbf{q}, \dot{\mathbf{q}}) / \partial \dot{\mathbf{q}}. \tag{4.75}$$

The canonical equations

$$\dot{\mathbf{q}} = +\partial H^{(n)} / \partial \mathbf{p}, \; \dot{\mathbf{p}} = -\partial H^{(n)} / \partial \mathbf{q}, \tag{4.76}$$

are equivalent to, and can be deduced from, Hamilton's principle of least action,

$$\begin{aligned} 0 &= \delta \int_{t_1}^{t_2} dt L^{(n)}(\mathbf{q}, \dot{\mathbf{q}}, \\ &= \delta \int_{t_1}^{t_2} dt [\mathbf{p} \cdot \dot{\mathbf{q}} - H^{(n)}(\mathbf{q}, \mathbf{p})]. \end{aligned} \tag{4.77}$$

This can be written in terms of arc length ds as

$$\delta \int_{s_1}^{s_2} ds (\mathbf{p} \cdot \mathbf{q}' - H^n(\mathbf{q}, \mathbf{p}) t') = 0. \tag{4.78}$$

where the prime indicates differentiation with respect to s. Now define

$$q_{n+1} = t, \; p_{n+1} = -H^{(n)}, \tag{4.79}$$

so that the system has $n+1$, rather than just n, dimensions. Hamilton's principle becomes

$$\begin{aligned} 0 &= \delta \int_{t_1}^{t_2} dt L^{(n)}(\mathbf{q}, \dot{\mathbf{q}}), \\ &= \delta \int_{s_1}^{s_2} ds[\mathbf{p} \cdot \mathbf{q}' + p_{n+1} q'_{n+1}], \\ &= \delta \int_{s_1}^{s_2} ds L^{(n+1)}(\mathbf{q}, \mathbf{q}'). \end{aligned} \tag{4.80}$$

Thus Hamilton's principle in n dimensions can be expressed in terms of Hamilton's principle in $n+1$ dimensions by suitable definition of q_{n+1} and p_{n+1}, with the Lagrangian $L^{(n+1)}(\mathbf{q}, \mathbf{q}') = L^{(n)}(\mathbf{q}, \dot{\mathbf{q}})t'$ in the system in which s, rather than t, is the independent variable. The corresponding Hamiltonian is

$$\begin{aligned} H^{(n+1)}(\mathbf{q}, \mathbf{p}) &= \sum_{i=1}^{n+1} p_i q'_i - L^{(n+1)}(\mathbf{q}, \mathbf{q}'), \\ &= \sum_{i=1}^{n} p_i q'_i + p_{n+1} q_{n'+1} - L^{(n)} t'. \end{aligned} \tag{4.81}$$

But

$$H^{(n)}(\mathbf{q}, \mathbf{p}) = \sum_{i=1}^{n} p_i \dot{q}_i - L^{(n)}(\mathbf{q}, \dot{\mathbf{q}}).$$

Hence

$$\begin{aligned} H^{(n+1)} &= p_{n+1} q_{n'+1} + t' H^{(n)}, \\ &= q'_{n+1}(p_{n+1} + H^{(n)}). \end{aligned} \tag{4.82}$$

The essential point is that $H^{(n+1)}$ is independent of q_{n+1} and so does not depend explicitly on time. Hence there are two integrals of the system, i.e., two 3D integral hypersurfaces in the case $n = 1$ of present interest.

Consider a magnetic field $\mathbf{B}(\mathbf{r}) = (B_1, B_2, B_3)$ with $\partial B_3/\partial x^3 = 0$ while B_1 and B_2 depend on all three coordinates x^i. Write $x^1 = q_1, x^2 = p_1, x^3 = q_2$. Then on some hypersurface $p_2 = f(q_1, p_1, q_2) = constant$ we wish to establish the Hamiltonian $H^{(2)}$ such that the phase space flow corresponds to $\mathbf{B}$. Since q_2 is an ignorable coordinate, p_2 is a constant and can just as well be put equal to zero. The hypersurface is then $f(q_1, p_1, q_2) = 0$. The Hamiltonian is time independent, so there is the energy integral $H^{(2)} = constant$, which can as well be taken to be $H^{(2)} = 0$. The phase flow is then restricted to the intersection of the two 3D level surfaces $f(q_1, p_1, q_2) = 0$ and $H^{(2)}(q_1, p_1, q_2) = 0$. The phase flow lies on a family of 2D surfaces, i.e., the field lines lie on flux surfaces. Note that if there were no second integral (e.g., $p_2 = constant$), then the phase flow would ergodically fill the hypersurface $H^{(2)} = 0$, and there would be no flux surfaces.

The phase space flow is

$$\mathbf{v} = (q'_1, p'_1, q'_2, p'_2), \tag{4.83}$$

and we identify the contravariant components of $\mathbf{B}$ with the first three components of $\mathbf{v}$ since $p_2' = 0$ on the hypersurface $p_2 = 0$. Hence from Hamilton's equations the contravariant field components are

$$B^i = \left(\frac{\partial H^{(2)}}{\partial p_1}, -\frac{\partial H^{(2)}}{\partial q_1}, \frac{\partial H^{(2)}}{\partial p_2} \right), \tag{4.84}$$

evaluated at $p_2 = 0$. But if $\partial B^3/\partial x^3 = 0$, then with $x^3 = q_2$, it follows that $\partial^2 H^{(2)}/\partial q_2\, \partial p_2 = 0$, and $H^{(2)}$ must be of the form

$$H^{(2)} = f(q_1, p_1, q_2) + g(q_1, p_1, p_2). \tag{4.85}$$

Therefore

$$B^i = \left(\frac{\partial f}{\partial p_i} + \frac{\partial g}{\partial p_i}, -\frac{\partial f}{\partial q_1} - \frac{\partial g}{\partial q_1}, \frac{\partial g}{\partial p_2} \right). \tag{4.86}$$

To establish the relation between the Hamiltonian construction and the more familiar forms of equations (4.10)–(4.14), note that $\mathbf{B}$ can be written as

$$\mathbf{B} = \left[\frac{1}{h_2 h_3} \frac{\partial A_3}{\partial x^2}, -\frac{1}{h_1 h_3} \frac{\partial A_3}{\partial x^1}, \frac{1}{h_1 h_2} \frac{\partial A_2}{\partial x^1} \right] \tag{4.87}$$

in terms of the vector potential $[0, A_2(x^1, x^2), A_3(x^1, x^2, x^3)]$, where $\partial A_2/\partial x_3 = 0$ in order that $\partial B_3/\partial x_3 = 0$. Write

$$A \equiv A_3(x^1, x^2, x^3),\ G \equiv A_2(x^1, x^2). \tag{4.88}$$

and let

$$H^{(2)} = a(q_1, p_1, q_2) + p_2 m(q_1, p_1), \tag{4.89}$$

with $x^1 = q_1$, $x^2 = p_1$, $x^3 = q_2$. Then from Hamilton's equations the contravariant components of the field are

$$B^i = \left(\frac{\partial a}{\partial p_1}, -\frac{\partial a}{\partial q_1}, m \right) \tag{4.90}$$

on $p_2 = 0$. The physical components of $\mathbf{B}$ are just $B^1/h_2 h_3$, $B^2/h_3 h_1$, $B^3/h_1 h_2$, so that comparing the two expressions for $\mathbf{B}$ we obtain

$$\frac{\partial a}{\partial x^2} = \frac{\partial A}{\partial x^2}, \tag{4.91}$$

$$\frac{\partial a}{\partial x^1} = \frac{\partial A}{\partial x^1}, \tag{4.92}$$

$$m = \frac{\partial G}{\partial x_1}. \tag{4.93}$$

It is sufficient to put

$$A = a,\ G = \int dx^1\, m, \tag{4.94}$$

from which it is clear that $H^{(2)}$ can be calculated for a given **B** and vice versa. The 2D Hamiltonian and the vector potential are two equivalent schemes for representing a 3D magnetic field (with $\partial B_3/\partial x_3 = 0$ in the present example).

4.11.2 Destruction of Flux Surfaces

Consider the effect of adding a small perturbation $\varepsilon H_1^{(2)}(q_1, p_1, q_2, p_2)$ of nonintegrable form to the Hamiltonian $H^{(2)}(q_1, p_1, p_2)$. When $\varepsilon = 0$, there are two integrals of the system, viz $H^{(2)} = constant$ and $p_2 = constant$. The system is confined to a nested set of integral (2D) surfaces. The surfaces can be labeled by some parameter, e.g., A. The Kolmogoroff–Arnold–Moser theorem (Moser, 1966, 1973; Arnold, 1978) states that a finite portion of the integral surfaces are destroyed by the nonintegrable perturbation no matter how small is ε. That is to say, the complete set of nested integral surfaces exists only if ε is identically equal to zero. The basis for the theorem is simply the nonintegrability of the equations providing the generating function for the surfaces (conveniently expressed in terms of action angle variables. See the outline in Appendix B of Tsinganos, Distler, and Rosner, 1984). The absence of integral surfaces in any region of space means that the field lines are ergodic throughout that portion of space. In such circumstances the requirement $\mathbf{B} \cdot \nabla p = 0$, or $\mathbf{B} \cdot \nabla \alpha = 0$, can be fulfilled only for the degenerate cases $P = constant$ and $\alpha = constant$, respectively. These special cases have already been discussed, in §4.2. In the general circumstance, then, the perturbation **b** does not have the same special form as **B**, so **b** is a nonintegrable perturbation. Hence it destroys the possibility of static equilibrium throughout a finite portion of the perturbed region. This result was shown in §4.8 for infinitesimal ε. The Hamiltonian formulation and the Kolmogoroff–Arnold–Moser theorem extend these results for infinitesimal ε into the realm of $\varepsilon = O(1)$.

Consider, then, the implication that there is no static equilibrium of a magnetic field with a topology that is nonintegrable, in the sense just defined. The field is assumed to be embedded in a fluid with infinite electrical conductivity, so the field topology is preserved, whatever the field may do. The essential point is that the magnetic field cannot remain in a state of nonequilibrium indefinitely. Addition of a small viscosity would see to that. The field relaxes asymptotically into the lowest available energy state, which means magnetostatic equilibrium. Simple physical considerations allow no alternative. The field does this by developing tangential discontinuities, i.e., contact surfaces, between finite regions of field in which there are integral surfaces. But the integral surfaces are now split open in places, in a manner that is illustrated by the optical analogy formulated in Chapter 7. The integral surfaces contain gaps, i.e., they have edges, and the foregoing development of the Hamiltonian formulation of the field becomes inapplicable.

The physics is simply that already described in §§4.5–4.7. The addition of a small nonintegral topological alteration of the field, represented by the nonintegrable field perturbation $\varepsilon b(x^1, x^2, x^3)$ causes the individual field lines to wander away from the integral surfaces of the zero-order field. So the flux surfaces of the perturbed field extend along the field into increasing convolution, illustrated in Fig. 4.1.

5

Direct Integration of Equilibrium Equations

5.1 Description of the Field

It is instructive to write out exact integrals of the complete force-free equations (3.3) and (3.4) to illustrate those general aspects of the mathematical structure that lead to discontinuities in the equilibrium field. In special cases the integration can be carried through to provide analytic forms for the discontinuous equilibrium field. More generally the integration provides the algebraic sum of the strength of the discontinuities encountered by a closed contour. The exact calculations extended the perturbation calculations of §4.5–4.8. In the perturbation calculations the integration was along the field lines of the zero-order unperturbed field, which possessed an ignorable coordinate ($\partial/\partial z = 0$) so that the projections of the field lines onto the orthogonal surfaces ($z = constant$) formed closed curves. As one might expect, the exact calculations involve integration along the projection of the exact field lines onto the surfaces $z = constant$, and these projections generally do not form closed curves. On the other hand the family of curves orthogonal to the projection of the field lines often contains closed curves. So a study of the topology is an essential part of the development.

Consider, then, the formal description of the equilibrium field. Note that the continuous interwoven field, sketched in Fig. 1.5(b) and described by equation (3.11) is expressed by a scalar function ψ and a constant B_0. Quite generally, an arbitrary magnetic field $\mathbf{B}(\mathbf{r})$ in three dimensions is subject to one constraint, $\nabla \cdot \mathbf{B} = 0$, suggesting that its description permits two arbitrary scalar functions, e.g., the Euler potentials g and h, with $\mathbf{B} = \nabla g \times \nabla h$, throughout $0 < z < L$. The continuous field of equation (3.11) is described by the two functions ψ and B_0. When that field relaxes to equilibrium the z-component is no longer uniform and the second function is no longer merely a constant.

It is important to understand that no more than two independent scalar functions are available to describe the field. This fact figures prominently in the analysis (Parker, 1986a,b), where the restriction that the field be everywhere continuous imposes more than two conditions on the field. To demonstrate the restriction, consider the final equilibrium field obtained by allowing the field (3.11) to relax to force-free equilibrium while holding the footpoints of the field fixed at $z = 0$ and L. Denote the final equilibrium field by $b_i(x, y, z)$, extending from $z = 0$ to $z = L$. Consider what deformations (fluid motions) are necessary to restore it to the initially uniform $\mathbf{e}_z B_0$. We shall see that two suffice. Hence, if the process is reversed, the arbitrary final field b_i which may be continuous, or only piecewise continuous, is fully described in terms of two scalar functions, showing the limited freedom in satsifying such arbitrary constraints as continuity of the field between regions of different topology.

Starting with the arbitrarily interwoven $b_i(x, y, z)$ of Fig. 1.5(b), which contains internal discontinuities in most cases (i.e., which is piecewise continuous), apply an irrotational compressible deformation, carrying each point x_i to some other point $X_i(x_k)$. The deformation is designed to restore $b_z(x, y, z)$ to the uniform value B_0. Restrict the mapping to two dimensions so that

$$X = X(x, y, z),\ Y = Y(x, y, z),\ Z = z. \tag{5.1}$$

The footpoints of the field remain fixed at $z = 0, L$ during this mapping. The formal solution to the induction equation (2.30) can be written in terms of Cauchy's integral as

$$B_i(X_k) = b_j(x_k)\frac{\partial(x, y)}{\partial(X, Y)}\frac{\partial X_i}{\partial x_j} \tag{5.2}$$

(Lundquist, 1952; Roberts, 1967; Parker, 1979). Then if B_z is to equal B_0, it follows from the Z-component of equation (5.2) that the Jacobian is

$$\begin{aligned}\frac{\partial(X, Y)}{\partial(x, y)} &= \frac{b_z(x, y, z)}{B_0},\\ &\equiv 1 + \gamma(x, y, z),\end{aligned} \tag{5.3}$$

thereby expressing the Jacobian in terms of the scalar function $\gamma(x, y, z)$. Assume that $b_z > 0$ everywhere, so the Jacobian is nonsingular.[1]

In order that the mapping is irrotational, so that it cannot unwind any of the interweaving of the field lines (which would involve motion of the footpoints if there were a change in the overall topology), write the mapping in terms of the potential function $\Theta(x, y, z)$,

$$X = x - \partial\Theta/\partial x,\ Y = y - \partial\Theta/\partial y, \tag{5.4}$$

so that

$$\partial Y/\partial x - \partial X/\partial y = 0.$$

Note that Θ may contain discontinuities at which $\partial\Theta/\partial x$ and $\partial\Theta/\partial y$ are undefined. It is readily shown that equation (5.3) takes the form

$$\Theta_{xx} + \Theta_{yy} + \Theta_{xy}^2 - \Theta_{xx}\Theta_{yy} = -\gamma \tag{5.5}$$

where now the subscripts indicate partial differentiation.

Equation (5.5) is a nonlinear Poisson equation and has solutions for any bounded γ. Thus, for instance, if γ is a constant, the general solution is

$$\Theta = \frac{1}{2}a_1 x^2 + \frac{1}{2}a_2 y^2 + (a_1 a_2 - a_1 - a_2 - \gamma)^{\frac{1}{2}} xy + a_3 x + a_4 y + a_5$$

[1]With sufficient twisting a flux bundle can be made to kink back on itself to such a degree that the sign of b_z reverses at some locations. Such kinking automatically provides tangential discontinuities (cf. Rosenbluth, Dagazian, and Rutherford, 1973; Park, Monticello, and White, 1983; Strauss and Otani, 1988) so that case need not be pursued further.

where the a_n are arbitrary constants. Such solutions are unbounded on the xy-plane. In the present context, we are interested in quite a different class of "charge density" γ in which γ fluctuates about zero on a scale ℓ and averages to zero over any significantly larger scales, so that Θ is a bounded function. Recall that the original winding and interweaving of the field lines, described by the scalar function ψ through equation (3.11), is assumed to be without net displacement on any scale large compared to the dimension ℓ of the local swirls and eddies represented by ψ. Hence the restoration of the final equilibrium field involves no net x or y displacement on scales large compared to ℓ.

It is sufficient to note that the interweaving of the field b_i has a correlation length ℓ, beyond which the mean value of $b_z - B_0$ falls rapidly to zero. That is to say, the source term γ on the right-hand side of equation (5.5) fluctuates about zero with a characteristic transverse scale ℓ. Hence the mean value over any transverse area of scale h goes rapidly to zero as h exceeds ℓ. Note, then, that for small values of Θ_{ij} the terms Θ_{xy}^2 and $\Theta_{xx}\Theta_{yy}$ are small to second order in equation (5.5), which then reduces to the familiar Poisson equation, with a random source term. The mean "charge" density is zero, and the Poisson equation has bounded well-behaved solutions. Hence there is a bounded well-behaved solution to the nonlinear Poisson equation (5.5) at least for a suitably small upper bound γ_m on the fluctuating "charge density" γ. Therefore, there exists a compressive mapping Θ which restores any bounded $b_z(x, y, z)$ to B_0, just so long as γ averages to zero over any region much larger than $\ell \times \ell$ in the xy-plane. At this stage the mapping has restored the field to the form described by equation (3.11), with $B_z = B_0$.

There exists a mapping Ψ (which includes the reverse of ψ in equation (3.11), as well as the readjustments involved in the subsequent relaxation to equilibrium) restoring $B_x(X, Y, Z)$, $B_y(X, Y, Z)$ to the original $(0, 0, B_0)$. This follows from the fact that equation (3.11) with ψ replaced by Ψ, can be integrated to deduce Ψ for any given B_x and B_y, since the integrability requirement that $\partial^2\Psi/\partial X\partial Y = \partial^2\Psi/\partial Y\partial X$ is just the necessary divergence condition $\partial B_x/\partial X + \partial B_y/\partial Y = 0$. The integrability and existence of Ψ also follows directly from the physical fact that the final equilibrium field b_i arose as a result of a winding and interweaving of the field lines, by motion of the footpoints at $z = L$. Therefore running the deformation backward restores b_i to the initial uniform field $(0, 0, B_0)$. In conclusion, then, the potential function Θ and the stream function Ψ transform the initial arbitrarily wound equilibrium field b_i into the uniform field B_0 in the z-direction. It follows that the arbitrary equilibrium field b_i can be described by two scalar functions Θ and Ψ (Parker, 1986a).

Now turn this mapping or deformation around, taking the uniform field $\mathbf{e}_z B_0$ into the final interwoven force-free equilibrium field $b_i(x, y, z)$. Writing out the steps explicitly, some stream function $\varphi(x, y, kzt)$ carries the uniform field into

$$B_x(X, Y, Z) = +B_0 k t_1 \partial\varphi/\partial X, \tag{5.6}$$

$$B_y(X, Y, Z) = -B_0 k t_1 \partial\varphi/\partial Y, \tag{5.7}$$

$$B_z(X, Y, Z) = +B_0, \tag{5.8}$$

in a time t_1, where the coordinates are designated by (X, Y, Z). Then introduce the compressible irrotational mapping

$$X = x - \partial\theta/\partial x,\ Y = y - \partial\theta/\partial y,\ Z = z \tag{5.9}$$

carrying (X, Y, Z) into the final position (x, y, z). The Jacobian equation (equation 5.3) is

$$D = 1 - \theta_{xx} - \theta_{yy} + \theta_{xx}\theta_{yy} - \theta_{xy}^2, \tag{5.10}$$

with the subscripts indicating partial differentiation. The Cauchy integral of the induction equation is the inverse of equation (5.2), with the final field

$$b_i(x, y, z) = B_j(X, Y, Z)(\partial x_i/\partial X_j)D. \tag{5.11}$$

Differentiate equation (5.9) with respect to X, Y, and Z and solve for $\partial x_i/\partial X_j$. The six nonvanishing components are

$$\partial x/\partial X = (1 - \theta_{yy})/D,$$

$$\partial x/\partial Y = \partial y/\partial X = \theta_{xy}/D,$$

$$\partial x/\partial Z = [\theta_{xz}(1 - \theta_{yy}) + \theta_{yz}\theta_{xy}]/D,$$

$$\partial y/\partial Y = (1 - \theta_{xx})/D,$$

$$\partial y/\partial Z = [\theta_{yz}(1 - \theta_{xx}) + \theta_{xz}\theta_{xy}]/D.$$

It is convenient to express $\varphi(X, Y, kZt_1)$ in terms of the final coordinates (x, y, z), writing

$$\partial\varphi/\partial X_i = (\partial\varphi/\partial x_j)(\partial x_j/\partial X_i). \tag{5.12}$$

It can be shown from equations (5.6)–(5.8), (5.11), and (5.12) that

$$\begin{aligned} b_x(x, y, z) &= B_0\left[kt_1\left(\frac{\partial\varphi}{\partial Y}\frac{\partial x}{\partial X} - \frac{\partial\varphi}{\partial X}\frac{\partial x}{\partial Y}\right) + \frac{\partial x}{\partial Z}\right]D, \\ &= B_0\Big[\!\Big[kt_1 D^{-1}\Big\{[\varphi_x\theta_{xy} + \varphi_y(1 - \theta_{xx})](1 - \theta_{yy}) \\ &\quad - [\varphi_x(1 - \theta_{yy}) + \varphi_y\theta_{xy}]\theta_{xy}\Big\} + \theta_{xz}(1 - \theta_{yy}) + \theta_{yz}\theta_{xy}\Big]\!\Big], \\ &\equiv B_0 f_1(\theta, \varphi), \end{aligned} \tag{5.13}$$

$$\begin{aligned} b_y(x, y, z) &= B_0\left[kt_1\left(\frac{\partial\varphi}{\partial Y}\frac{\partial y}{\partial X} - \frac{\partial\varphi}{\partial X}\frac{\partial y}{\partial Y}\right) + \frac{\partial y}{\partial Z}\right]D, \\ &= B_0\Big[\!\Big[kt_1 D^{-1}\Big\{[\varphi_x\theta_{xy} + \varphi_y(1 - \theta_{xx})]\theta_{xy} \\ &\quad - [\varphi_x(1 - \theta_{yy}) + \varphi_y\theta_{xy}](1 - \theta_{xx})\Big\} + \theta_{yz}(1 - \theta_{xx}) + \theta_{xz}\theta_{xy}\Big]\!\Big], \\ &\equiv B_0 f_2(\theta, \varphi), \end{aligned} \tag{5.14}$$

$$b_z(x, y, z) = B_0 D. \tag{5.15}$$

For the record, then, equation (3.4) can be used to express α as

$$\alpha = \frac{f_1(\partial D/\partial y - \partial f_2/\partial z) + f_2(\partial f_1/\partial z - \partial D/\partial x) + D(\partial f_2/\partial x - \partial f_1/\partial y)}{f_1^2 + f_2^2 + D^2} \tag{5.16}$$

The essential point of this formal description of the field is that the interwoven equilibrium field $b_i(x, y, z)$ has sufficient freedom or arbitrariness that its description permits two and only two arbitrary mapping functions θ and φ. The function φ contains the information on the arbitrary topological interweaving of the field, as well as the information on the distribution of the interweaving along the field, while θ describes the local equilibrium compression and expansion of the field produced by the topological interweaving. The final equilibrium field shown in equations (5.13)–(5.15) satisfies the force-free equation (3.3), as well as being divergence free, which implies equation (3.5). The question is whether θ and φ have sufficient freedom to satisfy all these conditions while remaining continuous everywhere with φ constrained by the imposed arbitrary topology (interweaving) of the field lines. As has already been shown in Chapter 4, in most cases the topology allows only piecewise continuity, with one or more surfaces of discontinuity. Alternatively, requiring complete continuity greatly restricts the topology of the field lines. Since the topology is determined by the arbitrary physical winding and interweaving of the field lines, the equilibrium field contains discontinuities in almost all cases. As we will see again here in Chapter 5 the discontinuities are not introduced by a discontinuous mapping of the footpoints at $z = 0, L$, but by the necessary equilibrium conditions throughout the interior of the region $0 < z < L$.

5.2 The Equilibrium Equations for a Force-Free Field

Consider the circumstances under which the magnetic field b_i, described by equations (5.13)–(5.15), can satisfy the three components of the force-free equilibrium equation (3.3), given that b_i is expressible in terms of two scalar functions θ and φ, where φ is already constrained by the arbitrary winding and interweaving of the field. The force-free equation (3.3) becomes (Parker, 1986a)

$$\partial D/\partial y - \partial f_2/\partial z = \alpha f_1, \tag{5.17}$$

$$\partial f_1/\partial z - \partial D/\partial x = \alpha f_2, \tag{5.18}$$

$$\partial f_2/\partial x - \partial f_1/\partial y = \alpha D. \tag{5.19}$$

The condition in equation (3.5) is derived from the force-free equations with the additional constraint $\partial b_j/\partial x_j = 0$. It takes the form

$$f_1 \partial\alpha/\partial x + f_2\, \partial\alpha/\partial y + D\partial\alpha/\partial z = 0. \tag{5.20}$$

Define

$$f \equiv +(f_1^2 + f_2^2)^{\frac{1}{2}},$$

and note that the projection of the field $(B_0 f_1, B_0 f_2, B_0 D)$ onto any plane $z = constant$ has the direction cosines $(f_1/f, f_2/f)$ providing the unit vector $\mathbf{e}_p$ tangent to the projected field lines and the unit vector $\mathbf{e}_n$ with direction cosines $(+f_2/f, -f_1/f)$ perpendicular to the projected field lines. Note, then, that

$$\mathbf{e}_p \cdot \mathbf{e}_n = \mathbf{e}_n \cdot \mathbf{e}_z = \mathbf{e}_p \cdot \mathbf{e}_z = 0.$$

Denote by s the distance measured along a projected field line from some specified point (x_0, y_0), and denote by n the distance measured along an orthogonal curve. It follows that $\mathbf{e}_p \cdot \nabla = \partial/\partial s$ and $\mathbf{e}_n \cdot \nabla = \partial\partial n$.

Multiply equation (5.18) by f_1/f and equation (5.17) by f_2/f. Subtracting the former from the latter yields

$$\frac{\partial D}{\partial s} = \frac{\partial f}{\partial z}, \tag{5.21}$$

which is the counterpart of equation (4.32). On the other hand, multiply equation (5.18) by $-f_2/f$ and equation (5.17) by $-f_1/f$ and add, obtaining

$$\frac{\partial D}{\partial n} = -\alpha f + \frac{f_2}{f}\frac{\partial f_1}{\partial z} - \frac{f_1}{f}\frac{\partial f_2}{\partial z}. \tag{5.22}$$

for the orthogonal derivative of D. The divergence condition gives

$$\frac{\partial D}{\partial z} = -\frac{\partial f_1}{\partial x} - \frac{\partial f_2}{\partial y}. \tag{5.23}$$

Divide equation (5.20) by f, obtaining

$$\frac{\partial \alpha}{\partial s} = -\frac{D}{f}\frac{\partial \alpha}{\partial z}, \tag{5.24}$$

which is the counterpart of equation (4.47). Divide the z-component of equation (3.5) by f, obtaining

$$\frac{\partial \alpha}{\partial n} = -\frac{1}{f}(\nabla^2 + \alpha^2)D \tag{5.25}$$

for the orthogonal derivative of α.

These equations can be integrated along the projected field lines and along the orthogonal trajectories to determine D and α. Starting at (x_0, y_0), D and α are determined by integration of equations (5.21) and (5.24), respectively, along the projected field line through (x_0, y_0). Then integrate equations (5.22) and (5.25) along each orthogonal trajectory from the point of intersection of the trajectory with the projected field line. The essential point is that the specification of D and α at some point (x_0, y_0) determines D and α over the entire topological region (bounded by tangential discontinuities along the topological separatrices in almost all cases). The topology of the field lines in each region determines the general character of the field there (through constraints on the scalar function φ), so that the fields of neighboring topological regions generally do not match smoothly where they meet at the topological separatrices.

To go into a little more detail, suppose that the field (f_1, f_2, D) is specified along some line $(x = x_1, y = y_1)$ in the z-direction. Note that this line generally does not lie along the field, so it is not a characteristic of the field equations. It is evident that with all the field quantities prescribed on the line, the right-hand sides of equations (5.21) and (5.22) are known. Hence D may be computed throughout some small first-order neighborhood of (x_1, y_1) from the Taylor expansion

$$D(x_0 + \delta x, y_0 + \delta y, z) = D(x_0, y_0, z) + (\partial D/\partial s)_0\, \delta s + (\partial D/\partial n)_0\, \delta n + \ldots.$$

The torsion coefficient α can be extended in the same way using equations (5.24) and (5.25). Equations (5.19) and (5.23) provide the curl and divergence of the transverse field components f_1 and f_2, from which f_1 and f_2 can be extended outward. Equation (5.25) then provides $(\nabla^2 + \alpha^2)D$ over the same region, with α computed from equation (5.16). The integration proceeds uniquely outward in this way from the conditions at the line (x_1, y_1) for whatever topological winding and interweaving is prescribed by the field on (x_1, y_1). The two functions θ and φ are sufficient to provide a unique continuous field for some finite distance outward from (x_1, y_1) in most cases.

Suppose, then, that the same construction and integration is carried out for the line (x_2, y_2) at a distance $[(x_2 - x_1)^2 + (y_2 - y_1)^2]^{\frac{1}{2}} = O(\ell)$ in a neighboring topological region where the winding of the field has a different form, with different constraints on the scalar function φ, and different adjustments of θ and φ to balance the stresses. It is evident that the fields generally do not match where the two outward integrations meet at the topological separatrix, for each is determined uniquely by the arbitrarily different topological winding conditions in the two regions. The compressive adjustment θ can match field magnitudes, but φ is required to match field alignments, and φ is already largely determined by the arbitrarily different topology of the field line winding and interweaving in each separate topological region. Hence, there is generally a directional, i.e., tangential, discontinuity at the separatrix. The field magnitudes are continuous across the separatrices because the function θ is free to make that adjustment. Note too that while the mapping of the footpoints of the field at $z = 0, L$ is continuous, the discontinuity arises because the field lines on opposite sides of the tangential discontinuity are connected to regions of footpoints that are widely separated at $z = 0, L$. This will become clear in later chapters where the optical analogy is used to show the association of tangential discontinuities with gaps in the flux surfaces, through which otherwise separated flux bundles press against each other. That is to say, the fields on opposite sides of a local topological separatrix really are of different origin and form.

5.3 Topology of Field Lines

An examination of the projected field lines (on any plane $z = constant$) and the family of orthogonal trajectories shows that the orthogonal trajectories contain closed curves where the projected field lines contain neutral points. Integration of equations (5.22) and (5.25) on these closed curves provides a direct measure of the tangential discontinuities at the topological separatrices. Consider, then, the topology and connectivity of the field lines and of their dual formed by the orthogonal trajectories. The first point to be recognized is that the projections of the field lines onto $z = constant$ generally do not form closed curves. This is distinct from the invariant zero-order field ($\partial/\partial z = 0$) treated in Chapter 4, where the transverse field is expressed in terms of a vector potential $A(x, y)$, with projected field lines given by $A(x, y) = constant$ providing closed curves.

The essential features of the field lines in a topological region are captured by an idealized axisymmetric flux bundle, representing the near neighborhood of an O-type neutral point of the field projected onto a plane $z = constant$. There is no axisymmetry in the general case, of course, where the flux bundle is squashed out of round by the uneven pressures of the contiguous topological regions (flux bundles). The local field is not aligned with the z-axis, but generally veers first one way and then another as it threads its way among the adjacent flux bundles. The curvature of the flux bundle produces a field gradient across the bundle, because the field tension along the curved bundle is balanced by the magnetic pressure gradient across the bundle. However, these effects do not alter the basic local topology of the projected field lines. The axisymmetric field is sufficient for the present discussion.

Consider a twisted flux bundle ($\partial/\partial\varphi = 0$) in which the torsion coefficient α has a value $-q$ along the axis of the bundle (the z-axis) and varies only slowly with distance ϖ from the axis. Write the field components in terms of the flux function Φ, so that

$$B_{\varpi} = -\frac{1}{\varpi}\frac{\partial\Phi}{\partial z}, \; B_{\varphi} = \frac{F(\Phi)}{\varpi}, \; B_z = +\frac{1}{\varpi}\frac{\partial\Phi}{\partial\varpi}. \tag{5.26}$$

The ϖ and z components of the force-free equation (3.3) are automatically satisfied by putting $F'(\Phi) = \alpha$. The φ-component of equation (3.3) becomes

$$\varpi\frac{\partial}{\partial\varpi}\frac{1}{\varpi}\frac{\partial\Phi}{\partial\varpi} + \frac{\partial^2\Phi}{\partial z^2} + F'(\Phi)F(\Phi) = 0, \tag{5.27}$$

If now $\alpha = +q$ in the vicinity of the z-axis, write $F(\Phi) = +q\,\Phi$ in order that B_z vanish as ϖ declines to zero. Then equation (5.27) has solutions

$$\begin{aligned} \Phi &\propto \varpi J_1(\kappa\varpi)\exp(\pm kz), \\ &\cong \frac{1}{2}\kappa\varpi^2\exp(\pm kz) \end{aligned} \tag{5.28}$$

close the axis ($\kappa\varpi \ll 1$) where $\alpha \cong -q$ and $\kappa \equiv (k^2 + q^2)^{\frac{1}{2}}$. It follows that

$$B_{\varpi} \cong \frac{1}{2}B_0\, k\varpi\exp(-kz), \tag{5.29}$$

$$B_{\varphi} \cong \frac{1}{2}B_0\, q\varpi\exp(-kz), \tag{5.30}$$

$$B_z \cong B_0\exp(-kz). \tag{5.31}$$

The field lines are described by

$$\frac{d\varpi}{B_{\varpi}} = \frac{\varpi d\varphi}{B_{\varphi}} = \frac{dz}{B_z}. \tag{5.32}$$

The result is the spiral

$$\varpi = a\exp[(\varphi - \varphi_a)k/q] \tag{5.33}$$

for the field line crossing $\varpi = a$ at $\varphi = \varphi_a$. The z dependence is

$$qz = 2(\varphi - \varphi_0), \tag{5.34}$$

crossing $z = 0$ at $\varphi = \varphi_0$.

Starting at the origin ($\varpi = 0$, $\varphi = -\infty$) the length of the spiral in the plane $z = 0$ is

$$s = \int_{-\infty}^{\varphi} d\varphi[\varpi^2 + (d\varpi/d\varphi)^2]^{\frac{1}{2}}, \tag{5.35}$$

$$= a(\kappa/k)\exp[(\varphi - \varphi_a)k/q], \tag{5.36}$$

$$= \kappa\varpi/k, \tag{5.37}$$

out to a radius ϖ. Figure 5.1 is a plot of two such spiral field lines for the weak divergence $k = 0.1q$ of the field. The orthogonal lines are

$$\varpi = b\exp(-\varphi q/k), \tag{5.38}$$

where b is the radius at $\varphi = 0$, shown in Fig. 5.1. The orthogonal lines spiral outward from the origin, the distance along a line out to a radius ϖ being

$$n = \kappa\varpi/q. \tag{5.39}$$

Note, then, that the lengths of both the projected lines and the orthogonal lines are finite, $O(\varpi)$.

A similar (but more complicated) result obtains when the twisted flux bundle is at an angle ϑ to the z-direction. The field lines are described briefly in the Appendix at the end of this chapter for the interested reader.

Figure 5.2 is a sketch of the field lines and the orthogonal lines of two similar juxtaposed twisted flux bundles in a region of slowly diverging field. Note that the field lines of the projected field are discontinuous at the separatrix RP. The orthogonal lines converge to the center of each topological region, i.e., each twisted flux bundle. Any two orthogonal lines connecting the two centers form a closed contour around which the integration of equations (5.22) and (5.25) can be carried. Then

$$\oint dn\partial D/\partial n = -\oint dn\frac{\alpha}{f} + \oint dn\left(\frac{f_2}{f}\frac{\partial f_1}{\partial z} - \frac{f_1}{f}\frac{\partial f_2}{\partial z}\right), \tag{5.40}$$

$$\oint dn\frac{\partial\alpha}{\partial n} = -\oint\frac{dn}{f}(\nabla^2 + \alpha^2)B, \tag{5.41}$$

where the integration is carried along an orthogonal line from one center to the other and then back along another, different orthogonal line. The return line may be very near the outgoing line, or it may be farther away. The essential point is that if the field is continuous everywhere, then D and α are continuous and the line integrals on the left-hand sides of these two relations are identically zero. But this is

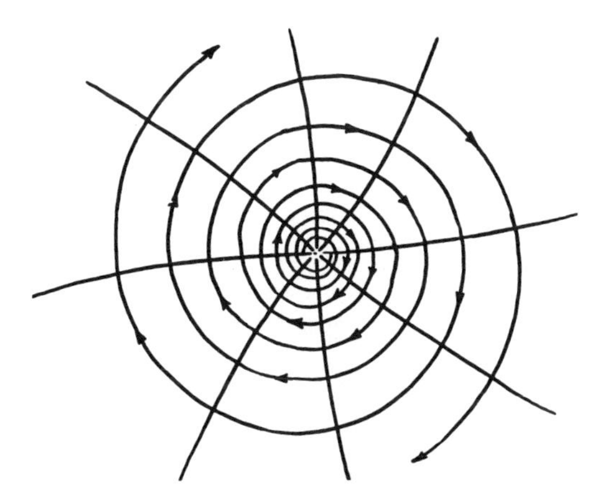

Fig. 5.1. A plot of the spiral field lines of equation (5.33) for the case $k = 0.1q$ of weak divergence of the field in the z-direction. The nearly radial curves represent members of the family of orthogonal curves, given by equation (5.38).

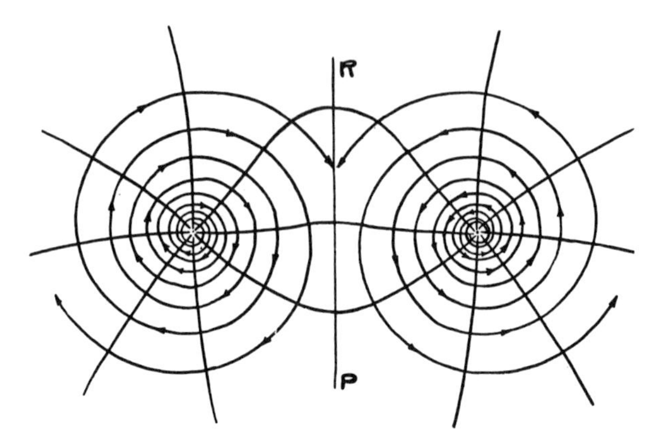

Fig. 5.2. A sketch of two neighboring diverging flux bundles, with separatrix RP, showing the projection of the field lines on a plane $z = constant$, with the orthogonal curves indicated.

possible only if the field is of such a form that the right-hand sides turn out to be zero for all choices of the outgoing and return orthogonal lines connecting the two centers. There is simply not enough freedom in θ and φ with φ constrained by the arbitrary topology of the field lines. As will be shown in §5.4 there is enough freedom in θ and φ to fit the boundary conditions and maintain a continuous field, but the field is then uniquely determined and the topology is fixed by the boundary conditions. Some other constraint must be removed if the topology is to be free, and that constraint is the continuity of the field. If we give up continuity, then the topology is free to be fitted to the requirements of the initial arbitrary winding prescribed by equation (3.11).

If ΔD and $\Delta\alpha$ represent the algebraic sums of the jump in D and α, respectively, at the tangential discontinuity (at the separatrix), then

$$\begin{aligned} \Delta D &= \oint dn \partial D/\partial n, \\ &= \oint \frac{dn}{f}\left(f_2 \frac{\partial f_1}{\partial z} - f_1 \frac{\partial f_2}{\partial z} - \alpha\right), \end{aligned} \tag{5.42}$$

and

$$\begin{aligned} \Delta\alpha &= \oint dn \partial\alpha/\partial n \\ &= -\oint \frac{dn}{f}(\nabla^2 + \alpha^2)D, \end{aligned} \tag{5.43}$$

providing a direct measure of the algebraic sum of the discontinuities encountered by the outbound and return orthogonal lines. Two nearby lines yield only a small ΔD and $\Delta\alpha$, of course, whereas widely separated orthogonal lines give a larger result.

5.4 The Primitive Torsional Field

The equilibrium equations (3.4) are nonlinear, but in simple cases can be integrated to provide analytic solutions illustrating the formation of tangential discontinuities by continuous large-scale deformation of the field. In particular the generic process of creation of discontinuities by a local field maximum can be carried though for the primitive force-free field

$$B_x = +B_0 \cos qz,\ B_y = -B_0 \sin qz,\ B_z = 0, \tag{5.44}$$

for which the torsion coefficient α is equal to q. This field represents the basic force-free field, in which the field lines are all straight and their direction rotates at a rate α (with the local value q) as one moves in some direction (chosen to be the z-direction) perpendicular to the field. This example exemplifies the general case, and a thorough analytic examination of the solutions is in order. There is, of course, an infinite family of continuous solutions, but with only limited topology of the field lines. The direct integration of the equations in a form that fits the boundary conditions (i.e., a solution with the topology of the field described by equation (5.44)) leads directly to tangential discontinuities (Parker, 1990). Now the form of the discontinuities follows in an elementary way from the optical analogy, developed in the succeeding chapters. However, we proceed by conventional formal mathematical methods, exploring the possibilities so as not to miss the opportunity for a continuous solution to the force-free equilibrium equations. The efficacy of the optical analogy in treating the topology and the discontinuities will become evident in the process.

Consider the formal integration of the force-free field equation (3.3) when the primitive force-free field of equation (5.44) is squeezed locally, in the vicinity of $x = y = 0$. Assume that the field in equation (5.44) is anchored on the distant rigid circular cylindrical boundary $\varpi = R$, and consider a slab of field $-h < z < +h$ where $qh = O(1)$ and $qR \gg 1$. The field within the slab is confined by the uniform pressure $B_0^2/8\pi$ exerted at $z = \pm h$ on an infinitely conducting passive membrane bounding the field. Suppose, then, that the applied pressure is increased by the amount $\nu(\varpi)B_0^2/8\pi$ inside a radius $\varpi = a$. The applied pressure has rotational symmetry and is written $[1 + \nu(\varpi)]B_0^2/8\pi$, locally squeezing and thinning the slab of field. The function $\nu(\varpi)$ is a smooth, differentiable, continuous function of ϖ, with $\nu(\varpi) \leqslant O(1)$ for $\varpi < a$ and $\nu(\varpi) = 0$ for $a < \varpi < R$.

The relative pressure enhancement $\nu(\varpi)$ is not small in $0 < \varpi < a$, so it produces a substantial change $\mathbf{b}(x, y, z)$ in the initial primitive force free field $\mathbf{B}(z)$. The problem is to compute the final total force-free field $\mathbf{B}(z) + \mathbf{b}(\mathbf{r})$. As we shall see, the equilibrium field equation (3.3) can be integrated over the cylindrical surface $\varpi = R$ by elementary analytical methods in the limit of large qR.

It is easiest to work in polar coordinates, writing the initial field of equation (5.44) as

$$B_\varpi = +B_0 \cos u, \tag{5.45}$$

$$B_\varphi = -B_0 \sin u, \tag{5.46}$$

$$B_z = 0, \tag{5.47}$$

with the uniform torsion coefficient $\alpha = q$ and $u \equiv qz + \varphi$. Denote the final torsion coefficient by $\alpha = q[1 + \tau(x, y, z)]$. Then the three components of the force-free equation (3.3) reduce to

$$\frac{1}{\varpi}\frac{\partial b_z}{\partial \varphi} - \frac{\partial b_\varphi}{\partial z} = q\tau B_0 \cos u + q(1 + \tau)b_\varpi, \tag{5.48}$$

$$\frac{\partial b_{\varpi}}{\partial z} - \frac{\partial b_z}{\partial \varpi} = -q\tau B_0 \sin u + q(1+\tau)b_{\varphi}, \tag{5.49}$$

$$\frac{1}{\varpi}\frac{\partial}{\partial \varpi}\varpi b_{\varphi} - \frac{1}{\varpi}\frac{\partial b_{\varpi}}{\partial \varphi} = q(1+\tau)b_z, \tag{5.50}$$

together with the divergence condition

$$\frac{1}{\varpi}\frac{\partial}{\partial \varpi}\varpi b_{\varpi} + \frac{1}{\varpi}\frac{\partial b_{\varphi}}{\partial \varphi} + \frac{\partial b_z}{\partial z} = 0. \tag{5.51}$$

These four equations suffice to compute the four quantities $b_{\varpi}, b_{\varphi}, b_z$, and τ.

The boundary condition on the deformed surfaces of the original slab of field is

$$(\mathbf{B}+\mathbf{b})\cdot(\mathbf{B}+\mathbf{b}) = [1+\nu(\varpi)]B_0^2. \tag{5.52}$$

The normal component of the field remains unchanged at the rigid boundary $\varpi = R$, so

$$b_{\varpi}(R, \varphi, z) = 0. \tag{5.53}$$

Hence equations (5.48)–(5.50) reduce to

$$\frac{1}{\varpi}\frac{\partial b_z}{\partial \varphi} - \frac{\partial b_{\varphi}}{\partial z} = +q\tau B_0 \cos u, \tag{5.54}$$

$$\frac{\partial b_z}{\partial \varpi} = +q\tau B_0 \sin u - q(1+\tau)b_{\varphi}, \tag{5.55}$$

$$\frac{1}{\varpi}\frac{\partial}{\partial \varpi}\varpi b_{\varphi} = +q(1+\tau)b_z, \tag{5.56}$$

at $\varpi = R$.

5.4.1 Elementary Integration

The nonlinear static equilibrium equations (5.54)–(5.56) can be reduced to elementary form on the basis of some simple physical considerations, providing the final solution with a minimum of mathematical obfuscation. The more complete mathematical solution, based only on the assumption of a power law for the asymptotic forms of b_{φ} and b_z at large $q\varpi$, is provided in the next section, §5.4.2, establishing that the elementary integration has overlooked nothing essential (Parker, 1990).

Consider, then, the relative magnitudes of the quantities in the equilibrium equations. We begin with the observation that the flux bundles passing close to the region of pressure enhancement $\varpi < a$ may conceivably be twisted by as much as a radian. However, this torsion is spread out uniformly along the full length $2R$ of the flux bundles. Hence the radian introduced at $\varpi = O(a)$ is spread out each way along the flux bundles over a distance R, contributing a net change $q\tau(\varpi, \varphi, z)$ in the torsion coefficient α that is small $O(1/R)$. As a matter of fact, it follows from equation (3.5) and the symmetry of each flux bundle about its midpoint that $q\tau$ is small $O(1/q^2R^2)$. For whatever the torsion introduced into a flux bundle by the perturbation, symmetry about the point of closest approach to $\varpi = a$ requires that the torsion have one sign on one half of the bundle extending out to $\varpi = R$ and the opposite sign on the other half out to $\varpi = R$. But equation (3.5) requires that torsion be the same on both legs. Hence $q\tau = 0$. As a matter of fact, the mathematical solution is carried out neglecting terms $O(1/q^2R^2)$ compared to one, and the result is $q\tau = O(1/q^2R^2)$ within the field.

To begin, then, neglect τ compared to one on the right-hand side of equation (5.56), obtaining

$$\frac{1}{\varpi}\frac{\partial}{\partial\varpi}\varpi b_\varphi = qb_z. \tag{5.57}$$

Now the perturbation has the asymptotic form (at large $q\varpi$) of a 2D dipole, so that b_ϖ and b_φ decline asymptotically as $(a/\varpi)^2$. It follows from equation (5.57) that b_z is asymptotically smaller than b_ϖ and b_φ by one power of a/ϖ. It also follows that $\partial b_z/\partial\varpi$ is smaller than $\partial b_\varpi/\partial z = O(qb_\varpi)$ by two powers of a/ϖ. Equations (5.54) and (5.55) reduce to

$$\frac{\partial b_\varphi}{\partial z} + q\tau B_0 \cos u = b_\varphi O(1/q^2R^2), \tag{5.58}$$

$$(1+\tau)b_\varphi - \tau B_0 \sin u = b_\varphi O(1/q^2R^2), \tag{5.59}$$

at $\varpi = R$. The right-hand sides of both these equations are small, second order in $1/qR$. Neglecting the right-hand side, equation (5.59) can be solved for τ, writing $\beta_\varphi = b_\varphi/B_0$ so that

$$\tau \cong \frac{\beta_\varphi}{\sin u - \beta_\varphi}, \tag{5.60}$$

while equation (5.58) becomes

$$\frac{\partial\beta_\varphi}{\partial z} + \frac{\beta_\varphi q \cos u}{\sin u - \beta_\varphi} \cong 0.$$

The applied pressure and the rigid boundary $\varpi = R$ both have rotational symmetry $(\partial/\partial\varphi = 0)$. Both apply across the thickness of the field at $\varpi = R$, i.e., they introduce no $\partial/\partial z$ at $\varpi = R$, to lowest order in $1/qR$. The only variation of **b** at $\varpi = R$ comes through the variation of the unperturbed field, described by equations (5.45)–(5.47), which depends only on u. Hence β_φ and τ can be functions only of u, to the order considered. It follows that

$$\frac{d\beta_\varphi}{du} + \frac{\beta_\varphi \cos u}{\sin u - \beta_\varphi} \cong 0,$$

which can be written as

$$\frac{d(\beta_\varphi - \sin u)}{d\sin u} = \frac{\sin u}{\beta_\varphi - \sin u}. \tag{5.61}$$

Integration yields

$$\beta_\varphi = \sin u \pm (\sin^2 u - C)^{\frac{1}{2}}, \tag{5.62}$$

where C is the integration constant, presumed to be real and positive for the moment.

Now β_φ at $\varpi = R$ is a small real quantity $O(1/q^2R^2)$ while sinu varies from –1 to +1, etc. Hence when $0 < u < \pi$, so that sin u > 0, the lower sign must be used on the right-hand side of equation (5.62), with C small $O(1/q^2R^2)$. It is also evident that the radical is imaginery and the solution is unphysical for $0 < \sin^2 u < C$. When $-\pi < u < 0$, and $\sin u \leqslant 0$, it is evident that the upper sign must be used. Thus when $0 < C \ll \sin^2 u \leqslant 1$,

$$\beta_\varphi = \frac{\pm C}{2\sin^2 u}\left[1 + \frac{C}{4\sin^2 u} - \frac{C^2}{8\sin^4 u} + \ldots\right],$$

where $\pm$ is chosen so that $\pm \sin u$ is positive. Thus β_φ is small $O(C) = O(1/q^2R^2)$ when $\sin^2 u = O(1)$.

For $\sin^2 u$ declining to C let

$$\sin^2 u = C(1+\xi). \tag{5.63}$$

Then

$$\beta_\varphi = \pm C^{\frac{1}{2}}[(1+\xi)^{\frac{1}{2}} - \xi^{\frac{1}{2}}], \tag{5.64}$$

$$\cong \pm C^{\frac{1}{2}}[1 - \xi^{\frac{1}{2}} + \frac{1}{2}\xi - \frac{1}{8}\xi^2 + \ldots]. \tag{5.65}$$

as ξ decreases to zero. It is evident that β_φ, which is small $O(C)$ for $\sin^2 u = O(1)$, increases to $O(C^{\frac{1}{2}}) = O(1/qR)$ as $\sin^2 u$ declines toward C. For $0 \leqslant \sin^2 u \leqslant C = O(1/q^2R^2)$ there is no physically acceptable solution, as already noted. The cause of this will appear shortly.

Note that the unphysical character of an imaginery value of the radical would be eliminated if C were chosen to be negative, say $C = -D$ with $D > 0$. For in that case $\sin^2 u + D$ is positive definite. However, it is immediately evident that the solution in equation (5.61) is unphysical in a more serious way, for the simple reason that, if $\sin^2 u + D$ has no zero, then $(\sin^2 u + D)^{\frac{1}{2}}$ has no branch point, and one sign or the other in equation (5.61) must apply with no chance to get from one to the other. Thus, if

$$\beta_\varphi = \sin u - (\sin^2 u + D)^{\frac{1}{2}}$$

for $0 < u < \pi$ when $\sin u \geqslant 0$, we have

$$\beta_\varphi \cong -D/2\sin u$$

for small D and $\sin u = O(1)$. The solution becomes

$$\beta_\varphi \cong 2\sin u,$$

when $-\pi < u < 0$, yielding $\beta_\varphi O(1)$ at $\varpi = R$. But we know that $\beta_\varphi \ll 1$ at $\varpi = R$. So it is necessary that $C > 0$. Hence it is not possible to avoid the small gap of width $\Delta u = 2C^{\frac{1}{2}} = O(1/qR)$ in the solution by a better choice of C.

The essential point is that equation (5.65) shows that β_φ jumps by

$$[\beta_\varphi] \cong \pm 2C^{\frac{1}{2}} = O(1/qR) \tag{5.66}$$

across the gap. The jump is $+2C^{\frac{1}{2}}$ at $u = 0, \pm 2\pi$, etc. and $-2C^{\frac{1}{2}}$ at $u = \pm\pi, \pi 3\pi$, etc.

The correction $q\tau$ to the torsion coefficient follows from equations (5.60) and (5.62), with

$$\tau = -1 \pm \frac{\sin u}{(\sin^2 u - C)^{\frac{1}{2}}}, \tag{5.67}$$

with the lower sign when $\sin u > 0$ and the upper sign when $\sin u < 0$, as with equation (5.62). The result is that τ is positive for all values of $\sin^2 u \geqslant C$. For $C \ll \sin^2 u < 1$,

$$\tau = \frac{C}{2\sin^2 u}\left(1 + \frac{C}{4\sin^2 u} + \ldots\right). \tag{5.68}$$

Thus in the present approximation τ has a small value $O(C) = O(1/q^2R^2)$ for $\sin u = O(1)$. But note that as $\sin^2 u$ declines toward C,

$$\tau \cong (1/\xi^{\frac{1}{2}})[1 - \xi^{\frac{1}{2}} + \frac{1}{2}\xi + \ldots] \tag{5.69}$$

upon using equation (5.63) to express $\sin^2 u$. The torsion coefficient becomes large without limit as $\sin^2 u$ declines to C and ξ declines to zero. The singularity is integrable, in association with the sharp downturn in β_φ, with

$$d\beta_\varphi/d\xi = \pm C^{\frac{1}{2}}/2\xi^{\frac{1}{2}}$$

from equation (5.65) as ξ declines to zero. The characteristic width of the spike in τ is $\frac{1}{2}C^{\frac{1}{2}}$.

The equilibrium equation (5.54) gives $(\nabla \times \mathbf{b})_\varpi$, and the current density j_ϖ, in terms of τ, so that integrating across the spike at $\sin u = +C^{\frac{1}{2}}$,

$$\begin{aligned}
\int dz(\nabla \times \mathbf{b})_\varpi &= B_0 \int_{\sin^{-1} C^{\frac{1}{2}}}^{\frac{\pi}{2}} du \cos u\,\tau(u), \\
&= B_0 \int_{C^{\frac{1}{2}}}^{1} dS[-1 + S/(S^2 - C)^{\frac{1}{2}}], \\
&= B_0[(1 - C)^{\frac{1}{2}} - 1 + C^{\frac{1}{2}}], \\
&\cong B_0[C^{\frac{1}{2}}(1 - \frac{1}{2}C^{\frac{1}{2}} + \frac{1}{6}C + \ldots)],
\end{aligned}$$

where $S = \sin u$. The spike at $\sin u = -C^{\frac{1}{2}}$ contributes equally, so that the total area under the curve of $(\nabla \times \mathbf{b})_\varpi$ is $2B_0C^{\frac{1}{2}}$ in the limit of large qR. Apart from the gap $-C^{\frac{1}{2}} < \sin u < +C^{\frac{1}{2}}$, the spike in the torsion can be represented by a delta function, with

$$\alpha(R, u) = q[1 + 2C^{\frac{1}{2}}\delta(u) - 2C^{\frac{1}{2}}\delta(u - \pi)], \tag{5.70}$$

This represents tangential discontinuities in the field, with b_φ jumping by $+2B_0C^{\frac{1}{2}}$ in passing across $u = 0$ and jumping by $-2B_0C^{\frac{1}{2}}$ in passing across $u = \pi$.

Now equation (3.5) asserts that the torsion $\alpha(R, u)$ computed at $\varpi = R$ extends uniformly along the field lines across the entire region $\varpi < R$. Hence the variation of α, given by equation (5.67) forms two spiral surfaces $u = qz + \varphi = 0, \pi$, representing current sheets and the associated tangential discontinuity in b_φ. The current is directed radially outward on $u = 0$ and inward on $u = \pi$.

The spiral surfaces of tangential discontinuity are deformed strongly in the neighborhood of $\varpi = a \ll R$, which the present elementary integration of the equilibrium equation (3.3) at $\varpi = R$ does not describe.

Note that the spikes in τ at $u = 0, \pi$, represented by the δ functions in equation (5.70) violate the approximation that τ is small $O(C^{\frac{1}{2}}) \leqslant O(1/qR)$, on which the calculation was based. Thus, the calculation indicates unambiguously that there is a localized spike in τ in the neighborhood of $u = 0, \pi$, and the area under the spike is given correctly by Amperes law, but the precise structure of the unbounded increase in τ is not correctly provided. The gap $-C^{\frac{1}{2}} < \sin u < +C^{\frac{1}{2}}$ in equations (5.62) and (5.67) in which a real solution does not exist is an artifact of this situation. The optical analogy, defined in Chapter 7, permits the field configuration to be treated precisely, and in Chapter 8 we return to the problem treated here, describing the structure of the field in detail.

In order to understand the physics of this result, note that the small value of τ, $O(1/q^2R^2)$, except at the singular surfaces, follows from the fact already noted, that the deformation of the field in $\varpi < a$ displaces field lines a distance $O(a)$, which varies over a scale $1/q$. Since $aq = O(1)$ this states that a net rotation or shear of the order of one radian may be introduced by the deformation in $\varpi < a$. This rotation is introduced in the central region, $\varpi < a$, of the field lines that extend to $\varpi = R$ on either side. But the field lines are anchored at each end in the rigid surface $\varpi = R$, so there is no net torsion between the end-points. Hence any rotation of the field introduced in $\varpi < a$ produces equal and opposite torsion in the two segments of field extending from a to R. But equation (3.5) decrees that the torsion is the same in the two segments, so the torsion must be negligible, i.e., $O(1/q^2R^2)$ in the limit of large R. But the substantial shear and rotation introduced by squeezing the field in $\varpi < a$ is real, and the only available option is for it to create surfaces of discontinuity, which contain the torsion but which escape the constraint of equation (3.5). The continuous solutions simply cannot accommodate this physical situation by themselves. So the regions of continuous solution make contact across surfaces of tangential discontinuity.

The purpose of the foregoing direct integration of equation (3.4), for a localized deformation of the primitive force-free field of equations (5.44) or (5.45)–(5.47), is to demonstrate by elementary mathematical methods the appearance of surfaces of tangential discontinuity, i.e., current sheets, and to see directly the contradiction that would arise without these surfaces of discontinuity. An obvious question is whether the foregoing elementary calculation, guided by simple physical considerations on the form of the field, may have missed something providing a continuous solution. This question is addressed in §5.4.2 where we undertake the integration of equations (5.54)–(5.56) without the simplification of discarding the right hand sides of equations (5.58) and (5.59). The final results are no different from those achieved in the present section, but the more elaborate mathematical solution demonstrates that the elementary solution worked out here has overlooked nothing essential. A second question is the nature of the continuous solutions, which surely exist for the boundary condition stated by equation (5.53). The continuous solutions are presented in §5.4.3, where it is shown that the boundary conditions at $\varpi = R$ permit continuous solutions but only with field line topologies qualitatively different from the topology of the basic field of equations (5.44) and (5.45)–(5.47). Thus, we can have continuous solutions if we require only that the solutions of the force-free field equations fit the boundary conditions. But if we also prescribe the internal topology of the field, as well as satisfying the boundary conditions, then we must give up continuity of the equilibrium fields. There is simply a limit to the number of conditions that can be imposed. The arbitrary winding and interweaving of the field lines introduced through equation (3.11) is too much to expect continuity as well. The arbitrary winding can be accommodated only by the torsion in the tangential discontinuities that escapes the invariance requirement of equation (3.5).

5.4.2 Asymptotic Solution

Consider the integration of equations (5.54)–(5.56) with only the assumption that the perturbation in $\varpi < a$ can be expressed in terms of multipoles (Parker, 1990) so that b_φ and b_z vary asymptotically as inverse powers of ϖ at large ϖ. We make no assumption about τ. Then if $b_\varphi \sim (a/\varpi)^\kappa$, it follows that

$$\frac{1}{q\varpi}\frac{\partial}{\partial\varpi}(\varpi\beta_\varphi) \sim -\frac{\kappa-1}{q\varpi}\beta_\varphi \tag{5.71}$$

at large $q\varpi$, where $\beta_\varphi = b_\varphi/B_0$ again. If $b_z \sim (a/\varpi)^\mu$, then

$$\frac{\partial\beta_z}{\partial\varpi} = -\frac{\mu\beta_z}{\varpi}, \tag{5.72}$$

where $\beta_z = b_z/B_0$. In this way we eliminate $\partial/\partial\varpi$ so that the calculation can be completed at $\varpi = R$. Use these asymptotic expressions in equations (5.54)–(5.56). Equation (5.56) becomes

$$\beta_z = -\frac{(\kappa-1)\beta_\varphi}{qR(1+\tau)}, \tag{5.73}$$

upon using equation (5.71). Equation (5.54) can now be written

$$\frac{\partial\beta_\varphi}{\partial\zeta} + p^2\frac{\partial}{\partial\varphi}\left(\frac{\beta_\varphi}{1+\tau}\right) + \tau\cos u = 0 \tag{5.74}$$

at $\varpi = R$, where $\zeta = qz$, $u = qz + \varphi$, and $p^2 = (\kappa - 1)/q^2R^2$. Use equation (5.73) to eliminate B_z from equation (5.72), obtaining

$$\frac{\partial\beta_z}{\partial\varpi} \sim +\frac{\mu(\kappa-1)\beta_\varphi}{qR^2(1+\tau)},$$

so that equation (5.55) can be written

$$(\tau+\tau_1)(\tau+\tau_2)\beta_\varphi = \tau(\tau+1)\sin u, \tag{5.75}$$

with

$$\tau_{1,2} = 1 \pm \mu^{\frac{1}{2}}p. \tag{5.76}$$

Use equation (5.75) to eliminate β_φ from equation (5.74), with the result

$$\begin{aligned}(\tau^2 + 2\tau_1\tau_2\tau + \tau_1\tau_2)\sin u\frac{\partial\tau}{\partial\zeta} + p^2(\tau_1\tau_2 - \tau^2)\sin u\frac{\partial\tau}{\partial\varphi}\\ + \tau(\tau+\tau_1)(\tau+\tau_2)(\tau+\tau_3)(\tau+\tau_4)\cos u = 0,\end{aligned} \tag{5.77}$$

where it is convenient to write τ_3 and τ_4 as

$$\tau_{3,4} = \frac{3}{2} \pm \frac{1}{2} E, \tag{5.78}$$

with

$$E \equiv [1 + 4(\mu - 1)p^2]^{\frac{1}{2}}. \tag{5.79}$$

Equation (5.77) is a quasi-linear first-order equation for τ. The characteristic equations are

$$\frac{d\zeta}{d\sigma} = (\tau^2 + 2\tau_1\tau_2\tau + \tau_1\tau_2) \sin u, \tag{5.80}$$

$$\frac{d\varphi}{d\sigma} = p^2(\tau_1\tau_2 - \tau^2) \sin u, \tag{5.81}$$

$$\frac{d\tau}{d\sigma} = -\tau(\tau + \tau_1)(\tau + \tau_2)(\tau + \tau_3)(\tau + \tau_4) \cos u, \tag{5.82}$$

in terms of the arbitrary parameter σ. Adding equations (5.80) and (5.81) gives

$$\frac{du}{d\sigma} = (1 - p^2)(\tau + \tau_5)(\tau + \tau_6) \sin u, \tag{5.83}$$

where

$$\tau_{5,6} = \frac{(1 - \mu p^2)}{(1 - p^2)} \left[1 \pm \frac{ip^2(\mu - p^2)^{\frac{1}{2}}}{(1 - \mu p^2)^{\frac{1}{2}}} \right]. \tag{5.84}$$

These two roots are complex upon recognizing that $\mu = O(1)$ while p^2 is small $O(1/q^2R^2)$. Divide equation (5.82) by equation (5.83). The result can be written

$$H(\tau)d\tau + d\ln \sin u = 0, \tag{5.85}$$

where

$$H(\tau) = \frac{(\tau + \tau_5)(\tau + \tau_6)(1 - p^2)}{\tau(\tau + \tau_1)(\tau + \tau_2)(\tau + \tau_3)(\tau + \tau_4)}, \tag{5.86}$$

$$= \frac{a}{\tau} - \frac{b}{\tau + \tau_1} - \frac{c}{\tau + \tau_2} + \frac{d}{\tau + \tau_3} + \frac{e}{\tau + \tau_4}, \tag{5.87}$$

where

$$\begin{aligned} a &= (1 - p^2)\tau_5\tau_6/\tau_1\tau_2\tau_3\tau_4, \\ &= \frac{1 + p^2}{2 - (\mu - 1)p^2}, \\ &\cong \frac{1}{2}\left[1 + \frac{1}{2}(\mu + 1)p^2 + \frac{1}{4}(\mu + 1)(\mu - 1)p^4 + \dots\right], \end{aligned} \tag{5.88}$$

$$\begin{aligned} b &= (1 - p^2)(\tau_5 - \tau_1)(\tau_6 - \tau_1)/\tau_1(\tau_2 - \tau_1)(\tau_3 - \tau_1)(\tau_3 - \tau_1), \\ &= 1, \end{aligned} \tag{5.89}$$

$$\begin{aligned} c &= (1 - p^2)(\tau_5 - \tau_2)(\tau_6 - \tau_2)/\tau_2(\tau_1 - \tau_2)(\tau_3 - \tau_2)(\tau_4 - \tau_2), \\ &= 1, \end{aligned} \tag{5.90}$$

$$d = -(1-p^2)(\tau_5-\tau_3)(\tau_6-\tau_3)/\tau_3(\tau_1-\tau_3)(\tau_2-\tau_3)(\tau_4-\tau_3),$$
$$= +2\frac{1-6(\mu-1)p^2-(10\mu-7)p^4-2(2\mu-1)p^6+E(1-p^2)[1+(2\mu-3)p^2]}{(1-p^2)E(3+E)(1-2p^2+E)},$$
$$\cong \frac{1}{2}[1+\frac{1}{2}(3\mu-1)p^2+O(p^4)], \tag{5.91}$$

$$e = -(1-p^2)(\tau_5-\tau_4)(\tau_6-\tau_4)/\tau_4(\tau_1-\tau_4)(\tau_2-\tau_4)(\tau_3-\tau_4),$$
$$= -2\frac{1+6(\mu-1)p^2-(10\mu-7)p^4+2(2\mu-1)p^6-E(1-p^2)[1+(2\mu-3)p^2]}{(1-p^2)E(3-E)(1-2p^2-E)},$$
$$\cong 1-\mu p^2+O(p^4). \tag{5.92}$$

The quantities a, b, c, d, e are all positive for small p.

Expanding the right-hand sides of equations (5.86) and (5.87) in ascending powers of τ and equating coefficients, it is readily shown that

$$a-b-c+d+e = a+d+c-2 = 0, \tag{5.93}$$

$$b\tau_1+c\tau_2-d\tau_3-e\tau_4 = \tau_1+\tau_2-d\tau_3-e\tau_4 = 0, \tag{5.94}$$

$$b\tau_1^2+c\tau_2^2-d\tau_3^2-e\tau_4^2 = \tau_1^2+\tau_2^2-d\tau_3^2-e\tau_4^2, = -1+p^2, \tag{5.95}$$

etc.

The variables are separated in equation (5.85), so that integration yields the quadrature

$$\ln\sin u + \int d\tau H(\tau) = -\frac{1}{2}\ln Q,$$

where Q is a real constant. Using equation (5.87) the integral of $H(\tau)$ can be performed, the final result being

$$\frac{\tau^a(\tau+\tau_3)^d(\tau+\tau_4)^e}{(\tau+\tau_1)(\tau+\tau_2)}\sin u = Q^{\frac{1}{2}}. \tag{5.96}$$

In the limit of large qR, for which p^2 declines to zero, equation (5.96) reduces to

$$\tau(\tau+2)\sin^2 u = Q(\tau+1)^2,$$

which can be solved for τ to provide equation (5.67). Thus, the earlier result, based on the obvious approximations associated with large qR, is formally correct.

Consider what happens when $\sin^2 u$ in equation (5.96) declines to zero. For $\sin u = O(1)$, with τ small $O(1/q^2R^2)$, equation (5.96) reduces to

$$\frac{\tau^a\tau_3^d\tau_4^e}{\tau_1\tau_2}\sin u \cong Q^{\frac{1}{2}},$$

where $\tau_1 \cong \tau_2 \cong 1$, $\tau_3 \cong 2$, $\tau_4 = 1$ and $a \cong \frac{1}{2}$, $d \cong \frac{1}{2}$, $e \cong 1$. Hence

$$\tau \cong Q/2\sin^2 u$$

which is just equation (5.68) again. Now if $\sin u$ declines from $O(1)$ toward zero, the coefficient of $\sin u$ on the left-hand side of equation (5.96) must become large in order that the product with $\sin u$ maintain the fixed value $Q^{\frac{1}{2}}$. For instance the denominator might become small because τ approaches either $-\tau_1$ or $-\tau_2$, i.e., τ approaches -1. But in that case, with $a \cong \frac{1}{2}$, the factor τ^a is imaginery and the solution is unphysical. The alternative is that τ becomes large without bound, so that equation (5.96) can be written

$$\frac{\tau^{a+d+e-2}(1+\tau_3/\tau)^d(1+\tau_4/\tau)^e}{(1+\tau_1/\tau)(1+\tau_2/\tau)} \sin u = Q^{\frac{1}{2}}.$$

It follows from equation (5.93) that the factor $\tau^{a+d+e-2}$ is equal to one. Expanding the other factors in descending powers of τ, and using equation (5.94), it can be shown that

$$(1 + \mu p^2/\tau^2 + \ldots) \sin u = Q^{\frac{1}{2}}.$$

Hence, $\sin u$ declines to $Q^{\frac{1}{2}}$ as τ becomes large without bound, and the solution has all the essential properties of the elementary result of equation (5.64). The solution is valid everywhere except as $\sin^2 u$ approaches Q and τ becomes large without bound. Evidently the assumed asymptotic forms $b_\varphi \sim \varpi^{-\kappa}$, $b_z \sim \varpi^{-\mu}$ are not applicable where τ becomes large, so that the precise form of τ and β_φ is not given properly there. The calculation shows that as $\sin^2 u$ becomes small, τ increases abruptly from the general value Q or C providing a tangential discontinuity in b_φ, i.e., a current sheet, but the form of the spike in τ cannot be physically correct, having a gap in the middle of it, $-Q^{\frac{1}{2}} < \sin^2 u < +Q^{\frac{1}{2}}$. This may arise from inadequacy in the limiting procedures for large τ and qR.

It is interesting, then, to consider the nature of the continuous solutions of equations (5.54)–(5.56) to see what is available.

5.5. Continuous Solutions

Consider the set of all continuous solutions of the static equilibrium equations (5.54)–(5.56) for the deformed primitive force-free field shown in equations (5.45)–(5.47). We shall see that the solutions are uniquely determined by the boundary conditions and, consequently, are restricted to special topologies that are unlike the topology of the deformed primitive field (Parker, 1990). Thus the deformed field is not to be found among the continuous solutions.

To effect a solution of the static force-free equations, we take advantage of the fact that the perturbation $q\tau$ to the torsion coefficient is small, second order in $1/qR$. The net twist across the strongly perturbed regions is essentially zero, of course, because the field lines all connect through the rigid, boundary of $\varpi = R$. Then $\alpha \cong q$ to a sufficient approximation and equation (3.6) reduces to

$$(\nabla^2 + q^2)B_i = 0$$

for each Cartesian component B_i. The essential point is that this equation is fully elliptic so that the continuous solutions are uniquely determined by the symmetry of B_i and by specification of the normal component of the field at the boundaries. However, in the real physical world the topology of the field throughout the region is not uniquely related to the normal component at the boundary, so the unique continuous solution generally does not possess the topology of the field under consideration. Hence the assumption that the field is continuous throughout the volume is incompatible with the physical situation, and discontinuous solutions such as treated in the foregoing sections of §5.4, are the appropriate representation of the field.

The problem, then, is to develop the continuous solutions in such a way as to establish their uniqueness in the context of the perturbed primitive force-free field of equations (5.45)–(5.47). The uniqueness is the essential part of the development, which must, therefore, explore every possible alternative solution in the process. For this reason the development of the solution is more detailed and tedious than would otherwise be the case.

It is convenient to work in cylindrical coordinates (ϖ, φ, z) again, so that the z-component of the field satisfies

$$\nabla^2 B_z + q^2 B_z = 0. \tag{5.97}$$

The ϖ and φ components of equation (3.3) are

$$\frac{1}{\varpi}\frac{\partial B_z}{\partial \varphi} - \frac{\partial B_\varphi}{\partial z} = qB_\varpi, \tag{5.98}$$

$$\frac{\partial B_\varpi}{\partial z} - \frac{\partial B_z}{\partial \varpi} = qB_\varphi. \tag{5.99}$$

There is no advantage in breaking up the field into the primitive force-free field given by equations (5.45)–(5.47) and the deformation **b** caused by application of the excess pressure in the neighborhood of $\varpi = 0$. We proceed instead to work with the complete field B_i in terms of the general continuous solutions to equations (5.87)–(5.99). The task is to fit the solutions to the appropriate boundary conditions, with the field fixed in the rigid infinitely conducting boundary $\varpi = R$, and generally of the primitive form of equations (5.45)–(5.47) except insofar as the field is deformed at $\varpi = a$.

Use equation (5.98) to eliminate B_ϖ from equation (5.98) and use equation (5.99) to eliminate B_φ from equation (5.98). The result is

$$\left(\frac{\partial^2}{\partial z^2} + q^2\right) B_\varphi = \left(\frac{1}{\varpi}\frac{\partial^2}{\partial \varphi \partial z} - q\frac{\partial}{\partial \varpi}\right) B_z, \tag{5.100}$$

$$\left(\frac{\partial^2}{\partial z^2} + q^2\right) B_\varpi = \left(\frac{\partial^2}{\partial \varpi \partial z} + \frac{q}{\varpi}\frac{\partial}{\partial \varphi}\right) B_z, \tag{5.101}$$

respectively, providing B_φ and B_ϖ in terms of B_z. Solving for B_φ and B_ϖ, we obtain

$$\begin{aligned} B_\varpi = &-\frac{1}{q}\int_{z_1}^{z} d\xi \sin q(\xi - z)\left(\frac{\partial^2}{\partial \varpi \partial \xi} + \frac{q}{\varpi}\frac{\partial}{\partial \varphi}\right) B_z(\varpi, \varphi, \xi) \\ &+ F_\varpi(\varpi, \varphi)\cos q(z - z_1) + G_\varpi(\varpi, \varphi)\sin q(z - z_1), \end{aligned} \tag{5.102}$$

$$\begin{aligned} B_\varphi = &-\frac{1}{q}\int_{z_1}^{z} d\xi \sin q(\xi - z)\left(\frac{1}{\varpi}\frac{\partial^2}{\partial \varphi \partial \xi} - q\frac{\partial}{\partial \varpi}\right) B_z(\varpi, \varphi, \xi) \\ &+ F_\varphi(\varpi, \varphi)\cos q(z - z_1) + G_\varphi(\varpi, \varphi)\sin q(z - z_1), \end{aligned} \tag{5.103}$$

where F_ϖ, G_ϖ, F_φ, G_φ are arbitrary functions of ϖ and φ at this point in the calculation, and z_1 is an arbitrary value of z.

These solutions must satisfy the divergence condition and the individual components of the field equations. To satisfy $\nabla \cdot \mathbf{B} = 0$ use equations (5.102) and (5.103) to eliminate B_ϖ and B_φ from $\nabla \cdot \mathbf{B}$, and then use equation (5.97) to eliminate the second derivatives of B_z with respect to ϖ and φ. Successive integrations by parts reduces the result to two terms, one in $\sin q(z - z_1)$ and the other in $\cos q(z - z_1)$. Equating the coefficients to zero, it follows that

$$\nabla \cdot \mathbf{G} = -(1/q)\partial^2 B_z(\varpi, \varphi, z_1)\partial z_1^2 \tag{5.104}$$

from the coefficient of $\sin q(z - z_1)$ and

$$\nabla \cdot \mathbf{F} = -\partial B_z(\varpi, \varphi, z_1)/\partial z_1. \tag{5.105}$$

from the coefficient of $\cos q(z - z_1)$. The "vector" $\mathbf{F}$ is made up of (F_ϖ, F_φ), while $\mathbf{G}$ is (G_ϖ, G_φ).

If B_ϖ and B_φ are eliminated in a similar manner form equation (5.98), the result is

$$G_\varphi + F_\varpi = (1/q\varpi)\partial B_z(\varpi, \varphi, z_1)/\partial\varphi, \tag{5.106}$$

$$G_\varpi - F_\varphi = (1/q)\partial B_z(\varpi, \varphi, z_1)/\partial\varpi. \tag{5.107}$$

Equation (5.99) yields the same. The z-component of equation (3.3), i.e.,

$$\frac{1}{\varpi}\frac{\partial}{\partial\varpi}\varpi B_\varphi - \frac{1}{\varpi}\frac{\partial B_\varpi}{\partial\varphi} = qB_z,$$

can be treated similarly, with the result

$$(\nabla \times \mathbf{F})_z = qB_z(\varpi\varphi, z_1), \tag{5.108}$$

$$(\nabla \times \mathbf{G})_z = \partial B_z(\varpi, \varphi, z_1)/\partial z_1. \tag{5.109}$$

These six conditions, equations (5.104)–(5.109), are not all independent, of course. Thus, equations (5.106) and (5.107) can be solved for (G_ϖ, G_φ), and used to eliminate G_ϖ and G_φ from the left-hand side of equation (5.104). The result is equation (5.108). Alternatively, solve equations (5.106) and (5.107) for (F_ϖ, F_φ) and eliminate F_ϖ and F_φ from equation (5.105), obtaining equation (5.109). So there are only four independent equations restricting the four functions $(F_\varpi, F_\varphi, G_\varpi, G_\varphi)$. It is convenient and sufficient to use equations (5.105) and (5.108) to specify the divergence and curl of $\mathbf{F}$, computing $\mathbf{G}$ from equations (5.106) and (5.107). Note, then, that $\mathbf{F}$ is determined from B_z, apart from an arbitrary gage transformation $\mathbf{F}' = \mathbf{F} + \nabla\Psi$, where $\nabla^2\Psi = 0$.

The general form of $B_z(\varpi, \varphi, z)$ follows from inspection of equation (5.97) in cylindrical polar coordinates, with

$$B_z = B_0 \sum_{k,v} C_{kv}\, Z_{\pm v}(\lambda\varpi)\exp(\pm ikz \pm iv\varphi), \tag{5.110}$$

where $\lambda^2 = q^2 - k^2$, Z_v is any solution of Bessel's equation of order v, and k may be real or complex. For the special case $k = q$ the parameter λ vanishes, and the solutions take the form

$$B_z = B_0 \sum_v C_{qv}\, \varpi^{\pm v} \exp(\pm iqz \pm iv\varphi). \tag{5.111}$$

The $\pm$ signs in these solutions are all independent of each other. The solutions form a complete orthogonal set over φ and z, if k is real, so that B_z is determined throughout the interior of a volume by specification of B_z over the surface $\varpi = R$. The parameter v must have integral values $n = 0, 1, 2, 3, \ldots$ in order that the solution be single valued and continuous over φ inside $\varpi = R$, on which the normal component B_ϖ is fixed with the value $+B_0 \cos(qz + \varphi)$ from equation (5.45).

5.5.1 Solution for Primitive Force-Free Field

To treat a simple physical example, suppose that $qh = \infty$ and the initial primitive force-free field, described by equations (5.45)–(5.47), is deformed by thrusting a superconducting needle of radius a and infinite length through the field along the z-axis, so that the field is excluded from $\varpi \leqslant a$ (Parker, 1990). Then $B_\varpi = 0$ on $\varpi = a$ while $B_\varpi = +B_0 \cos(qz + u)$ on $\varpi = R$. It is evident that the φ, z dependence of the deformed field arises only from the initial primitive field, described by equations (5.45)–(5.47), and depending upon φ and z only through the linear combination $u = qz + \varphi$. Thus, if $kz + n\varphi$ is to have the form u, it follows that $k = nq$ and

$$\lambda^2 = q^2(1 - n^2). \tag{5.112}$$

The lowest mode is $n = 1$, for which $\lambda = 0$, and the solutions must be of the form in equation (5.110). For the higher modes $n = 2, 3, 4, \ldots$ it follows that λ is imaginery and $Z_{\pm n}(\lambda\varpi)$ involves the modified Bessel functions $I_n(|\lambda|\varpi)$ and $K_n(|\lambda|\varpi)$.

First of all, note that all the modes $n = 1, 2, 3, \ldots$ of B_ϖ must vanish at $\varpi = a$, while the modes $n = 2, 3, \ldots$ must also vanish at $\varpi = R$. What is more, the right-hand sides of equations (5.101) and (5.102) must reduce to functions of ϖ and of the linear combination $u = qz + \varphi$. These two mathematical requirements allow one and only one continuous solution for the given boundary conditions. To see how this comes about, note that linear combinations of the monotonic functions $I_n[q(1 - n^2)^{\frac{1}{2}} \varpi]$ and $K_n[q(1 - n^2)^{\frac{1}{2}} \varpi]$ do not lend themselves to vanishing at both points $\varpi = a$ and $\varpi = R$. Now if B_z were chosen to vary as $\exp inu$ it is evident from equations (5.108) and (5.109) that $\mathbf{F}$ and $\mathbf{G}$ would vary as $\exp in\varphi$, and the terms, e.g., $F_\varpi \cos q(z - z_1)$ on the right-hand sides of equations (5.102) and (5.103) would vary as $\exp(\pm iqz + in\varphi)$. This does not provide the required $\exp(\pm iu)$ unless $n = 1$. Furthermore, the integrals on the right-hand sides of equations (5.102) and (5.103) introduce a variation $\exp[i(n \pm 1)qz + i\varphi]$ because of the factor $\sin q(\xi - z)$ in the integrands. This variation does not reduce to the linear combination $u = qz + \varphi$ even if $n = 1$. Hence the right-hand sides of equations (5.102) and (5.103) generally provide inappropriate solutions for B_ϖ and B_φ. Only the special case $n = 1$ and $B_z = 0$ avoids these difficulties, which we now explore.

With $n = 1$ and $B_z = 0$ it follows from equations (5.104), (5.105), (5.108), and (5.109) that $\mathbf{F}$ and $\mathbf{G}$ are expressible as gradients of scalar functions that satisfy Laplace's equation. Equations (5.106) and (5.107) show that $\mathbf{F} \cdot \mathbf{G} = 0$. The available solutions to Laplace's equation for $n = 1$ are $\varpi^{\pm 1}\exp(\pm i\varphi)$, where the two $\pm$ signs are independent of each other. The solution $\varpi^{-1}\exp(\pm i\varphi)$ for **F** yields

$$F_\varpi = -G_\varphi = +\varpi^{-2}\exp(\pm i\varphi),$$

$$F_\varphi = +G_\varpi = \pm i\varpi^{-2}\exp(\pm i\varphi),$$

with

$$B_\varpi = \varpi^{-2}\exp[\pm i(\varphi - qz)],$$

$$B_\varphi = \pm i\varpi^{-2}\exp[\pm i(\varphi - qz)],$$

from equations (5.102) and (5.103) upon setting $z, = 0$. This dipole field is excluded by the boundary conditions, which depend only on $\varphi + qz$, rather than $\varphi - qz$.

Consider, then, the solution $\varpi\exp \pm i\varphi$, for which

$$F_\varpi = \pm iF_\varphi = -G_\varphi = \pm iG_\varpi = \exp(\pm i\varphi).$$

Putting $z_1 = 0$ it is readily seen that equations (5.102) and (5.103) reproduce the initial primitive force-free field shown in equations (5.45)–(5.47).

To accommodate the exclusion of the field from $\varpi < a$, suppose that B_z is nonvanishing. For $n = 1$ the two available modes follow from equation (5.111) as

$$B_z = B_0 C_1 q\varpi \exp iu, \tag{5.113}$$

and

$$B_z = \frac{B_0 C_2}{q\varpi}\exp iu. \tag{5.114}$$

Consider the physical implications of these two different modes. The first mode substituted into equations (5.102) and (5.103) yields

$$\begin{aligned} B_\varpi = {} & qC_1 B_0\left\{\frac{1}{2}i[1 - \exp i2q(z_1 - z)] + (z - z_1)\right\}\exp i(qz + \varphi) \\ & + F_\varpi \cos q(z - z_1) + G_\varpi \sin q(z - z_1), \end{aligned} \tag{5.115}$$

$$\begin{aligned} B_\varphi = {} & qC_1 B_0\left\{-\frac{1}{2}[1 - \exp i2q(z_1 - z)] + i(z - z_1)\right\}\exp i(qz + \varphi) \\ & + F_\varphi \cos q(z - z_1) + G_\varphi \sin q(z - z_1), \end{aligned} \tag{5.116}$$

with equations (5.104) and (5.107) providing the conditions

$$\frac{1}{\varpi}\frac{\partial}{\partial\varpi}\varpi F_\varpi + \frac{1}{\varpi}\frac{\partial F_\varphi}{\partial\varphi} = -iq^2 C_1 B_0 \varpi \exp i(qz_1 + \varphi), \tag{5.117}$$

and

$$\frac{1}{\varpi}\frac{\partial}{\partial\varpi}\varpi F_\varphi - \frac{1}{\varpi}\frac{\partial F_\varpi}{\partial\varphi} = q^2 C_1 B_0 \varpi \exp i(qz_1 + \varphi) \tag{5.118}$$

on **F**. It is obvious by inspection that the term $z - z_1$ in the braces on the right-hand side of equations (5.115) and (5.116) is unbounded as $|z|$ becomes large, and it is evident that the various combinations of qz and φ in the exponentials cannot all be reduced to the combination $u = qz + \varphi$. Nor is there any available choice for **F** and **G** that can correct this difficulty. For if we eliminate F_φ between equations (5.117) and (5.118), the result is

$$\frac{1}{\varpi^2}\frac{\partial}{\partial\varpi}\varpi\frac{\partial}{\partial\varpi}\varpi F_\varpi + \frac{1}{\varpi^2}\frac{\partial^2 F_\varpi}{\partial\varphi^2} + 4iq^2 C_1 B_0 \exp i(qz_1 + \varphi) = 0.$$

The solutions proportional to $\exp i\varphi$ are

$$F_\varpi = +C_1 B_0\left[-\frac{1}{2}iq^2\varpi^2 + D_1 + D_2/q^2\varpi^2\right]\exp i(qz_1 + \varphi),$$

$$F_\varphi = -C_1 B_0[+2\varpi^2 q^2 + D_1 - D_2/q^2\varpi^2]\exp i(qz_1 + \varphi),$$

$$G_\varpi = +C_1 B_0[(1 - 2\varpi^2 q^2) + D_1 - D_2/q^2\varpi^2]\exp i(qz_1 + \varphi),$$

$$G_\varphi = +C_1 B_0\left[i\left(1 + \frac{1}{2}q^2\varpi^2\right) - D_1 - D_2/q^2\varpi^2\right]\exp i(qz_1 + \varphi),$$

where D_1 and D_2 are arbitrary constants. Substituting these expressions into the right-hand sides of equations (5.115) and (5.116) does not remove the linear terms in z, nor does it reduce the z and φ dependence to the linear combination $u = qz + \varphi$. It follows that the mode in equation (5.113) must be excluded from the solution.

Consider, then, the solution of equation (5.114). Substituting into the right-hand sides of equations (5.102) and (5.103), the integrands vanish identically, and the result reduces to

$$B_\varpi = F_\varpi \cos q(z - z_1) + G_\varpi \sin q(z - z_1), \tag{5.119}$$

$$B_\varphi = F_\varphi \cos q(z - z_1) + G_\varphi \sin q(z - z_1). \tag{5.120}$$

Equations (5.105) and (5.108) become

$$\frac{1}{\varpi}\frac{\partial}{\partial\varpi}\varpi F_\varpi + \frac{1}{\varpi}\frac{\partial F_\varphi}{\partial\varphi} = -i\frac{B_0 C_2}{\varpi}\exp i(qz_1 + \varphi),$$

$$\frac{1}{\varpi}\frac{\partial}{\partial\varpi}\varpi F_\varphi - \frac{1}{\varpi}\frac{\partial F_\varpi}{\partial\varphi} = +\frac{B_0 C_2}{\varpi}\exp i(qz_1 + \varphi),$$

respectively. Then eliminating F_φ between these two equations, it follows that

$$\frac{1}{\varpi}\frac{\partial}{\partial\varpi}\varpi\frac{\partial}{\partial\varpi}\varpi F_\varpi + \frac{1}{\varpi}\frac{\partial^2 F_\varpi}{\partial\varphi^2} + \frac{2B_0 C_2}{\varpi}\exp i(qz_1 + \varphi) = 0.$$

The solutions with the necessary dependence $\exp i\varphi$ are

$$F_\varpi = B_0 C_2\left[-i\ln\varpi + D_3 + D_4/q^2\varpi^2\right]\exp i(qz_1 + \varphi), \tag{5.121}$$

$$F_\varphi = iB_0 C_2\left[-i\ln\varpi + D_3 - D_4/q^2\varpi^2\right]\exp i(qz_1 + \varphi), \tag{5.122}$$

$$G_\varpi = B_0 C_2\left[\ln\varpi + iD_3 - (1 + iD_4)/q^2\varpi^2\right]\exp i(qz_1 + \varphi), \tag{5.123}$$

$$G_\varphi = iB_0 C_2\left[\ln\varpi + iD_3 + (1 + iD_4)/q^2\varpi^2\right]\exp i(qz_1 + \varphi), \tag{5.124}$$

where D_3 and D_4 are arbitrary constants. The field components follow from equations (5.114) and from equations (5.119) and (5.120) with the coefficients of equations (5.121)–(5.124). The unique result is

$$B_\varpi = B_0 C_2\{[-i \ln \varpi + D_3 + i/2q^2\varpi^2] \exp i(qz + \varphi) \\ + [(D_4 - i/2)/q^2\varpi^2] \exp[+i(-qz + \varphi)]\}, \tag{5.125}$$

$$B_\varphi = B_0 C_2\{[+\ln \varpi + iD_3 + 1/2q^2\varpi^2] \exp i(qz + \varphi) \\ - i[(D_4 - i/2)/q^2\varpi^2] \exp i(-qz + \varphi)\}. \tag{5.126}$$

Let $D_4 = \frac{1}{2}i$ so as to eliminate the terms in $\exp i(-qz + \varphi)$. Then since $B_\varpi = 0$ at $\varpi = a$, it follows that

$$D_3 = i(\ln a - 1/2q^2a^2). \tag{5.127}$$

Finally, write $C_2 \equiv D \exp i\gamma$. Then the real parts of equations (5.125) and (5.126) become

$$B_\varpi = B_0 D\left[\ln \frac{\varpi}{a} + \frac{1}{2q^2}\left(\frac{1}{a^2} - \frac{1}{\varpi^2}\right)\right] \sin(qz + \varphi + \gamma),$$

$$B_\varphi = B_0 D\left[\ln \frac{\varpi}{d} + \frac{1}{2q^2}\left(\frac{1}{a^2} + \frac{1}{\varpi^2}\right)\right] \cos(qz + \varphi + \gamma).$$

The boundary condition at $\varpi = R$ is $B_\varpi = B_0 \cos(qz + \varphi)$, which is satisfied with $\gamma = \frac{1}{2}\pi$ and

$$D = \left[\ln \frac{R}{a} + \frac{1}{2q^2}\left(\frac{1}{a^2} - \frac{1}{R^2}\right)\right]^{-1} \tag{5.128}$$

The result is

$$B_\varpi = +B_0 D\left[\ln \frac{\varpi}{a} + \frac{1}{2q^2}\left(\frac{1}{a^2} - \frac{1}{\varpi^2}\right)\right] \cos(qz + \varphi), \tag{5.129}$$

$$B_\varphi = -B_0 D\left[\ln \frac{\varpi}{a} + \frac{1}{2q^2}\left(\frac{1}{a^2} + \frac{1}{\varpi^2}\right)\right] \sin(qz + \varphi), \tag{5.130}$$

$$B_z = -\frac{B_0 D}{q\varpi} \sin(qz + \varphi). \tag{5.131}$$

This solution fits all the boundary conditions and is continuous throughout the entire volume $a < \varpi < R$, $z^2 < \infty$. It is unique. But the topology of the field lines is not at all like the connection of field lines straight across $\varpi = R$ of the field described by equations (5.45)–(5.47).

5.5.2 The Field Topology

Consider the field lines, described by the integrals of

$$\frac{d\varpi}{B_\varpi} = \frac{\varpi d\varphi}{B_\varphi} = \frac{dz}{B_z}.$$

It follows that

$$\frac{du}{d\varpi} = q\frac{dz}{d\varpi} + \frac{d\varphi}{d\varpi},$$

$$= \frac{q\varpi B_z + B_\varphi}{\varpi B_\varpi},$$

$$= -\frac{(dI/d\varpi)}{I}\frac{\sin u}{\cos u},$$

from equations (5.129)–(5.131), with

$$I(q\varpi) \equiv q\varpi\left[\ln\frac{\varpi}{a} + \frac{1}{2q^2a^2} - \frac{1}{2q^2\varpi^2}\right]. \tag{5.132}$$

Integration yields

$$I(q\varpi)\sin u = \Lambda, \tag{5.133}$$

where Λ is the constant of integration and $I(\varpi)\sin u$ is effectively the vector potential,

$$A = B_0 D I(q\varpi)\sin u, \tag{5.134}$$

with

$$B_\varpi = +\frac{1}{q\varpi}\frac{\partial A}{\partial u}, \tag{5.135}$$

$$B_\varphi = \frac{-1}{1+q^2\varpi^2}\left(q\varpi A + \frac{\partial A}{\partial q\varpi}\right), \tag{5.136}$$

$$B_z = \frac{+1}{1+q^2\varpi^2}\left(A - q\varpi\frac{\partial A}{\partial q\varpi}\right). \tag{5.137}$$

Then

$$\begin{aligned}\frac{d\varphi}{d\varpi} &= \frac{B_\varphi}{\varpi B_\varpi},\\ &= -\frac{q\varpi[q\varpi I(q\varpi) + I'(q\varpi)]\sin u}{(1+q^2\varpi^2)I(q\varpi)\cos u}.\end{aligned} \tag{5.138}$$

Note, then, that

$$x\,I(x) + I'(x) \equiv [1 + xI(x)](1+x^2)/x^2. \tag{5.139}$$

Use equation (5.133) to express $\sin u$ and $\cos u$ in terms of I, so that

$$\frac{d\varphi}{d\varpi} = \frac{\Lambda(1+q\varpi I)}{q^2\varpi^2 I(I^2-\Lambda^2)^{\frac{1}{2}}}. \tag{5.140}$$

Integrate this equation along a field line from its point of closest approach $\varpi = \varpi_0$ to $\varpi = a$ at $\varphi = \varphi_0$, obtaining

$$\varphi(\varpi) = \varphi_0 \pm \Lambda\int_{q\varpi_0}^{q\varpi}\frac{ds[1+sI(s)]}{s^2 I(s)[I(s)^2-\Lambda^2]^{\frac{1}{2}}}. \tag{5.141}$$

Noting that $I(x) > 0$ for $x > qa$, so that

$$I'(x) = \frac{I(x)}{x} + 1 + \frac{1}{x^2}, \qquad (5.142)$$
$$> 0,$$

it follows from equation (5.133) that the minimum $q\varpi (= q\varpi_0)$ occurs at $u = \frac{1}{2}\pi$, with $\Lambda = I(q\varpi_0)$. In a similar fashion it can be shown that

$$z(\varpi) = z_0 \pm \frac{\Lambda}{q} \int_{q\varpi_0}^{q\varpi} \frac{ds}{I(s)[I(s)^2 - \Lambda^2]^{\frac{1}{2}}}, \qquad (5.143)$$

where $z = z_0$ is the level at which the field line approaches closest ($\varpi = \varpi_0$) to $\varpi = a$.

To carry out the indicated integration, note from equation (5.142) that

$$[1 + xI(x)]/x^2 = I'(x) - 1,$$

so that equation (5.141) can be rewritten

$$\varphi(\varpi) = \varphi_0 \pm \Lambda \left[\int_{\Lambda}^{I(q\varpi)} \frac{dI}{I(I^2 - \Lambda^2)^{\frac{1}{2}}} - \int_{q\varpi_0}^{q\varpi} \frac{ds}{I(s)[I(s)^2 - \Lambda^2]^{\frac{1}{2}}} \right], \qquad (5.144)$$
$$= \varphi_0 \pm \cos^{-1}[\Lambda/I(q\varpi)] - q(z - z_0),$$

where $z - z_0$ is given by equation (5.143). It follows that

$$u = u_0 \pm \cos^{-1}[\Lambda/I(q\varpi)], \qquad (5.145)$$

reducing to equation (5.33) since $u_0 = \frac{1}{2}\pi$.

Consider the asymptotic location and direction of the field line passing closest to $\varpi = a$ at $\varpi = \varpi_0$, $\varphi = \varphi_0$, $z = z_0$. Since $I(q\varpi) \sim q\varpi \ln(\varpi/a)$ in the limit of large $q\varpi$, it is evident from equation (5.145) that

$$u \sim u_0 \pm \frac{1}{2}\pi. \qquad (5.146)$$

It is also evident from equation (5.143) that

$$z \sim z_0 \pm Z(q\varpi), \qquad (5.147)$$

where

$$Z(q\varpi) = \frac{1}{\Lambda q} \int_{q\varpi_0}^{q\varpi} \frac{ds}{W(s)[W^2(s) - 1]^{\frac{1}{2}}}, \qquad (5.148)$$

with $W(s) = I(s)/q$. The integral is bounded so that $Z(\infty)$ is finite. It follows that in passing outward in both directions along the line of force from the point $z = z_0$ of closest approach, the field line moves to different levels, approaching $z_0 + Z(\infty)$ in one direction and $z_0 - Z(\infty)$ in the other. It follows from equation (5.146) that

$$\varphi \sim u_0 - qZ(\infty) + \frac{1}{2}\pi \qquad (5.149)$$

in one direction and

$$\varphi \sim u_0 + qZ(\infty) - \frac{1}{2}\pi \qquad (5.150)$$

in the other. On the other hand, the topology of the field described by equations (5.44) or (5.45)–(5.47) requires

$$\varphi \sim \varphi_0 + \frac{1}{2}\pi = u_0 - qz_0 + \frac{1}{2}\pi, \tag{5.151}$$

and

$$\varphi \sim \varphi_0 - \frac{1}{2}\pi = u_0 - qz_0 - \frac{1}{2}\pi, \tag{5.152}$$

with the field line approaching $z = z_0$ in both directions along the line. That is to say, the field lines enter and leave (in the limit of large qR) in opposite directions at the same level z. But if the field is required to be continuous everywhere, the solution is uniquely prescribed by the boundary conditions (at $\varpi = R$) so that the field lines enter at one level and depart at another, the two levels differing by $2Z(\infty)$. The azimuths of arrival and departure differ by $\pi - 2qZ(\infty)$ rather than π. So the projection of a field line onto $z = z_0$ provides a V-shaped line, not at all like nearly straight lines of field of equation (5.44). The unique continuous field simply has a different topology from the primitive force-free field of equation (5.44).

It is evident, then, that restricting the solution to be continuous everywhere overconstrains the problem. As soon as that artificial constraint is removed, the solutions are no longer uniquely determined by the normal component on the boundaries and the mathematics readily accommodates the topology of the field, as in equations (5.62) and (5.67) or (5.96).

Appendix: Field Lines in a Canted Field

Consider the axisymmetric field described by equations (5.29)–(5.31), written in Cartesian coordinates as

$$B_x \cong kx - qy,\ B_y \cong ky + qx,\ B_z \cong 1 - 2kz,$$

in dimensionless form with $kz \ll 1$. Consider the oblique rectangular coordinate system (x', y', z'), rotated by an angle ϑ about the y-axis from the (x, y, z) system. Then $x = x'\cos\vartheta + z'\sin\vartheta$, $y = y'$, $z = -x'\sin\vartheta + z'\cos\vartheta$. The field components are

$$B'_x = B_x \cos\vartheta - B_z \sin\theta,$$

$$B'_y = B_y,$$

$$B^1_z = B_x \sin\vartheta + B_z \cos\vartheta.$$

It follows that

$$B'_x = k(1 - 3\sin^2\vartheta)x' - q\cos\vartheta y' + 3k\sin\vartheta\cos\vartheta z' - \sin\vartheta,$$

$$B'_y = q\cos\theta x' + ky' + q\sin\vartheta z',$$

$$B'_z = 3k\sin\vartheta\cos\vartheta x' - q\sin\vartheta y' + k(1 - 3\cos^2\vartheta)z' + \cos\vartheta.$$

There is a neutral point ($B'_x = B'_y = 0$) on $z' = 0$ at the position

$$x'_0 = \frac{k \sin \vartheta}{k^2(1 - 3\sin^2 \vartheta) + q^2 \cos^2 \vartheta},$$

$$y'_0 = -\frac{q \sin \vartheta \cos \vartheta}{k^2(1 - 3\sin^2 \vartheta) + q^2 \cos^2 \vartheta}.$$

Let $x' = x'_0 + \xi$, $y' = y'_0 + \eta$, with the result that

$$B'_x = k(1 - 3\sin^2 \vartheta)\xi - q \cos \vartheta \eta,$$

$$B'_y = q \cos \vartheta \xi + k\eta.$$

The projection of the field lines onto $z' = 0$ satisfies

$$\frac{d\xi}{\xi + b\eta} = \frac{e\, d\eta}{\xi + c\eta},$$

where

$$c = k/q \cos \vartheta,$$

$$e = -1/b = k(1 - 3\sin^2 \vartheta)/q \cos \vartheta.$$

The integral of this equation can be written as

$$[\xi + (c - \beta_1)\eta]^{1+\beta_1/\beta_2} = k[\xi + (c - \beta_2)\eta]^{1+\beta_2/\beta_1},$$

where k is the integration constant, and the quantities

$$\beta_{1,2} = \frac{1}{2}\{c + e \pm [(c - e)^2 - 4]^{\frac{1}{2}}\}$$

are the roots of the quadratic equation

$$\beta^2 - (c + e)\beta + ec + 1 = 0.$$

The special case $\vartheta = 0$ treated in the text follows immediately with $c = e = k/q$, $\beta_{1,2} = \pm i + k/q$. In the general case note that $c - e = 3(k/q)\sin^2 \vartheta/\cos \vartheta$, which can be larger than two, so that the roots $\beta_{1,2}$ are real, for modest $\vartheta(<1)$ if the torsion is weak ($q \ll k$). In that case the divergence dominates and the field lines extend more or less radially outward from the neutral point. Otherwise the roots $\beta_{1,2}$ are complex and the field lines form eccentric spirals outward from the neutral point giving essentially the result already described in §5.3.

6

Examples of Field Discontinuities

6.1 Introduction

It is instructive to complement the generally analytical study of the magnetostatic field equations of the preceding chapters with specific examples of the formation of tangential discontinuities by continuous deformation of simple field topologies. The examples illustrate the common conditions under which discontinuities appear with the deformation of a continuous field. In fact it appears that the familiar theoretical X-type neutral point is rarely to be found in pure form in nature. Instead one expects two Y-type neutral points with a surface of tangential discontinuity extending between them and usually weaker discontinuities extending out along both arms of each Y.

The exposition has concentrated so far on the spontaneous appearance of tangential discontinuities throughout a region of field far removed from rigid boundaries. In particular we have examined the interior of the volume $0 < z < L$, in the limit of large L. These circumstances are appropriate for the formation of discontinuities throughout the extended bipolar fields of active regions on the Sun and other stars, the inflated halo of a galaxy, etc. On the other hand, the formation of tangential discontinuities near a boundary is interesting for the onset of solar flares etc. caused by the collision of two lobes of bipolar field. The basic principles are the same in both cases, of course. The formation of discontinuities follows from the local squeezing of magnetic field so as to open gaps in layers of flux through which otherwise separate field regions come into contact (i.e., deforming X-type neutral lines into pairs of Y-type neutral lines). But the details are sufficiently different, and the eventual applications are sufficiently important, that both are treated in the course of the chapter. And both cases illustrate the ubiquitous nature of the tangential discontinuity.

We begin with an extension of the deformation of the primitive force-free form given by equations (1.5)–(1.7) where now the slab of field is compressed by application of an infinitely long ridge of pressure across the face of the slab. The next example treats a 2D field with a linear variation across a neutral sheet, which is subsequently compressed to provide a tangential discontinuity across the neutral sheet. Then we take up the construction of formal solutions to the Grad–Shafranov equations (1.8) and (1.9) by the physical manipulation of an initially uniform field, as described by equations (1.3) and (1.4), showing how the absence of mathematical precision in the physical manipulation leads to discontinuities. We describe also how any overall deformation of continuous solutions of the Grad–Shafranov equations produces discontinuities.

A number of numerical experiments are now available showing the rapid development of tangential discontinuities as the field deforms from uniformity. The experiments deal with the onset of discontinuities resulting from quasi-static twisting and massaging of an initially uniform field and also from the nonlinear development of an instability of a nonuniform field.

Several of the illustrations of the formation of discontinuities permit easy calculation of the free magnetic energy of the field. The free energy is the energy available for resistive dissipation from the initial discontinuous field to the final potential field with the same normal component on the boundaries. It is a direct measure of the magnetic energy available in flares and coronal heating.

6.2 Compression of a Primitive Force-Free Field

As a first example consider the effect of squeezing a slab of force-free field along a strip extending straight across the surface of the slab, thereby exhibiting what has been called the surface rotation effect (Parker, 1987). It is convenient to use the primitive force-free field of equation (5.44). Relocate and reorient the xyz coordinate system so that the field is

$$B_x = 0,\ B_y = +B_0 \sin qx,\ B_z = +B_0 \cos qx, \tag{6.1}$$

satisfying the force-free magnetostatic equation (1.5) with $\alpha = q$. The slab of field $-h < x < +h$ (where $qh = O(1)$) is confined by a uniform pressure $P_0 = B_0^2/8\pi$ and bounded by flexible infinitely conducting sheets at $x = \pm h$. The slab is then compressed by application of a simple symmetric (about $y = 0$) ridge of strongly enhanced pressure $\Delta P(y)(\Delta P(y) = \Delta P(-y))$ applied to the surfaces $x = \pm h$, compressing them to $x = \pm[h - \Delta h(y)]$. The applied pressure ΔP and the indentation Δh are both maximum on the centerline $y = 0$. The ridge of pressure extends uniformly and indefinitely in the z-direction and has a characteristic width b in the y-direction, going asymptotically to zero in the limit of large y. The result is a smooth groove in each face of the slab, extending in the z-direction so that $\partial/\partial z = 0$ in the deformed field as well as in the initial field. Hence the deformed field, like the initial field, is described by the Grad–Shafranov equations (1.8) and (1.9).

Symmetry decrees that $x = 0$ is a flux surface of the deformed field, on which there is no loss of generality in putting $A = 0$. Then in the deformed field write out the explicit form

$$\begin{aligned} B_z &= B_z(A), \\ &= \sum_{n=0}^{\infty} \beta_n A^n, \end{aligned} \tag{6.2}$$

for $B_z(A)$ where the constant coefficients β_n are given and where $B_z(A)$ is a continuous, well-behaved function of A on either side of $x = 0$. Hence the infinite series has a finite radius of convergence on either side of $x = 0$. Consider solutions to the Grad–Shafranov equation (1.9) of the form

$$A = \sum_{n=1}^{\infty} f_n(y)x^n, \tag{6.3}$$

which also has a finite radius of convergence on either side of $x = 0$. Use equations (6.2) and (6.3) to express A and B in equation (1.9). Equating the coefficients of like powers of x to zero, the result is easily shown to be

$$f_2(y) = -\frac{1}{2}\beta_0\beta_1, \tag{6.4}$$

$$f_3(y) = -\frac{1}{3 \times 2}[f''_1(y) + (\beta_1^2 + 2\beta_0\beta_2)f_1(y)], \tag{6.5}$$

$$f_4(y) = -\frac{1}{4 \times 3}[f''_2(y) + (\beta_1^2 + 2\beta_0\beta_2)f_2(y) + 3(\beta_1\beta_2 + \beta_0\beta_3)f_1(y)^2], \tag{6.6}$$

$$= -\frac{1}{4 \times 3}\left[-\frac{1}{2}\beta_0\beta_1(\beta_1^2 + 2\beta_0\beta_2) + 3(\beta_1\beta_2 + \beta_0\beta_3)f_1(y)^2\right], \tag{6.7}$$

etc. The solution is completely determined by the choice of the function $f_1(y)$ once the β_n are specified so as to determine the functional form of $B_z(A)$.

The field components are

$$B_x = f'_1(y)x + O(x^3), \tag{6.8}$$

$$B_y = -f_1(y) + \beta_0\beta_1 x + \frac{1}{2}[f''_1(y) + (\beta_1^2 + 2\beta_0\beta_1)f_1(y)]x^2 + \ldots, \tag{6.9}$$

$$B_z = \beta_0 + \beta_1 f_1(y)x + \left[\beta_2 f_1(y)^2 - \frac{1}{2}\beta_0\beta_1^2\right]x^2 + \ldots. \tag{6.10}$$

The magnetic pressure is

$$B^2/8\pi = \{\beta_0^2 + f_1^2(y) + [f'_1(y)^2 - f_1(y)f''_1(y)]x^2 + \ldots\}/8\pi. \tag{6.11}$$

Now the applied pressure $P_0 + \Delta P(y)$ declines monotonically from a maximum $P_0 + \Delta P(0)$ on the centerline $y = 0$, going asymptotically to P_0 at large y/b. In the neighborhood of $x = 0$, it follows from equation (6.11) that $B^2 \cong \beta_0^2 + f_1(y)^2$ to lowest order, while it follows from equation (6.9) that $B_y = -f_1(y)$. On the other hand the slab is undisturbed in the limit of large y^2/b^2, with the field described by equation (6.1) for which $B_y = 0$ at $x = 0$. Hence $f_1(y)$ and all its derivatives must fall asymptotically to zero at large y/b, where then $B = B_z = B_0$. It follows that $\beta_0 = B_0$. Comparing equations (6.9) and (6.10) with equation (6.1) at large y/b, it is obvious that $\beta_1 = q$.

To continue, note that B_y is an odd function of x in the undisturbed field of equation (6.1), and this symmetry is preserved in the deformed field. It follows that A is an even function of x in the deformed field, as is B_z as well. With the odd symmetry of B_y in mind, note that $f_1(y)$ has a maximum at $y = 0$ where $B^2 = B_0^2 + f_1(y)^2$ and the field is squeezed by the maximum applied pressure $P_0 + \Delta P(0)$. Obviously $f_1(0)^2$ is of the order of $8\pi\Delta P(0)$. But it is evident by inspection of equation (6.9) that B_y can be an odd function of x only if $f_1(y)$ jumps from $-[f_1^2(0)]^{\frac{1}{2}}$ across $x = 0$ to $+[f_1^2(0)]^{\frac{1}{2}}$ on the positive side. That is to say $f_1(y)$ is discontinuous in an amount $O\{[8\pi\Delta P(0)]^{\frac{1}{2}}\}$ across $x = 0$, so that B_y is discontinuous across $x = 0$ by the same amount. Note, then, that this discontinuity is in keeping with the field equations, which have the field lines as characteristics (discussed in §3.4). Figure 6.1 is a sketch of a field line on each side of the plane $x = 0$. This "surface rotation effect" is readily understood in terms of the optical analogy and the refraction of the field lines in a pressure maximum, taken up in Chapter 7. In effect $\Delta P(y)^{\frac{1}{2}}$ plays the role of the index of refraction, so that the field lines cross the pressure ridge at a finite critical angle. A field line crossing the ridge at less than the critical angle is subject to total internal reflection, so that it is trapped within the ridge all the way to $z = \pm\infty$, and does not participate in the topology of the field obtained from equation (6.1).

The essential point is that a simple localized compression of the primitive force-free field of equation (6.1) provides a surface of tangential discontinuity.

6.3 Tangential Discontinuities in Two Dimensions

The foregoing example illustrates the formation of a tangential discontinuity associated with a field maximum in a 3D space. The discontinuity can be understood in terms of the refraction (rotation) of the field lines, described by the optical analogy. However, the rotation effect does not exist in a 2D magnetic field. So it is of interest to consider an example of the formation of a discontinuity in the more constrained topology of two dimensions. Consider, then, the properties of a neutral sheet in a 2D magnetic field $b(x)$ in the y-direction confined to the slab $-h < x < +h$ and embedded in an infinitely conducting incompressible fluid. Following a special example by Kulsrud and Hahm (1982) and Hahm and Kulsrud (1985) consider the general case that $b(x)$ passes through zero at $x = 0$. It is convenient to describe $b(x)$ in terms of the vector potential, or flux function, $a(x)$, with $b(x) = da/dx = a'(x)$. Then $a(x)$ is an even function of x, increasing monotonically with x^2, and there is no loss of generality in assuming that $a(0) = 0$. The total flux on either side of $x = 0$ is $a(h)$. The fluid pressure p is a function of x and can be written $p(a)$ as a functional of $a(x)$. Static equilibrium dictates that

$$8\pi p(a) + a'(x)^2 = 8\pi p_0, \tag{6.12}$$

where p_0 is the uniform external pressure applied to the surfaces $x = \pm h$.

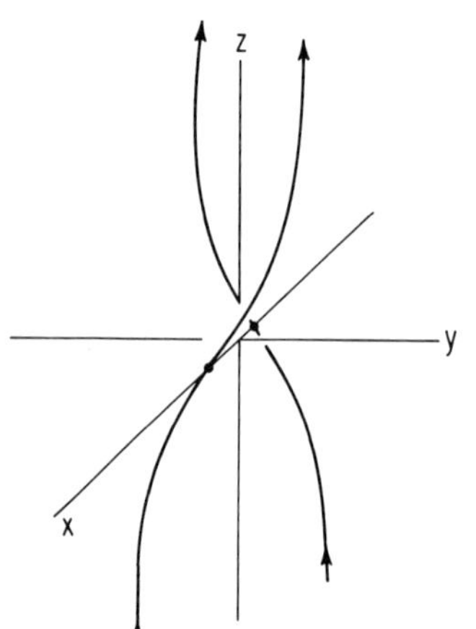

Fig. 6.1. A sketch of the field lines on opposite sides of the plane $x = 0$, showing the directional or tangential discontinuity created by the local maximum in the applied pressure $\Delta P(y)$.

Suppose, then, that the external pressure is changed from the initial p_0 to Π_1 over the interval $0 < y < 1$, and to Π_2 over $\ell_1 < y < \ell_1 + \ell_2$ (with $\Pi_2 < \Pi_1$) repeated periodically over the entire range of y. To make the problem tractable consider the limit that $\ell_1, \ell_2 \gg h$, so that the transition regions of scale $O(h)$ between ℓ_1 and ℓ_2 occupy a negligible fraction of the whole space, and the tension along the field plays a negligible role, its effect being small to first order in h/ℓ_1 and h/ℓ_2 compared to the effect of the magnetic pressure.

The essential point here is that the pressure difference $\Delta\Pi = \Pi_1 - \Pi_2$ causes local compression of the field, with displacement of the incompressible fluid along the field from the region of enhanced pressure Π_1 to the region of lower pressure Π_2. In particular, the fluid is squeezed out of $0 < y < \ell_1$ at $x = 0$ (where $b(0) = a'(0) = 0$) by the enhanced pressure Π_1, so that the squeezed field has finite rather than vanishing pressure in ℓ_1 to accommodate $\Delta\Pi_1$. But of course the direction of the field still reverses across $x = 0$, so there is a finite discontinuity in the field as a consequence of $\Delta\Pi$ and the redistribution of fluid.

So the effect of $\Pi_1 > \Pi_2$ is to compress the field in $0 < y < \ell_1$, and expand the field in $\ell_1 < y < \ell_1 + \ell_2$, and the problem is to calculate the field density as a function of x in each region. Denote by $X_1(x)$ the position of the elements of fluid in $0 < y < \ell_1$ that were initially at x, and denote by $X_2(x)$ the position of the elements of fluid in $\ell_1 < y < \ell_1 + \ell_2$ initially at x. Then conservation of fluid requires that the interval $(x, x + \Delta x)$ maps into the two intervals $X_1'(x)\Delta x$ and $X_2'(x)\Delta x$, sketched in Fig. 6.2, so that conservation of fluid is described by

$$\ell_1 X_1'(x)\Delta x + \ell_2 X_2'(x)\Delta x = (\ell_1 + \ell_2)\Delta x. \tag{6.13}$$

The magnetic field is displaced in the x-direction along with the fluid, so that the flux function $a(x)$ is transported without change $(da/dt = 0)$ from x to $X_1(x)$ and $X_2(x)$. Hence, the field densities are da/dX or

$$B_1(X_1) = a'(x)/X_1'(x), \tag{6.14}$$

$$B_2(X_1) = a'(x)/X_2'(x). \tag{6.15}$$

The fluid pressure, denoted by $P(a)$, is the same at X_1 and X_2 because X_1 and X_2 lie on the same field line. Magnetostatic equilibrium requires that

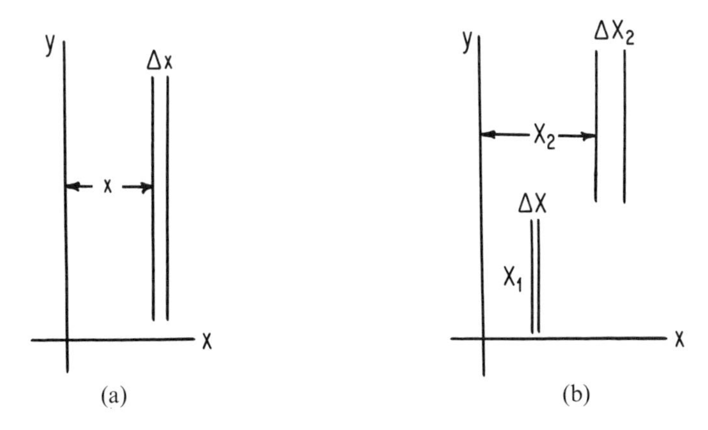

Fig. 6.2. (a) A sketch of a flux strip $(x, x + \Delta x)$ extending in the y-direction, and (b) deformed into a strip of width ΔX_1 in $0 < y < \ell_1$ and into ΔX_2 in $\ell_1 < y < \ell_1 + \ell_2$.

$$8\pi P(a) + [a'(x)/X_1'(x)]^2 = 8\pi\Pi_1, \tag{6.16}$$

and

$$8\pi P(a) + [a'(x)/X_2'(x)]^2 = 8\pi\Pi_2. \tag{6.17}$$

Subtract these two relations, with the result

$$(a'/X_1')^2 - (a'/X_2')^2 = 8\pi\Delta\Pi. \tag{6.18}$$

This can be written as

$$a'^2 X_1'^2 = X_2'^2(a'^2 - 8\pi\Delta\Pi\, X_1'^2). \tag{6.19}$$

Then solve equation (6.13) for $X_2'(x)$ and eliminate X_2' from equation (6.19). Let

$$\alpha = \ell_1/(\ell_1 + \ell_2),\ 1 - \alpha = \ell_2/(\ell_1 + \ell_2), \tag{6.20}$$

obtaining

$$(1 - \alpha X_1'^2)(a'^2 - 8\pi\Delta\Pi\, X_1'^2) = (1 - \alpha)^2 a'^2 X_1'^2. \tag{6.21}$$

Similarly

$$[1 - (1 - \alpha)X_2']^2(a'^2 + 8\pi\Delta\Pi\, X_2'^2) = \alpha^2 a'^2 X_2'^2. \tag{6.22}$$

These equations are quartics in X_1' and in X_2', respectively. Solution provides $X_1'(x)$ in terms of $a'(x)$, reducing the solution for $X_1(x)$ to a quadrature, and similarly with X_2.

The essential feature of the system is the compression of the field to finite values at $x = 0$ in $0 < y < \ell_1$ in any situation where the initial field $b(x) = a'(x)$ passes smoothly and continuously (presumably linearly) across zero at $x = 0$. For in that case $X_1'(x)$ also declines to zero at $x = 0$, and equation (6.21) can be written as

$$\frac{8\pi\Delta\Pi X_1'^2}{a'^2} = 1 - \frac{(1 - \alpha)^2 a'^2 X_1'^2}{(1 - \alpha X_1')^2}, \tag{6.23}$$

$$\cong 1 + O(a'^2 X_1'^2). \tag{6.24}$$

To lowest order

$$X_1'(x) = \pm a'(x)/(8\pi\Delta\Pi)^{\frac{1}{2}}, \tag{6.25}$$

where the positive sign is appropriate for $x > 0$ and the negative sign for $x < 0$. It follows that

$$X_1(x) = \pm\frac{a(x)}{(8\pi\Delta\Pi)^{\frac{1}{2}}}. \tag{6.26}$$

It follows from equation (6.14) that

$$B_1(X_1) = \pm(8\pi\Delta\Pi)^{\frac{1}{2}}, \tag{6.27}$$

which is nonvanishing as x declines to zero, with the upper sign appropriate for $x > 0$ and the negative sign for $x < 0$. The essential point is that the external squeezing of $0 < y < \ell_1$ can be balanced only by compressing the field so that the field has a pressure $B_1^2/8\pi = \Delta\Pi$ in the immediate vicinity of $x = 0$. The field passed smoothly across zero before the pressure was applied. The enhanced magnetic pressure is achieved by expelling fluid along the field lines from $0 < y < \ell_1$ into $-\ell_2 < y < 0$ and into $\ell_1 < y < \ell_1 + \ell_2$, etc. The discontinuity in the field strength across $x = 0$ is

$$[B_1] = 2(8\pi\Delta\Pi)^{\frac{1}{2}}, \tag{6.28}$$

as a result.

The fluid expelled from ℓ_1 inflates the field in ℓ_2, with the result from equation (6.22) that

$$X_2'(0) = 1/(1 - \alpha). \tag{6.29}$$

It follows from equation (6.15) that

$$B_2(X_2) \cong (1 - \alpha)a'(x) \tag{6.30}$$

to lowest order, where

$$X_2(x) \cong x/(1 - \alpha). \tag{6.31}$$

Thus $B_2(X_2)$ vanishes at $x = 0$ where the initial field vanished. If the initial field varies linearly across $x = 0$, so that

$$a'(x) = B_0\,x/h, \tag{6.32}$$

then at a position X_2 in the near vicinity of zero,

$$B_2(X_2) \cong (1 - \alpha)^2 B_0\,X_2/h. \tag{6.33}$$

One factor of $(1 - \alpha)$ on the right-hand side arises from the inflation of the field by fluid expelled from ℓ_1 and the second factor arises from the outward displacement of the field by the inflation so that the final field $B_2(X_2)$ is compared with the initial field $B_0\,X_2/h$ which is stronger than the field $B_0 x(X_2)/h$ at the initial location x.

6.4 Free Energy of a Discontinuity

A simple quantitative illustration of the field and the free energy associated with the deformation of an X-type neutral point to form a tangential discontinuity was pointed out by Sheeley (1991), starting with two antiparallel magnetic dipoles in a space filled with a passive ($\nabla p = 0$) infinitely conducting fluid. To fix ideas, suppose that each dipole consists of a spherical volume of rigid infinitely conducting material of small radius ε in which there is embedded a uniform magnetic field B_0, thereby providing an exterior dipole moment $M = \frac{1}{2}B_0\varepsilon^3$. The entire space outside the two small spheres ε is filled with a passive infinitely conducting fluid, exerting no force on the magnetic field. The topology of the magnetic fields of the two dipoles is assumed to be free of torsion, with the result that

$$\mathbf{B} = -\nabla\phi$$

everywhere outside the two small spheres. The vacuum field of two antiparallel dipoles is sketched in Fig. 6.3.

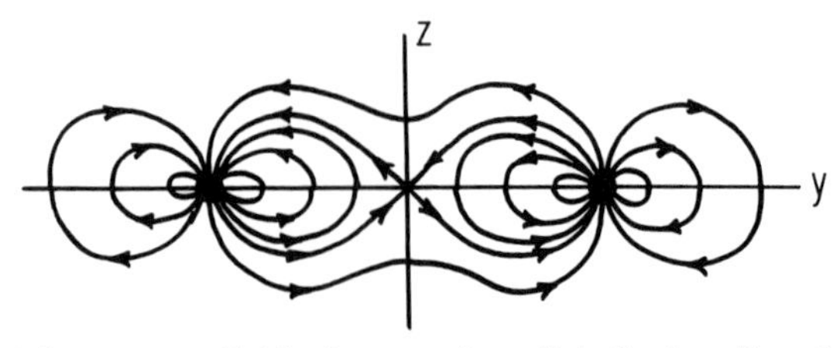

Fig. 6.3. A sketch of the vacuum field of two antiparallel dipoles aligned with the z-direction and located symmetrically on the y-axis at $y = \pm H$.

The vacuum field of two parallel dipoles is sketched in Fig. 6.4. The two fields repel each other and the force of repulsion is just equal to the magnetic pressure $B^2/8\pi$ integrated over the midplane. On the other hand, the force of repulsion can be computed with less effort from the fact that the force on one dipole M pointing in the z-direction in the field B_y of the other is

$$F = M\partial B_y/\partial z, \tag{6.34}$$

so suppose that the two widely separated (by a distance $2H$) antiparallel dipoles sketched in Fig. 6.3 are embedded in a passive infinitely conducting fluid and then pushed to some small separation $2h(h \ll H)$. The fields are frozen into the fluid, so they do not interpenetrate and the final field configuration is that shown in Fig. 6.4. The only difference (for $H/h \to \infty$)) is that the field B_i on one side of the midplane is reversed from that shown in Fig. 6.4 for two parallel dipoles. So there is a discontinuity in the field, i.e., a current sheet, on the midplane. But the magnetic pressure is the same in both cases because it depends only on the square of the field components B_i. Hence the work done in pushing the two antiparallel dipoles together in the presence of an infinitely conducting fluid is the same as for two parallel dipoles in vacuo, with the force (see equation 6.35) opposing the mutual approach of the dipoles. It is an elementary exercise to integrate equation (6.34) over y from H to h.

Define

$$r_{1,2} = [x^2 + (y \pm h)^2 + z^2]^{\frac{1}{2}} \tag{6.35}$$

as the distance of the point (x,y,z) from the dipoles at $y = \pm h$, respectively. Then note that the potential ϕ of the vacuum field of two parallel dipoles is just the sum of the potentials ϕ_1, and ϕ_2 of the individual dipoles at $y = \pm h$, respectively, where

$$\phi_{1,2} = Mz/r_{1,2}^3. \tag{6.36}$$

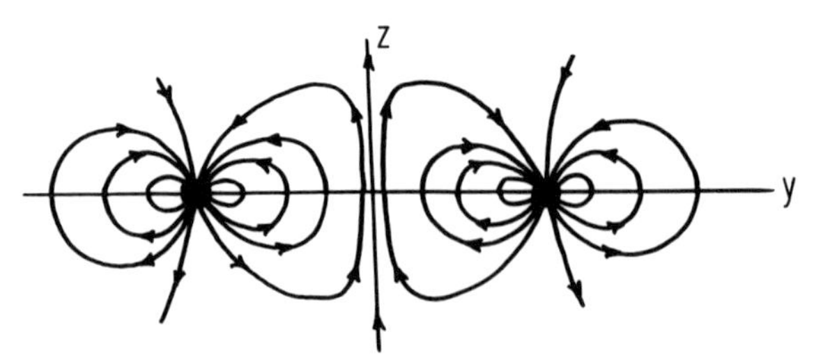

Fig. 6.4. A sketch of the field of two parallel dipoles pushed together from initial widely separated positions at $y = \pm H$, with or without the presence of an infinitely conducting fluid.

The field of the dipole at $y = -h$ is

$$B_{1x} = +3Mxz/r_1^5,$$

$$B_{1y} = +3M(y+h)z/r_1^5,$$

$$B_{1z} = +M[2z^2 - (y+h)^2 - x^2]/r_1^5,$$

so that

$$\frac{\partial B_{1y}}{\partial z} = \frac{3M(y+h)[x^2 + (y+h)^2 - 4z^2]}{r_1^7}.$$

Hence, at $x = z = 0$, $y = h$, the force F follows from equation (6.34) as

$$F(h) = 3M^2/16h^4. \tag{6.37}$$

Needless to say, the same result is readily obtained by the more laborious task of integrating the pressure $B^2/8\pi$ of the total field over the midplane $y = 0$, where

$$B_x = 6Mxz/(x^2 + h^2 + z^2)^{\frac{5}{2}},$$

$$B_y = 0,$$

$$B_z = 2M(2z^2 - h^2 - x^2)/(x^2 + h^2 + z^2)^{\frac{5}{2}}.$$

The work, W, required to push both parallel dipoles from the large separation $2H$ to the small separation $2h$ is

$$W = 2\int_h^H dh' F(h')$$

$$= M^2/8h^3, \tag{6.38}$$

in the limit of large H/h. The energy of the field is increased by this amount as the two antiparallel dipoles are pushed together to the final separation h.

In the presence of a passive infinitely conducting fluid the field of two antiparallel dipoles has the same form, with the direction of the field reversed in $y < 0$, so that the force is the same. Therefore W is a precise measure of the increase of the field energy when two antiparallel dipoles are pushed together in the presence of an infinitely conducting fluid.

The next step is to consider the energy released from the discontinuous field of two antiparallel dipoles (separated by $2h$) if a small resistivity and viscosity are introduced into the fluid. The dissipation of magnetic energy eventually reduces the initially discontinuous field to the final vacuum field configuration of two antiparallel dipoles separated by $2h$, whose form is sketched in the original Fig. 6.3. The magnetic energy lost in this way is the free energy $\mathcal{E}$ associated with the tangential discontinuity, sketched in Fig. 6.5. The energy of the vacuum field of two antiparallel dipoles is easily determined from the fact that the field of either dipole at the position of the other is reversed from the case of two parallel dipoles. So the antiparallel dipoles attract rather than repel and the field energy is now diminished rather than increased by the amount W given by equation (6.38). It follows that the free energy is 2W, or

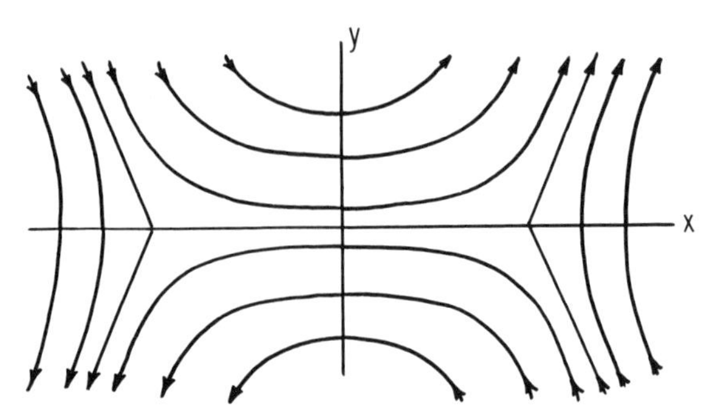

Fig. 6.5. A sketch of the field lines $U(x, y) = constant$ from equation (6.44), showing the discontinuity in $-b < x < b$, $y = 0$ given by equation (6.51) and the form of the field around the two Y-type neutral points, given by equation (6.53).

$$\mathcal{E} = M^2/4h^3. \tag{6.39}$$

This is the free energy of a strongly deformed X-type neutral point, with a current sheet extending asymptotically to $x^2, z^2 = \infty$ (in the limit of large H/h), comparable to the magnetic energy in the deformed region of characteristic dimension h. For the field strength at $x = z = 0$ on either side of the surface of discontinuity $y = 0$ is $B_z = 2M/h^3$. Over a characteristic volume h^3 the energy of such a field is $M^2/2\pi h^3$, which is closely comparable to the result $\mathcal{E}$ listed above. This simple result, then, provides a general idea of the free energy available in any squashed X-type neutral point. It is, of course, only the asymptotic value in the limit of large deformation, to the dipole separation $2h$ from the much larger initial separation $2H$. In this limit the surface of discontinuity is extremely broad, $O(2H)$. In §6.6 we take up the case where the deformation is modest, producing a tangential discontinuity of arbitrary width.

It becomes clear at this point that Sheeley's method can be applied to the calculation of the free energy of the magnetic discontinuities formed by a number of magnetic multipole fields in an infinitely conducting fluid. The essential requirement is that the multipoles be oriented so that their symmetry prevents any vacuum field lines from one multipole connecting to another. That is to say, the symmetry must be such that each multipole occupies its own separate topological domain, i.e., it must have its own topologically isolated "magnetosphere," so that its direction can be reversed without changing the shape of its magnetosphere and without changing the magnitude of the field at the boundary of the magnetosphere. For that is how the surface of discontinuity is produced. However, the simple case of the two antiparallel dipoles is sufficient for present purposes and we pursue the method no further.

6.5 The Y-Type Neutral Point in a Potential Field

In preparation for a more detailed study of the formation of tangential discontinuities consider the formation of Y-type neutral points at each end of the tangential discontinuity, or current sheet formed when an X-type neutral point is either compressed or stretched in one direction. The potential field in the vicinity of a 2D X-type neutral point is represented by the scalar potential

$$V = B_0\, xy/a, \tag{6.40}$$

and by the vector potential

$$U = B_0(y^2 - x^2)/2a, \tag{6.41}$$

so that

$$B_x = -\partial V/\partial x = +\partial U/\partial y = B_0\, y/a,$$

$$B_y = -\partial V/\partial y = -\partial U/\partial x = B_0\, x/a.$$

The field magnitude is B_0 at a distance $\varpi \equiv (x^2 + y^2)^{\frac{1}{2}} = a$. The energy density of the field is

$$B^2/8\pi = (B_0^2/8\pi)\varpi^2/a^2,$$

providing a total magnetic energy per unit length in $\varpi < R$ of

$$\begin{aligned} \mathcal{E}(R) &= 2\pi \int_0^R d\varpi\varpi B^2/8\pi \\ &= B_0^2 R^4/16a^2. \end{aligned} \tag{6.42}$$

This field can be represented by the complex potential

$$F(\zeta) = -B_0\,\zeta^2/2a,$$

where ζ is the complex variable $\zeta = x + iy$. Then if $F(\zeta) = U + iV$, the real part U is the vector potential and the imaginery part V is the scalar potential.

Now suppose that the field is compressed in the y-direction. We may imagine that the field (see equation 6.41) is anchored in the infinitely conducting boundaries $y = \pm L$ and the boundaries are moved closer together. As Syrovatskii (1971, 1978, 1981) and Bobrova and Syrovatskii (1979) have emphasized, the result is to squash the X-type neutral point into two Y-type neutral points with a tangential discontinuity, or current sheet, between. If the Y-type neutral points are at $x = \pm b$, $y = 0$ the complex potential function can be written

$$F(\zeta) = -\frac{B_0}{2a}[[\zeta(\zeta^2 - b^2)^{\frac{1}{2}} - b^2 \ln\{[\zeta + (\zeta^2 - b^2)^{\frac{1}{2}}]/b\}]] \tag{6.43}$$

in the neighborhood of the origin, which Syrovatskii constructed using a conformal mapping. The real part of $F(\zeta)$ is again the vector potential, with

$$\begin{aligned} U = -\frac{B_0}{2a}\Bigg[\Bigg[& x\mathrm{Re}(\zeta^2 - b^2)^{\frac{1}{2}} - y\mathrm{Im}(\zeta^2 - b^2)^{\frac{1}{2}} \\ & - \frac{1}{2}\ln\{[x + Re(\zeta^2 - b^2)^{\frac{1}{2}}]^2 + [y + \mathrm{Im}(\zeta^2 - b^2)^{\frac{1}{2}}]^2\}\Bigg]\Bigg], \end{aligned} \tag{6.44}$$

where

$$\mathrm{Re}(\zeta^2 - b^2)^{\frac{1}{2}} = 2^{-\frac{1}{2}}\{[(x^2 - y^2 - b^2)^2 + 4x^2 y^2]^{\frac{1}{2}} + x^2 - y^2 - b^2\}^{\frac{1}{2}}, \tag{6.45}$$

$$\mathrm{Im}(\zeta^2 - b^2)^{\frac{1}{2}} = 2^{-\frac{1}{2}}\{[(x^2 - y^2 - b^2)^2 + 4x^2 y^2]^{\frac{1}{2}} - x^2 + y^2 + b^2\}^{\frac{1}{2}}. \tag{6.46}$$

The field is computed most directly from

$$\begin{aligned} B_x - iB_y &= iF'(\zeta) \\ &= -i\frac{B_0}{a}(\zeta^2 - b^2)^{\frac{1}{2}}, \end{aligned} \tag{6.47}$$

so that

$$B_x = +\frac{B_0}{a}\,\mathrm{Im}(\zeta^2 - b^2)^{\frac{1}{2}}, \tag{6.48}$$

$$B_y = +\frac{B_0}{a}\,\mathrm{Re}(\zeta^2 - b^2)^{\frac{1}{2}}. \tag{6.49}$$

The energy density is

$$\frac{B^2}{8\pi} = \frac{B_0^2}{8\pi a^2}\left[(x^2 - y^2 - b^2)^2 + 4x^2y^2\right]^{\frac{1}{2}}. \tag{6.50}$$

The change in the total field energy from the X-type neutral point can be computed only when the boundaries are fully and properly specified so that the field at large ϖ is determined. Rigid infinitely conducting boundaries at $y = \pm L(t)$ would be particularly appropriate, but the present complex potential function involves a readjustment of the field at large ϖ which affects the total energy to about the same degree as the deformation of the neutral point at the origin. Hence it is not an appropriate example. The next section §6.6 treats the energy change when the X-type neutral point formed between two line dipoles is compressed or extended, so that the boundary conditions are fully specified.

The important aspect of the present calculation is simply the form of the field in the general neighborhood of the two Y-type neutral points and the surface of discontinuity that connects the neutral points. Figure 6.5 is a plot of the field lines, given by $U = constant$. The discontinuity across $-b < x < +b$, $y = 0$ is readily shown from equations (6.45), (6.46), (6.48), and (6.49), yielding

$$B_x = \pm B_0(b^2 - x^2)^{\frac{1}{2}}/a \tag{6.51}$$

for y approaching zero ($0 < y^2 \ll b^2 - x^2$) from above and below, respectively, while B_y declines to zero as

$$B_y = \frac{B_0\,xy}{a(b^2 - x^2)^{\frac{1}{2}}}. \tag{6.52}$$

In the neighborhood of the neutral point at $x = b, y = 0$ write $\zeta = b + r\exp i\theta$ where $r \ll b$. It follows that

$$U \cong -\frac{2^{\frac{3}{2}}B_0\,b^{\frac{1}{2}}r^{\frac{3}{2}}}{3a}\cos\frac{3}{2}\theta. \tag{6.53}$$

Then

$$B_r = +\frac{1}{r}\frac{\partial U}{\partial\theta} = B_0\left(\frac{2br}{a^2}\right)^{\frac{1}{2}}\sin\frac{3}{2}\theta,$$

$$B_\theta = -\frac{\partial U}{\partial r} = B_0\left(\frac{2br}{a^2}\right)^{\frac{1}{2}}\cos\frac{3}{2}\theta,$$

with

$$B_x = B_0\left(\frac{2br}{a^2}\right)^{\frac{1}{2}}\sin\frac{1}{2}\theta, \tag{6.54}$$

$$B_y = B_0\left(\frac{2br}{a^2}\right)^{\frac{1}{2}}\cos\frac{1}{2}\theta. \tag{6.55}$$

The field line $U = 0$ provides the Y configuration with $\theta = \pm\frac{1}{3}\pi$, and $\theta = \pi$ for $r > 0$, the neutral point $r = 0$ lying at the center of the Y at $x = b$, $y = 0$.

Syrovatskii (1981) provides a discussion and some wave and similarity solutions to the magnetohydrodynamic equations, illustrating special forms of the onset of the squeezing that leads to bifurcation of the X-type neutral point to form the discontinuity across $-b < x < +b$, $y = 0$. For our purposes it is sufficient to know that the deformation progresses in a finite time (measured by the sound and Alfven transit times), so that with a slow deformation the field is always close to the equilibrium form described above, with b an increasing function of time. It is evident from equation (6.48) that the magnitude of the discontinuity in the field has a maximum value of $B_0 b/a$ at $x = 0$, increasing linearly with b.

The simple tangential discontinuity described here has been discussed in more formal detail by Priest and Raadu (1975) and Hu and Low (1982), and, in the context of the larger 2D field in which the discontinuity is embedded, by Aly and Amari (1989) and Amari and Aly (1990).

6.6 Free Energy Above a Plane Boundary

The free magnetic energy produced by quasi-static deformation of a 2D X-type neutral point depends upon the form of the field throughout the entire region of deformation and upon the width of the current sheet or discontinuity. It is not without interest, therefore, to supplement the examples of the previous sections with another in which the field is anchored in an infinitely conducting plane boundary and the neutral point is deformed to only a limited degree. The field configuration can be represented by two colinear two dimensional dipoles, sketched in Fig. 6.6. The analytic form of the field has been worked out by Priest and Raadu (1975) and Hu and Low (1982) in two dimensions using the properties of analytic complex functions. They treat the case where the colinear dipoles are initially separated by a finite distance $2a_0$ in the x-direction and that distance is subsequently decreased to $a < a_0$, squashing the X-type neutral point in much the same manner as described in §6.4. Hu and Low (1982) also have worked out the discontinuous field when the dipoles are pulled apart to a larger separation, described in the next section. The field configuration is sketched in Fig. 6.7 when the dipoles are pushed together, showing the vertical surface of discontinuity.

The introduction of a small resistivity provides dissipation of magnetic energy at

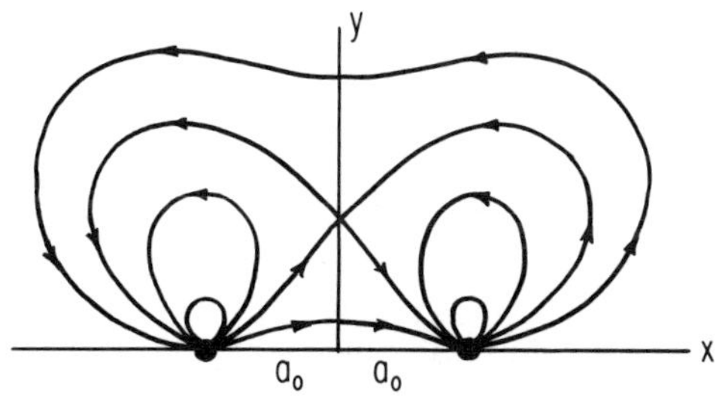

Fig. 6.6. A sketch of the field lines above the x-axis of the continuous potential field of two colinear 2D dipoles separated by a distance $2a_0$.

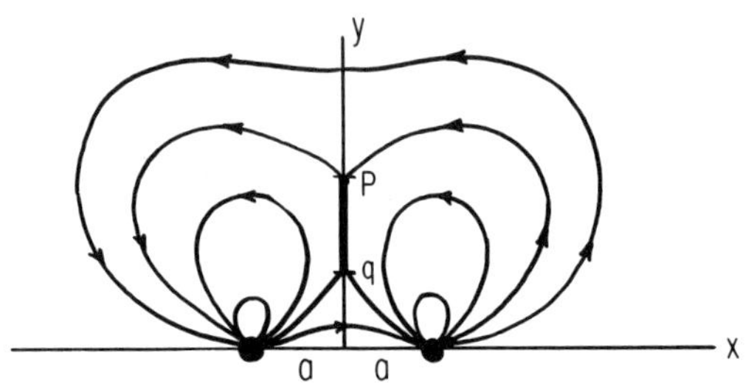

Fig. 6.7. A sketch of the field lines of the colinear dipoles of Fig. 6.6. after being pushed toward each other, reducing their separation from $2a_0$ to $2a$, and showing the discontinuity (heavy line) that forms from the original X-type neutral point.

either discontinuity, eventually reducing the field configuration to the continuous form sketched in Fig. 6.6 but now with the reduced dipole separation a. The current sheets or discontinuities responsible for the dissipation decline to zero only as the field takes up the final continuous form of Fig. 6.6.

The discontinuous potential field, sketched in Fig. 6.7, is described by

$$B_x - iB_y = \frac{2D(\zeta^2 + a_0^2)}{(\zeta^2 - a_0^2)^2}, \tag{6.56}$$

where $\zeta = x + iy$ and D is the dipole moment in the x-direction per unit length (in the z-direction, perpendicular to the xy-plane). The two dipoles point in the x–direction and are located at $x = \pm a_0$, $y = 0$. The boundary plane $y = 0$ is rigid and anchors all field lines. We are concerned only with the space $y \geqslant 0$. It is readily shown that

$$B_x = \frac{2D[(x^2 - y^2 - a_0^2)^2 (x^2 - y^2 + a_0^2) + 4x^2y^2(x^2 - y^2 - 3a_0^2)]}{[(x^2 - y^2 - a_0^2)^2 + 4x^2y^2]^2}, \tag{6.57}$$

$$B_y = \frac{4Dxy[(x^2 - y^2 - a_0^2)(x^2 - y^2 + 3a_0^2) + 4x^2y^2]}{[(x^2 - y^2 - a_0^2)^2 + 4x^2y^2]^2}. \tag{6.58}$$

At the boundary $y = 0$, the normal component B_y vanishes everywhere except at $x = \pm a_0$, while

$$B_x = \frac{2D(x^2 + a_0^2)}{(x^2 - a_0^2)^2}. \tag{6.59}$$

On the y-axis the y component vanishes ($B_y = 0$) and

$$B_x = \frac{2D(a_0^2 - y^2)}{(a_0^2 + y^2)^2}.$$

There is an X-type neutral point at $x = 0$, $y = a$. The magnetic energy density is

$$\begin{aligned} \frac{B^2}{8\pi} &= \frac{4D^2(\zeta^2 + a_0^2)(\zeta^{*2} + a_0^2)}{(\zeta^2 - a_0^2)^2(\zeta^{*2} - a_0^2)^2} \\ &= 4D^2\, T(a_0)^2/S(a_0)^4, \end{aligned} \tag{6.60}$$

where

$$T^2(a) = (x^2 - y^2 + a^2)^2 + 4x^2 y^2, \tag{6.61}$$

$$S^2(a) = (x^2 - y^2 - a^2)^2 + 4x^2 y^2. \tag{6.62}$$

This field is deformed, maintaining infinite electrical conductivity throughout the entire region, by moving the two dipoles closer together, from $x = \pm a_0$ to $x = \pm a$, where $a < a_0$. The total flux through each dipole, i.e., the dipole moments, are conserved, and a tangential discontinuity, or current sheet is formed along the y-axis in the interval $q < y < p$. The deformed field is given by

$$B_x - iB_y = \frac{E(\zeta^2 + p^2)^{\frac{1}{2}}(\zeta^2 + q^2)^{\frac{1}{2}}}{(\zeta^2 - a^2)^2}, \tag{6.63}$$

where the real quantities p, q, and E are determined by the three relations

$$E = 4Da^2/(a^2 + p^2)^{\frac{1}{2}}(a^2 + q^2)^{\frac{1}{2}}, \tag{6.64}$$

$$pq = a^2, \tag{6.65}$$

$$D = a_0 E \int_0^q dy \frac{(p^2 - y^2)^{\frac{1}{2}}(q^2 - y^2)^{\frac{1}{2}}}{(a^2 + y^2)^2}. \tag{6.66}$$

in terms of D, a_0, and a.

It is readily shown that

$$B_x - iB_y = \frac{EPQ}{S^z} \exp i(\alpha + \beta - 2\gamma), \tag{6.67}$$

where

$$P^4 = (x^2 - y^2 + p^2)^4 + 4x^2 y^2, \tag{6.68}$$

$$Q^4 = (x^2 - y^2 + q^2)^4 + 4x^2 y^2, \tag{6.69}$$

with

$$2^{\frac{1}{2}} \sin\alpha = [1 - (x^2 - y^2 + p^2)/P^2]^{\frac{1}{2}},$$

$$2^{\frac{1}{2}} \cos\alpha = [1 + (x^2 - y^2 + p^2)/P^2]^{\frac{1}{2}},$$

$$2^{\frac{1}{2}} \sin\beta = [1 - (x^2 - y^2 + q^2)/Q^2]^{\frac{1}{2}},$$

$$2^{\frac{1}{2}} \cos\beta = [1 + (x^2 - y^2 + q^2)/Q^2]^{\frac{1}{2}},$$

$$\sin\gamma = 2xy/S,$$

$$\cos\gamma = (x^2 - y^2 - a^2)/S.$$

It follows from equation (6.64) that $B_y = 0$ at the boundary $y = 0$, where

$$B_x = \frac{E(x^2 + p^2)^{\frac{1}{2}}(x^2 + q^2)^{\frac{1}{2}}}{(x^2 - a^2)^2},$$

exhibiting the two dipoles at $x = \pm a$. On the midline $x = 0$, it follows form equation (6.63) that

$$B_x - iB_y = \frac{E(p^2 - y^2)^{\frac{1}{2}}(q^2 - y^2)^{\frac{1}{2}}}{2(y^2 + a^2)^2}.$$

Then with $y^2 < q^2 < p^2$ or with $y^2 > p^2 > q^2$, the right-hand side is real and $B_y = 0$, with B_x varying as indicated by the right-hand side. On the other hand, if $q^2 < y^2 < p^2$ it follows that $(p^2 - y^2)^{\frac{1}{2}}$ is real while $(q^2 - y^2)^{\frac{1}{2}}$ is pure imaginery. So $B_x = 0$ with

$$B_y = \pm\frac{E(p^2 - y^2)^{\frac{1}{2}}(y^2 - q^2)^{\frac{1}{2}}}{2(y^2 + a^2)^2}. \tag{6.70}$$

The upper sign applies on the $x > 0$ side and the lower sign applies on the $x < 0$ side. The jump ΔB_y across the discontinuity at $x = 0$, $q < y < p$ is twice the right-hand side. The strength of the discontinuity falls to zero at each end ($y = q, p$) with a maximum between, located nearer to q than to p. Figure 6.8 provides a plot of $a^2\Delta B_y/E$ for $0 < q/a \ll 1$ and for $q/a = 0.6$, 0.8 and 1.0, for which $E/D = 4a/p$, 1.765, 1.951, and 2.00, respectively.

The free energy ΔW of the field is represented by the energy of the deformed field in equation (6.60) in excess of the corresponding continuous potential field of equation (6.56) (with $a = a_0$), representing the minimum energy for the given distribution of the normal component of the field on the boundary $y = 0$. Thus the free energy above some arbitrary level $y = h > 0$ is

$$\begin{aligned}\Delta W(a,p,q,h) &= \frac{1}{8\pi}\int_{-\infty}^{+\infty} dx \int_h^{\infty} dy[B_x^2(a,p,q) + B_y^2(a,p,q) - B_x^2(a,a,a) - B_y^2(a,a,a)] \\ &= \frac{1}{4\pi}\int_0^{\infty} dx \int_h^{\infty} dy\left(\frac{E^2P^2Q^2 - 4D^2T^2}{S^4}\right),\end{aligned} \tag{6.71}$$

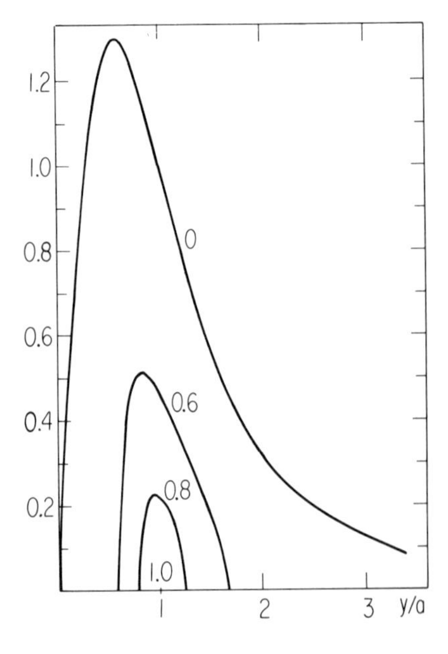

Fig. 6.8. A plot of the strength $a^2\Delta B_y/D$ over the breadth of the discontinuity shown in Fig. 6.7. for $q/a = 0, 0.6, 0.8$ and 1.0.

where T, S, E, D, P and Q are given by equations (6.61), (6.62), (6.64), (6.66), (6.68) and (6.69), respectively. Note that both fields diverge at $\zeta = \pm a$ and the integral of each separately diverges. Hence to evaluate the integral it is necessary to expand both fields to second order in the distance from the origin, cancelling the lower order terms which cause the divergence of the integral. The calculation is tedious because the integrals are not of elementary form. Low and Hu (1983) carry it through using the result that $B^2 = \mathbf{B} \cdot \nabla \times \mathbf{A} = \mathbf{A} \cdot \nabla \times \mathbf{B} + \nabla \cdot (\mathbf{A} \times \mathbf{B})$. Then

$$\begin{aligned} \int dVB^2 &= \int dV\mathbf{B} \cdot \nabla \times \mathbf{A} \\ &= \int dV\mathbf{A} \cdot \nabla \times \mathbf{B} + \int d\mathbf{S} \cdot \mathbf{A} \times \mathbf{B} \\ &= \frac{4\pi}{c} \int dV\mathbf{j} \cdot \mathbf{A} + \int d\mathbf{S} \cdot \mathbf{A} \times \mathbf{B}, \end{aligned}$$

where both Gauss' theorem and Ampere's law (see equation 2.32) have been used. This approach simplifies the computation but involves some considerations on the work done by $\mathbf{j}$, etc.

Figure 6.9 is taken from Low and Hu (1983, Fig. 2), given $\Delta W(a, p, q, h)$ in units of the free interaction energy $W = D^2/4a^2$ of two coaxial dipoles separated by a distance $2a$. W is the work required to separate the two dipoles to infinity. The value of q/a is indicated on each curve, plotting $\Delta W/W$ as a function of h/a. For the uppermost curve ($q/a = 0.6$, $p/a = 1.667$) the width of the sheet of discontinuity is $(p - q)/a = 1.067$, representing a substantial compression with $\Delta W = 0.060\,\mathrm{W}$. The negative slope of the individual curves shows the distribution of free energy with height h. Thus with $q/a = 0.6$, the free energy above $h = 0.5a$ is $0.052\,\mathrm{W}$, compared to $0.018\,\mathrm{W}$ below $h = 0.5a$. This distribution expresses the fact that most of the free energy is in the neighborhood of the surface of tangential discontinuity.

6.7 Free Energy of an Extended Neutral Point

Consider the tangential discontinuity and the associated free energy when the separation of the dipoles is increased, rather than decreased (i.e., $a > a_0$). The result is sketched in Fig. 6.10, showing the horizontal arc formed by the discontinuity. Hu and Low (1982) give

$$B_x - iB_y = \frac{2D}{\cos\phi} \frac{(\zeta^2 - \zeta_A^2)^{\frac{1}{2}}(\zeta^2 - \zeta_B^2)^{\frac{1}{2}}}{(\zeta^2 - a^2)^2}, \tag{6.72}$$

where $\zeta = x + iy$ again, D is the 2D dipole moment per unit length in the z-direction and ϕ is a quantity to be adjust to conserve flux passing below the surface of discontinuity. The points $\zeta_A = a \exp i(\frac{1}{2}\pi - \phi)$ and $\zeta_B = a \exp i(\frac{1}{2}\pi + \phi)$ represent the end-points of the discontinuity, which can be shown to be the arc of a circle,

$$\zeta = a \exp i\left(\frac{1}{2}\pi - \theta\right),$$

bisected by the y-axis at $\theta = 0$ and with end points at $\theta = \pm\phi$. Thus the angles ϕ and θ are measured from the imaginery axis.

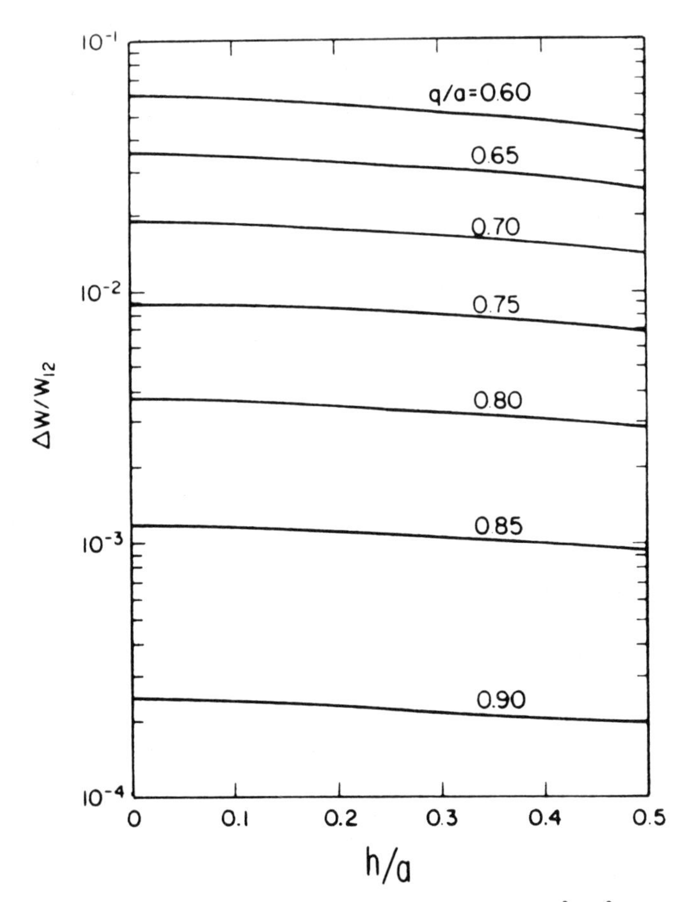

Fig. 6.9. A plot of the free energy ΔW above $y = H$ in units of $D^2/4a^2$ as a function of h/a with the value of q/a indicated on each curve, reproduced from Low and Hu (1983) with the kind permission of the authors, Fig. 2.

Now ϕ is defined by the preservation of the total flux between $y = 0$ and the arc of discontinuity, requiring that

$$\frac{D}{a_0} = \frac{2D}{a\cos\phi}\int_0^1 \frac{d\xi(\xi^4 - 2\xi^2\cos 2\phi + 1)^{\frac{1}{2}}}{(1+\xi^2)^2}, \tag{6.73}$$

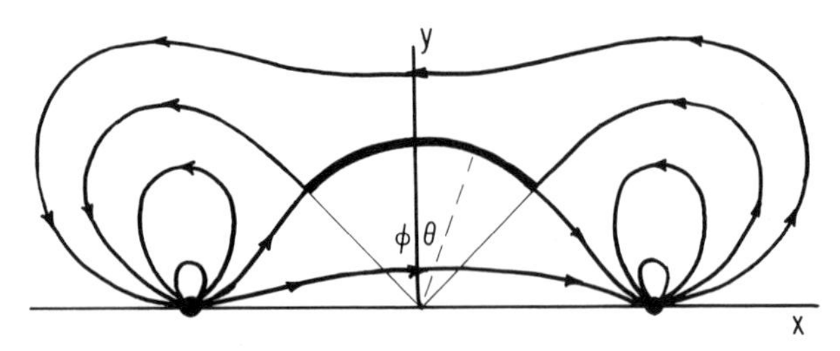

Fig. 6.10. A sketch of the field and the discontinuity (heavy arc) formed when the two colinear dipoles shown in Fig. 6.6. are pulled apart, increasing their separation from $2a_0$ to $2a$, where now $a > a_0$.

which relates ϕ to a and a_0 when $a > a_0$. The angular width of the current sheet is 2ϕ, as already noted, and it follows from equation (6.73) that

$$\frac{d}{d\phi}\left(\frac{a}{a_0}\right) = \frac{2\sin\phi}{\cos^2\phi}\int_0^1 \frac{d\xi(\xi^4 - 2\xi^2\cos 2\phi + 1)^{\frac{1}{2}}}{(1+\xi^2)^2}$$
$$+ \frac{2}{\cos\phi}\int_0^1 \frac{d\xi \quad 2\xi^2 \sin 2\phi}{(1+\xi^2)(\xi^4 - 2\xi^2\cos 2\phi + 1)^{\frac{1}{2}}}.$$

The right-hand side is obviously greater than zero for $0 < \phi < \frac{1}{2}\pi$, showing that ϕ is a monotonically increasing function of a/a_0 for a given initial a_0. The farther apart the dipoles are moved, the longer is the arc of discontinuity, with ϕ increasing to $\frac{1}{2}\pi$ in the limit of large a/a_0.

The field at the arc of discontinuity is entirely azimuthal, of course, so it is convenient to convert to polar coordinates, with

$$B_x - iB_y = -\exp i\theta(B_\theta - iB_\varpi).$$

Then with $\zeta = a\exp i(\frac{1}{2}\pi - \theta) = ia\exp(-i\theta)$, it follows from equation (6.72) that

$$\begin{aligned} B_\theta - iB_\varpi &= -\frac{2D\exp(-i\theta)[\exp(-i4\theta) - 2\cos 2\phi\exp(-i2\theta) + 1]^{\frac{1}{2}}}{a^2\cos\phi[(1+\exp(-i2\theta)]^2}, \\ &= -\frac{2D(\sin^2\phi - \sin^2\theta)^{\frac{1}{2}}}{a^2\cos\phi\cos^2\theta}. \end{aligned} \tag{6.74}$$

The right-hand side is real for $-\phi < \theta < +\phi$, so $B_\varpi = 0$. Thus, the right-hand side provides B_θ, with the proper choice of sign for the square root on either side of the boundary, giving $B_\theta < 0$ above and $B_\theta > 0$ below the discontinuity. There is a branch point at each end, $\theta = \pm\phi$. Beyond the discontinuity ($\theta^2 > \phi^2$) the right-hand side is pure imaginery, giving $B_\theta = 0$ and

$$B_\varpi = \frac{D(\sin^2\theta - \sin^2\theta)^{\frac{1}{2}}}{a^2\cos\phi\cos^2\theta},$$

where the sign of the square root is to be taken so that B_ϖ is negative for $\theta > \phi$ and positive for $\theta < -\phi$. It is interesting, then, that on the circle $\zeta = ia\exp(-i\theta)$ the field is precisely radial for $\theta^2 > \phi^2$ and precisely azimuthal for $\theta^2 < \phi^2$.

The strength of the discontinuity follows at once from equation (6.74) as

$$\Delta B_\theta = \frac{2D(\sin^2\phi - \sin^2\theta)^{\frac{1}{2}}}{a^2\cos\phi\cos^2\theta}. \tag{6.75}$$

for $-\phi < \theta < +\phi$. The surface current density is $c\Delta B_\theta/4\pi$, of course. For $\phi \ll 1$,

$$\Delta B_\theta \cong \frac{2D(\phi^2 - \theta^2)^{\frac{1}{2}}}{a^2}.$$

It follows from equation (6.74) that

$$a \cong a_0\left(1 + \frac{1}{2}\phi^2 + \dots\right).$$

The strength of the discontinuity ΔB_θ is plotted in Fig. 6.11 for various values of ϕ, showing how it evolves from a convex to a concave shape across $\theta = 0$ as ϕ increases across $\frac{1}{4}\pi$. Note that $a/a_0 = 1.40$, 2.94, and 8.98 for $\phi = 30°$, 60°, and 80°, respectively, so that the extension of the field for $\phi = 80°$ is quite large, with the strongly double peaked current density shown in Fig. 6.11. It is readily shown that

$$B_x - iB_y = \frac{2D\,\mathrm{MN}}{S^2 \cos\phi} \exp i(\nu + \mu - 2\gamma), \tag{6.76}$$

where

$$M^4 = (x^2 - y^2 + a^2 \cos 2\phi)^2 + (2xy + a^2 \sin 2\phi)^2, \tag{6.77}$$

$$N^4 = (x^2 - y^2 + a^2 \cos 2\phi)^2 + (2xy - a^2 \sin 2\phi)^2, \tag{6.78}$$

with

$$2^{\frac{1}{2}} \sin\nu = \left[1 - (x^2 - y^2 + a^2 \cos 2\phi)/M^2\right]^{\frac{1}{2}},$$

$$2^{\frac{1}{2}} \cos\nu = \left[1 + (x^2 - y^2 + a^2 \cos 2\phi)/M^2\right]^{\frac{1}{2}},$$

$$2^{\frac{1}{2}} \sin\mu = \left[1 - (x^2 - y^2 + a^2 \cos 2\phi)/N^2\right]^{\frac{1}{2}},$$

$$2^{\frac{1}{2}} \cos\mu = \left[1 + (x^2 - y^2 + a^2 \cos 2\phi)/N^2\right]^{\frac{1}{2}}.$$

The free energy ΔW associated with the increase of a is the excess energy of the deformed field given by equation (6.75) minus the energy of the potential field of equation (6.56) sketched in Fig. 6.6 with a_0 put equal to a so that the free energy in $y > h$ is

$$\Delta W(a, \phi, h) = \frac{D^2}{\pi} \int_0^{+\infty} dx \int_h^{\infty} dy \frac{1}{S^4} \left(\frac{M^2 N^2}{\cos^2\phi} - T^2 \right),$$

where T, S, M, and N are given by equations (6.61), (6.62), (6.77), and (6.78), respectively. Figure 6.12 is a plot of ΔW as a function of h for various values of ϕ, with ΔW expressed in units of the free interaction energy $W = D^2/4a^2$ of two coaxial dipoles separated by a distance $2a$.

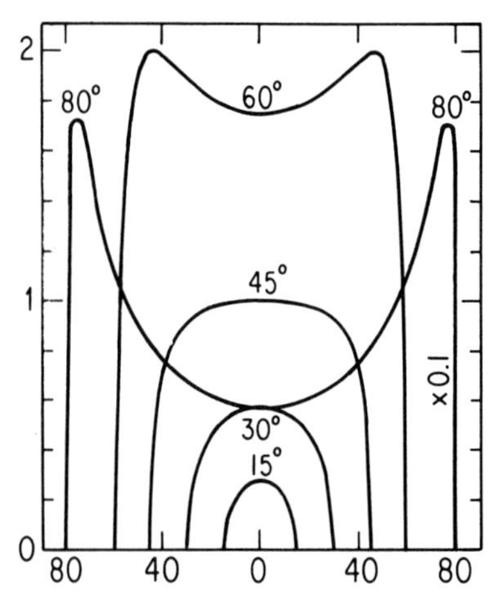

Fig. 6.11. A plot of the strength $a^2 \Delta B_0/D$ of the discontinuity versus position θ along the arc for different values of ϕ, from equation (6.74). The curve for $\phi = 80°$ is reduced by a factor of 10.

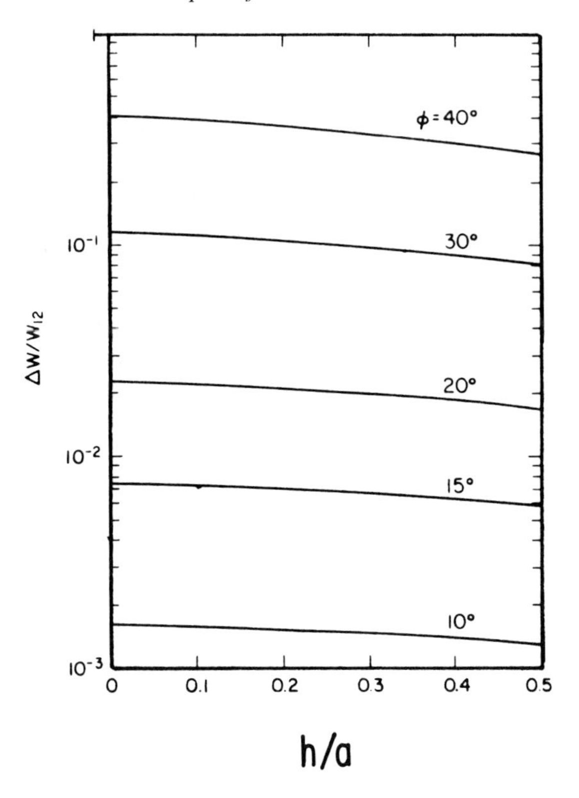

Fig. 6.12. A plot of the free energy ΔW above $y = H$ in units of $W = D^2/4a^2$ as a function of h/a with the value of ϕ indicated on each curve, reproduced from Low and Hu (1983) with the kind permission of the authors, Fig. 4.

It is evident by inspection of Fig. 6.12 that the free energy produced by increased separation of the colinear dipoles is comparable in magnitude to the free energy produced by decreased separation, plotted in Fig. 6.9, as one would expect. The essential physical features are the width of the current sheet and the characteristic field strength.

6.8 The Creation of X-Type Neutral Points

The examples provided so far illustrate the creation of a tangential discontinuity by deforming an X-type neutral point. It is generally the case that tangential discontinuities are formed in this way, and even an infinitesimal deformation of an X-type neutral point may produce a discontinuity, albeit a discontinuity of infinitesimal strength and width. However, the presence of an initial X-type neutral point is not a necessary condition for the deformation of the field to produce a tangential discontinuity. A field without neutral points may develop one or more X-type neutral points as a result of a finite deformation. Any further deformation squashes the neutral point into a tangential discontinuity in the manner described in the foregoing sections.

Generally speaking X-type neutral points are formed when separate lobes of field (i.e., separate and distinct topological regions) come into contact. In the simplest case (Moffatt, 1987; Low, 1987; Low and Wolfson, 1988) imagine a potential field configuration of the form shown in Fig. 6.13(a), consisting of two bipolar lobes of field above the surface $y = 0$ overlain by a field of larger scale. Such a potential field is easily constructed, in the simplest case by imagining two colinear horizontal dipoles somewhere beneath the surface, but of course, the dipoles are virtual. The essential feature is the normal component of the field at the surface $y = 0$. Then suppose that the boundary $y = 0$ is compressed in the neighborhood of $x = 0$, carrying the footpoints of the field inward toward the y-axis. The region $y > 0$ is assumed to be filled with an infinitely conducting passive fluid, so the field remains a potential field as the two lobes are pushed together by the displacement of the footpoints toward $x = 0$.

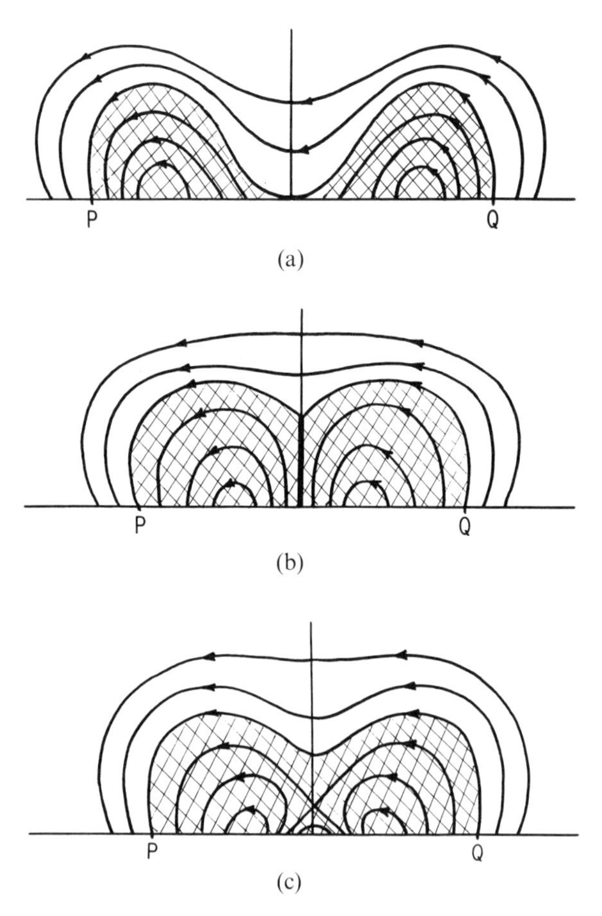

Fig. 6.13. (a) A sketch of a potential field with two separate lobes surrounded by a large-scale field. The cross-hatching indicates the area occupied by the two lobes with outer edges intersecting the boundary $y = 0$ at the points labeled P and Q. (b) The deformed field of (a) following compression of the footpoints with the discontinuity indicated by the heavy vertical line up the middle. (c) The continuous potential field with the same normal component at $y = 0$ as in (b).

The topology of the field lines is preserved by the infinite conductivity, so the continuous potential field for the same distribution of the normal component B_y at $y = 0$, sketched in Fig. 6.13(a), is not an available option. Instead the approaching lobes expel the larger scale field from the region between them, coming into contact to form first an X-type neutral point and then a surface of tangential discontinuity, sketched in Fig. 6.13(b). Only the discontinuous field of Fig. 6.13(b) preserves the topology. In the presence of a small but nonvanishing resistivity the discontinuous field configuration (b) eventually decays into the continuous field (c). The discontinuity declines to zero as the field approaches the form (c).

The problem can be treated formally. We turn to an example worked out by Low (1989), treating an octupole field external to a spherical boundary $r = 1$. The external field is deformed by continuous motion of the footpoints on the boundary $r = 1$. Compression of the footpoints towards the equator ($\theta = \frac{1}{2}\pi$) suppresses the equatorial lobe, opening up the surrounding space so that the lobes at high positive and negative latitude bulge equatorward, coming into contact across the equatorial plane to produce a surface of tangential discontinuity that extends outward from finite $r(>1)$ to $r = \infty$.

The demonstration is in two stages. The first stage demonstrates the motion of the footpoints that provides the coming together of the high latitude lobes at the equatorial plane. The calculation is carried out for a vacuum field so that an X-type neutral line is formed at $r = r_0 > 1$ in the equatorial plane, when the lobes come into contact, showing the change in topology of the field lines in the two lobes if no infinitely conducting fluid is present. This demonstrates that there would be a surface of discontinuity formed at the X-type neutral point in the equatorial plane in the presence of an infinitely conducting fluid (which would preserve the topology).

The second stage provides a formal analytical illustration of the tangential discontinuity for fields of high latitude lobes that come into contact at the equatorial plane.

Consider, then, the axisymmetric poloidal field

$$B_r = +\frac{1}{r^2 \sin\theta}\frac{\partial A}{\partial \theta}, \tag{6.79}$$

$$B_\theta = -\frac{1}{r\sin\theta}\frac{\partial A}{\partial r}, \tag{6.80}$$

$$B_\varphi = 0, \tag{6.81}$$

expressed in spherical coordinates (r, θ, φ) in terms of the flux function A. The field lines lie in meridional planes and are given by $A(r, \theta) = constant$, while the magnetic flux through an annulus with edges at (r, θ) and $(r + dr, \theta + d\theta)$ is $2\pi dA$, where dA is the difference in A between the edges of the annulus.

In the presence of a passive infinitely conducting fluid the magnetic field **B** is a potential field and

$$\frac{\partial^2 A}{\partial r^2} + \frac{\sin\theta}{r^2}\frac{\partial}{\partial\theta}\frac{1}{\sin\theta}\frac{\partial A}{\partial\theta} = 0. \tag{6.82}$$

The variables are separable and provide solutions of the form

$$W_n = \frac{1}{nr^n}(1-\mu^2)^{\frac{1}{2}}P_n^1(\mu), \tag{6.83}$$

where n is a positive integer and $\mu = \cos\theta$, with the usual notation $P_n^1(\mu)$ for the associated Legendre polynomial of the first kind.

Consider, then,

$$W_3 = \frac{1}{3r^3}(1-\mu^2)(1-5\mu^2), \tag{6.84}$$

representing three lobes, each lobe extending to $r = \infty$ between $\mu = \pm 1$ and $\mu = \pm 1/5^{\frac{1}{2}} = 0.4472$, i.e., the equatorial lobe extends out to $r = \infty$ between latitudes $\pm 26°33'(63°27' < \theta < 116°33')$ and the high latitude lobes extend out to $r = \infty$ between latitudes $\pm 26°33'$ and $\pm 90°$, respectively, $(0 < \theta < 63°27'$, $116°33' < \theta < 180°)$. The field is sketched in Fig. 6.14. The flux function is plotted in Fig. 6.15. The minimum value of $(1-\mu^2)(1-5\mu^2)$ is -0.8, at $\mu^2 = 0.6$, rising to zero at the poles $(\mu^2 = 1)$ and to one at the equator $(\mu^2 = 0)$.

Now suppose that the footpoints of the field on the infinitely conducting sphere $r = 1$ are displaced in latitude, i.e., a given footpoint moves from the polar angle θ to the new position ϑ. If the resulting flux function for the deformed field is $A_d(r, \vartheta)$, conservation of magnetic flux requires that

$$A_d(1, \vartheta) = A(1, \theta), \tag{6.85}$$

providing ϑ in terms of θ for any specified $A_d(1, \vartheta)$. It is convenient to treat the case that

$$A_d(1, \theta) = \frac{(1-\mu^2)(\mu^4 + h\mu^2 + h + k)}{h+k}, \tag{6.86}$$

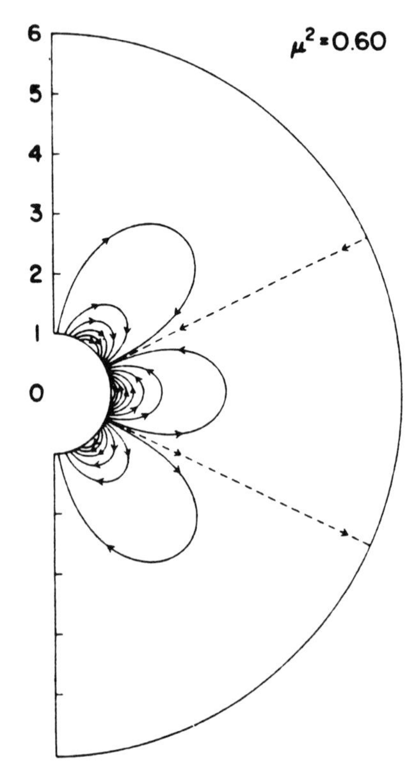

Fig. 6.14. A sketch of the initial field W_3 from equation (6.84), reproduced from Low (1989), Fig. 1(b), with the kind permission of the author.

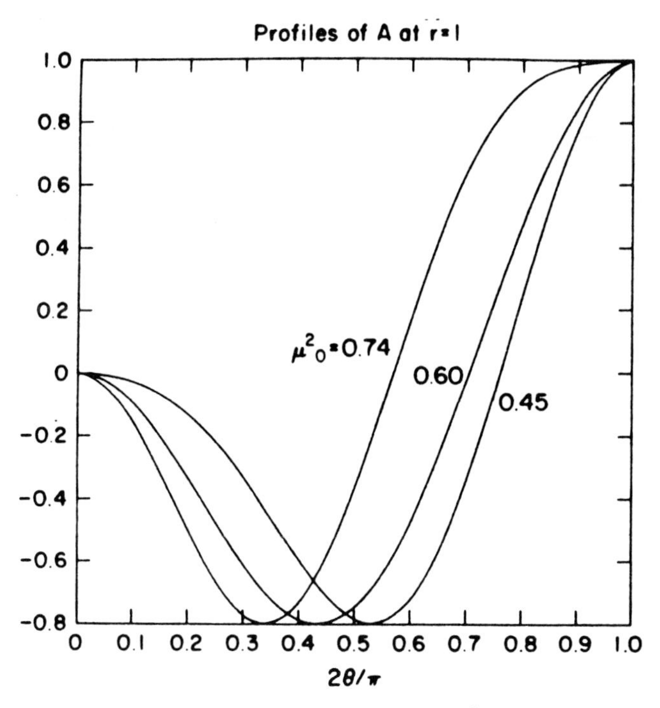

Fig. 6.15. The flux function W_3 from equation (6.84) for $\mu_0^2 = 0.6$, and from equation (6.89) for $\mu_0^2 = 0.45$ representing a compression of the footpoints toward low latitudes, reproduced from Low (1989), Fig. 1(a), with the kind permission of the author.

where $\mu = \cos\theta$, and h and k are defined in terms of the single parameter μ_0,

$$h = -2\mu_0^2 \frac{5\mu_0^4 - 16\mu_0^2 + 9}{5\mu_0^4 - 18\mu_0^2 + 9}, \tag{6.87}$$

$$k = +\mu_0^2 \frac{5\mu_0^6 - 27\mu_0^2 + 18}{5\mu_0^4 - 18\mu_0^2 + 9}. \tag{6.88}$$

Note then that if $\mu_0^2 = 0.6 + \varepsilon$ we have $h \cong +0.1200/\varepsilon$ and $k = -0.1440/\varepsilon$ with the result that $A_d(1, \theta)$ reduces to $+3W_3$, representing the initial octupole field. More generally, with $A_d(1, \theta)$ given by equation (6.86) it can be shown that

$$A_d(r, \theta) = \frac{-1}{21(h+k)}\left\{\frac{8}{3}W_5 + \frac{14}{5}(3h+2)W_3 + \left[\frac{21}{5}(6h+5k) + \frac{9}{5}\right]W_1\right\} \tag{6.89}$$

represents the field for $r > 1$. The field topology is sketched in Fig. 6.16 for $\mu_0^2 = 0.45$, representing a displacement of the footpoints toward the equator, as may be seen from the plot of $A_d(1, \theta)$ in Fig. 6.16.

The essential point to be seen in Fig. 6.16 is the X-type neutral point at $r \cong 4.5$ at the equator, $\theta = \frac{1}{2}\pi$, formed by the two high latitude lobes coming into contact and freely reconnecting their field lines in the vacuum for which the calculation was carried out. Thus the topology of the two lobes shown in Fig. 6.16 is different from the topology shown in Fig. 6.14 where there is no connection between the two high

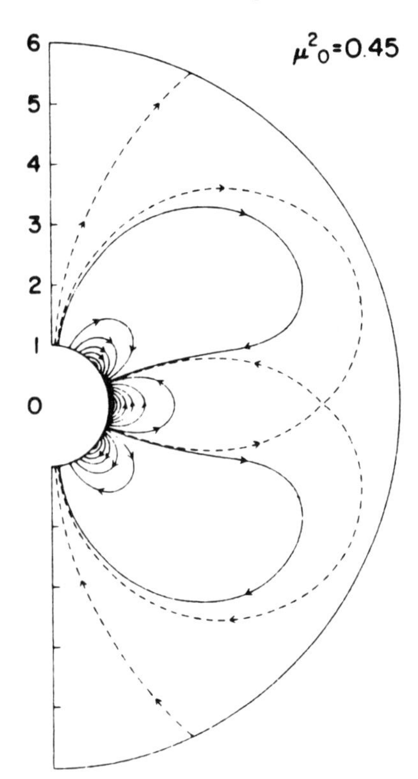

Fig. 6.16. A sketch of the continuous potential field described by the flux function A_d, equation (6.89) for $\mu_0^2 = 0.45$, representing a compression of the footpoints into low latitudes, of the field shown in Fig. 6.14, reproduced from Low (1989), Fig. 1(d), with the kind permission of the author. The flux function is plotted in Fig. 6.15.

latitude lobes. In the presence of an infinitely conducting fluid there would have been no reconnection of field lines, and a surface of tangential discontinuity would have formed in the equatorial plane, extending out to infinity from somewhere in the vicinity of the neutral point. In that case the field would be equivalent to an octupole field at the origin at the center of a circular hole of radius b in an impenetrable sheet in the equatorial plane (representing the current sheet or surface of tangential discontinuity), sketched in Fig. 6.17. The field of the lobe at large positive latitude fills the distant space $(0 < \theta < \frac{1}{2}\pi)$ above the sheet and the other lobe fills the distance space $(\frac{1}{2}\pi < \theta < \pi)$ below the sheet.

The second stage of the demonstration consists of the construction of a potential field of the form shown in Fig. 6.17, which reduces to the flux function $A_d(1, \theta)$ from equation (6.89) with $\mu_0^2 = 0.45$ at $r = 1$ while retaining the separate topologies of the three lobes of the initial field $\mu_0^2 = 0.60$. Oblate spheroidal coordinates (ξ, η, φ) are appropriate, with φ measuring azimuth, as in spherical coordinates, and ξ, η related to r, θ and z, ϖ by

$$\xi^2 = -\frac{1}{2}\left(1 - \frac{r^2}{b^2}\right) + \frac{1}{2}\left[\left(1 - \frac{r^2}{b^2}\right)^2 + 4\frac{r^2}{b^2}\cos^2\theta\right]^{\frac{1}{2}}, \tag{6.90}$$

$$\eta^2 = -\frac{1}{2}\left(\frac{r^2}{b^2} - 1\right) + \frac{1}{2}\left[\left(\frac{r^2}{b^2} - 1^2\right)^2 + 4\frac{r^2}{b^2}\cos^2\theta\right]^{\frac{1}{2}}, \tag{6.91}$$

and

$$z = b\xi\eta, \; \varpi = b(\xi^2 + 1)^{\frac{1}{2}}(1 - \eta^2)^{\frac{1}{2}}, \tag{6.92}$$

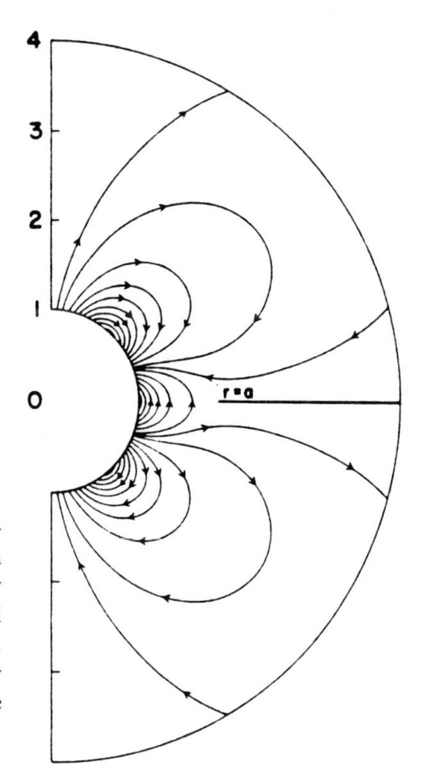

Fig. 6.17. A sketch of the potential field with footpoints compressed toward low latitudes, as in Fig. 6.16, but with the field topology preserved by the infinite electrical conductivity of the fluid and described by the flux function of equation (6.102). The equatorial current sheet is indicated by the heavy line, reproduced from Low (1989), Fig. 2(c), with the kind permission of the author.

in terms of the spherical coordinates (r, θ, φ) and cylindrical coordinates (ϖ, φ, z). The surfaces $\xi = constant$ represent a family of confocal oblate spheroids with focus at $r = b$, $\theta = \frac{1}{2}\pi$. The surfaces $\eta = constant$ represent a family of confocal hyperboloids of revolution with the same focus as the spheroids. The spheroidal coordinates have ranges $0 \leqslant \xi \leqslant \infty$ and $0 \leqslant \eta \leqslant 1$. The impenetrable equatorial sheet corresponds to $\eta = 0$ with the edge of the hole $r = b$ at the focus. The z-axis lies on $\eta = 1$. The basic solutions of equation (6.82) for the vector potential A are

$$w_n = (1 + \xi^2)^{\frac{1}{2}} (1 - \eta^2)^{\frac{1}{2}} Q_n^1(i\xi) P_n^1(\eta), \tag{6.93}$$

where Q_n^1 represents the associated Legendre function of the second kind. The $n = 3$ (octupole) solution is

$$w_3 = (1 - \eta^2)(1 - 5\eta^2)[3(1 + \xi^2)(1 + 5\xi^2)\cot^{-1}\xi - 15\xi^3 - 13\xi], \tag{6.94}$$

and the $n = 1$ (dipole) solution is

$$w_1 = (1 - \eta^2)[(1 + \xi^2)]\cot^{-1}\xi - \xi]. \tag{6.95}$$

Low (1986a) defines the functions

$$u^2 = -\frac{1}{2}\left(1 - \frac{b^2}{r^2}\right) + \frac{1}{2}\left[\left(1 - \frac{b^2}{r^2}\right)^2 + \frac{4b^2}{r^2}\cos^2\theta\right]^{\frac{1}{2}}, \tag{6.96}$$

$$v^2 = -\frac{1}{2}\left(\frac{b^2}{r^2} - 1\right) + \frac{1}{2}\left[\left(\frac{b^2}{r^2} - 1\right)^2 + \frac{4b^2}{r^2}\cos^2\theta\right]^{\frac{1}{2}}, \tag{6.97}$$

which are obtained from ξ and η, respectively, by inversion with respect to the sphere $r = b$. The inversion transformation converts the disk $r \leqslant b$, $\theta = \frac{1}{2}\pi$ into an infinite sheet $r \geqslant b$, $\theta = \frac{1}{2}\pi$ with a circular hole at the center. The transformed solutions take the form

$$S_n = r(1+u^2)^{\frac{1}{2}}(1-v^2)^{\frac{1}{2}}Q_n^1(iu)P_n^1(v). \tag{6.98}$$

With these basic solutions Low constructs the potentials

$$\begin{aligned} Z_1 &= S_1 - \pi b^2 W_1 + 2b\eta \\ &= r(1-v^2)\left[(1+u^2)\cot^{-1}u - u\right] - \frac{\pi b^2}{2}\frac{(1-\mu^2)}{r} + 2b\eta, \end{aligned} \tag{6.99}$$

and

$$\begin{aligned} Z_3 &= \frac{3}{4}S_3 - \frac{135\pi b^4}{8}W_3 + \frac{9\pi b^2}{2}W_1 + 12b\eta \\ &= \frac{3}{4}r(1-v^2)(1-5v^2)[3(1+u^2)(5u^2+1)\cot^{-1}u - 15u^3 - 13u] \\ &\quad - \frac{45b^4}{8}\frac{(1-u^2)(1-5\mu^2)}{r^3} + \frac{9\pi b^2}{2}\frac{(1-\mu^2)}{r} + 12b\eta, \end{aligned} \tag{6.100}$$

so that Z_1 and Z_3 reduce to W_1 and W_3 as r declines to zero, and W_1 and W_3 are given by equation (6.83). On the equator $\theta = \frac{1}{2}\pi$ there is a cut extending from $r = b$ to infinity across which there is no field. The tangential component of the field reverses sign across the equator. The terms $2a\eta$ and $12a\eta$ have been added to keep the field from diverging at the inner edge $r = b$ of the current sheet or cut. Starting with the initial field

$$A_i(r,\theta) = b^5\frac{75}{8}W_5 - w_3, \tag{6.101}$$

with $b = 5/8$, the flux function $A_i(1,\theta)$ is plotted in Fig. 6.18 and the form of the field is essentially that of Fig. 6.14. The footpoints of the field on $r = 1$ are pushed toward the equator, providing the final flux function A_f plotted in Fig. 6.18. The flux function is then

$$A_f = (6Z_1 - Z_3)/b, \tag{6.102}$$

sketched in Fig. 6.18, where the high latitude lobes have pushed together, meeting at the equatorial plane to create a tangential discontinuity, or current sheet, extending from $r = b = 1.796$ to infinity.

Suppose, then, that a small resistivity is introduced so that there is dissipation of magnetic energy at the tangential discontinuity. The result is a gradual weakening and shrinking of the discontinuity until it disappears altogether upon reaching the continuous form

$$A_p = (6H_1 - H - 3)/b, \tag{6.103}$$

where

$$H_1 = w_1 - \frac{\pi b^2}{2}W_1,$$

$$H_3 = \frac{3}{4}w_3 - \frac{135}{8}\pi b^4 W_3 + \frac{9\pi b^2}{2}W_1.$$

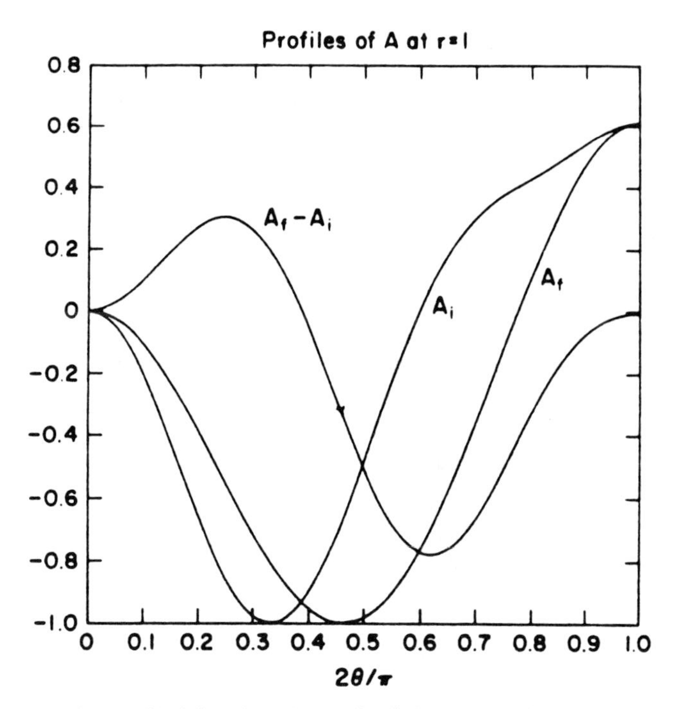

Fig. 6.18. A plot of the initial flux function $A_i(1, \theta)$ from equation (6.101) and the final flux function $A_f(1, \theta)$ from equation (6.102) at the boundary $r = 1$, reproduced from Low (1989), with the kind permission of the author.

The field then has same the form as the field shown in Fig. 6.16 with only an X-type neutral point instead of a tangential discontinuity. Low (1989) makes the general point that any bipolar magnetic lobe tends to retract when its footpoints are compressed in the direction of the axis of the bipole. Hence such longitudinal compression permits neighboring lobes to expand into the space left by the retraction, which leads them into contact to form a surface of discontinuity in the manner depicted by the foregoing example. He also points out that the transverse compression, perpendicular to the axis of a bipolar lobe, increases the field strength causing the lobe to expand transversely beyond the area of its footpoints into the surrounding region of weaker field, perhaps coming into contact in this way with other lobes to form a surface of discontinuity. Two such expanded lobes are sketched in Fig. 6.19.

Further examples of the formation of tangential discontinuities are available in the literature. Jensen (1989) treats the formation of surfaces of tangential discontinuity within a spherical region as a consequence of the topology of the field. Low and Wolfson (1988), Aly and Amari (1989), and Amari and Aly (1989) provide formal calculations of the surfaces of tangential discontinuity in 2D potential fields. Low (1986a) provides examples involving partially open magnetospheres with equatorial current sheets associated with the expansion of fields whose footpoints are subject to ongoing shear (Low 1986b). Linardatos (1992) develops an interesting and quite general variational approach to magnetostatic equilibria and steady Euler flows in two dimensions. He works out several examples to illustrate the method, including cases where the relaxation squashes an X-type neutral point to provide a

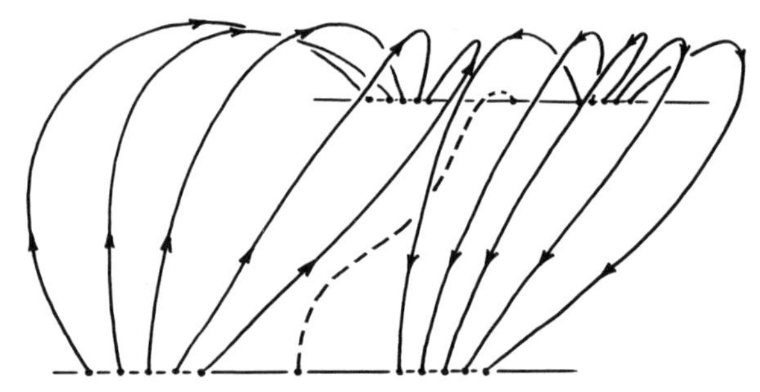

Fig. 6.19. A sketch of two oppositely oriented bipolar lobes of field whose footpoints, indicated by the dots, have been compressed perpendicular to the axis of the bipoles, showing their tendency to crowd out the weaker uncompressed magnetic field between (dashed line) thereby making contact and forming a current sheet at the midplane.

tangential discontinuity. Titov (1992) describes a general analytical approach to constructing 2D potential fields with arbitrary current sheets, starting with a specification of the field topology and the normal component of the field on the enclosing boundaries.

6.9 Direct Numerical Simulation

Mikic, Schnack, and Van Hoven (1989) have taken a direct approach to the problem of field deformation with a numerical experiment showing the rapid development of current sheets in response to footpoint displacement of the initially uniform field B_0 extending from $z = 0$ to $z = L$. The field is deformed from this uniform state by moving the footpoints at $z = 0$ while holding fixed the footpoints at $z = L$. The footpoints at $z = 0$ are then moved by a sequence of alternately perpendicular shear flows, along the lines employed by Van Ballegooijen (1986). The first and all subsequent odd-numbered flows are in the y-direction with

$$v_x^{(2i-1)} = 0,$$

$$v_y^{(2i-1)} = v_0 \sin(ky + \phi_{2i-1}),$$

where $i = 1, 2, 3, \ldots$, The even-numbered flows are in the x-direction, with

$$v_x^{(2i)} = v_0 \sin(kx + \phi_{2i}),$$

$$v_y^{(2i)} = 0.$$

The successive phases ϕ_1 are chosen at random. Each flow runs for a time T such that $kv_0 T = 1$, so that the field is strongly deformed. The field is allowed to relax toward static equilibrium between successive flows, with a small viscosity included to damp the associated kinetic energy. The calculation is carried out in a 64^3 mesh in a cube of side L. The characteristic dynamical time is the Alfven transit time $\tau_A = L/C$ where C is the Alfven speed $B_0/(4\pi\rho)^{\frac{1}{2}}$. They use $v_0 = C/10\pi$. Then $T = 5\tau_A$ and $k = 2\pi/L$ so that there is one wavelength across the cubical cell and $kv_o T = 1$.

The magnitude of the computation can be seen from the seven million words in the code, requiring a Cray-2 computer. With $N = 64$ grid points in each direction the code can resolve Fourier components up to $\frac{1}{3}N = 21$. Thus with wave numbers $2\pi m/L$ and $2\pi n/L$ in the x- and y-directions, it follows that $-21 \leqslant mn \leqslant +21$.

The computational scheme behaves well through $i = 12$ and thereafter (it was carried to $i = 15$) begins to show jitter indicating the accumulation of truncation error. The results recover the earlier smaller simulation of Van Ballegooijen (1988) and go on to show the approximate exponential growth of the peak current density once the deformation gets underway (the initial current density is zero, of course). The growth is approximately $\exp(0.1i)$ at ith step in the sequence.

Figure 6.20 is reproduced from Mikic, Schnack, and Van Hoven (1989), showing the configuration of 16 field lines (all initially straight and uniformly spaced) after the twelfth step. It is evident that the deformation of the field is substantial, but it has not progressed to a state that could be described as strongly wrapped and wound. This makes the concentration of the current density at $z = \frac{1}{2}L$ shown in Fig. 6.21 from Mikic, Schnack, and Van Hoven (1986) to be all the more impressive. The rms current $\langle j_z^2 \rangle^{\frac{1}{2}}$ has increased by a factor of about $e = 2.73$ from $i = 2$ to $i = 12$, and the concentration toward thin sheets (large wave numbers) is conspicuous. They have continued to upgrade their numerical code and a recent simulation has been successfully carried out with a magnetic Reynolds number of the order of 10^4, providing a vivid illustration of the rapid development of thin current sheets.

Van Ballegooijen (1988) and Mikic, Schnack, and Van Hoven (1989) refer to the concentration of current toward thin filaments as a "cascade toward high wave numbers," in the spirit of turbulence theory where there is a continual dynamical transfer (cascade) of energy to large wave numbers. They view their results in that perspective rather than as an approach to tangential discontinuities in static equilibrium. It seems to be a different vocabulary to describe the same effect, because the tangential discontinuity is the formal end point of the "cascade" to large wave numbers.

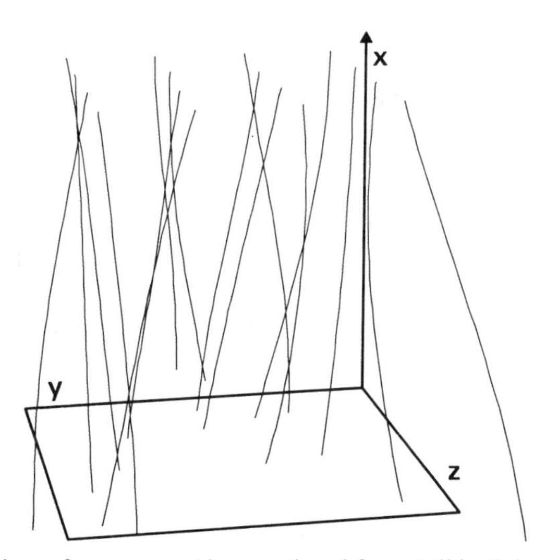

Fig. 6.20. The field lines after step $i = 12$, reproduced from Mikic, Schnack, and Van Hoven (1989) with the kind permission of the authors.

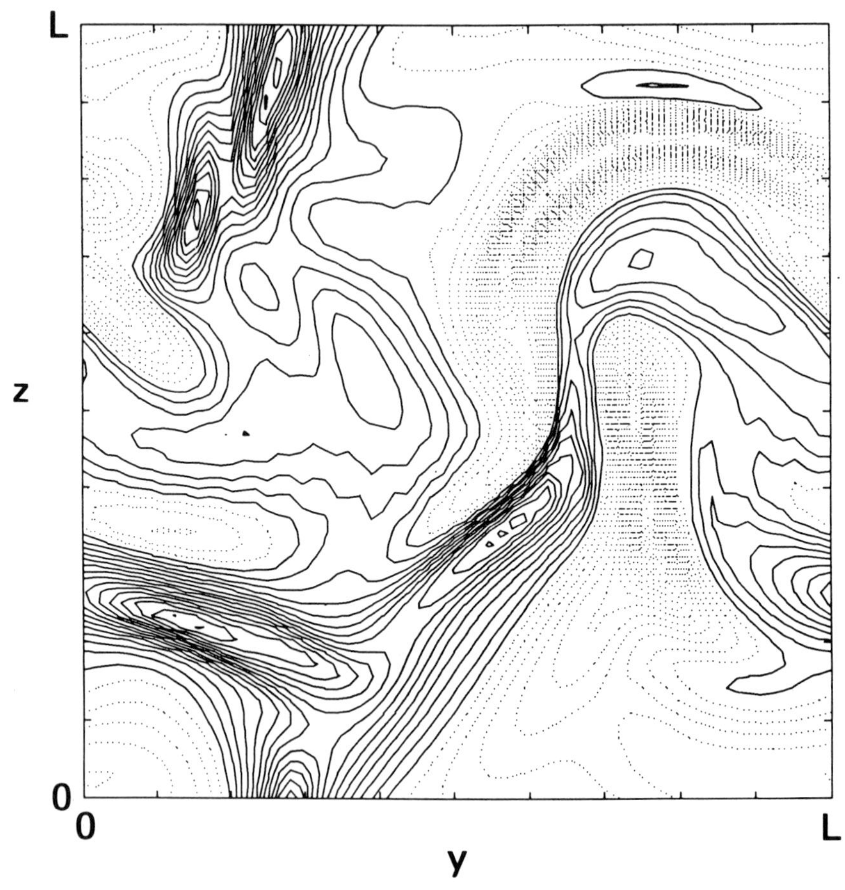

Fig. 6.21. The contours of constant current density after step $i = 12$, reproduced from Mikic, Schnack, and Van Hoven (1989), with the kind permission of the authors.

6.10 Discontinuity Through Instability

The foregoing examples illustrate the formation of one or more internal surfaces of tangential discontinuity by the continuous deformation of an equilibrium magnetic field. The present section takes up a variation of this general effect, treating a field that is continuous in magnetostatic equilibrium, but the equilibrium is an unstable one. That is to say, the equilibrium is not the lowest energy state available and in seeking a lower energy state the field becomes discontinuous. As a simple example we imagine a uniform field B_0 extending from $z = 0$ to $z = L$. Then a region of radius a about the z-axis is subject to twisting, maintaining strict rotational symmetry about the z-axis, so that the field remains in equilibrium during the rotation of the footpoints,

$$v_\varphi = v(\varpi/a)\exp(-\varpi^2/a^2),$$

at $z = L$, while $v_\varphi = 0$ at $z = 0$. The field remains continuous as the twisting proceeds, but the twisted region becomes increasing unstable as the winding accumulates.

Suppose, then, that the flux in the vicinity of the z-axis becomes very strongly twisted after a time, so that the safety factor $\varpi B_z/RB_\varphi(\varpi)$ is substantially less than one. Then turn off v_φ and perturb the system slightly, so that the unstable twisted field develops one or more strong kinks, sketched in Fig. 6.22, much as a tightly twisted rubber band kinks when insufficient tension is maintained to prevent it from doing so. The loops and kinks of the stronger field push against each other as well as out into the surrounding field, producing surfaces of strong tangential discontinuity. In fact only a weak kinking is sufficient to produce discontinuities, as we shall see.

An extreme example is provided by Low (1986b) who shows that shearing the footpoints of a bipolar lobe of force-free field beyond a certain limit may cause the lobe to thrust outward if there is nothing to impede its expansion, forming a current sheet between the two extended legs. In general, then, the expanding lobe collides with other unrelated fields in the surrounding space, forming tangential discontinuities at the surfaces of contact.

A particularly simple example is noted at the end of §6.12, where it is pointed out that a close packed rectangular array of alternately twisted flux bundles is unstable to slipping into a close packed hexagonal array. All vertices in a hexagonal array are three-way (Y-type) and the unavoidable contact between flux bundles of the same torsion produces surfaces of tangential discontinuity at each vertex.

Rosenbluth, Dagazian, and Rutherford (1973) consider the onset of the $m = 1$ magnetohydrodynamic kink instability in a cylindrical geometry (ϖ, φ, z) in terms of the safety factor $\varpi B_z/RB_\varphi(\varpi)$ of the helical field. The mode is unstable where the safety factor is less than one, with the boundary $\varpi = R$ fixed or free. The linear analysis assumes a radial displacement of the form $\xi(\varpi)\cos(m\varphi + k_z z)$ and exhibits the resonant surface $\varpi = s$ at which $\mathbf{k} \cdot \mathbf{B} = k_z B_z + B_\varphi/\varpi = 0$, with $\xi(\varpi) = \xi_0$ for $\varpi < s$ and $\xi(\varpi) = 0$ for $\varpi > s$. The discontinuity in the displacement at the resonant surface carries over into the nonlinear regime, providing a jump in B_φ across $\varpi = 0$ so that there is a current sheet at the singular surface $\varpi = s$. The helical field in $\varpi < s$ presses against one side $(\varphi = \varphi_0)$ of $\varpi = s$ and pulls away from the opposite side $(\varphi = \varphi_0 + \pi)$, creating an O-type neutral point and an X-type neutral point, respectively, when resistivity is introduced. The essential point here is that even a very small spiralling produces surfaces of weak discontinuity.

Park, Monticello, and White (1984) develop a numerical simulation of the $m = 1$ resistive internal kink instability to study the rate of reconnection of field in the presence of a small resistivity. The current sheet shows up strongly, dying away only when finally dissipated by resistive reconnection of the field. They perform a similar simulation of the $m = 2$ tearing mode, with only a weak current spike

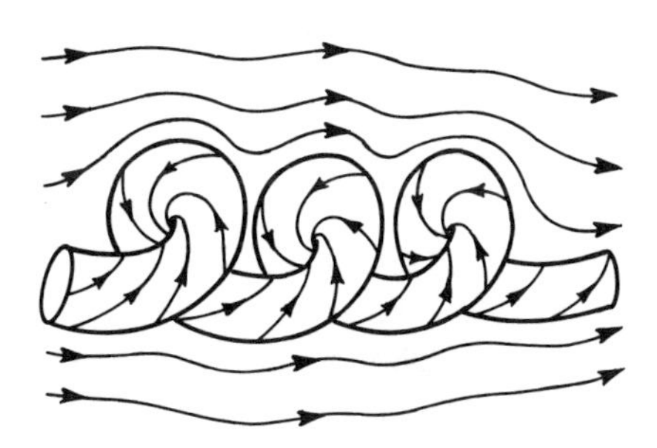

Fig. 6.22. A sketch of a strongly twisted and kinked flux bundle embedded in an ambient field.

appearing in the radial profile. Evidently it is the $m = 1$ mode that produces the principal tangential discontinuities.

Strauss and Otani (1988) examine the $m = 1$ kink instability in a field extending between infinitely conducting end plates at $z = 0$ and $z = L$. The field lines are then twisted by moving the footpoints in circles at $z = L$ while holding the footpoints fixed at $z = 0$. The twisting is over a transverse scale ℓ small compared to L, and the transverse field components (B_x, B_y) are small compared to $B_z \cong B_0$, even after many revolutions of the footpoints, because $L \gg \ell$. They consider the dynamical stability of the resulting field as a function of line twist or rotation transform. In this approximation the dynamical equations reduce to (Strauss, 1976)

$$B_x = +\partial A/\partial y,\ B_y = -\partial A/\partial x,\ B_z = B_0,$$

$$v_x = +\partial U/\partial y,\ v_y = -\partial U/\partial x,\ v_z = 0,$$

with

$$\frac{\partial A}{\partial t} = \frac{\partial(U, A)}{\partial(x, y)} + \frac{\partial U}{\partial z} + \eta\nabla_\perp^2 A,$$

$$\frac{\partial}{\partial t}\nabla_\perp^2 U = \frac{\partial(U, \nabla_\perp^2 A)}{\partial(x, y)} + \frac{\partial(\nabla_\perp^2 A, A)}{\partial(x, y)} + \frac{\partial}{\partial z}\nabla_\perp^2 A + \nu\nabla_\perp^4 U,$$

where $\partial(f, g)/\partial(x, y) \equiv \nabla f \times \nabla g \cdot \mathbf{e}_z$ and $\nabla_\perp^2$ is the Laplacian in x and y, with $\partial^2/\partial z^2$ small in comparison. The resistive diffusion coefficient η and the kinematic viscosity ν provide dissipation and diffusion.

The field is anchored in the end plates, where $\mathbf{v}$ is prescribed for the purpose of twisting the field. The axis of the twisting is taken to lie along the z-axis and a cylindrical boundary at $\varpi = R$ is imposed, fixing the field there with a time-independent radial component admitted to accommodate noncircular flux surfaces. The most interesting case treated by Strauss and Otani involves noncircular twisting of the field, with a radial component at $\varpi = R$ of the form $\cos 2\theta$. The elliptical contours of constant A and $\nabla_\perp^2 A$ are shown in Fig. 6.23 from their paper. The rotational transform, defined as

$$\frac{1}{q} = \frac{L}{4\pi^2\varpi}\int_0^{2\pi} d\varphi \frac{B_\varphi(\varpi, \varphi)}{B_0},$$

is a function of ϖ. This rotational transform represents the number of rotations around the z-axis of the average azimuthal field at a fixed radius ϖ, with the average field defined as

$$\langle B_\varphi \rangle = \frac{1}{2\pi}\int_0^{2\pi} d\varphi B_\varphi(\varpi, \varphi).$$

Thus the mean pitch angle θ between the field and the z-direction is given by

$$\tan\theta = \langle B_\varphi \rangle / B_0.$$

The field line makes one revolution around the z-axis in a distance $2\pi\varpi\cot\theta$, so that $1/q$ represents the number of revolutions in a distance L at a radius ϖ. The variation of $1/q$ with ϖ is shown in Fig. 6.24, reproduced from Strauss and Otani (1988), declining from a value $1/q_0 = 12$ on the axis. Instability occurs for $1/q_0 \gtrsim 9$, which depends on

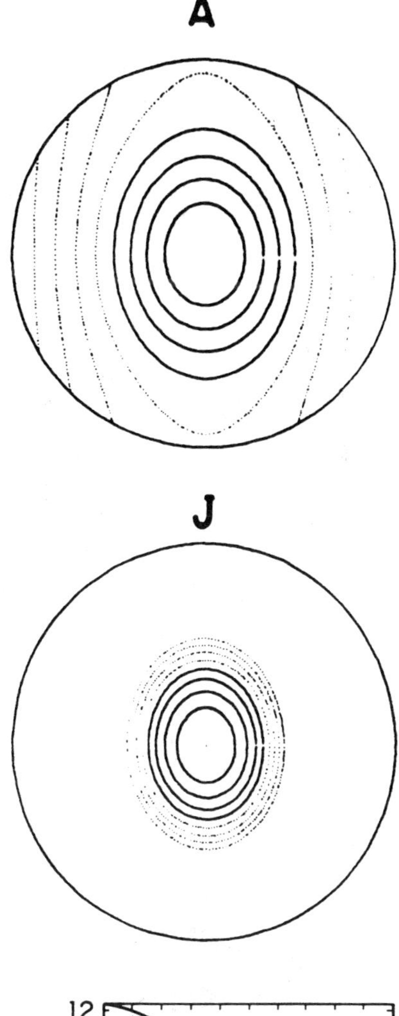

Fig. 6.23. The contours of A and $\nabla_{\perp}^2 A$ for the 2D equilibrium providing the dynamical instability, reproduced from Strauss and Otani (1988) with the kind permission of the authors.

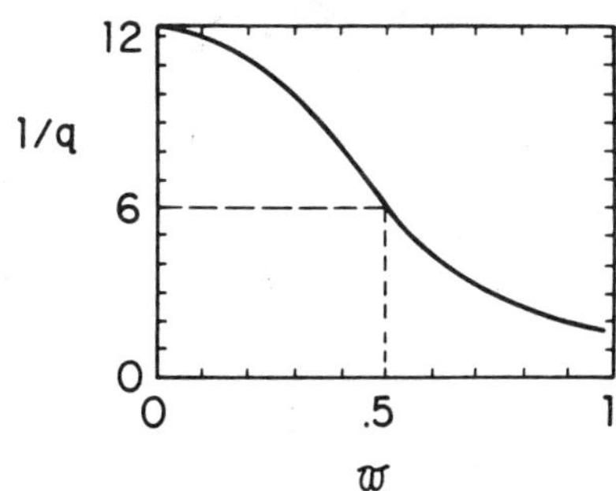

Fig. 6.24. The rotation transform $1/q$ employed in the calculations, reproduced from Strauss and Otani (1988) with the kind permission of the authors.

the ellipticity of the flux surfaces shown in Fig. 6.23, of course, declining with increasing eccentricity. The characteristic growth time of the instability for $1/q_0 = 12$ is approximately 10^{-1} of the Alfven transit time over the length L of the field.

Strauss and Otani integrate the dynamical equations numerically, following the instability into the nonlinear regime. Current sheets develop and their thickness is limited by the resistive diffusivity η. Figure 6.25, from Strauss and Otani (1988), shows the contours of constant $\nabla_{\perp}^2 A$, i.e., the current density or torsion in the field at $z = 0, \frac{1}{4}L, \frac{1}{2}L, \frac{3}{4}L$ from the numerical calculation at the time when the instability has reached its maximum kinetic energy. The current sheets are conspicuous in Fig. 6.25, while maintaining finite thickness as a consequence of diffusion. The basic result of the calculation is a detailed illustration of production of current sheets by the $m = 1$ kink-ballooning instability, indicated by earlier calculations already cited.

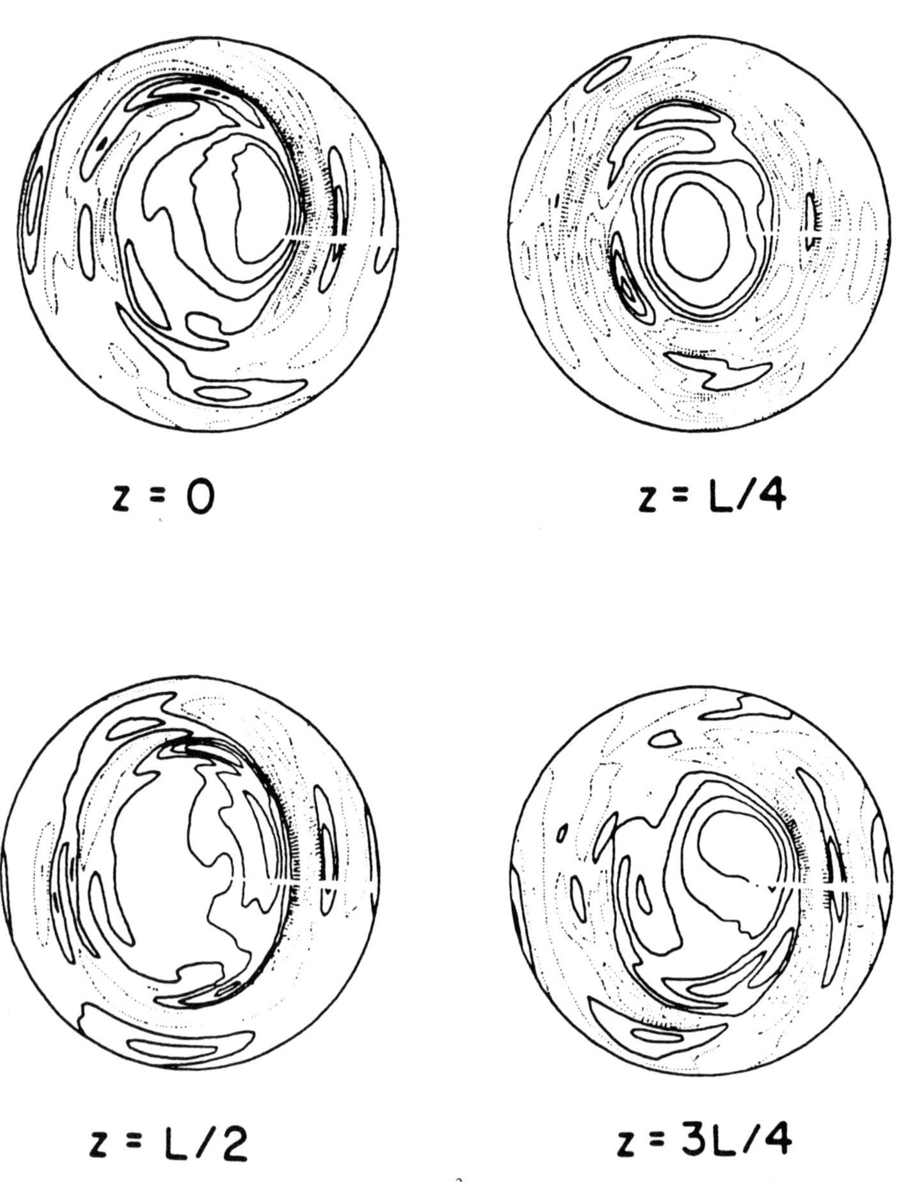

Fig. 6.25. A plot of the contours of constant $\nabla_{\perp}^2 A$, i.e., current density, at the time the kinetic energy of the instability has reached its maximum, reproduced from Strauss and Otani (1988) with the kind permission of the authors.

6.11 Discontinuities Between Twisted Flux Bundles

A flux bundle increases its radius as it is increasingly twisted, with the result that it pushes outward into other regions where it may come into contact with other flux bundles, forming a surface of tangential discontinuity where they meet (Parker, 1990) even without kinking. This simple phenomenon is not widely appreciated, so the present section provides a formal example to illustrate the expansion with increasing twisting. As a case in point consider a regular hexagonal array of twisted flux bundles, created from an initially uniform field B_0 extending in the z-direction from $z = 0$ to $z = L$. The twisted flux bundles come into being when the fluid (with infinite electrical conductivity) is locally rotated out to a radius b about the lines (x_{mn}, y_{mn}) in the z-direction, which become the axes of the twisted flux bundles. The cross section $z = constant$ is sketched in Fig. 6.26(a) for a regular hexagonal array, for which $y_{mn} = 3^{\frac{1}{2}}na$ with $x_{mn} = 2ma$ when n is even, and $(2m + 1)a$ when n is odd. The distance between the axes of nearest neighbors is $2a$. Specifically at time $t = 0$, the fluid throughout $0 \leqslant z \leqslant L$ is set into the circular motion

$$v_{\varphi}(\varpi) = \omega kzf(\varpi)$$

about each axis (x_{mn}, y_{mn}) where ϖ represents distance from the axis (x_{mn}, y_{mn}) of each region,

$$\varpi = \left[(x - x_m)^2 + (y - y_m)^2\right]^{\frac{1}{2}}.$$

For rigid rotation out to a radius b, put $f(\varpi) = \varpi \exp[-(\varpi/b)^{2q}]$ where q is an integer large compared to one so that $f(\varpi)$ cuts off rapidly as ϖ increases beyond the radius b. The result of the rigid rotation is the magnetic field (see equations (3.7) and (3.11)),

$$B_{\varpi} = 0,\ B_{\varphi} = B_0\, k\varpi\tau f(\varpi),\ B_z = B_0,$$

after a time τ when the fluid motion is abruptly halted. The resulting twisted flux bundles are not in hydrostatic equilibrium, so when the field and fluid are released from the constrained motion v_{φ} of the fluid, the flux bundles expand with their outer radius increasing from b to R. If $R \geqslant a$, then the flux bundles press together, as sketched in Fig. 6.26(b), forming tangential discontinuities where neighboring flux bundles meet. We calculate, in the paragraphs below, the degree of twisting of the individual flux bundle that is necessary and sufficient to provide contact between nearest neighbors. That depends, of course, on what happens to the initial field B_0 in the interstices when the twisted flux bundles expand and the interstitial area is decreased. In the simplest case, the interstices are compressed by the expanding flux bundles so that B_z is increased above the initial value B_0. The enhanced magnetic pressure opposes the expansion of the twisted flux bundles. On the other hand, one can imagine that, if only a few twisted flux bundles are present, instead of an infinite array, then the interstitial field is squeezed out of the region so that the interstitial magnetic pressure remains constant. Both cases are treated.

It should be noted that if the individual flux bundles are strongly twisted with $kL\varpi\tau > 2\pi$, the individual flux bundles are subject to the kink instability, with the axis of each bundle becoming helical, the more so the stronger the twist. For sufficient twisting, depending upon the ratio a/b, the kinking spiralling flux bundles

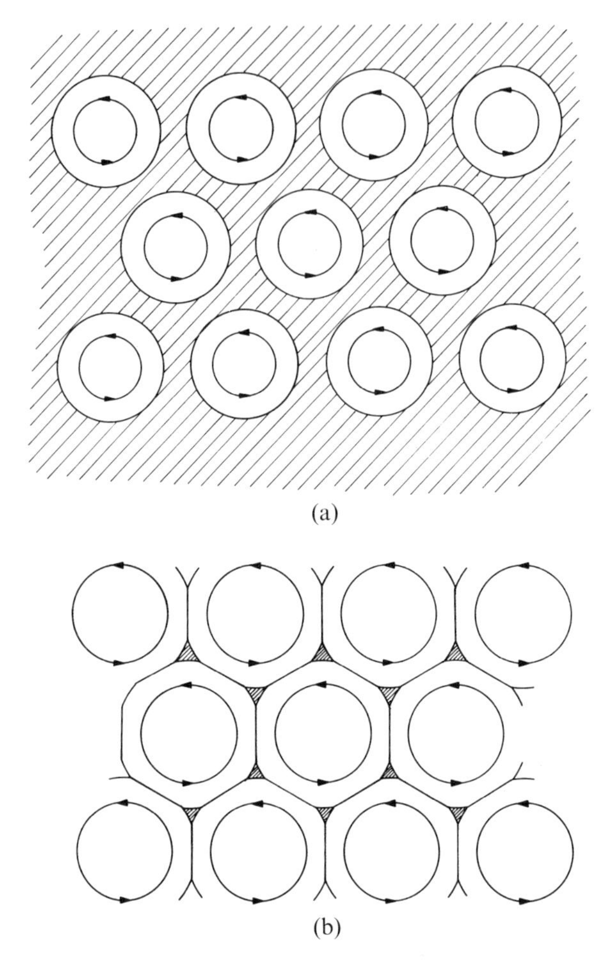

Fig. 6.26. (a) Schematic representation of the cross-section (z = constant) of the hexagonal array of isolated twisted flux bundle when $R < a$. The hatching represents the interstice where the field is not twisted. (b) Further twisting causes the bundles to expand so that they come into contact and are squashed out of the rotational symmetry. Note the extreme compression of the interstices.

make large excursions and come into contact with each other at various points, where they create a tangential discontinuity (see §6.9). Further, they become misaligned with respect to the interstitial field, which has no twist, so that their entire surface becomes a surface of tangential discontinuity. Some of the numerical experiments described in §6.10 show detailed examples of this. But even without kinking, the neighborhood flux bundles expand so as to come into contact to form discontinuities. So suppose that in the perfect world of mathematics, there are no perturbations to excite the kink instability, and the twisted flux bundles remain straight as they relax to magnetostatic equilibrium. In that case the twisted flux bundle maintains its rotational symmetry (Vainshtein and Parker, 1986), and the flux surface initially at the radius ϖ expands to a radius $\Pi(\varpi)$. The initial field $(0, B_0 k\omega\tau\varpi, B_0)$ in $\varpi \leqslant b$ becomes $[0, B_\varphi(\Pi), B_z(\Pi)]$ in $\Pi < R$, where R is the outer boundary of the twisted flux bundle. The footpoints of the field are fixed in the end plates $z = 0, L$ and do not expand, so the individual twisted flux bundle is constricted at each end, with the constriction extending a characteristic distance $1/b$ from each end. Far from either end, throughout the broad in interior of $0 < z < L$ $(L \gg a, b)$, the flux bundle is uniform $(\partial/\partial z = 0)$, so that the equilibrium can be described by a suitable generating function

$$F(\Pi) = B_\varphi^2(\Pi) + B_z^2(\Pi),$$

with

$$B_z^2(\Pi) = F(\Pi) + \frac{1}{2}\Pi F'(\Pi),$$

$$B_\varphi^2(\Pi) = -\frac{1}{2}\Pi F'(\Pi).$$

Conservation of longitudinal magnetic flux in the annulus $(\varpi, \varpi + d\varpi)$ that maps into $(\Pi, \Pi + d\Pi)$ requires that

$$B_0 \varpi d\varpi = B_z(\Pi)\Pi d\Pi. \tag{6.104}$$

Conservation of azimuthal flux per unit length requires that

$$B_0\, k\omega\tau\varpi d\varpi = B_\varphi(\Pi) d\Pi. \tag{6.105}$$

The ratio of these two expressions can be written

$$B_\varphi(\Pi) = \omega\tau k \Pi B_z(\Pi).$$

Squaring both sides and introducing the generating function $F(\Pi)$ leads to differential equation

$$F'(\Pi)(1 + Q^2\Pi^2) = -2Q^2\Pi F(\Pi),$$

where $Q = \omega\tau k$ is a wave number proportional to the degree of twisting, representing the angle of rotation per unit length along the flux bundle. The integral of the differential equation is

$$F(\Pi) = \frac{B_z^2(0)}{1 + Q^2\,\Pi^2}, \tag{6.106}$$

and

$$B_\varphi(\Pi) = \frac{Q\Pi B_z(0)}{1 + Q^2\,\Pi^2}, \tag{6.107}$$

$$B_z(\Pi) = \frac{B_z(0)}{1 + Q^2\,\Pi^2}. \tag{6.108}$$

The magnetic pressure $B^2/8\pi$ is continuous across the outer surface of the region of twisting, $\Pi = R$, in the presence of a uniform fluid pressure. Hence if the field is $B_z = B_2$ in the interstitial region $(\varpi > R)$, it follows that $F(R) = B_z^2$, or

$$B_z^2(0) = B_2^2(1 + Q^2R^2) \tag{6.109}$$

The total longitudinal flux, as well as the azimuthal flux per unit length, are conserved within the twisted bundle, requiring that

$$B_0\, Q^2b^2 = B_z(0) \ln(1 + Q^2R^2) \tag{6.110}$$

in both cases.

Now the initial interstitial field was B_0, filling the initial interstitial area $(2 \times 3^{\frac{1}{2}}a^2 - \pi b^2)$ per flux bundle. As the flux bundles are twisted, expanding to a radius $R > b$, the interstitial area decreases to $(2 \times 3^{\frac{1}{2}}a^2 - \pi R^2)$ per bundle, compressing the longitudinal field to

$$B_z = B_0 \frac{\nu a^2 - b^2}{\nu a^2 - R^2}, \tag{6.111}$$

where ν is the number $2 \times 3^{\frac{1}{2}}/\pi = 1.10266$. Eliminating $B_z(0)$ and B_z from equations (6.109)–(6.111) the result can be written

$$Q^4 b^4 = \left(\frac{\nu a^2 - b^2}{\nu a^2 - R^2}\right)^2 (1 + Q^2R^2)[\ln(1 + Q^2R^2)]^2. \tag{6.112}$$

It is evident by inspection that in the limit of large twist, i.e., large Q, the left-hand side increases as Q^4 while the right-hand side increases only as $Q^2R^2 \ln Q^2R^2$. Hence R increases to maintain the equality, but of course, when R becomes equal to a, neighboring twisted flux bundles come into contact, forming a current sheet where they touch because of the opposite sense of B_φ on either side of the line of contact. If $R > a$, then the bundles squash together, in the manner sketched in Fig. 6.26(b) and the present analysis, based on rotational symmetry, no longer provides an adequate description.

It should be noted that the thin smooth transition layer from rigid rotation in $\varpi < b$ to no rotation beyond b remains in place so long as $R < a$, so that the field is everywhere continuous. However, when neighboring bundles come into contact this thin layer, of characteristic thickness $b/2q$, is squeezed, so that it is subject to the surface rotation effect, described in §§6.2, 7.3 and 7.9. The surface of tangential discontinuity forms at the contact surface between the two transition layers of continuous flux bundles. The transition layers in no way cushion or soften the mathematical discontinuity across which B_φ reverses sign.

Putting $R = a$, it follows from equation (6.112) that contact occurs when

$$\Upsilon(Qa) = \frac{b^4}{a^4}\left(\frac{\nu - 1}{\nu - b^2/a^2}\right)^2, \tag{6.113}$$

where Υ is the function

$$\Upsilon(x) = (1 + x^2)[\ln(1 + x)]^2/x^4, \tag{6.114}$$

$$\cong 1 - x^4/12 + x^6/12 + 0(x^8), \tag{6.115}$$

$$\sim (\ln x^2)^2/x^2 + [(\ln x^2)^2 + 2\ln x^2]/x^4 + \dots. \tag{6.116}$$

Solving equation (6.113) for b^2/a^2 results in the relation

$$\frac{b^2}{a^2} = \frac{\nu \Upsilon(Qa)^{\frac{1}{2}}}{\nu - 1 + \gamma(Qa)^{\frac{1}{2}}}, \tag{6.117}$$

$$\cong 1 - \frac{\nu - 1}{24\nu}(Qa)^4 + O^6(Qa), \tag{6.118}$$

$$\sim \frac{2\nu}{\nu - 1}\frac{\ln Qa}{Qa}\left[1 - \frac{2\ln Qa}{\nu Qa} + O^2\left(\frac{\ln Qa}{Qa}\right)\right]. \tag{6.119}$$

The quantity Qa is plotted in Fig. 6.27 as a function of b/a. As an example, an initial radius $b = 0.949a$ expands to make contact ($R = a$) when Qa reaches 10. An initial radius $b = 0.8a$ expands to make contact when $Qa \cong 60$, and $b = 0.52a$ when $Qa \cong 500$.

Consider the form $\Pi(\varpi)$ of the expansion. Use equation (6.106) to replace $B_z(\Pi)$ in equation (6.108), with $B_z(0)$ given by equation (6.110), to integrate equation (6.104). The result is the relation

$$\frac{\varpi^2}{b^2} = \frac{\ln(1 + Q^2\Pi^2)}{\ln(1 + Q^2R^2)} \tag{6.120}$$

between the initial radius $\varpi(\leqslant b)$ and the final radius $\Pi(\leqslant R)$ of a given flux surface. Note that for weak twist ($QR \ll 1$),

$$\varpi/b \cong \Pi/R, \tag{6.121}$$

and the expansion is uniform across the flux bundle.

In the limit of strong twisting ($Qb \gg 1$) the bundle contracts at small radius ($Q\Pi \ll 1$), with

$$\begin{aligned} \frac{\Pi^2}{\varpi^2} &\cong \frac{\ln(1 + Q^2R^2)}{Q^2b^2} \\ &\cong \frac{(\nu^2a^2 - R^2)}{(\nu^2a^2 - b^2)(1 + Q^2R^2)^{\frac{1}{2}}}, \end{aligned} \tag{6.122}$$

where the second expression follows with the aid of equation (6.112). Since $b < R$, the right-hand side is obviously less than one. On the other hand, at larger radius, the flux bundle expands outward to provide $R > b$.

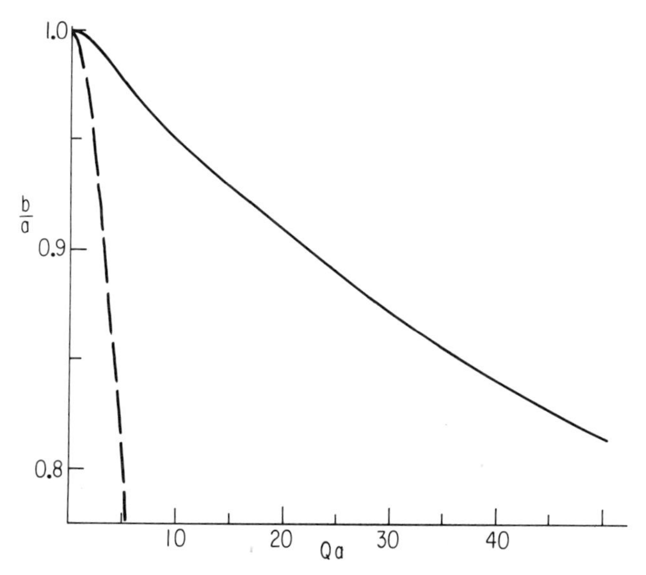

Fig. 6.27. The solid line is a plot of the twisting $Qa = ka\omega\tau$ required to expand a flux bundle with initial radius b to a radius a such that it makes contact with its nearest neighbors, from equation (6.117). The dashed line is a plot of the same from equation (6.124).

The essential point is that flux bundles expand when twisted, bringing otherwise separate bundles into contact to produce tangential discontinuities where they touch. If Q is increased beyond the value (for a given b/a) shown in Fig. 6.27, the bundles flatten against each other, as sketched in Fig. 6.26(b).

The enormous compression of the interstitial field should be noted when b is significantly less than a. The increased interstitial pressure strongly inhibits the expansion of the flux bundle, so that the large values of Qa indicated in Fig. 6.27 are required, even when b is only a little smaller than a. This is, of course, a consequence of working with an infinite array, equivalent to enclosing the region with rigid boundaries. The simple situation in which the interstitial pressure $B_0^2/8\pi$ remains constant is probably closer to the conditions in stellar and galactic magnetic fields. So consider that case briefly, putting $B_z = B_0$ in equation (6.109) and

$$B_z(0) = B_0(1 + Q^2R^2)^{\frac{1}{2}}$$

in place of equation (6.111). Then equation (6.112) reduces to

$$Q^4b^4 = (1 + Q^2R^2)[\ln(1 + Q^2R^2)]^2, \tag{6.123}$$

with R reaching a when

$$\frac{b^4}{a^4} = \frac{(1 + Q^2a^2)[\ln(1 + Q^2a^2)]^2}{Q^4a^4}, \tag{6.124}$$

in place of equation (6.113). The value of Qa providing contact for a given b/a is plotted by the dashed curve in Fig. 6.27. The expansion is much stronger, with contact for $b/a \cong 0.46$ at $Qa = 10$, instead of the value $b/a = 0.949$ when the expansion of the bundle compresses the interstitial field. This simple case shows more clearly the expansion of an individual tube. Relatively little twisting expands the flux bundle so that it pushes its way out into new territory where it presses against other flux bundles to produce current sheets.

6.12 Physical Construction of Continuous Fields

The foregoing examples illustrate the production of surfaces of tangential discontinuity by quasi-static deformation of magnetostatic fields that are initially continuous. The examples show that even if a field is continuous, almost every continuous deformation produces discontinuities.

There is another approach to the problem, presented in this section, which starts from a somewhat different premise but leads to the same conclusion. And that is the improbability of ever constructing an initially continuous field in the first place. For recall that in the real world a magnetic configuration is constructed by physical manipulation of the "infinitely conducting" fluid in which the magnetic field is embedded. The construction of the continuous field shows the unlikelihood that the vertices of four different topological regions can be physically juxtaposed with the mathematical precision that is required to form an X-type neutral point. Any error at all provides two Y-type neutral points with a surface of tangential discontinuity stretching between them.

A number of illustrative examples are provided elsewhere (Parker, 1982, 1983) but one will suffice here (Parker, 1990). Consider the initially uniform field B_0 extending in the z-direction through an infinitely conducting fluid between the boundaries $z = 0$ and $z = L$ in a suitable laboratory experimental setup. Then suppose that the fluid is set in the motion described by the stream function $\psi(x, y)$ in equation (3.7) in §3.3 for a time $t = \tau$. The laboratory apparatus guides ψ so that the field is deformed from the initial uniform state to the final state described by equation (3.11). Hence $\partial/\partial z = 0$ because $\partial\psi/\partial z = 0$. Then hold the footpoints fixed at $z = 0, L$ and release the fluid so that the field relaxes to the lowest available energy state. Imagine that the fluid motion has been guided so that the final field relaxes to the form

$$B_x = +\partial A/\partial y,\ B_y = -\partial A/\partial x,\ B_z^2(A) = B_0^2 + 2k^2A^2, \tag{6.125}$$

with the vector potential given by

$$A = (B_1/k)\sin kx \sin ky \tag{6.126}$$

throughout the interior of the region, i.e., except for a transition layer of characteristic thickness $1/k$ at each end, $z = 0, L$ of the region $(1/k \ll L)$. Note, then, that the final field throughout the interior satisfies the equilibrium equations (4.9)–(4.11) with A restricted by the linear form

$$\nabla^2 A + 2k^2 A = 0. \tag{6.127}$$

The field represents a rectangular array of close packed, alternately twisted flux bundles, sketched in Fig. 6.28. Each bundle is bounded by the field lines $A = 0$. The field is without discontinuities.

Now, as a matter of fact, this ideal field configuration is only approximately achieved, even by the most carefully controlled laboratory manipulation of the fluid, because there is always some slight error in the physical manipulation. The errors involve all forms, from $\partial/\partial z \neq 0$, to imprecise location of the topological separatrices $kx, ky = \pm n\pi\ (n = 0, 1, 2, \ldots)$, to a small mixing of field lines across the separatrices, to slightly more twisting in some domains (flux bundles) than in others, etc. Each of these many small errors contributes to the mismatch of neighboring flux bundles and to their improper positioning, with the result that the precise positioning of the four identical vertices at each X-type neutral point is not achievable.

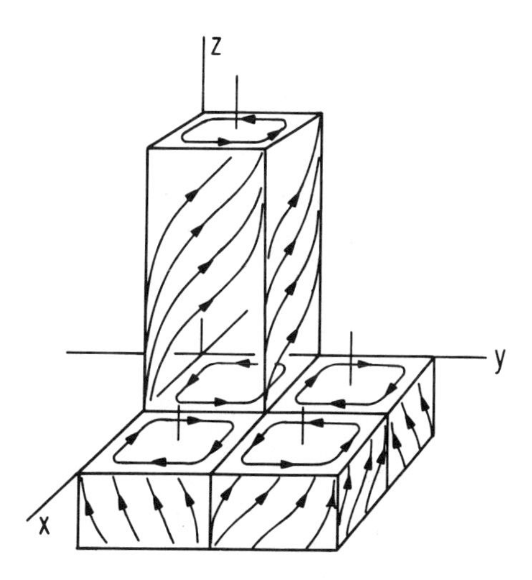

Fig. 6.28. A sketch of the ideal rectangular array of close packed alternately twisted flux bundles, described by the vector potential given in equation (6.104).

Consider, for instance, the single error that the twisting of the individual flux bundles (topological domains) is biased slightly with the right-hand helical fields twisted more than the left-hand fields. The result of the error is that the right-hand bundles are slightly expanded compared to the left-hand bundles, so that they do not pack together to form X-type neutral points at their vertices. Figure 6.29 is a sketch of the result, with the ideal continuous configuration sketched in (a) and the actual result shown in (b).

To treat this more precisely, note that the torsion coefficient α of the ideal field of equation (6.106) is

$$\begin{aligned}\alpha &= B_z'(A) \\ &= 2k^2 A / B_z^2(A) \\ &= 2kB_1 \sin kx \sin ky / B_z^2(A),\end{aligned} \tag{6.128}$$

which vanishes at the separtrices kx, $ky = 0$, i.e., at $A = 0$. Suppose that the actual manipulation achieves this condition precisely but errs in the manner already described, with too much twisting of each right-handed flux bundle (e.g., $0 < kx$, $ky < \pi$) and too little of each left-handed bundle (e.g., $-\pi < kx < 0$, $0 < ky < \pi$), providing the configuration sketched in Fig. 6.29(b), where the vertices are deformed into two Y-type neutral points.

To treat the deformed vertices analytically, consider the ideal case that the error is introduced equally throughout the array of flux bundles, so that in place of equation (6.125), we have $B_z(A)$ in the functional form

$$B_z^2(A) = B_0^2 + 2k^2 A^2 + \varepsilon f(A), \tag{6.129}$$

where $\varepsilon f(A)$ represents the error. Suppose that $f(A)$ is an analytic function of its argument in the vicinity of $A = 0$, so that it can be expanded about $A = 0$ in ascending powers of A. For if $f(A)$ is not analytic, it provides for discontinuities in the field gradients as well as in the field itself. Now it is clear that the asymmetry between right- and left-handed flux bundles involves terms in odd powers of A. If the lowest order term is A^{2m+1}, write

$$f(A) = f(0) + \frac{1}{(2m+1)!} f^{(2m+1)}(0)\, A^{2m+1} + \dots \tag{6.130}$$

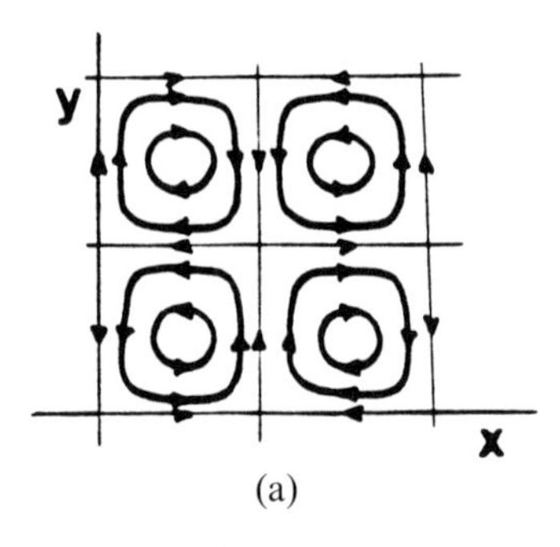

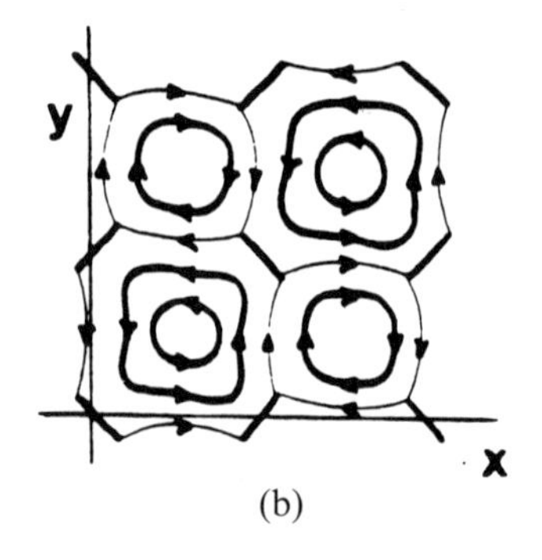

Fig. 6.29. (a) A section $z = constant$ through the rectangular array shown in Fig. 6.28, and (b) the same section with the counterclockwise cells (right-hand helicity) more strongly twisted than the clockwise cells (left-hand helicity). The discontinuities (heavy lines) appear where the corners of the counterclockwise cells are squashed together.

where m is a positive integer. Then

$$\alpha = \left[2k^2 A + (\varepsilon/(2m)!) f^{(2m+1)}(0) A^{2m} + \ldots\right] / B_z^2(A), \tag{6.131}$$

and the asymmetry between right- and left-handed flux bundles (where $A > 0$ in the right-handed domains and $A < 0$ in the left-handed domains) is obvious in the factor A^{2m}. In the gauge for which $A = 0$ on the separatrices, the torsion $\alpha = B_z'(A)$ remains zero there. The essential point is that $B_z'(A)$ falls to zero in proportion to A as the separatrices are approached, so that the Grad–Shafranov equation $\nabla^2 A + B_z'(A) B_z(A) = 0$ reduces to Laplace's equation,

$$\nabla^2 A \cong 0,$$

in some sufficiently small neighborhood of $A = 0$. The field is a potential field and can be represented by the analytic function

$$B_x - iB_y = (B_1/b)\left\{\zeta^2 - \left[b \exp\left(-i\frac{\pi}{4}\right)\right]^2\right\}^{\frac{1}{2}}$$

close to the separatrix, $A = 0$, where $\zeta \equiv x + iy$, and the origin of the xy coordinate system is displaced slightly to lie at the center of the small current sheet. Figure 6.30 is a magnified view of the current sheet into which the vortex at the origin in Fig. 6.29(a) is deformed. The width of the current sheet, or discontinuity, is $2b(kb \ll 1)$ inclined at an angle $\frac{1}{4}\pi$ to the x-axis. The cut (discontinuity) in the complex plane lies along the current sheet between the two points $\pm b(1-i)/2^{\frac{1}{2}}$. The field immediately to the "northeast" of the current sheet is

$$B_x = +(B_1/2^{\frac{1}{2}})(1 - \varpi^2/b^2)^{\frac{1}{2}} = -B_y, \tag{6.132}$$

with the opposite field on the "southwest" side, where $\varpi = (x^2 + y^2)^{\frac{1}{2}}$ represents distance from the origin at the midpoint. The strength of the discontinuity is, then, $2^{\frac{1}{2}} B(1 - \varpi^2/b^2)^{\frac{1}{2}}$. The Y-type neutral point at each end is made up of separatrices with directions differing by $2\pi/3$, illustrated in Fig. 6.30.

This idealized example is intended only to emphasize that a continuous field requires mathematical precision in its construction, whereas in the physical world

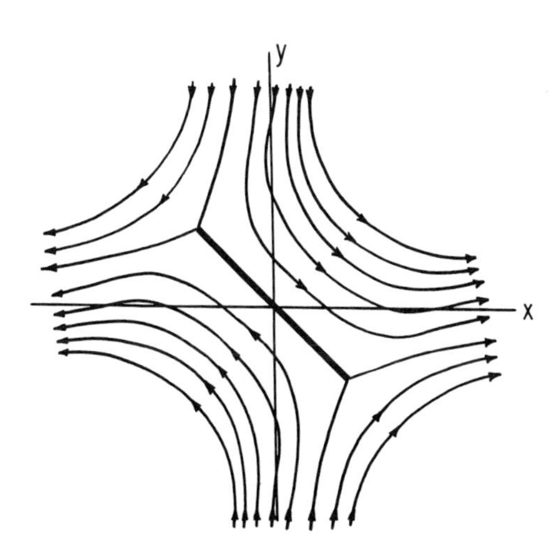

Fig. 6.30. A close up of the discontinuity (heavy line) in the neighborhood of the origin.

that precision is not available in the mechanical deformations that go into forming the topology of the field.

In fact there is a further difficulty in the physical world, besides the one of forming each twisted flux bundle with mathematical precision. And that is the positioning of the individual flux bundles once they are formed. For the simple fact is that the rectangular array of alternately twisted flux bundles is unstable to slipping into a close packed hexagonal array, sketched in Fig. 6.31. The hexagonal array has slightly lower energy because the field is able to expand into the obtuse vertices more readily than into the right-angled vertices of the rectangular array. In a right-angled corner, the field declines as $r \exp i2\varphi$ into the corner, whereas in the obtuse angle $2\pi/3$ of the hexagon the field declines only as $r^{\frac{1}{2}} \exp i3\varphi/2$, with r representing radial distance from the vertex. It follows that all vertices are three-way Y-type vertices, which cannot avoid a tangential discontinuity on at least one of the three separatrices.

In view of these simple considerations, it would appear that separatrices always meet in three–way vertices and never in four-way (X-type) vertices in the physical world where neither precise formation of separate lobes of field occurs, nor the necessary precise positionings of the lobes.

6.13 Time-Dependent Fields

The foregoing examples of the formation of tangential discontinuities have concentrated on equilibrium field configurations. We must not be misled, however, into thinking that the tangential discontinuities are limited to stationary or static fields. For it will be shown in Chapter 7 that, irrespective of time dependence, the component of any vector field $\mathbf{F}(\mathbf{r})$ in the flux surfaces of $\nabla \times \mathbf{F}$ may show discontinuity of one form or another if the magnitude of the component is sufficiently inhomogeneous. Thus, for instance, the formation of vortex sheets in hydrodynamic turbulence is associated with local maxima in the velocity component in the vortex sheet. Similarly the formation of current sheets in magnetohydrodynamic turbulence is associated with the local maxima in the magnetic field. These are geometrical rather than dynamical associations, of course, based on the topological construction called the *optical analogy* (Parker, 1981, 1989a,b, 1991) described in Chapter 7.

The essential point is that numerical simulations of magnetohydrodynamic turbulence provide vivid illustration of the dynamical formation of tangential discontinuities or current sheets by the ejection of fluid from between approaching lobes of field. The

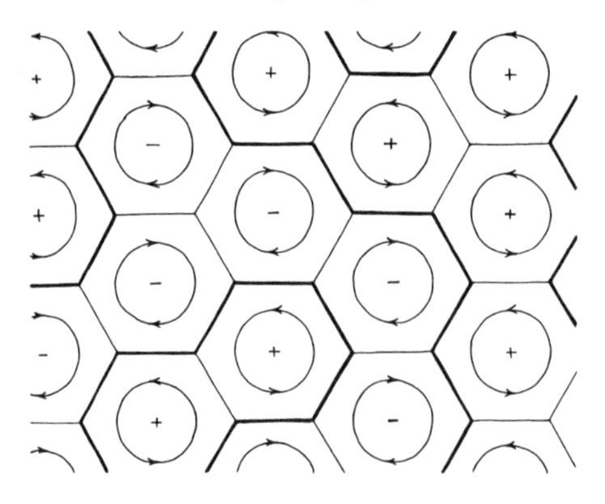

Fig. 6.31. A sketch of the hexagonal close packed array, with slightly lower energy than the rectangular array. The discontinuities are indicated by the heavy lines.

hydrodynamics of the ejection of fluid is explored in Chapter 9, but as an example of the trend toward tangential discontinuities it is appropriate to show here the results of some of the numerical simulations of magnetohydrodynamic turbulence.

There is an extensive literature on numerical simulation of magnetohydrodynamic turbulence, at magnetic Reynolds numbers up to about 10^3 (Hayashi and Sato, 1978; Sato, et al. 1978; Sato and Hayashi, 1979; Cheng, 1980; Kraichnan and Montgomery, 1980; Matthaeus, 1982; Tajima, Brunel, and Sakai, 1982; Biskamp and Welter, 1989). Matthaeus and Montgomery (1980) provide a numerical simulation of the relaxation of an initial 2D magnetic field ($\partial/\partial z = 0$) composed of a close-packed array of twisted flux bundles of irregular size and shape. The initial field is not in equilibrium, so it evolves rapidly into a dynamical state of 2D magnetohydrodynamic turbulence, with a magnetic Reynolds number of about 10^3. The contact between twisted flux bundles with the same sign of torsion provides concentrated current sheets almost immediately. Resistive dissipation is rapid as a consequence of the declining thickness of the current sheets so that the topology of the field evolves continually with the passage of time. The thinning of the current sheets is limited by the dissipation, of course, remaining approximately uniform once the sheets are formed. Figure 6.32 is reproduced from Matthaeus and Montgomery (1980), showing the contours of constant current density. It is evident by inspection of the figure that the characteristic thickness of the current sheets, once formed, remains approximately 0.02 of the dimension of the entire region. The small thickness is maintained in opposition to resistive diffusion by the expulsion of fluid out both ends of each current sheet. Matthaeus and Montgomery (1981), Matthaeus (1982), and Biskamp and Welter (1989) provide a detailed numerical simulation of the jetting of fluid from current sheets, and the hydrodynamics of this phenomenon is taken up in Chapter 9 (Parker, 1982, 1983).

It should be emphasized again that 2D turbulence simulates the formation of tangential discontinuities in a close packed array of parallel twisted flux bundles in the presence of a small resistivity. For the formation of tangential discontinuities cannot be avoided when flux bundles are packed together unless the initial flux bundles are precisely tailored to fit together and then precisely positioned (Parker, 1982, 1983) so that their boundaries form X-type neutral points without Y-type neutral points. Most initial states lack these precise conditions, forming tangential discontinuities. In the absence of infinite electrical conductivity, the discontinuities give way to rapid reconnection of field and jetting of fluid, which is precisely 2D magnetohydrodynamic turbulence.

Riyopoulos, Bondeson, and Montgomery (1982) illustrate the process with a numerical simulation of the relaxation of an axisymmetric ($\partial/\partial\varphi = 0$) 3D field. The initial nonequilibrium field quickly forms current sheets which then exhibit resistive reconnection of the field. The dynamical evolution passes through four distinct phases. The initial field is not in complete static equilibrium, of course, so the first phase represents the immediate adaption of the field to unbalanced Maxwell stresses. This phase is essentially a readjustment in terms of Alfven wave propagation across the region of field. The largest wave numbers are dissipated by resistivity (the Lundquist number is of the order of 10^3) during this first phase, which then evolves into a quiescent phase with fluid velocities v only of the order of $10^{-3}C$, where C is the Alfven speed. The quiescent phase lasts for about 10 Alfven crossing times. By that time the current sheets have thinned down to where resistive reconnection

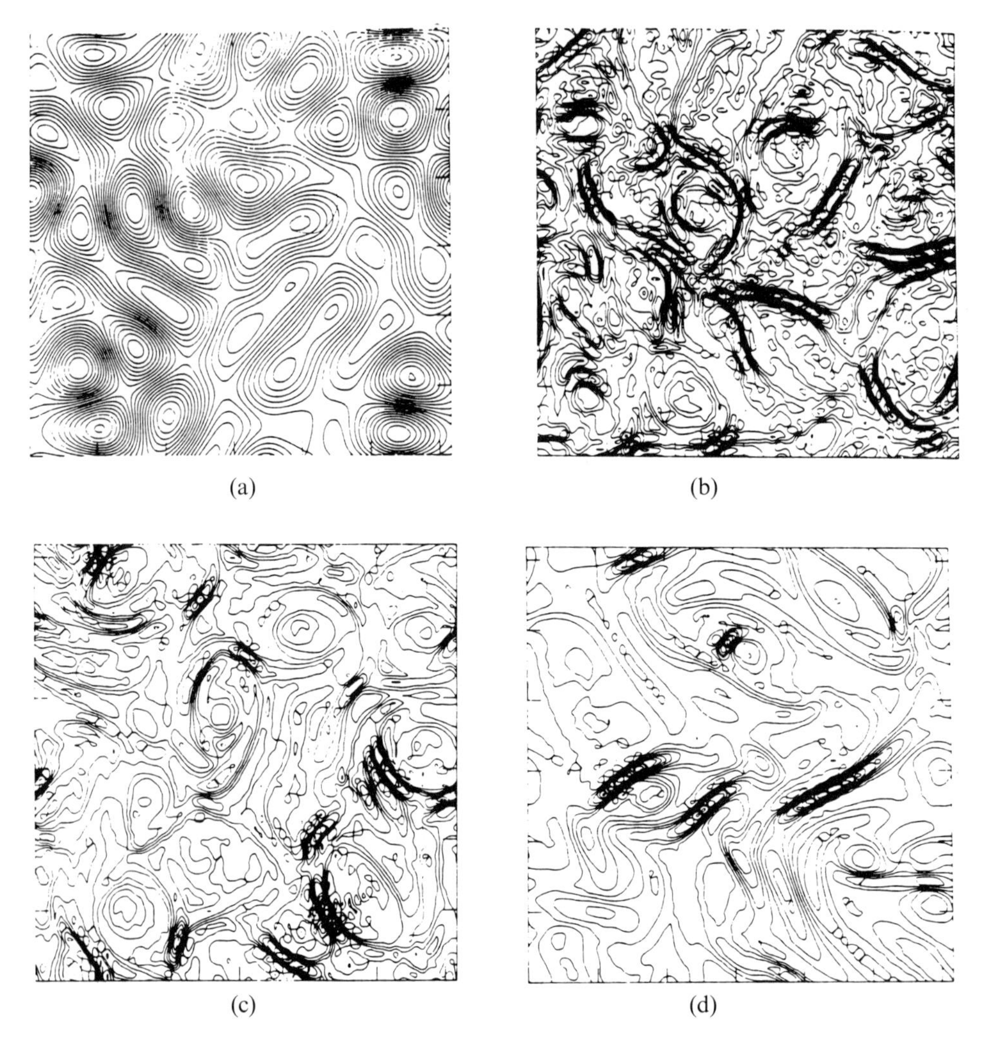

Fig. 6.32. Contours of constant current density at successive times (a) $t = 0$, (b) $t = 500$, (c) $t = 1000$, and (d) $t = 2500$ time steps, reproduced from Matthaeus and Montgomery (1980), Fig. 8, with the kind permission of the authors.

between opposite fields appears, ushering in a simplification of the field topology through coalescence of separate lobes of field (coalescence of islands). The magnetic energy declines rapidly during the reconnection phase, of course, in which fluid velocities are typically 0.5 C. The final state arises when neutral point reconnection has reduced the field topology to two separate regions of opposite rotational sense. A quasi-static equilibrium then prevails with the relatively slow resistive relaxation of these last two large-scale islands.

The essential point of these numerical experiments is the dynamical trend toward formation of thin current sheets in time dependent fields, in association with local maxima in the fields. The spontaneous appearance of discontinuities in hydrodynamics and in magnetohydrodynamics is as potent in time varying fields as in static fields.

7

The Optical Analogy

7.1 The Basic Construction

There is a basic geometrical relation between the magnitude and the pattern of a vector field $\mathbf{F}(\mathbf{r})$ in three, or more dimensions, called the *optical analogy* (Parker, 1981a, 1983, 1989a, 1991). The relation is based on the fact that the field lines of a potential field $-\nabla\phi$ coincide with the optical ray paths computed in an index of refraction $|\nabla\phi|$. The optical analogy is applicable to every potential field and to every well-behaved nonpotential vector field $\mathbf{F}(\mathbf{r})$ with nonvanishing helicity $(\mathbf{F}\cdot\nabla\times\mathbf{F})$ because there are always two of the three or more components of a nonpotential $\mathbf{F}(\mathbf{r})$ that form a potential field. In a variety of special cases, such as the steady flow of an inviscid fluid, or a magnetostatic field, there is a family of flux surfaces of $\mathbf{F}(\mathbf{r})$ in which the entire $\mathbf{F}(\mathbf{r})$ is expressible as the gradient of some scalar function ϕ in the individual flux surface. Indeed, in Beltrami flows and force-free magnetic fields, $\mathbf{F}(\mathbf{r})$ is expressible as $-\nabla\phi$ in every flux surface. The optical analogy provides an effective tool for treating the field topology and the formation of tangential discontinuities. The power of the optical analogy derives from the fact that the magnitude of the field, which determines the field pattern, is often available from relatively simple considerations, e.g., boundary conditions. In particular, the optical analogy shows that any sufficiently localized maximum in the field magnitude $|\nabla\phi|$ is associated with a gap (a birfurcation) in the field pattern, producing a discontinuity in the field, or a discontinuity in the spatial derivatives, or both. For magnetostatic fields and stationary hydrodynamic flows of an ideal fluid the result of a localized maximum in the field is a tangential discontinuity, i.e., a current sheet and a vortex sheet, respectively, in the absence of nearby rigid boundaries. In fact the bifurcation in the field in a flux surface is simply a global view of the squashing of an X-type neutral point to form a current sheet, as originally described by Syrovatskii (1971, 1978, 1981).

Consider, then the fact that the projection $\mathbf{F}_s$ of $\mathbf{F}(\mathbf{r})$ onto any surface locus, or flux surface, S of $\nabla\times\mathbf{F}$ is expressible as $\nabla\phi$ in the surface S (cf. Brand, 1947, p. 225). We begin with the construction of a surface locus, or flux surface, of a vector field $\mathbf{F}(\mathbf{r})$. The surface locus is locally defined by a curve C, noting that any curve C drawn across (i.e., not parallel to) a vector field $\mathbf{F}(\mathbf{r})$ in a 3D space intersects a family of field lines of $\mathbf{F}(\mathbf{r})$ which defines a local flux surface, or surface locus, Σ_c. The same curve C presumably also cuts across the field lines of $\nabla\times\mathbf{F}$ and defines a surface locus, or flux surface, S_c of $\nabla\times\mathbf{F}$. Figure 7.1 is a sketch of a curve C and the local flux surfaces Σ_c and S_c that it defines. The surface loci, or flux surfaces, Σ_c and S_c

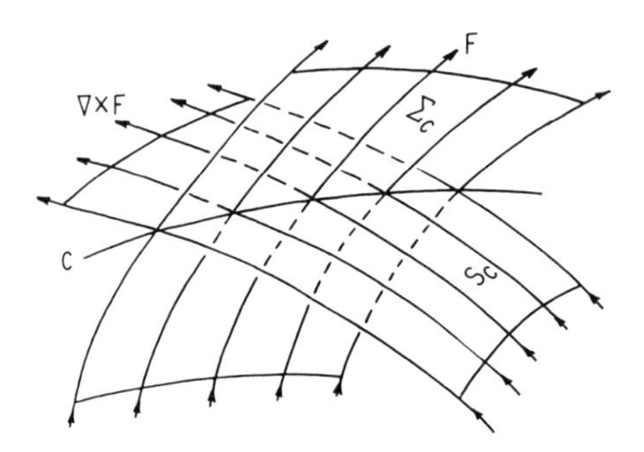

Fig. 7.1. A sketch of the flux surfaces Σ_c and S_c generated in the vector field **F** and in $\nabla \times \mathbf{F}$ by the transverse curve C.

project away from C in both directions along **F** and $\nabla \times \mathbf{F}$, respectively. In a global toroidal topology the field lines, and hence Σ_c and S_c, circle endlessly, and in most cases incommensurably and ergodically, around the toroidal space, filling at least a portion of the space. However, the optical analogy is generally for local application, so the global topology is not a concern. We concentrate on the local surfaces Σ_c and S_c, in the general neighborhood of their generating curve C.

Denote by $\mathbf{F}_s$ the projection of the vector **F** at S_c onto the surface locus S_c of $\nabla \times \mathbf{F}$, sketched in Fig. 7.2. The essential fact is that $\mathbf{F}_s$ in any surface locus S_c is expressible as the gradient of a scalar function ϕ_s of position in the surface S_c. For consider the line integral of $\mathbf{F}_s$ around any closed curve Γ in S_c, applying Stokes' theorem to obtain

$$\oint_\Gamma d\mathbf{s} \cdot \mathbf{F}_s = \int_{S_c} dS(\nabla \times \mathbf{F})_\perp$$

where the surface integral is over S_c interior to Γ, and $(\nabla \times \mathbf{F})_\perp$ is the component of $\nabla \times \mathbf{F}$ perpendicular to S_c. But $(\nabla \times \mathbf{F})_\perp = 0$ everywhere on the surface locus S_c, with the result that $\oint d\mathbf{s} \cdot \mathbf{F}_s = 0$ around every closed contour Γ. Hence $\mathbf{F}_s$ is irrotational and can be written

$$\mathbf{F}_s = -\nabla \phi_s \tag{7.1}$$

in each flux surface S_c of $\nabla \times \mathbf{F}$.

Now in general a surface locus S_c is not a flat space, i.e., it is a non-Euclidean 2D space, so the metric tensor cannot be represented by a constant. Denote by $h_1(\xi_1, \xi_2)$ and $h_2(\xi_1, \xi_2)$ the distance measure of the two orthogonal coordinates (ξ_1, ξ_2), so that length ds in S_c is given by

$$ds^2 = h_1^2\, d\xi_1^2 + h_2^2\, d\xi_2^2. \tag{7.2}$$

Then $\phi_s = \phi_s(\xi_1, \xi_2)$ and

$$F_{s_1} = -\frac{1}{h_1}\frac{\partial \phi_s}{\partial \xi_1}, \; F_{s_2} = -\frac{1}{h_2}\frac{\partial \phi_s}{\partial \xi_2}. \tag{7.3}$$

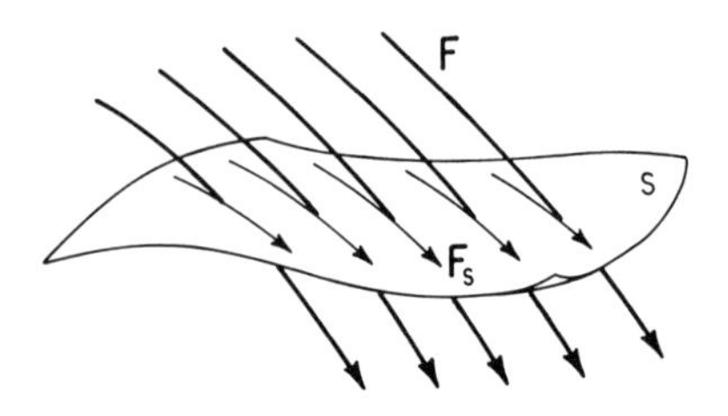

Fig. 7.2. A sketch of the projection $\mathbf{F}_s$ of the vector field **F** on a surface locus S of $\nabla \times \mathbf{F}$.

The field lines of $\mathbf{F}_s$ are described by

$$F_s h_1 \frac{d\xi_1}{ds} = -\frac{1}{h_1}\frac{\partial \phi_s}{\partial \xi_1}, \tag{7.4}$$

$$F_s h_2 \frac{d\xi_2}{ds} = -\frac{1}{h_2}\frac{\partial \phi_s}{\partial \xi_2}. \tag{7.5}$$

The magnitude F_s is given by (dropping the subscript s on the potential ϕ),

$$F_s^2 = \left(\frac{1}{h_1}\frac{\partial \phi}{\partial \xi_1}\right)^2 + \left(\frac{1}{h_2}\frac{\partial \phi}{\partial \xi_2}\right)^2, \tag{7.6}$$

recognizeable as the eikonal equation in a medium with index of refraction F_s with the scalar potential ϕ playing the role of the eikonal (see §7.11).

Equations (7.4) and (7.5) are recognizable as the equations for the optical ray path $x_i = x_i(s)$ in a medium with index of refraction $F_s(\xi_1, \xi_2)$. Fermat's principle applies, with the ray paths given by

$$\delta \int_{P_1}^{P_2} d\tau \left(h_1^2 \dot{\xi}_1^2 + h_2^2 \dot{\xi}_2^2\right)^{\frac{1}{2}} F_s(\xi_1, \xi_2) = 0, \tag{7.7}$$

where $d\tau$ is a parametric measure of distance along the ray path between fixed end points P_1 and P_2. The dot indicates differentiation with respect to τ. The Euler equations are,

$$\frac{d}{d\tau}(h_1^2 \dot{\xi}_1 F_s) = \frac{\partial F_s}{\partial \xi_1} + F_s\left(h_1 \frac{\partial h_1}{\partial \xi_1}\dot{\xi}_1^2 + h_2 \frac{\partial h_2}{\partial \dot{\xi}_1}\dot{\xi}_2^2\right) \tag{7.8}$$

$$\frac{d}{d\tau}(h_2^2 \dot{\xi}_2 F_s) = \frac{\partial F_s}{\partial \xi_2} + F_s\left(h_1 \frac{\partial h_1}{\partial \xi_2}\dot{\xi}_1^2 + h_2 \frac{\partial h_2}{\partial \xi_2}\dot{\xi}_2^2\right) \tag{7.9}$$

if $d\tau$ is put equal to arc length ds so that $h_1^2\, \xi_1^2 + h_2^2\, \xi_2^2 = 1$. These two equations can be rewritten as

$$F_s h_1^2 \ddot{\xi}_1 + \dot{\xi}_1^2 F_s h_1 \frac{\partial h_1}{\partial \xi_1} + \dot{\xi}_1 \dot{\xi}_2 \frac{\partial}{\partial \xi_2} h_1^2 F_s - \dot{\xi}_2^2 h_2 \frac{\partial}{\partial \xi_2} h_2 F_s = 0 \tag{7.10}$$

and

$$F_s h_2^2 \ddot{\xi}_2 + \dot{\xi}_2^2 F_s h_2 \frac{\partial h_2}{\partial \xi_2} + \dot{\xi}_1 \dot{\xi}_2 \frac{\partial}{\partial \xi_1} h_2^2 F_s - \dot{\xi}_1^2 h_1 \frac{\partial}{\partial \xi_1} h_1 F_s = 0 \tag{7.11}$$

The same result can be obtained by multiplying equations (7.4) and (7.5) by h_1 and h_2, respectively, and differentiating with respect to s. Then use equations (7.4) and (7.5) to eliminate $d\xi_2/ds$ from the right-hand side. The manipulation proceeds from equation (7.4) as

$$\begin{aligned}
\frac{d}{ds}\left(F_s h_1^2 \frac{d\xi_1}{ds}\right) &= -\frac{d\xi_j}{ds}\frac{\partial}{\partial \xi_j}\frac{\partial \phi}{\partial \xi_1} \\
&= \frac{1}{2F_s}\left[\frac{\partial}{\partial \xi_1}\left(\frac{1}{h_1}\frac{\partial \phi}{\partial \xi_1}\right)^2 + \frac{\partial}{\partial \xi_1}\left(\frac{1}{h_2}\frac{\partial \phi}{\partial \xi_2}\right)^2\right] \\
&\quad + \frac{1}{F_s}\left[\left(\frac{\partial \phi}{\partial \xi_1}\right)^2 \frac{1}{h_1^3}\frac{\partial h_1}{\partial \xi_1} + \left(\frac{\partial \phi}{\partial \xi_2}\right)^2 \frac{1}{h_2^3}\frac{\partial h_2}{\partial \xi_1}\right] \\
&= \frac{1}{2F_s}\frac{\partial}{\partial \xi_1}F_s^2 + F_s\left(h_1 \frac{\partial h_1}{\partial \xi_1}\dot{\xi}_1^2 + h_2 \frac{\partial h_2}{\partial \xi_1}\dot{\xi}_2^2\right),
\end{aligned} \tag{7.12}$$

where equation (7.6) has been used to simplify the first term in brackets, and equation (7.3) was used to eliminate $(\partial\phi/\partial\xi_1)^2$ and $(\partial\phi/\partial\xi_2)^2$ from the second term in brackets. The result is recognizable as equation (7.8).

The physical implications of the Euler equations are readily exhibited in a flat space ($h_1 = h_2 = 1$), writing Fermat's principle as

$$\delta \int dx(1 + y'^2)^{\frac{1}{2}} F_s(x, y) = 0.$$

The Euler equation can then be written as

$$\frac{y''}{(1 + y'^2)^{\frac{3}{2}}} = \frac{1}{(1 + y'^2)^{\frac{1}{2}}} \frac{\partial \ln F_s}{\partial y} - \frac{y'}{(1 + y'^2)^{\frac{1}{2}}} \frac{\partial \ln F_s}{\partial x} \tag{7.13}$$

in terms of the coordinates x, y in the flux surface. Define the unit vector

$$\mathbf{e}_s = \frac{1}{(1 + y'^2)^{\frac{1}{2}}}, \frac{y'}{(1 + y'^2)^{\frac{1}{2}}} \tag{7.14}$$

tangent to the ray path and note the curvature

$$K = \frac{y''}{(1 + y'^2)^{\frac{3}{2}}}.$$

The equation for the ray path is then just

$$K = \mathbf{e}_s \times \nabla \ln F_s, \tag{7.15}$$

showing that the ray, or field line, is concave toward a local maximum in F_s. The same geometrical relation holds if the surface is not flat, of course, but the expression is more complicated because ∇h_1 and ∇h_2 are nonvanishing.

An alternative form of equation (7.13) can be written in terms of the change $d\theta$ in the angle θ of inclination with respect to the x-axis along the ray path $y = y(x)$, so that

$$\tan\theta = y'(x). \tag{7.16}$$

Then in place of equation (7.13) we have

$$\mathbf{e}_n \, d\theta = d\mathbf{s} \times \nabla \ln F_s(\mathbf{r}), \tag{7.17}$$

where $\mathbf{e}_n$ is the unit vector normal to the flux surface and $d\mathbf{s}$ is the element of displacement $(dx, y'(x)dx, 0)$ along the field line. If $\partial F_s/\partial x = 0$, then equation (7.13) or (7.17) integrates to

$$y'(x) = \pm\{[F_s(y)/F_s(y_0)]^2 - 1\}^{\frac{1}{2}}$$

for total internal reflection at $y = y_0$. The ray path exists only where $F_s(y) > F_s(y_0)$, with the inclination increasing with $F_s(y)/F_s(y_0)$.

In summary, the optical analogy provides the pattern of the field lines of $\mathbf{F}_s$ in terms of the optical ray paths in the index of refraction F_s. An optical ray is concave toward a local maximum in the index of refraction. In particular, a sufficiently localized maximum in the index of refraction provides a shorter optical path length around the maximum than over the top of the maximum, with the result that there is a gap in the field line pattern of $\mathbf{F}_s$ in the surface locus S where the field lines pass around, rather than through, the local maximum, sketched in Fig. 7.3. A gap extending through a stack of flux surfaces of finite thickness, i.e., a gap extending through a slab of field, sketched in Fig. 7.4, permits the otherwise separate fields on either side of the stack to come into contact through the gap. The fields from the separate regions are generally not parallel where they meet in the gap, so their contact in the gap creates a tangential discontinuity in almost all cases.

The gap, then, is not empty, as might appear from Fig. 7.3, but is filled with field from the other surface loci pressed in from either side. That is to say, when the gap is created by enhanced pressure of the field on either side, the fields on either side poke into the gap and come into contact with each other, forming a tangential discontinuity where they meet. Thus, a gap in the field pattern of $\mathbf{F}_s$ means discontinuity of some form in $\mathbf{F}$. It is evident, then, that the gap in the field pattern of $\mathbf{F}_s$ in the surface locus S is also a gap in the surface locus S itself, for a discontinuity in $\mathbf{F}$ and, presumably therefore in $\nabla \times \mathbf{F}$, means a discontinuity in the surface loci of $\nabla \times \mathbf{F}$.

The gap in the field pattern does not constitute an internal inconsistency in the optical analogy, where we declared there was a localized maximum in F_s, creating the gap. A more precise statement would be that F_s increases toward the region, from all sides, so that the ray paths or field lines are concave toward the region and are refracted around the region, leaving the gap in the middle. Hence the field $\mathbf{F}_s$ and the index of refraction F_s are well defined wherever the refraction is to be computed. Their absence in the gap means only that there is nothing there to which the optical analogy might be applied in the context of S. As already noted, other fields have poked into the gap, with their own flux surfaces.

As a final remark, the optical analogy corresponds to a minimum energy principle. To show this consider an elemental flux bundle of a conserved field $\mathbf{H}$ (i.e., $\nabla \cdot \mathbf{H} = 0$). The cross-section of the bundle has the small value $\Delta A(s)$ where s represents distance

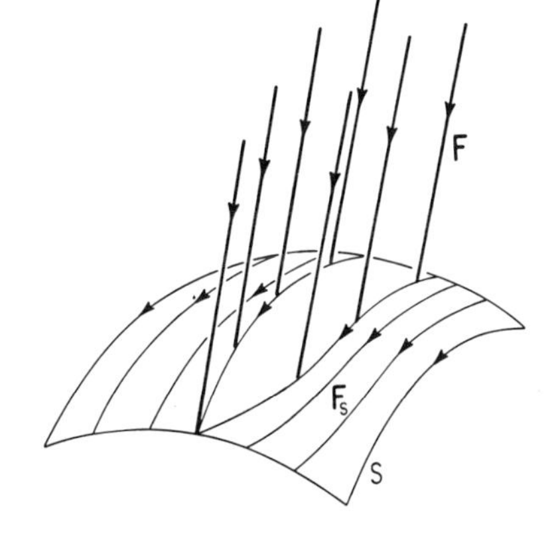

Fig. 7.3. A sketch of a gap in the field pattern of $\mathbf{F}_s$ as a consequence of a local maximum in F_s.

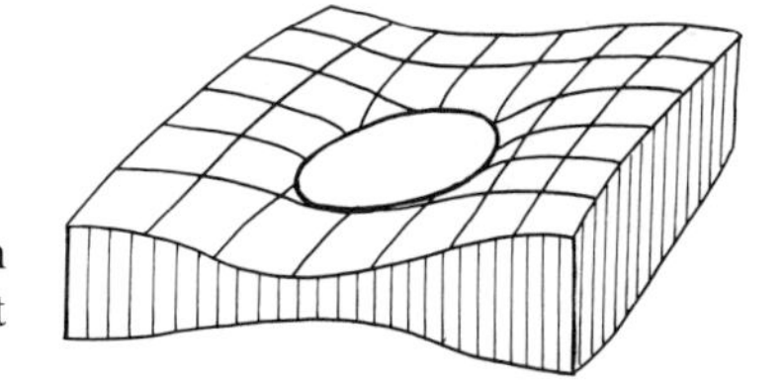

Fig. 7.4. A sketch of a finite stack of surface loci S with a common gap through the middle, permitting contact between the fields on opposite sides of the stack.

along the bundle. The total flux in the bundle is then $\Delta\Phi = H(s)\Delta A(s)$ and is constant along the bundle by virtue of $\nabla \cdot \mathbf{H} = 0$. Consider the volume integral of $H^2(s)$ over the bundle, which may be called the total "energy" ΔW. Then for the bundle connecting from point P to point Q, the total energy is

$$\begin{aligned}\Delta W(P,Q) &= \int_P^Q ds \Delta A(s) H^2(s),\\ &= \Delta\Phi \int_P^Q ds H(s).\end{aligned}$$

The energy $\Delta W(P,Q)$ is then an extremum, and presumably a minimum, when the first variation of the integral vanishes

$$\delta \int_P^Q ds H(s) = 0$$

This is precisely the optical analogy, of course, given by equation (7.7).

It should be noted in this general discussion that although there is a formal similarity of the field line equations (7.4) and (7.5) to the dynamical equations of a particle, the optical analogy in practice is distinct from the general global Hamiltonian formulation of the field, based on representation of the field lines of $\mathbf{F}$ as particle trajectories in phase space (Kerst, 1962; Boozer, 1982; Cary and Littlejohn, 1983; Tsinganos, Distler, and Rosner, 1984). The Hamiltonian formulation is usually applied to toroidal geometries with the azimuthal position φ around the torus corresponding to the time in Hamiltonian mechanics. It is particularly useful when there is an ignorable coordinate, e.g., φ. The Hamiltonian representation of a toroidal magnetic field was described in §§4.10 and 4.10.1 in the form applied by Tsinganos, Distler, and Rosner (1984). In contrast, the optical analogy is applied locally to the 2D subspaces formed by the field lines of $\nabla \times \mathbf{F}$, or to the full 3D space of a potential field, allowing computation of the local pattern of the field lines from a specification of the field magnitude F_s.

7.2 Special Cases

There are circumstances in which a family of curves C exists such that the associated surface loci Σ_c and S_c are coincident. In those cases the optical analogy applies to the entire vector $\mathbf{F}$ (i.e., $\mathbf{F}_s = \mathbf{F}$), and a gap around a maximum in $F = |\mathbf{F}|$ indicates a tangential discontinuity in $\mathbf{F}$. The optical analogy is more restrictive and hence more powerful in this case.

To explore the possible coincidence of Σ_c and S_c, consider the vector

$$\mathbf{G} \equiv (\nabla \times \mathbf{F}) \times \mathbf{F}.$$

Then obviously

$$\mathbf{F} \cdot \mathbf{G} = (\nabla \times \mathbf{F}) \cdot \mathbf{G} = 0,$$

and both $\mathbf{F}$ and $\nabla \times \mathbf{F}$ are perpendicular to $\mathbf{G}$ at each point P in space. Together $\mathbf{F}$ and $\nabla \times \mathbf{F}$ define a surface perpendicular to $\mathbf{G}$ and that surface is a surface locus of both $\mathbf{F}$ and $\nabla \times \mathbf{F}$ at the point P. That is to say, there are surface loci of $\mathbf{F}$ and

$\nabla \times \mathbf{F}$ that are tangent to each other at any point in space. The question is whether the equations for the local surface perpendicular to $\mathbf{G}$ can be integrated to provide continuous surfaces, so that there exist families of Σ_c and S_c that are coincident, rather than merely tangent at a point.

The differential equation for a surface everywhere orthogonal to the vector field $\mathbf{G}(\mathbf{r})$ is

$$d\mathbf{r} \cdot \mathbf{G}(\mathbf{r}) = 0 \tag{7.18}$$

with $d\mathbf{r}$ representing an infinitesimal displacement with components (dx, dy, dz). This equation is integrable only if there exists an integrating factor $\lambda(\mathbf{r})$ such that $\lambda\mathbf{G} \cdot d\mathbf{r}$ is a perfect differential, i.e.,

$$\lambda\mathbf{G} \cdot d\mathbf{r} = d\Phi.$$

That is to say, the equation is integrable only if

$$\frac{\partial\Phi}{\partial x_i} = \lambda G_i.$$

Then since $\partial^2\Phi/\partial x_i \partial x_j = \partial^2\Phi/\partial x_j \partial x_i$, it follows that $\nabla \times (\lambda\mathbf{G}) = 0$, which is

$$\lambda\nabla \times \mathbf{G} + \nabla\lambda \times \mathbf{G} = 0.$$

The scalar product with $\mathbf{G}$ yields

$$\mathbf{G} \cdot \nabla \times \mathbf{G} = 0. \tag{7.19}$$

This is a necessary (and also sufficient) condition for integrability. The requirement is that $\mathbf{G}$ be free of torsion, i.e., $\nabla \times \mathbf{G} \perp \mathbf{G}$ with no parallel component of $\nabla \times \mathbf{G}$. It is easy to see that this is essential, because torsion in $\mathbf{G}$ means that the element of area dA perpendicular to $\mathbf{G}$ at some point P forms a spiral surface if extended outward from P, so that it is everywhere perpendicular to $\mathbf{G}$. A spiral does not meet itself in passing once around the field line through P.

In summary, then, there exists a family of curves C such that the associated surface loci Σ_c and S_c coincide if $\mathbf{F}$ is such that the vector $\mathbf{G} = (\nabla \times \mathbf{F}) \times \mathbf{F}$ is without torsion. That is to say, if $\mathbf{G}$ can be written $u\nabla v$ where u and v are two scalar functions of position, then $\nabla \times \mathbf{G} = \nabla u \times \nabla v$ and $\mathbf{G} \cdot \nabla \times \mathbf{G} = 0$, so that $\mathbf{G}$ is without torsion. In that case, there exists a family of curves C for which the flux surfaces Σ_c and S_c coincide so that the optical analogy applies to the entire vector $\mathbf{F}$ in those coincident surface loci.

In the situation where $\mathbf{F}$ itself has no torsion, i.e., if $\nabla \times \mathbf{F}$ is perpendicular to $\mathbf{F}$, so that there exist two scalar functions U and V such that $\mathbf{F} = U\nabla V$, then $\mathbf{F}/U$ is a potential field and the optical analogy applies to the vector $\mathbf{F}/U$ in 3D space. The field lines are given by

$$\frac{dx_i}{ds} = \frac{U}{F}\frac{\partial V}{\partial x_i} \tag{7.20}$$

$$= \frac{1}{|\nabla V|}\frac{\partial V}{\partial x_i} \tag{7.21}$$

and may be computed using the index of refraction $|\nabla V|$ in any flux surface of $\mathbf{F}$ regardless of whether $\nabla \times \mathbf{F}$ has a nonvanishing component perpendicular to the flux surface. The only requirement is that $\nabla \times \mathbf{F}$ has no component parallel to $\mathbf{F}$.

On the other hand, suppose that $\nabla \times \mathbf{F}$ is precisely parallel to $\mathbf{F}$, as in a force-free magnetostatic field. In that case

$$(\nabla \times \mathbf{B}) \times \mathbf{B} = 0,$$

so that equation (1.5),

$$\nabla \times \mathbf{B} = \alpha \mathbf{B},$$

is applicable. Therefore, every flux surface of $\mathbf{B}$ is also a surface locus of $\nabla \times \mathbf{B}$, and the optical analogy applies to the whole $\mathbf{B}$ in every flux surface of $\mathbf{B}$. The same follows for Beltrami flows of an ideal inviscid incompressible fluid, for which

$$(\nabla \times \mathbf{v}) \times \mathbf{v} = 0,$$

and

$$\nabla \times \mathbf{v} = \beta \mathbf{v}.$$

The optical analogy is particularly effective in these cases because it applies to every flux surface, and therefore to two flux surfaces that are locally perpendicular, intersecting along a given field line. The optical analogy determines the field pattern in three dimensions then, with interesting results for solenoidal fields such as $\mathbf{B}$ and $\mathbf{v}$, described in §7.9.

Two cases of particular interest are the magnetostatic field $\mathbf{B}(\mathbf{r})$, satisfying the equation

$$(\nabla \times \mathbf{B}) \times \mathbf{B}/4\pi = \nabla p + \rho \nabla \Phi \tag{7.22}$$

in a fluid with pressure p and density ρ in a gravitational potential Φ, and the stationary flow of an inviscid fluid, satisfying the time-independent Euler equation

$$(\nabla \times \mathbf{v}) \times \mathbf{v} = -\nabla\left(\frac{1}{2}v^2 + \Phi\right) - \frac{1}{\rho}\nabla p. \tag{7.23}$$

Then if $p = p(\rho)$ and ρ is constant on level surfaces $\Phi = constant$, so that $\rho = \rho(\Phi)$ in equation (7.22), and if $\rho = \rho(p)$ in equation (7.23), these equations can be written

$$(\nabla \times \mathbf{B}) \times \mathbf{B} = 4\pi \nabla P, \tag{7.24}$$

$$(\nabla \times \mathbf{v}) \times \mathbf{v} = -\nabla \Pi, \tag{7.25}$$

where

$$P = p + Q(\Phi),$$

$$\Pi = \frac{1}{2}v^2 + \Phi + R(p),$$

and the functions $Q(\Phi)$ and $R(p)$ are defined by

$$Q'(\Phi) = \rho(\Phi),$$

$$R'(p) = 1/\rho(p).$$

It is obvious from equations (7.24) and (7.25) that the curls of $\mathbf{G} = (\nabla \times \mathbf{B}) \times \mathbf{B}$ and $\mathbf{G} = (\nabla \times \mathbf{v}) \times \mathbf{v}$ are zero. Hence $\mathbf{G} \cdot \nabla \times \mathbf{G} = 0$ in both cases and $\mathbf{G}$ is without torsion. So the isobaric and level surfaces $P = constant$ in the magnetostatic field are surface loci of both $\mathbf{B}$ and $\nabla \times \mathbf{B}$. The optical analogy applied to the isobaric surfaces, then, involves the complete magnetic field and not merely some projection of the field. Similarly the Bernoulli surfaces $\Pi = constant$ are surface loci of both the fluid velocity $\mathbf{v}$ and the vorticity $\nabla \times \mathbf{v}$. The optical analogy applied to the Bernoulli surfaces, then, involves the complete velocity field $\mathbf{v}$ and not merely some projection of the field.

The optical analogy applies to electromagnetic radiation in important cases, e.g., circularly polarized waves. It does not apply to a plane linearly polarized wave, e.g.,

$$E_x = B_y = D \cos \omega(t - z/c)$$

for which $\mathbf{E}$ and $\mathbf{B}$ are each perpendicular to their respective curls. However, for the circularly polarized wave, with the additional components

$$E_y = -B_x = D \sin \omega(t - z/c),$$

it follows that

$$\nabla \times \mathbf{E} = \frac{\omega}{c}\mathbf{E},\ \nabla \times \mathbf{B} = \frac{\omega}{c}\mathbf{B}.$$

Hence every flux surface of $\mathbf{E}$ is also a flux surface of $\nabla \times \mathbf{E}$, with the same for $\mathbf{B}$, similar to the force-free magnetohydrodynamic field and the stationary Beltrami flow of an ideal inviscid incompressible fluid.

Finally note that a potential field

$$\mathbf{F} = -\nabla\phi$$

is subject to the optical analogy in 3D space in any and all flux surfaces. In this case the optical analogy applies to any flux bundle or field line regardless of its orientation relative to the ambient field (Parker, 1981a).

Now there may be an occasion when it is desired to compute the path of an infinitesimal flux bundle connecting between two given points P_1 and P_2 in an arbitrary ambient field $\mathbf{B}$. It is necessary to specify the particular flux surface $S(P_1, P_2)$ of $\mathbf{B}$ along which the test bundle passes, there being infinitely many flux surfaces containing the two points P_1 and P_2. We suppose that the ambient field $\mathbf{B}$ contains a fluid with nonuniform pressure p (but with $\mathbf{B} \cdot \nabla p = 0$ in the absence of gravity). The test bundle contains its own internal fluid pressure $\wp$, taken to be uniform across the infinitesimal cross-section of the test bundle. It follows that the field $\mathbf{b}$ in the test bundle is a potential field $\mathbf{b} = -\nabla\phi$. The total pressure $\wp + b^2/8\pi$ is determined by the total ambient pressure $p + B^2/8\pi$ confining the bundle, from which it follows that the effective index of refraction is

$$|\nabla\phi| = [B^2 + 8\pi(p - \wp)]^{\frac{1}{2}},$$

where B and p are prescribed functions of position on the chosen flux surface $S(P_1, P_2)$ of $\mathbf{B}$.

Suppose, then, that P_1 and P_2 are chosen to lie on the same field line of $\mathbf{B}$ in $S(P_1, P_2)$. The test bundle in $S(P_1, P_2)$ generally does not follow the same path between P_1 and P_2 as the field line of $\mathbf{B}$ in $S(P_1, P_2)$ because the net tension $b^2/4\pi$ in the test bundle is different from the tension $B^2/4\pi$ in the ambient field.

One can ask, then, what is the path of a test bundle within which the fluid pressure $\wp$ varies substantially across the infinitesimal width of the bundle. The total pressure $\wp + b^2/8\pi$ on each field line in the test bundle is equal to the ambient pressure $p + B^2/8\pi$, of course, so that the field in the bundle is

$$b = [B^2 + 8\pi(p - \wp)]^{\frac{1}{2}}$$

which varies substantially across the width of the bundle. The net tension in the test bundle is $b^2/4\pi$, which varies across the bundle because of the variation of $\wp$. It follows that each elemental flux bundle – essentially each field line – within the test bundle has a different curvature, depending upon its $b^2/4\pi$, which opposes the gradient in the total ambient pressure. Thus an elemental part of the test bundle with a given pressure $\wp$ follows the path of a test bundle with uniform internal fluid pressure, viz. the case described in the foregoing paragraph. Hence, the individual elemental parts of the test bundle, each with their own $\wp$, fan out into a ribbon of finite width, each end of which converges to the point P_1 or P_2. A unique answer for the path of a test bundle depends, therefore, on specification of a single internal fluid pressure value $\wp$ over the entire infinitesimal cross-section.

7.3 The Field Rotation Effect

Consider the quantitative relations between a localized maximum in the index of refraction $F_s = |\mathbf{F}_s|$ and the refraction of the field pattern of $\mathbf{F}_s$. Write $\mathbf{H}$ in place of $\mathbf{F}_s$, or in place of $\mathbf{F}$ in the case that Σ_c coincides with S_c (see §7.1). Then, as a first example, consider the compression of the vector field

$$H_x = 0,\ H_y = H_0 \sin qx,\ H_z = H_0 \cos qx \tag{7.26}$$

treated in §6.2 to produce the surface rotation effect (Parker, 1987). The field is initially confined to the slab $-h < x < +h$, so that the field lines lie in the planes $x = constant$ with the inclination $dy/dz = \tan qz$ to the z-direction in each plane. The surfaces $x = \pm h$ bounding the region of field are then squeezed locally in the neighborhood of $y = 0$ and deformed inward to form a shallow groove $x = \pm hf(y)$ where $f(y) \leqslant 1$ is an even function of y, increasing monotonically with increasing y^2 and going asymptotically to one in the limit of large y^2. That is to say, $f(y)$ has a broad symmetric minimum at $y = 0$.

It is assumed that $f(y)$ varies slowly so that $f'(y)^2 \ll q^2$ and $h^2 f'(y)^2 \ll 1$. Then the x-component of the field, of the order of $Hhf'(y)$, is small compared to H. The field magnitude H is increased by the reduced thickness in the vicinity of $y = 0$, but the x component can be neglected and H^2 is adequately approximated by the sum of the squares of the y and z-components of the deformed field. The magnetic pressure, and hence the field magnitude $H(y)$, is uniformly distributed across the slowly varying width $2hf(y)$. In this way the initial flux surface $x = x_0$ $(-h < x_0 < +h)$ is deformed into $x = x_0 f(y)$, and $\mathbf{H}$ is subject to the optical analogy in these deformed flux surfaces.

Consider the pattern of the field in the flux surface $x = x_0 f(y)$ $(x_0^2 < h^2)$ with $f(y)$ a given function of y. The surface is Euclidean and distance in the y-direction is directly measured by y upon neglecting terms second order in $hf'(y)$ compared to one. The field magnitude is now $H(y)$ and the Euler equation (7.13) becomes

$$\frac{z''}{z'(1 + z'^2)} = \frac{d \ln H}{dy}, \tag{7.27}$$

where $z = z(y)$ represents the trajectory or ray path. Integration yields Snell's law

$$H_z^2 = H^2(y)\sin^2\theta(y) = H_0^2\sin^2\theta_0 \tag{7.28}$$

for the ray inclined at an angle θ_0 to the y-direction at $y = \pm\infty$ where $H = H_0$. It follows from equation (7.26) that $\theta_0 = \frac{1}{2}\pi - qx_0$. Equation (7.28) reduces to

$$H(y)\sin\theta = \pm H_0\cos qx_0 \tag{7.29}$$

with the sign chosen to match the sign of qx_0, as may be seen from equation (7.26). The y-component of the field is

$$\begin{aligned} H_y &= H(y)\cos\theta \\ &= \pm[H^2(y) - H_0^2\cos^2 qx_0]^{\frac{1}{2}}. \end{aligned} \tag{7.30}$$

The discontinuity $\Delta\theta$ in field direction across $qx_0 = 0$ is, then, the difference in field direction as qx_0 declines to zero from above and below,

$$\Delta\theta = 2\sin^{-1}[H_0/H(y)]. \tag{7.31}$$

This is precisely the result obtained in §6.2 by formal solution of the Grad–Shafranov equation. It has been obtained here directly from the optical analogy by elementary means. That is to say, the flux rotation effect computed formally from the Grad–Shafranov equation is just the refraction of the field in the enhanced index of refraction of the constricted field.

As a matter of fact, it is now straightforward to proceed with calculation of the field magnitude, and hence the individual field components, using the field pattern deduced from the optical analogy and the condition of flux conservation during the compression. Thus, for instance, the layer of flux $(x_0, x_0 + \delta x_0)$ has a thickness $\delta x(x_0, y)$ given by the condition for conservation of flux in the y-direction, so that

$$\delta x(x_0, y)H(y)\cos\theta = \delta x_0\, H_0\sin qx_0 \tag{7.32}$$

With equation (7.30) this yields,

$$\delta x(x_0, y) = \frac{\delta x_0 H_0\sin qx_0}{[H^2(y) - H_0^2\cos^2 qx_0]^{\frac{1}{2}}} \tag{7.33}$$

providing an expression for the thickness $\delta x(x_0, y)$. This expression may be integrated over x_0 to give the position $x(x_0, y)$, remembering the symmetry requirement that the center plane is not displaced, $x(0, y) = 0$. It is readily shown that

$$qx = \sin^{-1}\frac{H_0}{H(y)} - \sin^{-1}\left(\frac{H_0}{H(y)}\cos qx_0\right), \tag{7.34}$$

or

$$\frac{H(y)}{H_0} = \frac{[1 - 2\cos qx\cos qx_0 + \cos^2 qx_0]^{\frac{1}{2}}}{|\sin qx|}. \tag{7.35}$$

The boundary $x = hf(y)$ follows for $x_0 = h$, so that

$$\frac{H(y)}{H_0} = \frac{[1 - 2\cos[qhf(y)]\cos qx_0 + \cos^2 qx_0]^{\frac{1}{2}}}{\sin qhf(y)} \tag{7.36}$$

7.4 Refraction Around a Maximum

We come now to the refraction of a field **H** around a local maximum in the field magnitude H. As in §7.3 it is presumed that **H** is subject to the optical analogy in Euclidean flux surfaces, which can be flattened to form the xy-plane. Consider, then, a localized maximum of characteristic scale $1/k$ in the field magnitude $H(x,y)$ centered on the origin, and work out its effect on the ray path extending across the xy plane from $y=a$ at $x=-L_1$ to $y=b$ at $x=+L_2$, sketched in Fig. 7.5. It is sufficient for a preliminary investigation (Parker, 1989b) to assume a weak localized maximum with $kL_1, kL_2 \gg 1$. It follows that

$$c-a \cong L_1\vartheta_1,\ c-b \cong L_2\vartheta_2 \tag{7.37}$$

The total deflection $\vartheta(c)$ at $y=c$ is

$$\vartheta(c) = \vartheta_1+\vartheta_2 \tag{7.38}$$

$$= \frac{c}{L}-\frac{a}{L_1}-\frac{b}{L_2} \tag{7.39}$$

where

$$\frac{1}{L} \equiv \frac{1}{L_1}+\frac{1}{L_2}$$

The refraction of the ray depends only on the gradient $\partial H/\partial y$ at $y=c$. Neglecting y' compared to one, equation (7.13) for the ray path reduces to

$$y'' \cong \frac{\partial \ln H}{\partial y} \tag{7.40}$$

with F_s replaced by H. For a weak maximum write

$$H = H_0+\varepsilon H_1(x,y), \tag{7.41}$$

and write the ray path as

$$y = c+\varepsilon\xi(x). \tag{7.42}$$

Then

$$H_0\xi''(x) = \frac{\partial H_1(x,c)}{\partial c} \tag{7.43}$$

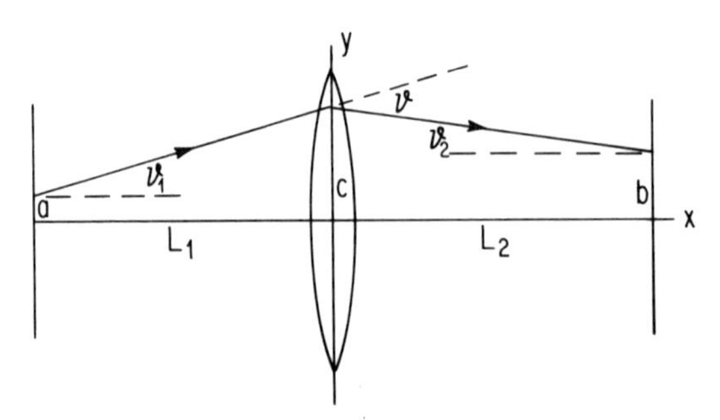

Fig. 7.5. A schematic drawing of the ray path extending from $(-L_1, +a)$ to $(+L_2, +b)$ and crossing the maximum at $(0, c)$.

The change ϑ in the slope of the ray is the integral of this expression across the region of nonvanishing $H_1(x, c)$, or

$$\vartheta = -\varepsilon f'(c) \tag{7.44}$$

where the function $f(c)$ is defined as

$$f(c) = \int_{-\infty}^{+\infty} dx \frac{H_1(x, c)}{H_0} \tag{7.45}$$

Thus equation (7.39) becomes

$$\frac{c}{L} + \varepsilon f'(c) = \frac{a}{L_1} + \frac{b}{L_2}.$$

and for the symmetric case that $a = b$,

$$a = c + \varepsilon L f'(c). \tag{7.46}$$

The next step is explore the refraction patterns represented by equation (7.44) for a variety of forms of $f(c)$. The investigation is limited to single smooth symmetric maxima, represented by a smooth bell-shaped form for $f(y)$, which refracts an otherwise uniform field. As already noted, the rays are concave toward the maximum, tending to pass around it, rather than over the top. Figure 7.6(a) is a sketch of the field for weak refraction. Fig. 7.6(b) shows stronger refraction, producing a gap where the optical path length around the maximum is shorter than across the maximum. Figure 7.6(c) sketches the refraction by a flat topped maximum, showing a gap on either flank. Finally, Fig. 7.6(d) exhibits some of the overlapping ray paths that come out of the mathematics for a peak sufficiently sharp as to produce a gap in the field pattern. For it is always possible to move a ray into an equilibrium position at any value of $y = c$ by the proper choice of the footpoint positions, $y = a = b$. This may involve a decrease of a and b for increasing c, resulting in overlapping ray paths. This situation is unstable, of course, because there is also a smaller stable value of c for the reduced a and b. The unstable $(da/dc < 0)$ values of c are not realized in nature so those values of c are left uncovered by rays, resulting in the gaps shown in Fig. 7.6(b) and (c). That is to say, there are rays only for those values of c for which the rays are stable. Hence the criterion for overlapping is also the criterion for a gap.

Suppose, then, that the footpoints of the field line, at $y = a, b$, are displaced slightly to $a + \delta a$ and $b + \delta b$, respectively. The result is a small change δc in c and a small change $\vartheta'(c)\delta c$ in the deflection $\vartheta(c)$. It follows from equation (7.39) that

$$\left[\vartheta'(c) - \frac{1}{L}\right]\delta c + \frac{\delta a}{L_1} + \frac{\delta b}{L_2} = 0 \tag{7.47}$$

Overlapping occurs if δc is negative when δa and δb are positive, so the criterion for a single valued monotonic relation between δa and δb on the one hand and δc on the other is that equation (7.44) be satisfied for δa, δb, and δc all of the same sign, which is possible only if the coefficient of δc is negative. Hence

$$L\vartheta'(c) < 1 \tag{7.48}$$

is the condition for the absence of gaps in the ray pattern. Conversely, if

$$L\vartheta'(c) > 1,$$

there is a gap.

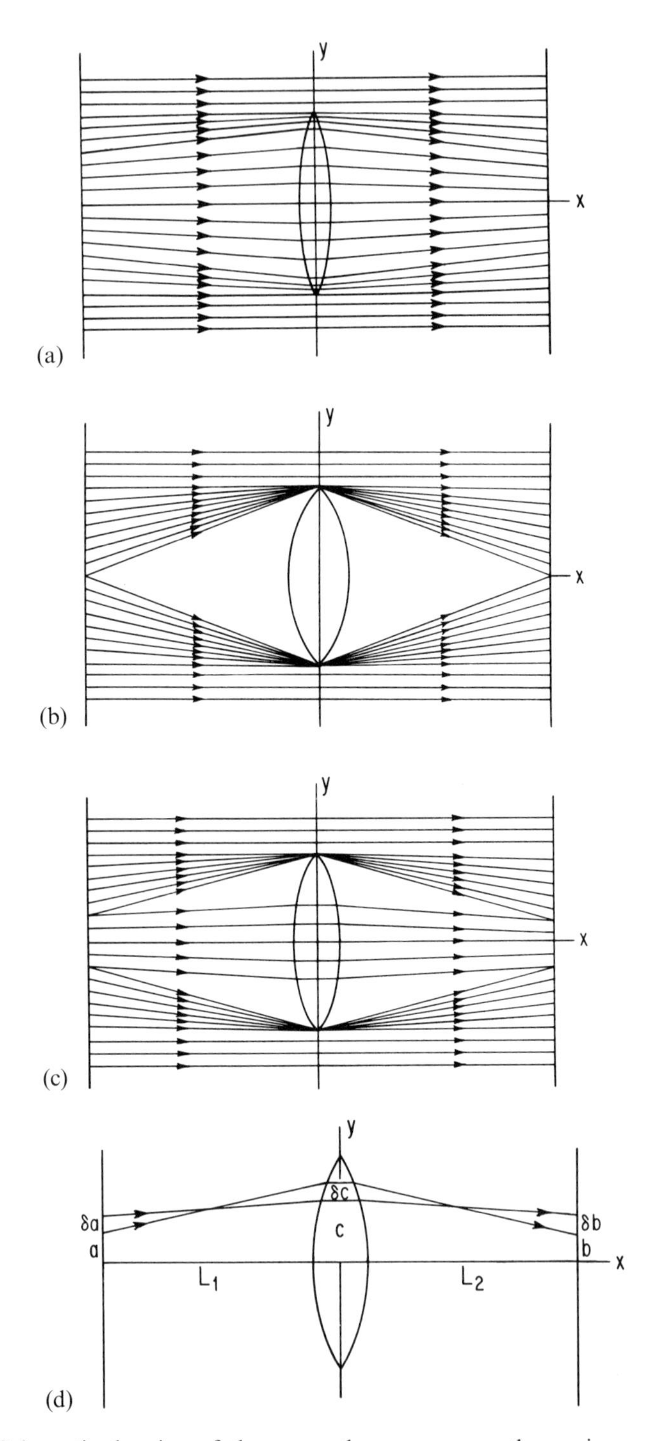

Fig. 7.6. (a), Schematic drawing of the ray paths across a gentle maximum, (b), across a maximum sufficiently sharp as to produce a gap in the field pattern, (c), across a broader maximum with steep sides, and (d), theoretical overlapping ray paths that arise at a maximum sufficiently steep as to form a gap.

For the symmetric field pattern $a = b$, the condition for the absence of a gap, or zone of exclusion, is simply $da/dc > 0$. This is readily seen from the fact that with $\delta a = \delta b$, equation (7.47) reduces to

$$\frac{da}{dc} = 1 - L\vartheta'(c) \tag{7.49}$$

$$= 1 + \varepsilon L f''(c) \tag{7.50}$$

so that the inequality of equation (7.48) yields $da/dc > 0$.

It is clear in principle that a sufficiently strong sharp maximum in $H(x, y)$, or in $f(y)$, produces a gap in the field pattern (Fig. 7.6b), whereas a sufficiently broad weak maximum does not. It is evident that a combination can be formed if $f(c)$ has the form of a mesa, or butte, with a nearly flat top and steep sides, so that there is a continuous band of field across the top with a gap on each side (Fig. 7.6c). The following subsections provide formal examples of these cases for symmetric geometry $a = b$.

7.4.1 General Parabolic Maximum

It is simplest, and adequate for present purposes, to treat a flat flux surface using cartesian coordinates (x, y) oriented so that the field lines extend in the x-direction (i.e., a uniform field) in the absence of any refraction so that $a = b$. Equations (7.39) and (7.44) yield the result

$$L\vartheta(c) = c - a \tag{7.51}$$

$$= -\varepsilon L f'(c) \tag{7.52}$$

for the deflection $\vartheta(c)$ of the ray path across $x = 0$ at $y = c$. Differentiating equations (7.51) and (7.52) with respect to a leads to

$$\frac{dc}{da}[1 + \varepsilon L f''(c)] = 1$$

Hence the criterion $dc/da > 0$ for a stable continuous field pattern becomes

$$\varepsilon L f'' > -1. \tag{7.53}$$

There is an absence of field lines, i.e., a gap in the field pattern, where $\varepsilon L f'' < -1$.

Consider the effect of various analytic forms of the maximum $f(y)$. In general, one expects that a symmetric maximum in $f(y)$ located at $y = 0$ can be expanded in ascending powers of $(ky)^2$, so that

$$kf(y) = 1 - p(ky)^2 - q(ky)^4 + \ldots$$

where $1/k$ is the characteristic width of the maximum and the coefficients p and q satisfy $p^2 + q^2 = O(1)$. If the coefficient p is $O(1)$ and $0 \leqslant q \ll p$, the refraction opens a gap extending along the field and centered on $y = 0$. On the other hand, if $0 \leqslant p \ll q$, there is a band of continuous field extending along $y = 0$ with gaps beyond (Parker, 1989b).

As a first example, then, consider the general parabolic form

$$kf(y) = \begin{cases} 1 - (ky)^{2n} & \text{for } 0 < k^2y^2 < 1 \\ 0 & \text{for } k^2y^2 > 1 \end{cases} \tag{7.54}$$

for the net refraction, given by equation (7.52), where n is a positive integer. This simple form provides a local maximum in the index of refraction with the flatness of the maximum increasing with n. It follows from equation (7.52) that a and c are related by

$$a = c[1 - 2\varepsilon kLn(kc)^{2n-2}]. \tag{7.55}$$

The criterion of equation (7.53) for a continuous distribution of field becomes

$$2\varepsilon kLn(2n-1)(kc)^{2n-2} < 1. \tag{7.56}$$

For $n = 1$ the relation in equation (7.55) becomes $a = c(1 - 2\varepsilon kL)$ and the inequality in equation (7.56) reduces to

$$2\varepsilon kL < 1 \tag{7.57}$$

for all kc. If the condition is fulfilled, the ray paths or field lines provide a continuous field across the region, sketched in Fig. 7.6(a). If, on the other hand $2\varepsilon kL > 1$, there is a gap across the entire region, from $ky = +1$ to $ky = -1$, sketched in Fig. 7.6(b).

Suppose, then, that $n = 2$, so that the maximum has a flatter top. Equation (7.56) becomes

$$12\varepsilon kL(kc)^2 < 1, \tag{7.58}$$

which is satisfied for all $kc(0 < k^2c^2 < 1)$ if $12\varepsilon kL < 1$. This is a more restrictive limit on the maximum strength εkL than (7.57) for $n = 1$ because the top of the maximum is flatter while the sides are correspondingly steeper. If $12\varepsilon kL$ is not less than one, the inequality of equation (7.58) is satisfied, and the field is continuous only over the width $0 \leqslant c^2 < c_g^2$, where

$$k^2c_g^2 = 1/12\varepsilon kL. \tag{7.59}$$

There is a gap beyond c_g, where the inequality of equation (7.58) is not satisfied. It follows from equation (7.55) that the inner edge of the gap, for which c_g is given by equation (7.59), connects into $a_g = \frac{2}{3}c_g$ at either end of the region ($x = -L_1, +L_2$). The configuration is sketched in Fig. 7.6(c), showing the band of continuous field across the middle, and the gap extending from kc_g to $k^2c^2 = 1$, with continuous field in $k^2c^2 > 1$. Figure 7.7 is a plot of ka versus kc from equation (7.55) for $n = 2$ for the values of $2\varepsilon kL$ indicated on each curve. Continuity requires that a increases monotonically with c throughout $0 < k^2c^2 < 1$, which is not satisfied for $2\varepsilon kL$ larger than $1/6$.

In the general case ($n \geqslant 2$) it follows from equation (7.56) that there is no gap in $0 < k^2c^2 < 1$ provided that

$$2\varepsilon kL < 1/n(2n-1) \tag{7.60}$$

If the maximum is not sufficiently weak as to satisfy this requirement, there is a gap extending from $c = c_g$ out to $k^2c^2 = 1$, where

$$(kc_g)^{2(n-1)} = 1/2n(2n-1)\varepsilon kL, \tag{7.61}$$

corresponding to

$$a = c_g(n-1)/(n-1/2). \tag{7.62}$$

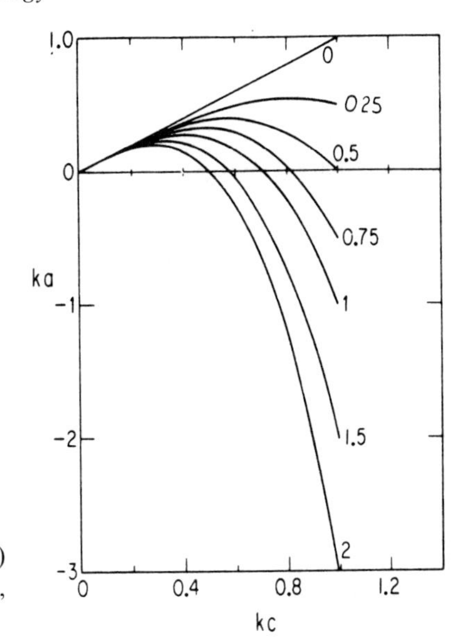

Fig. 7.7. A plot of ka versus kc from equation (7.55) for $n = 2$ and the values of $2\varepsilon kL = 0, 0.25, 0.5, 0.75, 1.0, 1.5, 2$ indicated on each curve.

Figure 7.8 is a plot of the half–width c_g of the band of continuous field as a function of $2\varepsilon kL$ for $n = 2, 3, 4$. The interesting feature in Fig. 7.8 is the cross-over of the curves for progressively larger n. The gap on either side of the central band of continuous field (see Fig. 7.6c) appears at progressively smaller $2\varepsilon kL$ as n increases, but the width $2c_g$ of the band of continuous field declines less rapidly with increasing $2\varepsilon kL$ for larger n so that for $2\varepsilon kL \gtrsim 0.5$, the band of continuous field is progressively wider, and the gap is progressively narrower, for the larger values of n.

As a final example here, suppose that with $0 < p < 1$,

$$kf(y) \equiv 1 - p(ky)^2 - (1 - p)(ky)^4$$

for $0 < k^2y^2 < 1$, and equal to zero for $k^2y^2 > 1$, combining the two cases $n = 1$ and $n = 2$. It follows from the inequality of equation (7.53) that there is a continuous distribution of field for

$$2\varepsilon kL[p + 6(1 - p)(kc)^2] \leqslant 1$$

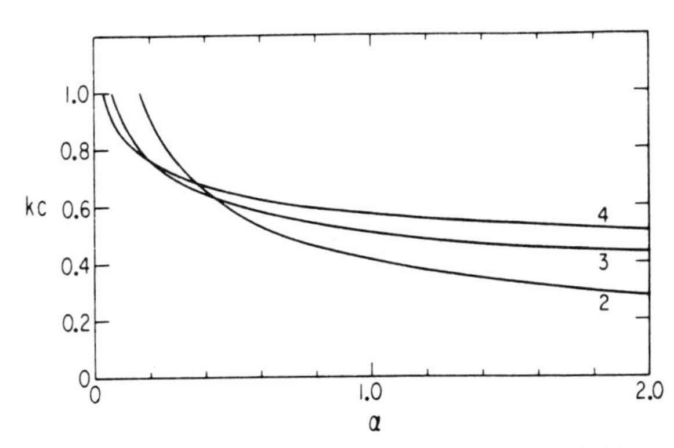

Fig. 7.8. A plot of the half width kc_g of the band of continuous field across the top of the maximum as a function of $2\varepsilon kL$ for $n = 2, 3, 4$ indicated for each curve.

in the region $(0 < k^2c^2 < 1)$ of refraction. It is evident by inspection that the field is continuous everywhere throughout $0 < k^2c^2 < 1$ if

$$2\varepsilon kL(6 - 5p) \leqslant 1.$$

On the other hand, if

$$\frac{1}{6 - 5p} < 2\varepsilon kL < \frac{1}{p},$$

the field is continuous over $0 < k^2c^2 < k^2c_g^2$, with

$$k^2c_g^2 = \frac{1/2\varepsilon kLp - 1}{6(1/p - 1)} < 1,$$

but a gap begins at kc_g and extends to $k^2c^2 = 1$. This is the configuration sketched in Fig. 7.6(c). Finally, if $2\varepsilon kLp > 1$, the band of continuous field $0 < c^2 < c_g^2$ vanishes and the entire region $0 < k^2c^2 < 1$ is devoid of field, with continuous field beyond $k^2c^2 = 1$, sketched in Fig. 7.6(b).

To summarize this result, $2\varepsilon kL(6.5p) < 1$ provides a continuous field across the entire region of refraction $0 < k^2c^2 < 1$. Since $6 - 5p > 1$ for $p < 1$, this requires $2\varepsilon kL$ somewhat smaller than one. As $2\varepsilon kL$ increases, a gap appears at $k^2c^2 = 1$ and extends inward to $k^2c^2 = k^2c_g^2 < 1$, reaching the x-axis when $2\varepsilon kLp = 1$. For $2\varepsilon kLp \geqslant 1$ the gap covers the entire refractive region $0 < k^2c^2 < 1$. This shows the competition between the $n = 1$ and $n = 2$ cases in determining the form of the gap, or gaps, created by a single symmetric maximum. This addresses the question of the form of the gaps expected in the flux surfaces of equilibrium fields in nature, where the effective value of the parameters $2\varepsilon kL$ and p are not well known. For simplicity of discussion in subsequent chapters we adopt the working hypothesis that p is sufficiently large that most gaps open up first along the center line, as sketched in Fig. 7.6(b). But it must be recognized that the double gap in Fig. 7.6(c) may sometimes occur. The creation of suprathermal effects in nature, through resistive dissipation at the associated discontinuities, generally does not depend critically on the form of the gap, however, so the issue seems not to be a primary concern at the present state of development of the theory.

7.4.2 *Gaussian Maximum*

The foregoing examples provide the qualitative picture of the refraction of field lines by a local maximum in the field magnitude, but they possesses the unattractive feature of a kink in the profile of the field magnitude $f(y)$ at $k^2y^2 = 1$, causing the field lines to pile up at each end, $k^2c^2 = 1$, as indicated in Fig. 7.6(b) and (c). Hence it is not without interest to treat the general gaussian form

$$kf(y) = \exp[-(k^2y^2)^n], \tag{7.63}$$

where n is again a positive integer and $f(y)$ varies smoothly and continuously over all y. Then it follows from equation (7.52) that a and c for an individual ray are related by

$$a = c\{1 - 2\varepsilon kLn(kc)^{2n-2}\exp[-(kc)^{2n}]\} \tag{7.64}$$

while the criterion (7.53) for continuous coverage by ray paths $(da/dc > 0)$ is

$$2\varepsilon kLn(2n - 1)(kc)^{2n-2}\left[1 - \frac{2n}{2n - 1}(kc)^{2n}\right]\exp[-(kc)^{2n}] \leqslant 1. \tag{7.65}$$

from $k^2c^2 = 0$ out to the first value of k^2c^2 at which the equality sign obtains. For the gaussian distribution $n = 1$, this gives

$$a = c\left[1 - 2\varepsilon kL\exp(-k^2c^2)\right], \tag{7.66}$$

and, in order that there are no gaps, the inequality of equation (7.65) requires

$$2\varepsilon kL(1 - 2k^2c^2)\exp(-k^2c^2) \leqslant 1. \tag{7.67}$$

The left-hand side of this inequality has a maximum value of $2\varepsilon kL$ at $kc = 0$. So if $2\varepsilon kL < 1$, there are no gaps. If $2\varepsilon kL > 1$, there is evidently a gap extending outward from $c = 0$.

The relation of equation (7.66) between a and c is plotted in Fig. 7.9 for $n = 1$ and the indicated values of $2\varepsilon kL$. Note that with $2\varepsilon kL > 1$, the coordinate a goes negative as c increases from zero, as may be seen directly from equation (7.66) for $k^2c^2 \ll 1$, so that

$$a \cong -c(2\varepsilon kL - 1)$$

to lowest order. Continuous and symmetric coverage across $c = 0$ requires that both $da/dc \geqslant 0$ and $a/c \geqslant 0$. The lower curve in Fig. 7.10 is a plot of the value of kc as a function of $2\varepsilon kL$ at which a reaches its minimum value when $2\varepsilon kL > 1$ (i.e., the value of kc at which $da/dc = 0$), given by the equality sign in equation (7.67). The upper curve in Fig. 7.10 gives the value of kc at which a increases back up to zero, obtained directly from equation (7.66) as $kc = (\ln 2\varepsilon kL)^{\frac{1}{2}}$. Continuous coverage occurs only for values of kc above the upper curve

$$k^2c^2 > \ln(2\varepsilon kL),$$

for which $2\varepsilon kL\exp(-k^2c^2) < 1$ so that equation (7.67) reduces to $k^2c^2 > 0$. Hence, the gap extends out to $kc = \pm(\ln 2\varepsilon kL)^{\frac{1}{2}}$, and the field is continuous beyond.

For $n \geqslant 2$ the situation is qualitatively different, in that for any finite $2\varepsilon kL > 0$ there is a range of $k^2c^2 > 0$ for which $da/dc \leqslant 0$, i.e., for which the inequality of equation (7.65) is satisfied and for which $a/c > 0$. This is evident from the fact that, starting from $k^2c^2 = 0$, equation (7.64) has the form

$$a \cong c[1 - 2\varepsilon kLn(kc)^{2n-2} + 2\varepsilon kLn(kc)^{4n-2} + \ldots] \tag{7.68}$$

while the inequality of equation (7.65) reduces to

$$2\varepsilon kLn(2n-1)(kc)^{2n-2} \leqslant 1,$$

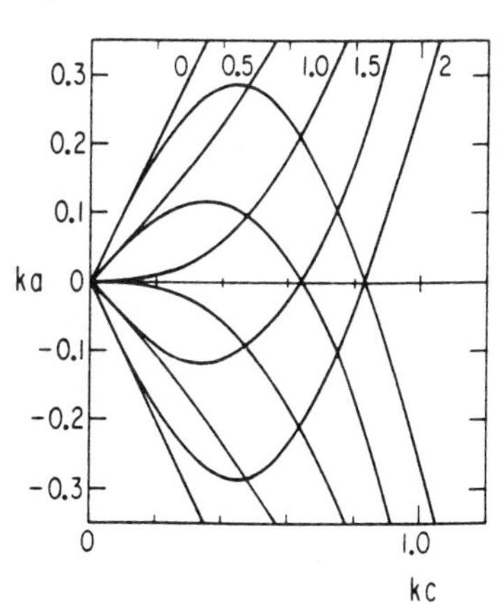

Fig. 7.9. A plot of a versus c from equation (7.66) for $n = 1$, for $2\varepsilon kL = 0, 0.5, 1, 1.5, 2$, indicated on each curve.

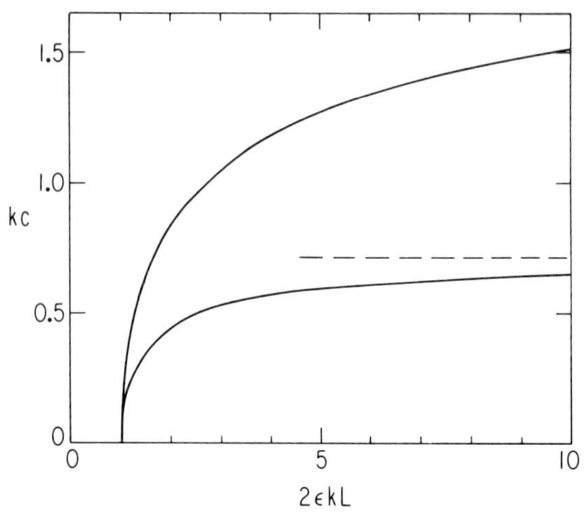

Fig. 7.10. The lower curve is a plot of the value of kc as a function of $\alpha \equiv 2\varepsilon kL$ at which ka passes through a minimum when $2\varepsilon kL > 1$, given by the equality sign in equation (7.67). The dash line represents the asymptotic value $kc \sim 2^{-1/2}$. The upper curve is the value of kc as a function of $2\varepsilon kL$ at which a returns to zero, given by equation (7.66) as $k^2c^2 = \ln 2\varepsilon kL$.

so there is always a band of continuous field of finite width lying across $k^2c^2 = 0$. If $2\varepsilon kL$ is sufficiently small, $da/dc > 0$ for all $k^2c^2 > 0$. This may be seen directly by inspection of equation (7.65), and it is also evident from the fact that with increasing k^2c^2 the left-hand side of (7.65) goes through a maximum, falls to negative values, passes through a minimum, and finally increases monotonically and asymptotically to zero as k^2c^2 becomes large. Hence if the inequality is to be violated, it is at the maximum of the left-hand side, at $c^2 = c_m^2$, where kc_m is the smaller of the two roots

$$(kc)^{2n} = \frac{1}{4n}\{3(2n-1) \pm [(2n-1)(10n-1)]^{\frac{1}{2}}\} \tag{7.69}$$

of the quadratic equation

$$2n^2(k^nc^n)^4 - 3n(2n-1)(k^nc^n)^2 + (n-1)(2n-1) = 0.$$

It follows that there is no break in the coverage of the field when $2\varepsilon kL$ is small enough that the inequality of equation (7.65) is satisfied at $c^2 = c_m^2$.

If $2\varepsilon kL$ is large enough that the inequality of equation (7.65) is violated for $c^2 = c_m^2$, then there is a gap in the field pattern extending from $c^2 = c_1^2 < c_m^2$ out to $c^2 = c_2^2 > c_m^2$, where c_1 is the value of c at which the left-hand side of the inequality (7.65) first becomes equal to one, and c_2 is the value of c at which the left-hand side reaches once again beyond c_m. The value a_1 of a associated with $c = c_1$, follows from equation (7.64). In fact equation (7.64) can be simplified at $c = c_1$ using the equal sign in equation (7.65) to eliminate $2\varepsilon kL$. The result is

$$a_1 = c_1 \frac{1 - (kc_1)^{2n} - 1/n}{1 - (kc_1)^{2n} - 1/2n} \tag{7.70}$$

Now for $c_1 < c < c_2$, the value of a associated with c is less than a_1, so that if such field lines existed, they would overlap the field for $c < c_1$. Such lines are unstable, because $da/dc < 0$, and so are not expected to occur in nature. The point is that as c increases to c_2, a increases back to a_1, so that the field is continuous again for $a > a_1$. That is to say, $a_2 = a_1$, and the relation in equation (7.69) gives the same a for both $c = c_1$ and $c = c_2$.

Consider, then, the special case that $n = 2$, which illustrates the qualitative features of $n > 2$. Then the two roots given by equation (7.69) are $k^4c^4 = 0.1813$ and 2.0687, so that the smaller gives $kc_1 = 0.6525$ for the inner edge of the gap. It follows that the inequality of equation (7.65) is satisfied for all kc if $2\varepsilon kL \leqslant 0.619$. If $2\varepsilon kL$ is larger, there is a gap whose inner boundary lies at $a = a_1$, $c = c_1$ with c_1 given by the equality sign in equation (7.65) and $kc_1 < 0.6525$. Then

$$12\varepsilon kL(kc_1)^2\left[1 - \frac{4}{3}(kc_1)^4\right]\exp[-(kc_1)^4] = 1 \tag{7.71}$$

The coordinate kc_1 is plotted as a function of $2\varepsilon kL$ in Fig. 7.11. The coordinate ka_1 of the ends of the field line follows from equation (7.64) and from equation (7.70) and is plotted in Fig. 7.11. For $2\varepsilon kL \gtrsim 3$ we have

$$kc_1 \cong \frac{1}{(12\varepsilon kL)^{\frac{1}{2}}}\left[1 - \frac{7}{6(12\varepsilon kL)^2} + \ldots\right] \tag{7.72}$$

and

$$ka_1 \cong \frac{2}{3(12\varepsilon kL)^{\frac{1}{2}}}\left[1 - \frac{7}{2(12\varepsilon kL)^2} + \ldots\right] \tag{7.73}$$

so that $ka_1 \sim \frac{2}{3}kc_1$ in the limit of large $12\varepsilon kL$. This defines the inner edge of the gap.

The outer edge of the gap connects to $y = a_1$ at each end, of course, and, therefore, is given by the larger value of kc satisfying equation (7.64) for $a = a_1$. Thus, denoting c by c_2 for the outer boundary, kc_2 is given by the larger root for kc in

$$ka_1 = kc\{1 - 4\varepsilon kL(kc)^2\exp[-(kc)^4]\} \tag{7.74}$$

wherein the second term within the braces is smaller than one because kc_2 exceeds one sufficiently that the exponential term is small. The resulting kc_2 is plotted

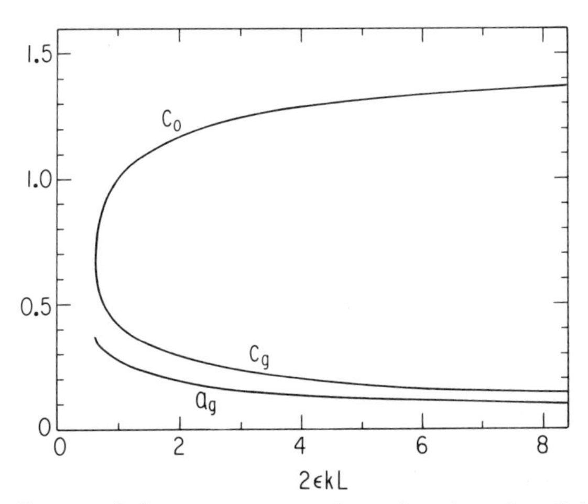

Fig. 7.11. The coordinates of the gap a_g, c_g, and c_o plotted against $2\varepsilon kL$ from equations (7.64), (7.71), and (7.74), respectively.

against $2\varepsilon kL$ in Fig. 7.11. An approximate expression for kc_2 can be deduced for large $2\varepsilon kL$. Write equation (7.64) as

$$1 - a/c = 2Q(kc)^2 \exp(-k^4c^4)$$

where $Q \equiv 2\varepsilon kL$ and $a = a_1$. Then for $c \gg a$ the logarithm of this relation gives

$$(kc)^4 \cong \ln 2Q + \ln (kc)^2 + a/c + 0(a^2/c^2).$$

Then write

$$k^4c^4 = (1 + \Delta) \ln 2Q$$

from which it follows to lowest order that

$$\Delta \cong \frac{\ln \ln 2Q + 2ka/(\ln 2Q)^{\frac{1}{4}}}{2 \ln 2Q - 1 + ka/2(\ln 2Q)^{\frac{1}{4}}}$$
$$\cong \frac{\ln \ln 2Q}{2 \ln 2Q}.$$

Thus the outer boundary has coordinates $a = a_1$ and

$$kc = (\ln 2Q)^{\frac{1}{4}} \left[1 + \frac{\ln \ln 2Q}{8 \ln 2Q} + \ldots \right] \tag{7.75}$$

in the limit of large $Q = 2\varepsilon kL$. The field is continuous beyond the outer boundary.

These examples provide a simple illustration of the symmetric refraction of the field by a local maximum in the field magnitude, i.e., in the index of refraction. As one can see, the parameter $2\varepsilon kL$ must be at least of the order of one to produce a gap in the field pattern, which is the basis for a tangential discontinuity. Whether the gap is a single hole in the flux surface, or whether it is two holes separated by a strip of continuous field, depends upon the flatness of the summit of the local maximum. As a guess, the double gap, produced when the fourth-order (or higher) term in the expansion of $f(y)$ dominates the second-order term, is less common in nature than the single gap, produced when the expansion is dominated by the second-order term.

7.5 Non-Euclidean Surfaces

The illustrative examples of the refraction of field lines have been restricted so far to Euclidean flux surfaces, i.e., flux surfaces that roll out flat without deformation. In particular, we have been concerned with the refraction of field lines around a local maximum in the field magnitude, producing a gap in the field, i.e., a hole in the flux surface, when the maximum is sufficiently concentrated. On the other hand the flux surfaces in 3D magnetic fields are generally not flat, i.e., not Euclidean, and it is important to understand the consequences (Parker, 1989b).

It follows from simple geometrical considerations that positive curvature (e.g., the surface of a sphere) enhances the formation of gaps, while negative curvature (e.g., a saddle-shaped surface) inhibits their formation. This may be seen from the fact that in the presence of a uniform index of refraction, the ray path is simply a geodesic of the flux surface. Figure 7.12 is an idealized sketch of a central region of

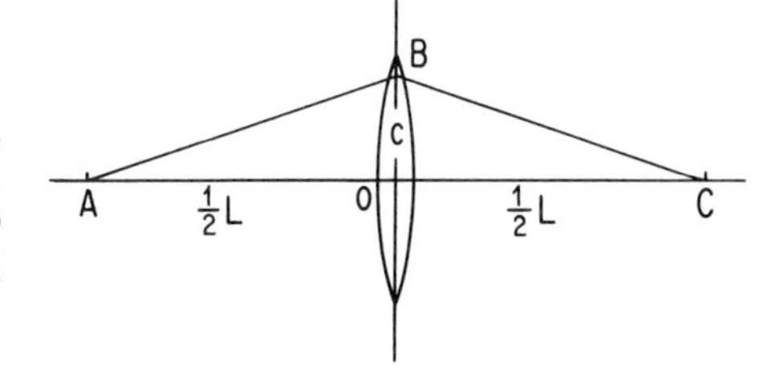

Fig. 7.12. A sketch of the geodesics AB, BC, and AOC on a surface of uniform curvature R. The oval drawn across the line AOC at O indicates a highly localized region of enhanced index of refraction.

enhanced index of refraction (indicated by the thin oval) and the two ray paths AOC and ABC connecting the points A and C separated by a distance L. The straight line AOC is the geodesic, representing the shortest geometrical path length from A to C. The path deflected around the end of the maximum, offset from the geodesic by a distance c, has a greater geometrical length, obviously, but by avoiding the enhanced index of refraction, it may have a shorter optical path length. A gap arises when the increased geometrical length of the deflected path is more than offset by the avoidance of the enhanced index of refraction, so that the deflected path ABC has a shorter optical path length. The essential point is that on a surface with positive curvature, the increased geometrical path length for a small offset $c(\ll L)$ is smaller than the value $2c^2/L$ for a flat surface, which in turn is smaller than the increased geometrical length on a surface of negative curvature. To take an extreme example, if A and B represent two diametrically opposite points on the surface of a sphere, the deflected path lengths are not increased at all.

If follows that a gap is produced on a surface of positive curvature with a relatively smaller increase in the index of refraction, because the path length penalty for avoiding the enhanced index of refraction is less. Similarly, on a surface of negative curvature, the increase in geometrical path length to avoid the enhanced index of refraction is greater and so provides a shorter optical path length only for a relatively stronger focal enhancement of the index of refraction.

The effect of the curvature of the flux surface may be demonstrated formally by computing the necessary deflection of the ray path at the point B in Fig. 7.12 as a function of the offset c on surfaces of positive and negative curvature. To keep the calculation at an elementary level, it is assumed that the index of refraction is uniform everywhere except in a thin region of enhancement along the midline. Thus, AB and BC are both geodesics. The deflection of the ray path at B takes place over a region of thickness h that is so small $(h \ll L, R)$ that the curvature R of the surface does not influence the local refraction. It is a straightforward procedure to compute the angular deflection of AB into BC at the point B.

The calculation begins by noting that the line element ds on the surface of a sphere of radius R can be written as

$$(ds)^2 = R^2\left[(d\theta)^2 + \sin^2\theta(d\varphi)^2\right] \tag{7.76}$$

in terms of the polar angle θ and the azimuthal angle φ relative to some arbitrary point O on the surface of the sphere. For negative curvature the line element is

$$(ds)^2 = R^2[(d\theta)^2 + \sin h^2\theta(d\varphi)^2]. \tag{7.77}$$

With reference to Fig. 7.12, the angle θ measures distance $R\theta$ from the origin O. It is convenient to measure azimuthal φ from the direction OB, so that the point B has coordinates (θ, φ) given by $\varphi = 0$ and $\theta_m = c/R$. The points A and C have $\varphi = \pm\frac{1}{2}\pi$ respectively, with $\theta_n = L/2R$. Fermat's principle requires that

$$\delta \int d\theta[1 + \Sigma(\theta)^2\varphi'^2]^{\frac{1}{2}} H(\theta, \varphi) = 0 \tag{7.78}$$

for an index of refraction $H(\theta, \varphi)$, where $\varphi' \equiv d\varphi/d\theta$ and $\Sigma(\theta) = \sin\theta$ for positive curvature and $\sin h\theta$ for negative curvature. The Euler equation is

$$\frac{d}{d\theta}\frac{\varphi'\Sigma^2 H}{(1 + \Sigma^2\varphi'^2)^{\frac{1}{2}}} = (1 + \Sigma^2\varphi'^2)^{\frac{1}{2}}\frac{\partial H}{\partial \varphi} \tag{7.79}$$

We are interested in the geodesic paths AB and BC, along which $H = constant$. The Euler equation integrates to

$$\varphi'\Sigma^2 = D(1 + \Sigma^2\varphi'^2)^{\frac{1}{2}}$$

where D is the constant of integration. This differential equation reduces to the quadrature

$$\varphi - \varphi_0 = \pm D\int \frac{d\theta}{\Sigma(\Sigma^2 - D^2)^{\frac{1}{2}}}$$

where φ_0 is the constant of integration. For a surface of uniform radius of curvature R the result is

$$\sin 2(\varphi - \varphi_0) = \frac{(D^2 + 1)\sin^2\theta - D^2}{(D^2 - 1)\sin^2\theta}. \tag{7.80}$$

Then for the path BC the polar angle of B is $\theta_n = c/R$ at $\varphi = 0$, and the polar angle of C is $\theta_n = L/2R$ at $\varphi = \frac{1}{2}\pi$, with the result that

$$\sin 2(\varphi - \varphi_0) = \frac{(\sin^2\theta_m + \sin^2\theta_n)\sin^2\theta - 2\sin^2\theta_m \sin^2\theta_n}{(\sin^2\theta_m + \sin^2\theta_n - 2\sin^2\theta_m\sin^2\theta_n)\sin^2\theta} \tag{7.81}$$

with

$$\sin 2\varphi_0 = \frac{\sin^2\theta_m - \sin^2\theta_n}{\sin^2\theta_m + \sin^2\theta_n - 2\sin^2\theta_m\sin^2\theta_n)},$$

$$\cos 2\varphi_0 = \pm\frac{2\sin\theta_m\cos\theta_m\sin\theta_n\cos\theta_n}{\sin^2\theta_m + \sin^2\theta_n - 2\sin^2\theta_m\sin^2\theta_n},$$

$$D^2 = \frac{\sin^2\theta_m\sin^2\theta_n}{\sin^2\theta_m + \sin^2\theta_n - \sin^2\theta_m\sin^2\theta_n}.$$

It is then a simple exercise to show that

$$\frac{d\theta}{d\varphi} = \pm\frac{\sin^2\theta_m\cos\theta_n}{\sin\theta_n} \tag{7.82}$$

at the point B, $(\varphi = 0,\ \theta = \theta_m = L/2R)$.

For negative curvature $\Sigma = \sin h\theta$ the corresponding results are

$$\sin 2(\varphi - \varphi_0) = \frac{(1 - D^2)\sin h^2\theta - 2D^2}{(1 + D^2)\sin \mathrm{h}^2\theta} \tag{7.83}$$

so that for the same end points

$$\sin 2(\varphi - \varphi_0) = \frac{(\sin h^2\theta_m + \sin h^2\theta_n)\sin h^2\theta - 2\sin h^2\theta_m \sin h^2\theta_n}{(\sin h^2\theta_m + \sin h^2\theta_n + 2\sin h^2\theta_m \sin h^2\theta_n)\sin h^2\theta}$$

with

$$\sin 2\varphi_0 = \frac{\sin h^2\theta_n - \sin h^2\theta_m}{\sin h^2\theta_m + \sin h^2\theta_n + 2\sin h^2\theta_m \sin h^2\theta_n},$$

$$\cos 2\varphi_0 = \pm\frac{2\sin h\theta_m \cos h\theta_m \sin h\theta_n \cos h\theta_n}{\sin h^2\theta_m + \sin h^2\theta_n + 2\sin h^2\theta_m \sin h^2\theta_n},$$

$$D^2 = \frac{\sin h^2\theta_m \sin h^2\theta_n}{\sin h^2\theta_m + \sin h^2\theta_n + \sin h^2\theta_m \sin h^2\theta_n}.$$

It is then a simple exercise to show that

$$\frac{d\theta}{d\varphi} = \pm\frac{\sin h^2\theta_m \cos h\theta_n}{\sin h\theta_n}. \tag{7.84}$$

To compare these results with a flat space, consider the slope of AB or BC at B. The geometrical slope $S = |Rd\theta/cd\varphi|$ relative to the line AOC is

$$S_+ = \frac{R\sin^2(c/R)\cos(L/2R)}{c\sin(L/2R)} \tag{7.85}$$

from equation (7.82) for a surface of positive curvature, and

$$S_- = \frac{R\sin h^2(c/R)\cos h(L/2R)}{c\sin h(L/2R)} \tag{7.86}$$

from equation (7.84) for a surface of negative curvature. For $c^2 \ll L^2 \ll R^2$ these two expressions reduce to

$$S_\pm \cong \frac{2c}{L}\left[1 \pm \frac{L^2}{12R^2} + \ldots\right] \tag{7.87}$$

to lowest order. In a flat space ($R = \infty$) the slope is just $S \cong 2c/L$. The refraction of the ray path at the point B is required to be $2S_\pm$, which is smaller by the factor $(1 - L^2/12R^2)$ for the surface of positive curvature and larger by $(1 + L^2/12R^2)$ for negative curvature.

Examples of ray paths in surfaces of nonvanishing curvature in a variable index of refraction $H(\theta)$ are available in the literature (Parker, 1989a,b), illustrating the point made here, that positive curvature is conducive to the formation of gaps by local maxima in H and negative curvature is inhibitory. The flux surfaces in the quasi-equilibrium magnetic fields of stars and galaxies may be expected to have curvature of either sign, so at present it is not apparent that there is any net effect of the generally non-Euclidean nature of the flux surfaces.

7.6 Refraction in a Slab and Field Line Topology

Consider the refraction in a slab of laminar force-free field $-h < z < +h$ described in equation (5.44), by the application of a local pressure maximum in the vicinity of the origin. The field is anchored in the distant circular boundary $\varpi = R$. This problem was treated by formal integration of the force-free field equations in §§ 5.4–5.5.2. It will now be treated with the optical analogy, permitting the solution to be constructed by inspection of Fig. 7.6(c).

The undisturbed field is (equations 5.44 and 5.45)

$$B_x = +B_0 \cos qz, \ B_y = -B_0 \sin qz, \ B_z = 0,$$

throughout the slab $-h < z + h$, so that the field is uniform with a strength B_0 in each flux surface $z = \lambda$ and is inclined at an angle qz to the x-axis. The field is confined by a uniform pressure $B_0^2/8\pi$ applied to the boundaries $z = \pm h$. The field is subsequently deformed by the applied pressure $[1 + F^2(\varpi)]B_0^2/8\pi$ in the circular region $\varpi \leqslant a$, with the uniform pressure $B_0^2/8\pi$ in $a < \varpi < R$. Assuming that $qR \gg qa \gg qh = O(1)$, it follows that the z component of the deformed field is small $O(h/a)$ so that the field magnitude is determined directly by the applied pressure. Then

$$B \cong B_0\left[1 + F^2(\varpi)\right]^{\frac{1}{2}}.$$

It follows from the optical analogy in the limit of large R that the field is excluded from $\varpi < a$, and the field lines are straight in the region $\varpi > a$. Hence the field pattern contains a diamond-shaped gap, in each flux surface, shown schematically in Fig. 7.13. The flux surfaces are deformed from the strictly planar $z = \lambda$ to a plane surface indented slightly in the vicinity of $\varpi = a$, to $z = \lambda[1 - \Lambda(\varpi, \varphi, \lambda)]$ where $\Lambda(\varpi, \varphi, \lambda) \sim 0$ for $\varpi \gg 4a$.

The precise nature of the discontinuity arising in $a < \varpi < R$ from the gap in each flux surface follows immediately from a superposition of flux surfaces, shown schematically in Fig. 7.14, in which the vertices of the gap in the uppermost surface are labeled PY and in the lowermost surface TU. Consider either of the points

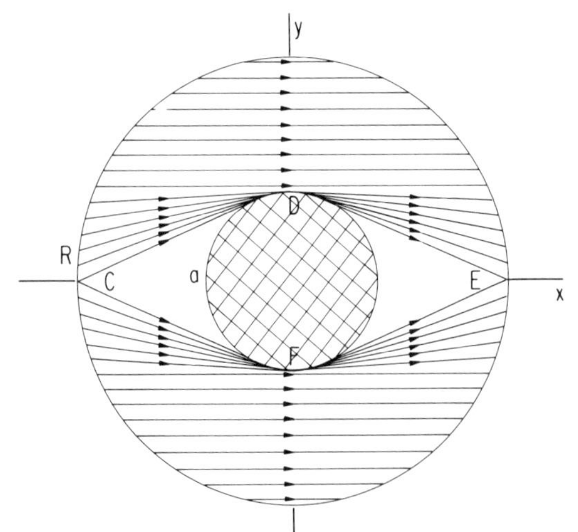

Fig. 7.13. A schematic drawing of the lines of force in the flux surface $z = 0$ fixed at $\varpi = R$ and deformed by application of an enhanced pressure within $\varpi = a$ (cross-hatched), producing the diamond-shaped gap CDEF in the flux surface.

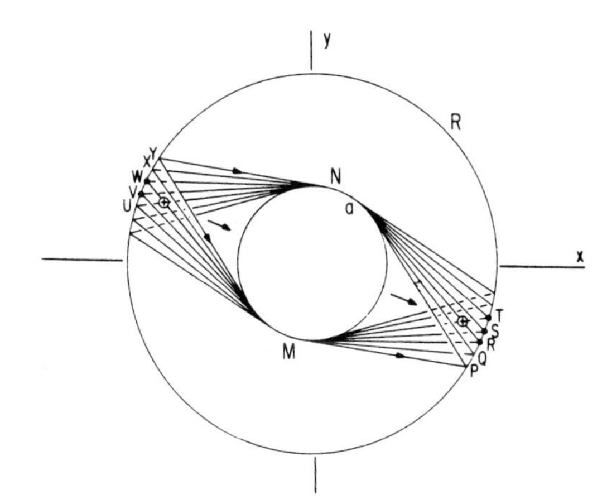

Fig. 7.14. A schematic drawing of the lines of force at the boundaries of the gaps in successive initial flux surfaces $z = \lambda > 0$, sketched in Fig. 7.13. at $z = 0$.

marked by the + sign in Fig. 7.14, at a radial distance ϖ. The essential point is that the field line RN crosses over the field line TM at the + on the right-hand side and WM crosses over UN at the + on the left hand side. The cross-over is at an angle 2θ with $\sin\theta = a/\varpi$. For $\varpi \gg a$ the field strength is B_0, so there is a jump from TM to RN in the azimuthal component in the amount $-2B_\perp$ where $B_\perp = B_0 \sin\theta$. Hence the field discontinuity is $2B_0 a/\varpi$ and the surface current density associated with the tangential discontinuity follows from Ampere's law. The surface current is in the outward radial direction with magnitude

$$J = (cB_0/2\pi)a/\varpi. \tag{7.88}$$

These results are precisely those obtained earlier in equations (5.62)–(5.70) at the outer boundary $\varpi = R$. They follow here from the optical analogy in more general form and with substantially less computational effort.

Note that the jump in the azimuthal field component at the + point on the left hand side of Fig. 7.14, in passing from UN to WM, is $+2B_\perp$ so that the surface current density is again given by (7.88) except that the current is now radially inward, as indicated by the arrows in Fig. 7.14.

The surface of tangential discontinuity, i.e., the current sheet, lies in the spiral surface in which the unperturbed field is radial ($B_\varphi = 0$), $\varphi + qz \cong 0, \pi$. It is evident from the right-hand rule and from Fig. 7.14 that the current is directed radially outward in the spiral surface $\varphi + qz = 0$ and radially inward in $\varphi + qz = \pi$. It is also evident from Fig. 7.14 that the individual field line, e.g., RN, that passes across the upper side of the spiral surface of discontinuity $\varphi + qz = 0$ on the right-hand side, then passes across the under side of the other surface $\varphi + qz = \pi$ where it is labeled WN on the left hand side. Conversely, the field line, e.g., RM, lying on the underside of $\varphi + qz = 0$ on the right-hand side passes across the upper side of $\varphi + qz = \pi$, where it is labeled WM on the left-hand side. The essential point is that any given field line passes from one side to the other of different surfaces of tangential discontinuity. A field line does not lie along the surface of only a single surface of discontinuity. This point was established by the work of Van Ballegooijen (1988) and Field (1990), who showed that when a line is confined to a single surface, conservation of electric current ($\nabla \cdot \mathbf{j} = 0$) is incompatible with the current being everywhere parallel to the field $\mathbf{B}$, as it is in a force-free equilibrium field. They interpreted their calculation as proof that there can be no surfaces of tangential discontinuity as a consequence of continuous deformation of an initially continuous field, whereas their work proves instead that surfaces of discontinuity do not have the simple form that they imagined. The optical analogy applied to the slab of laminar force-free field shows the surfaces to be of more complex topology, for which current is conserved in force-free equilibrium.

To prove that the field lines on either side of a surface of tangential discontinuity pass off onto other surfaces of discontinuity, suppose that the field lines do not pass over other surfaces of discontinuity but remain on only a single surface. In particular, consider the geometry of two field lines on opposite sides of a single surface with footpoints separated by only an infinitesimal distance, sketched in Fig. 7.15. To fix ideas suppose that one end of the surface, or current sheet, lies at the boundary $z = 0$ where the footpoints of the two field lines are labeled P and P'. The other end of the surface lies at the boundary $z = L$, where the footpoints of the field lines through P and P' are labeled Q and Q', respectively. The separation of P and P' is infinitesimal, and, with the assumption that the motion ψ (in equations 1.3 and 1.4) that deforms the field is continuous, it follows that the separation of Q and Q' is also infinitesimal. The field line from P to Q in Fig. 7.15 is labeled PMQ and is shown by the solid curve, while the line connecting $P'Q'$ is the solid curve $P'NQ'$. The mirror images of PMQ and $P'NQ'$ in the surface are shown by the dashed curves and are labeled $P'MQ'$ and PNQ, respectively.

Now the field immediately opposite PMQ across the surface on $P'MQ'$ has the same magnitude, in order that there be pressure balance across the surface, but, of course, the direction of the field on $P'MQ'$ differs from the direction on PMQ by some finite amount θ, because of the presence of the discontinuity. Similarly the field on PNQ immediately opposite $P'NQ'$ has the same magnitude but a different direction in some amount θ. Consider, then, the line integral of $\mathbf{B}$ around the closed contour $P'MQ'QMP$, along which distance from $z = 0$ is measured by s, and around the closed contour $P'NQ'QNP$, along which distance is measured by t. These line integrals measure the total current I through the respective contours, with

$$I_M = \frac{c}{4\pi}\left[\int_{P'MQ'} dsB(s)\cos\theta(s) - \int_{QMP} dsB(s)\right]$$

and

$$I_N = \frac{c}{4\pi}\left[\int_{P'MQ'} dtB(t) - \int_{QNP} dtB(t)\cos\theta(t)\right]$$

The integrals across PP' and QQ' are infinitesimal, if not identically zero, so they can be neglected. The current is everywhere parallel to the field in force-free equili-

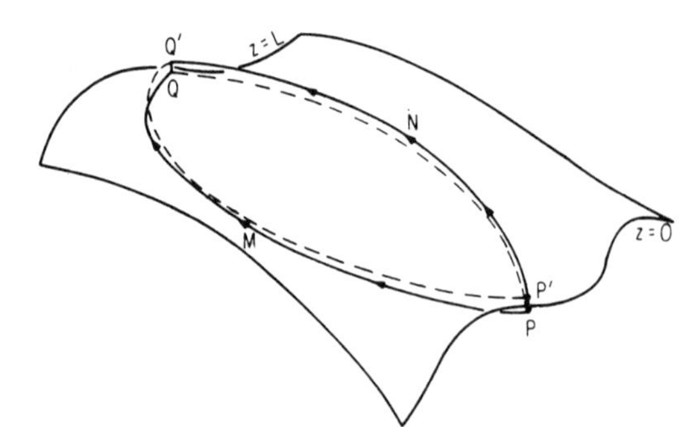

Fig. 7.15. A sketch of a hypothetical current sheet and two field lines of force, PMQ and P′NQ′, on opposite sides of the current sheet with footpoints separated by the infinitesimal distances PP′ and QQ′. The dashed curves P′MQ′ and PNQ represent the mirror images of the lines of force in the current sheet.

brium, so there can be no current across the surface bounded by the contour $PMQNP$ and across the surface bounded by $P'MQ'NP'$. Hence the current I_N flowing into the region from the right must be carried out by I_M on the left, i.e., I_N and I_M must be equal if current is conserved in force-free equilibrium. However, writing the integrals as

$$I_M = -\frac{c}{4\pi}\int_{PMQ} ds B(s)[1 - \cos\theta(s)],$$

$$I_N = +\frac{c}{4\pi}\int_{PNQ} dt B(t)[1 - \cos\theta(t)],$$

it is evident that I_M is negative definite and I_N is positive definite so long as there is a discontinuity so that $\theta \neq 0$. It follows that I_N and I_M cannot be equal and current cannot be conserved, from which it follows that surfaces of discontinuity forming spontaneously in a magnetostatic field subject only to continuous deformation have a more complicated structure than the simple surface shown in Fig. 7.15. The double spiral surface of the foregoing example, of the deformation of the laminar force-free field of equation (5.44), provides an illustration of a physically realizable case. Recall from the foregoing discussion that the individual field lines on each side of the surfaces of discontinuity pass over two different surfaces, below one and above the other. In fact, considering the hypothetical inhomogeneous field external to the slab, that is responsible for compressing the slab in the neighborhood of the origin, it is evident that there are additional current sheets branching out and flowing along the surfaces $z = \pm h$ of the slab of laminar force-free field. Figure 7.16 is a schematic drawing of the current sheets, omitting one of the spiral sheets for simplicity. So the topology of the current sheets in any real situation is not at all simple, with the currents flowing in and out and around on the connected surfaces. Chapter 8 takes up the topology of the current sheets in detail.

In summary, then, the example of the local compression of a slab of laminar force-free field shows that the resulting surfaces of tangential discontinuity spread out radially from the region of compression, their strength, i.e., the surface current density, declining inversely with distance so that radial current is conserved. The field lines on opposite sides of the current sheet that cross-over each other at any point on the sheet connect to widely different locations at the boundaries $\varpi = R$ or $z = 0, L$. The field lines from points close together at the boundary, but on opposite sides of a surface of discontinuity, pass off onto other branches of the current sheet before returning to neighboring footpoints on opposite sides of a surface of discontinuity at the other boundary.

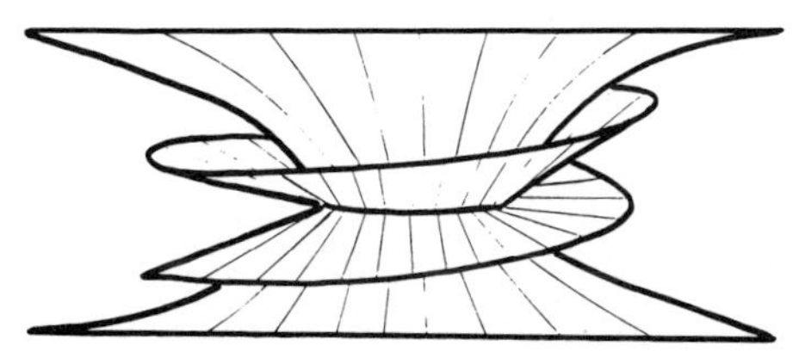

Fig. 7.16. A schematic drawing of the surfaces of discontinuity associated with the deformation of a slab of force-free field. Only one spiral surface is shown, the vertical dimension is greatly exaggerated, and the upper and lower surfaces represent the deformed boundaries of the slab. The light lines are included only to provide perspective.

7.7 Relative Motion of Index of Refraction

The field patterns treated in the foregoing sections are symmetric about their midline, but this need not be the situation in general. For suppose that the local field maximum moves slowly across the field with a small velocity u. The field lines flow over the top of the maximum in a continuous stream unless the maximum is sufficiently strong as to produce a gap. When there is a gap the field lines flow smoothly and continuously up to the leading edge of the gap, but once they cross the edge there is no equilibrium path until they reach the far edge. So they jump across the gap with the Alfven speed (computed in the field component perpendicular to the mean field direction). Upon reaching the far side they again join the continuous stream and flow away from the gap (Parker, 1990a).

To treat a specific case, suppose that the local maximum, defined in equation (7.45), has the specific form

$$f(y) = 1/k(1 + k^2y^2).$$

Then, with reference to Fig. 7.5 the end-points at $x = \pm 2L$, $y = a$ and the midpoint at $x = 0$, $y = c$ are related by equation (7.46), so that

$$a = c[1 - 2\varepsilon kL/(1 + k^2c^2)^2], \tag{7.89}$$

and

$$\frac{da}{dc} = 1 - \frac{2\varepsilon kL(1 - 3k^2c^2)}{(1 + k^2c^2)^3}. \tag{7.90}$$

Suppose, then, that the footpoints of the field at $x = \pm\frac{1}{2}L$ are set into uniform slow motion $da/dt = u > 0$, so that the field lines, or ray paths, remain in quasi-static equilibrium across the region while they drift in the positive y-direction. The motion dc/dt of the midpoint of a field line is related to the motion $da/dt = u$ of the end-points by

$$\frac{dc}{dt} = u\frac{dc}{da} \tag{7.91}$$

Figure 7.17 is a plot of $a(c)$ from equation (7.89) for various $2\varepsilon kL$. Figure 7.18 provides da/dc, from equation (7.90). Figure 7.19 plots dc/da, which gives the rate of motion dc/dt of the field line in units of $u = da/dt$.

To treat a specific example, let $2\varepsilon kL = 4$ and refer to Fig. 7.17, with $da/dt = +u$. A field line starting with large negative ka moves slowly in the positive y direction along the curve for ka as a function of kc. Note from the figure that when ka reaches zero, the field line has been pushed back by the local maximum to $kc = -1$. As ka increases further, kc increases only to about -0.438 at $ka \cong +0.795$. The quantity dc/da increases without bound as ka increases to 0.795 and becomes negative beyond. Thus $kc = -0.438$ represents the lower edge of the gap, with the field line slipping over the top of the maximum when ka increases beyond 0.795. Thus kc increases at the Alfven speed with ka remaining at $ka = 0.795$ until kc reaches a value approximately 1.42 (see Fig. 7.17) where there is an equilibrium again on the upper half of the curve, representing the upper side of the gap. The nonequilibrium path of kc across the gap is indicated by the arrows, from the lower side to the upper side. With u taken to be arbitrarily small compared to the Alfven

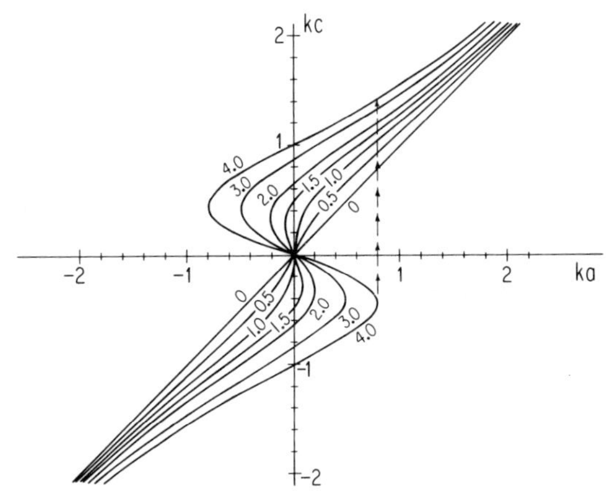

Fig. 7.17. A plot of kc versus ka for the values $2\varepsilon kL = 0, 0.5, 1, 1.5, 2, 3, 4$ indicated on each curve, from equation (7.89). The arrows indicate the path of kc from the lower side to the upper side of the gap for the case $2\varepsilon kL = 4$.

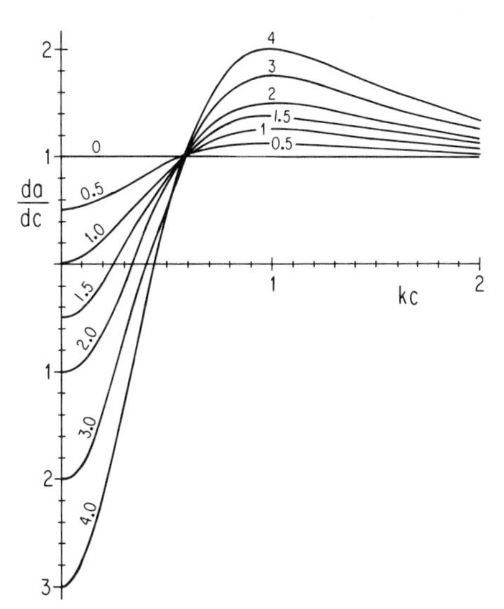

Fig. 7.18. A plot of da/dc for the values of $2\varepsilon kL$ indicated on each curve, from equation (7.90).

speed, the speed dc/dt with which the jump occurs is essentially infinite in units of u. This is reflected in the divergence of dc/da (plotted in Fig. 7.19) at the appropriate values of kc when there is a gap, $2\varepsilon kL > 1$. It is evident that the gap is asymmetric about $y = 0$. The asymmetry increases monotonically with increasing $2\varepsilon kL$, starting with a symmetric gap of vanishing width at $2\varepsilon kL = 1$. Figure 7.20 is a plot of the coordinates kc of the edges of the gap, with the position ka of the ends. In the limit of large $2\varepsilon kL$, it is readily shown from equation (7.90) that da/dc vanishes at $k^2c^2 = \frac{1}{3} - \zeta$, so that

$$kc \cong -\frac{1}{3^{\frac{1}{2}}}\left[1 - \frac{3}{2}\zeta - \frac{9}{8}\zeta^2 + \ldots\right]$$

where, with $2\varepsilon kL \equiv N$,

$$\zeta \cong \frac{64}{81N}\left(1 - \frac{16}{9N} + \ldots\right).$$

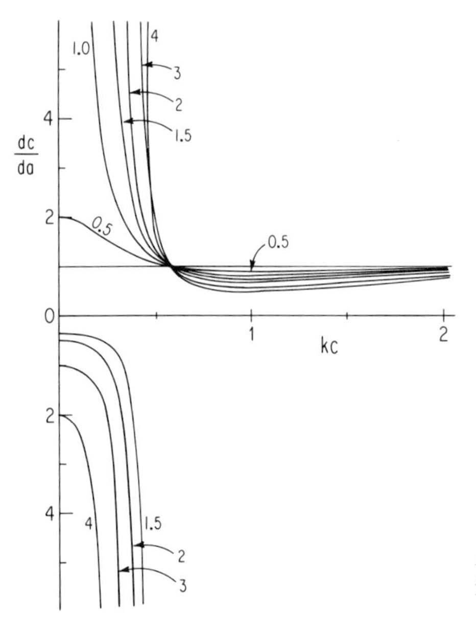

Fig. 7.19. A plot of dc/da for the values of $2\varepsilon kL$ indicated on each curve, from equation (7.90).

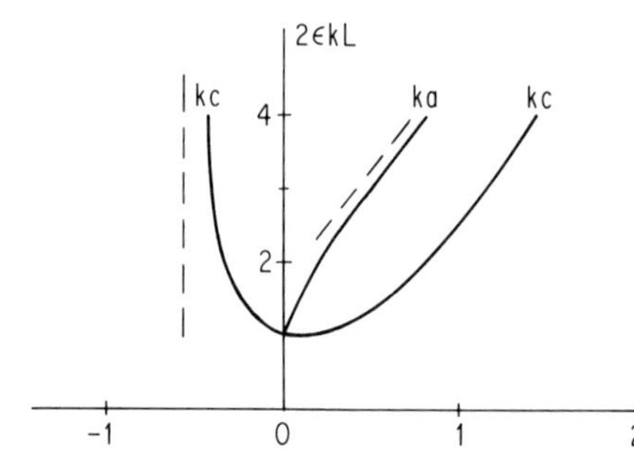

Fig. 7.20. A plot of kc on each side of the gap, with the location ka of the end-points, as a function of $2\varepsilon kL$. The dashed lines represent the asymptotes at large $2\varepsilon kL$.

This value of kc represents the lower edge of the gap,

$$kc \cong -\frac{1}{3^{\frac{1}{2}}}\left(1 - \frac{32}{27N} + \frac{1024}{729N^2} + \ldots\right). \tag{7.92}$$

The coordinate of the end-point follows from equation (7.89) as

$$ka \cong \frac{3^{\frac{3}{2}}N}{16}\left(1 - \frac{16}{9N} + \frac{256}{243N^2} + \ldots\right). \tag{7.93}$$

The upper edge of the gap follows from equation (7.89), solving for kc with this value of ka. Then it is readily shown that

$$kc \cong ka\left[1 + \frac{65536}{729N^3}\left(1 + \frac{64}{9N} + \ldots\right)\right] \tag{7.94}$$

Several notable features appear in these asymptotic forms. First of all, it is evident from equation (7.92) that the lower edge of the boundary approaches the fixed value $kc = -3^{-\frac{1}{2}} = -0.577$ in the limit of a strong maximum $2\varepsilon kL$. One might have thought that the gap edge would recede toward $-\infty$ as the strength of the maximum increased. Instead, the lower edge becomes fixed and ka increases in direct proportion to the strength, N. That is to say, the field lines bend increasingly sharply around the maximum before they are pulled across. This property depends on the precise form of the maximum in $f(c)$, of course.

The next point is that the upper edge of the gap, given by equation (7.94), becomes asymptotically straight in the limit of a strong maximum, as kc approaches ka. The field lines are increasingly sharply bent, into a hairpin form as $N = 2\varepsilon kL$ increases, until they slip over the top of the maximum. Their new equilibrium position is then essentially straight across between footpoints $y = a$ at $x = \pm 2L$.

The physically important aspect of the asymmetric gap created by a slow quasi-equilibrium drift of the field across the maximum (or vice versa) is the high speed motion of the thin sheet of field and fluid sliding across the gap. This layer of high speed fluid moves at the Alfven speed computed in the transverse component of the magnetic field, of the order of $(a - c)B/L$ if B is the local mean field, and a and c represent the coordinates of the lower edge of the gap. This velocity is independent of the very small rate of drift u of the applied maximum relative to the field in the flux surface under consideration. The thickness of the high speed layer diminishes in proportion to u, of course, and in the real world resistivity, viscosity, and Helmholtz and resistive instabilities must dominate the scene at sufficiently small u. However, the transverse component of the Alfven speed is estimated to be hundreds of km/sec for $2\varepsilon kL \cong 2$ in such common places as the X-ray corona of the Sun. So the effect may be of interest in the field dissipation and fluid acceleration. Formal examples of the fluid motion may be found in Chapter 9 and in Parker (1989c). For the present note that if the characteristic Alfven speed is $C = B/(4\pi\rho)^{\frac{1}{2}}$, then the transverse component $C_\perp$ is of the order of $(a - c)C/L$, computed for the upstream edge of the gap. Thus, if a strong field maximum drifts with a small velocity u across a flux layer of thickness h, the thickness δ of the high speed sheet of fluid streaming across the maximum is of the order of $hu/C_\perp$. As an example, then, C may be 2000 km/sec in the X-ray corona of the Sun, so that $C_\perp$ might be 300 km/sec for $2\varepsilon kL = 2$ (see Fig. 7.20). If h is 300 km, a slow drift $u = 1$ km/sec provides a thickness $\delta = 1$ km.

The dynamics of such a thin high speed layer is complex and, so far as we are aware, has not been developed under appropriate conditions. It is, of course, of the nature of a 2D jet, on which there is an extensive literature, but it is a jet containing, and confined by, transverse magnetic fields and driven by its own internal field. What is more, it is a jet across a region where rapid neutral point reconnection of field would otherwise (if $u = 0$) occur. The presence of the jet, or high speed fluid sheet, presumably reduces, or stops, reconnection. The jet is itself affected by the reconnection conditions.

7.8 Bifurcation of Fields

The illustration of the formation of gaps around local field maxima raises the question of how the conventional continuous solutions of field equations, such as the Grad–Shafranov equation, Laplace's equation, the Helmholtz equation, etc. avoid the gaps. The answer is, of course, that the continuous solutions are made up of properly balanced combinations of field maxima and minima such that the clustering effect of the minima is sufficient to draw the field lines smoothly over the maxima, thereby avoiding any gaps. Thus, for instance, the X-type neutral point,

$$H_x = x,\ H_y = -y,$$

involves the refraction of the field minimum

$$\begin{aligned} H &= (H_x^2 + H_y^2)^{\frac{1}{2}}, \\ &= \varpi, \end{aligned} \tag{7.95}$$

where $\varpi = (x^2 + y^2)^{\frac{1}{2}}$, so that there are no gaps in the field. Note, however, that the X-type neutral point represents a bifurcation of the field. The field lines approaching the neutral point, from either the x- or y-directions, are deflected through 90° into the y- or x-directions, respectively.

To see how the X-type neutral point and its local minimum in H fit into the bifurcation of a field at a local maximum in H, consider the potential field

$$\phi = \left(\varpi + \frac{a^2}{\varpi}\right) \cos \varphi \tag{7.96}$$

where φ is azimuth measured from the positive x-axis. The field $\mathbf{H} = -\nabla\phi$ has components

$$H_\varpi = -(1 - a^2/\varpi^2) \cos \varphi, \tag{7.97}$$

$$H_\varphi = +(1 + a^2/\varpi^2) \sin \varphi, \tag{7.98}$$

so that the magnitude is

$$H = [(1 + a^4/\varpi^4) - 2(a^2/\varpi^2) \cos 2\varphi]^{\frac{1}{2}}. \tag{7.99}$$

Then

$$\frac{\partial H^2}{\partial \varpi} = -\frac{4a^4}{\varpi^5}\left(1 - \frac{\varpi^2}{a^2} \cos 2\varphi\right),$$

from which it is obvious that H^2 declines with increasing radial distance from the origin for all $\varpi < a$. Figure 7.21 is a plot of the isobars ($H^2 = constant$) (light lines) showing a central maximum that is elongated in the $\pm\, y$-directions, and shortened in the $\pm\, x$-directions by the appearance of a local minimum on each side of the maximum. The field lines are refracted by the maximum and the two minima to the form

$$(\varpi/a - a/\varpi) \sin \varphi = \Lambda, \tag{7.100}$$

where Λ is a constant for each line, shown by the heavy lines in Fig. 7.21. Noting from equation (7.96) that $\nabla^2\phi = 0$, it is obvious that ϕ represents a potential field that is uniform and in the x-direction except for the fact that it is excluded from the interior of the circle $\varpi = a$. Thus there is a circular gap $\varpi \leqslant a$ in the exterior field, shown in Fig. 7.21, caused by the pressure maximum around the origin. The gap is circular rather than pointed at each end (as in Fig. 7.6), because the minima in H on each side of the maximum (centered at $x = \pm a$, $y = 0$) pull the field lines together into the two neutral points at $x = \pm a$, $y = a$. It is readily shown that equation (7.98) reduces to

$$H^2 \cong 4[x \pm a)^2 + y^2]$$

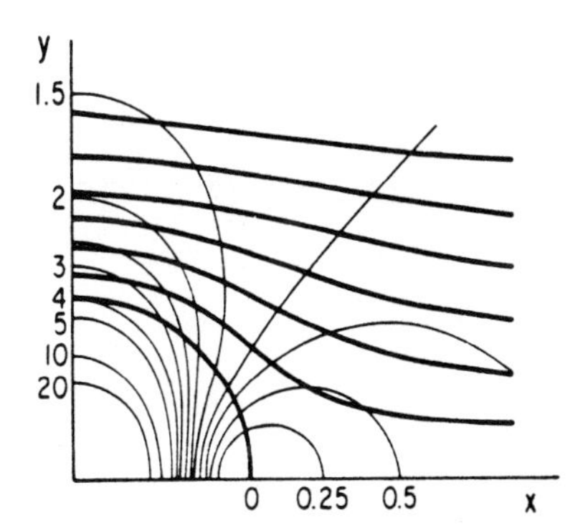

Fig. 7.21. The light lines represent contours of constant H^2, from equation (7.99), while the heavy lines represent the field lines as they are refracted in H, from equation (7.100).

in the vicinity of the two neutral points at $x = \pm a$. This is precisely the linear increase in H with distance from the neutral point that is required to produce the standard X-type neutral point, of course.

Note that the form of H inside the circle $\varpi = a$ is now immaterial, as the field is excluded from that region by the inward increasing field strength around the boundary $\varpi = a$ of the gap.

Now the bifurcation of a field can take place at an X-type neutral point, as when H is proportional to distance from the point, or the bifurcation can be at any arbitrary angle $\pi - \beta$, sketched in Fig. 7.22. To achieve the deflection of the field through an arbitrary angle β, let H vary as $\varpi^{\alpha-1}$. Then since $\partial H/\partial\varphi = 0$, the Euler equation is

$$\frac{d}{d\varpi}\frac{\varpi^2\varphi' H}{(1+\varpi^2\varphi'^2)^{\frac{1}{2}}} = 0,$$

where $\varphi' = d\varphi/d\varpi$. This can be integrated twice to

$$\varphi^{2\alpha}\sin^2\alpha(\varphi - \varphi_0) = \Lambda^2$$

where Λ is a constant on each ray path. A field line parallel to the x-axis in the limit of large φ requires $\varphi_0 = 0$. The line veers off to infinity again at $\varphi = \beta$ if $\alpha = \pi/\beta$, requiring $H \sim \varphi^{\pi/\beta-1}$. An appropriate potential is

$$\phi = (\beta/\pi)\varpi^{\pi/\beta}\cos(\pi\varphi/\beta),$$

providing the field

$$H_\varpi = -\varpi^{(\pi/\beta-1)}\cos\frac{\pi\varphi}{\beta},$$

$$H_\varphi = +\varpi^{(\pi/\beta-1)}\sin\frac{\pi\varphi}{\beta},$$

so that $H = \varpi^{\pi/\beta-1}$ as required by the deflection β. The field can as well be described by the vector potential

$$A = \varpi^{\pi/\beta}\sin\pi\varphi/\beta.$$

The field lines are given by $A = constant$. The line $A = 0$ lies on $\varphi = 0$, and on $\varphi = \pm\beta$. Hence the field has the form sketched in Fig. 7.22.

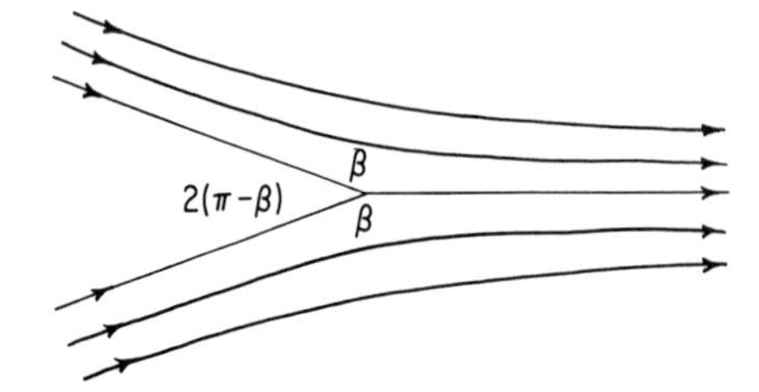

Fig. 7.22. A sketch of a bifurcation of angle $2(\pi - \beta)$ in a field $\mathbf{H}(\varpi, \varphi)$.

However, a bifurcation need not appear as an angle in the field direction but may, in fact, be in the form of a cusp, appearing in Fig. 7.23. To explore the conditions under which this occurs, consider the simple case that the index of refraction $F_s = H$ in the Euler equation (7.13) depends only on $y(\partial H/\partial x = 0)$ and reduces to the form

$$\ln H \cong Q + (ky)^{2(1-\alpha)} \tag{7.101}$$

in the limit that y declines to zero, where Q and $k(>0)$ are constants. Considering a field line tangent to the x-axis at $x = 0$ and splitting with increasing x to form the boundaries of a gap, equation (7.13) reduces to

$$y'' = d \ln H/dy \tag{7.102}$$

in the vicinity of the origin where $y'^2 \ll 1$. Multiply by y' and integrate, obtaining

$$y'^2 = 2[E + Q + (ky)^{2(1-\alpha)}] \tag{7.103}$$

where E is an arbitrary constant, put equal to $-Q$ for the field line under consideration. It follows that

$$ky = \pm(2^{\frac{1}{2}}\alpha kx)^{\frac{1}{\alpha}}, \tag{7.104}$$

representing a general "parabola" tangent to the x-axis at $x = 0$ for any $\alpha > 0$. In other words, the field line bifurcates at $x = 0$, each branch veering away from the

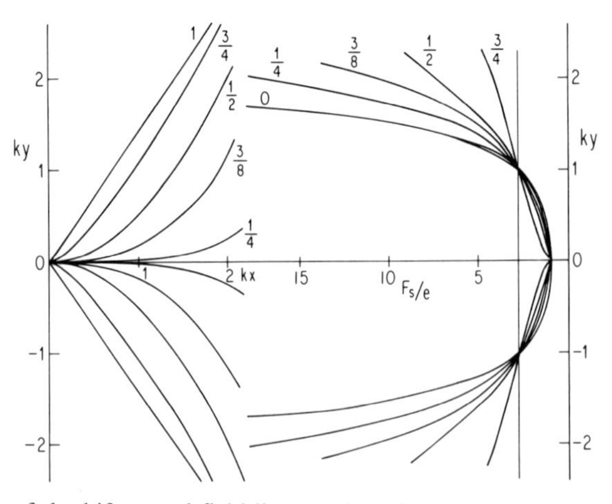

Fig. 7.23. A plot of the bifurcated field line on the left and the profile of the field magnitude F_s in units of $\exp Q$ on the right, from equations (7.104) and (7.101), respectively, for the values of $\alpha = 0, \frac{1}{4}, \frac{3}{8}, \frac{1}{2}, \frac{3}{4}, 1$ indicated on each curve.

x-axis. Figure 7.23 is a plot of the bifurcated field line from $\alpha = 0, \frac{1}{4}, \frac{3}{8}, \frac{1}{2}, \frac{3}{4}, 1$ on the left-hand side, while the right-hand side plots $H\exp(-Q)$ from equation (7.101) to show the variation of $H(y)$ associated with each form of the bifurcated field line.

The field line pattern on either side of the bifurcation is readily calculated from equation (7.103) with $E + Q \neq 0$. Write $E + Q = -(ky_0)^{2(1-\alpha)}$ so that the field line is parallel to the x–axis at $y = y_0$, i.e., $y = y_0$ is the distance of closest approach. The field line follows from the quadrature

$$\int_1^{y/y_0} \frac{ds}{\left[s^{2(1-\alpha)} - 1\right]^{\frac{1}{2}}} = \frac{2^{\frac{1}{2}}kx}{(ky_0)^{\alpha}} \tag{7.105}$$

if symmetry about $x = 0$ is assumed. Thus, if $\alpha = \frac{1}{4}$, let $s = w^2$ so that

$$x = 2^{\frac{1}{2}}(ky_0)^{\frac{1}{4}} \int_1^{(y/y_0)^{\frac{1}{2}}} \frac{dw\, w}{(w^3 - 1)^{\frac{1}{2}}}:$$

This elliptic integral can be expressed as

$$kx = \frac{2^{\frac{1}{2}}(ky_0)^{\frac{1}{4}}}{3^{\frac{1}{4}}} \left\{ (3^{\frac{1}{2}} + 1)\, F(\varphi, k) + 2 \times 3^{\frac{1}{2}} \left[\frac{sn\, u\, dnu}{1 + cnu} - E(\varphi, k) \right] \right\}, \tag{7.106}$$

(Byrd and Friedman, 1954) where

$$k^2 = \frac{1}{2} - \frac{3^{\frac{1}{2}}}{4} \tag{7.107}$$

and

$$cnu = \cos\varphi = \frac{\sqrt{3} + 1 - (y/y_0)^{\frac{1}{2}}}{\sqrt{3} - 1 + (y/y_0)^{\frac{1}{2}}}. \tag{7.108}$$

If $\alpha = \frac{1}{2}$, then the field lines are all parabolas, with

$$ky = ky_0 + \frac{1}{2}(kx)^2.$$

If $\alpha = \frac{3}{4}$, the field lines are given by

$$kx = \frac{1}{3}(4ky_0)^{\frac{3}{4}} \left[\left(\frac{y}{y_0}\right)^{\frac{1}{2}} - 1 \right]^{\frac{1}{2}} \left[\left(\frac{y}{y_0}\right)^{\frac{1}{2}} + 2 \right]. \tag{7.109}$$

The essential point of the calculations presented in this section is to illustrate the bifurcation of a field so as to pass around a maximum in the field magnitude. The splitting of the field is smooth and continuous for $0 < \alpha < 1$, and occurs in the presence of a field magnitude $F_s = H$ with an upward concavity sharper than quadratic, as may be seen from equation (7.101) with $0 < \alpha < 1$. The dependence $(ky)^{2(1-\alpha)}$ is not analytic in the sense that the second derivative becomes large without bound as ky approaches zero, but $(ky)^{2(1-\alpha)}$ is nonetheless a smooth continuous function as y passes across zero, so there is no obvious physical objection. If the variation in field magnitude, determined by the inhomogenous pressure applied to the field, does not have the necessary limiting form, then the bifurcation is not smooth and continuous, and further discontinuity is introduced in association with the gap in the field around a local maximum in the applied pressure.

7.9 Gaps and Various Discontinuities

The foregoing examples have illustrated the gaps that occur in the pattern of the projection $\mathbf{F}_s$ (of a vector $\mathbf{F}$) around a local maximum in F_s. It was pointed out in §7.1 that the gaps in successive flux surfaces provide a hole through a layer of magnetic flux of finite thickness, sketched in Fig. 7.4. The separate fields in the regions on opposite sides of the layer poke into the hole where they come into contact with each other. The separate fields are generally not precisely parallel, nor do they match perfectly to the fields around the sides of the hole, so the contact surface and the sides of the hole represent tangential discontinuities in almost all cases.

An obvious question is what exceptions there might be. That is to say, given a gap in the field pattern in a flux surface, is it possible to construct a continuous field that could be fitted into the gap to provide overall continuity in spite of the *topological* discontinuity represented by the gap. The answer seems to be that a continuous field can be constructed, but it involves second derivatives that are divergent, i.e., the field gradients are not continuous everywhere. The example of the cusp presented at the end of §7.7 is such a case. In magnetic and velocity fields the dissipation in the presence of nonvanishing resistivity and viscosity, respectively, depends on the second derivatives, which then would be unbounded. Therefore it appears that the topological discontinuity represented by the gap in the field pattern around a local maximum in F_s necessarily produces either a discontinuity in $\mathbf{F}_s$ or in its spatial derivatives, so that in any real situation there would be rapid dissipation of the field.

Consider, then, the minimum discontinuity in the field in the presence of a bifurcation of $\mathbf{F}_s$ associated with a gap around a local maximum in F_s. We arbitrarily fit an interior field $\mathbf{f}$ smoothly and continuously around the periphery of the gap sketched in Figs 7.3 and 7.6(c), and inquire how far it can be extended into the gap without producing a discontinuity in either the field or its spatial derivatives.

To begin, then, note that the external field immediately outside gap is not zero. Therefore its magnitude is bounded by F_0, where $F_s > F_0 > 0$, around the entire periphery. Therefore, a field $\mathbf{f}$ within the gap that fits smoothly to the external field is subject to the same lower bound F_0 at the boundary of the gap.

If $\mathbf{f}$ is smooth and continuous, with an upper bound on its spatial derivatives, then its magnitude is nonvanishing for some finite distance inward from the boundary, forming two strips of field around each side of the gap, the strips meeting at the ends of the gap. Suppose, then, that the two boundaries of the gap meet at a finite angle $2(\pi - \beta)$ in the manner sketched in Fig. 7.22. The meeting of the two strips of $\mathbf{f}$ around each boundary, sketched in Fig. 7.24, provides unbounded spatial

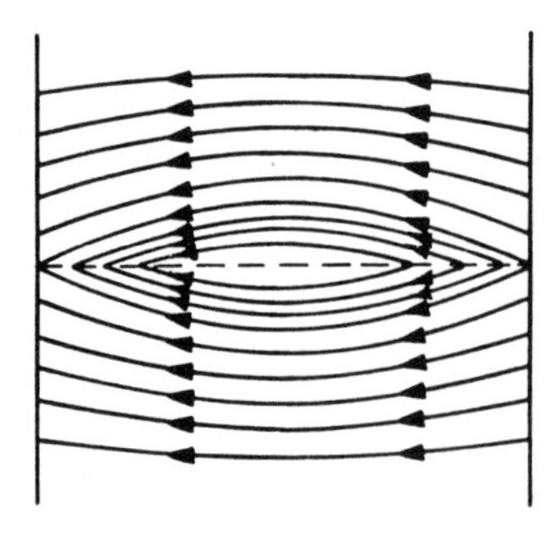

Fig. 7.24. A sketch of the field in the gap shown in the Fig. 7.6(c) tailored to fit smoothly and continuously to the field at the boundary.

gradients at the center line, where the field direction is discontinuous. Obviously this can be avoided if the field lines converge to a common source or sink at the apex. So use polar coordinates (ϖ, φ) with the origin at the apex of the left-hand end of the gap. In the simplest case the magnitude of the external field at the boundary is not only bounded but approaches some limiting value F_1 as the radial distance ϖ declines to zero at the apex. To avoid the discontinuity in field direction at the apex sketched in Fig. 6.24 suppose that the radial and azimuthal components of **f** are

$$f_\varpi = F_1, f_\varphi = 0.$$

Then

$$\nabla \cdot \mathbf{f} = F_1/\varpi,$$

which increases without bound as ϖ declines to zero. If z represents distance perpendicular to the flux surface, then the divergence of the original vector field **F** is

$$\begin{aligned} \nabla \cdot \mathbf{F} &= \nabla \cdot \mathbf{f} + \partial F_z/\partial z \\ &= F_1/\varpi + \partial F_z/\partial z \end{aligned}$$

Now $\nabla \cdot \mathbf{F}$ is bounded if there is no discontinuity at the apex, in which case $\partial F_z/\partial z$ is large without bound as ϖ declines to zero.

To see if this unbounded spatial derivative can be avoided in some way, introduce the cartesian coordinates (x, y) with the origin at the left-hand apex of the gap in Fig. 7.6(c) and the positive x-axis extending along the midline of the gap. Obviously, f_y is an odd function of y, passing across zero at $y = 0$. Close to the origin the lower bound $F_s \gtrsim F_0$ becomes $f_x \cong F_1$ while f_y passes smoothly across zero, at $y = 0$, if there is to be no singularity in the spatial derivatives. Hence the equation for the field lines is

$$\frac{dy}{dx} \cong \frac{f_y(x, y)}{F_1} \tag{7.110}$$

close to the origin. The boundary $y = \pm x \tan \beta$ must fit smoothly to this family of solutions. Suppose, then, that in the neighborhood of the origin

$$f_y(x, y) = F_1 K x^p y^q \tag{7.111}$$

to lowest order, where K is an arbitrary constant and p and q are arbitrary numbers. It follows that

$$\frac{dy}{dx} = K x^p y^q,$$

and integration yields

$$y = \left[\frac{1 - q}{1 + p} K\right]^{\frac{1}{1-q}} (x^{p+1} - x_0^{p+1})^{\frac{1}{1-q}}, \tag{7.112}$$

where x_0 is the integration constant. Now, it is necessary that $q < 1$ if y is to decline to zero, rather than increasing without bound, as x approaches x_0. But if $q < 1$, it follows that the derivative,

$$\frac{\partial f_y}{\partial y} = q K F_1 x^p y^{q-1},$$

diverges as y declines to zero. Further, if $y = x \tan \beta$ is to be a field line (for $x_0 = 0$), we must have $p + 1 = 1 - q$, i.e., $p = -q$, which leads to

$$\frac{\partial f_y}{\partial x} = -\frac{qK}{x} F_1 \tan^q \beta.$$

Hence $\partial f_y / \partial x$ diverges as x declines to zero along $y = x \tan \beta$.

Suppose, then, that the gap terminates at each end in a cusp, e.g., $y = \pm Q x^\gamma$ where $\gamma > 1$. The requirement that $q < 1$ remains, with the associated unbounded increase of $\partial f_y / \partial y$ as y declines to zero. To fit the boundary $y = Q x^\gamma$ (for $x_0 = 0$) it is necessary that $p = (\gamma - 1) - \gamma q$. Then $\partial f_y / \partial x$ varies as $x^{\gamma(1-q)-2}$, which is a constant if $\gamma = 2/(1 - q)$ and declines to zero with decreasing x if $\gamma > 2/(1 - q) > 2$ (for $q < 1$). Hence the cusp may avoid the unbounded increase in $\partial f_y / \partial x$, but the unbounded increase of $\partial f_y / \partial y$ with declining y refuses to go away.

The unbounded increase in the field gradient can be replaced by a discontinuity in the field itself, obviously. Thus, for instance, a uniform field can be put into the cusp, with a tangential discontinuity at the boundary of the gap, or a double ended field, sketched in Fig. 7.21, can be introduced with a discontinuity down the midline. The essential point is that a gap in the field pattern represents a fundamental topological discontinuity, and as such it implies either unbounded spatial derivatives, or what amounts to the same thing, tangential discontinuities of one form or another. Hence, a localized maximum and a gap in a magnetic field implies current sheets while in a velocity field it implies vortex sheets. Conversely, the appearance of current sheets and vortex sheets at the separatrices of distinct topological regions, which press together through a gap in the intervening field or flow, occurs only in the presence of local maxima in the respective fields.

7.10 Topology Around a Field Maximum

The maximum in field magnitude in a 2D surface locus has a close relation to the maximum in the field magnitude at the center of a twisted flux bundle in 3D space. The optical analogy provides an effective topological instrument for establishing the connection. To treat the simplest case, in a region far removed from rigid boundaries, consider a vector field with the properties of a force-free magnetic field **B**, for which $\nabla \cdot \mathbf{B} = 0$ and $\nabla \times \mathbf{B} = \alpha \mathbf{B}$ so that flux is conserved and every flux surface of **B** coincides with a surface locus of $\nabla \times \mathbf{B}$. Imagine, then, that there is a gentle maximum in the magnetic pressure $B^2/8\pi$, and therefore in the effective index of refraction B, centered on some point M in the 2D flux surface S. The result is a modest refraction of the field lines in S around the maximum in B, sketched in Fig. 7.6(a). The field lines spread out as they pass over the maximum.

Consider, then, an elemental flux bundle with total flux $\delta^2\Phi$ and cross-sectional area $\delta^2 A$, so that $\delta^2\Phi = B\delta^2 A$. It follows from conservation of $\delta^2\Phi$ that $\delta^2 A$ declines with increasing B where the flux bundle passes over the maximum in B, at the same place that the field spreads out and the width δw of the bundle in the surface S increases. This requires that the thickness δh of the bundle (in the direction perpendicular to S) decreases by a larger factor than the width δw increases, so that the product $\delta^2 A = \delta w \delta h$ decreases to provide the increase in B.

Now introduce the flux surface $S_\perp$ that is perpendicular to S along the field line

PMP' through the point M at the maximum of B in S. This is sketched in Fig. 7.25(a) with the simplification that the flux surfaces S and $S_\perp$ are shown as planes. In view of the minimum in δh in passing over the maximum in B, the field pattern in $S_\perp$ must be strongly concave in the direction away from S. This concavity requires that B increase across $S_\perp$ with increasing distance from S. But the increase in field magnitude cannot go on indefinitely, so there must be a local maximum in B somewhere nearby. The fact is that the force-free field has a true local maximum magnitude B only on the axis of a twisted bundle of field. Such a maximum is conveniently illustrated by the infinitely long twisted bundle with rotational symmetry about the z-axis, expressible in terms of a generating function $F(\varpi)$ as

$$B_\varpi = 0,\ B_\varphi^2(\varpi) = -\frac{1}{2}F'(\varpi),\ B_z^2(\varpi) = F(\varpi) + \frac{1}{2}\varpi F'(\varpi). \tag{7.113}$$

$F(\varpi)$ is an arbitrary function of distance ϖ from the z-axis, subject only to the conditions that

$$-\frac{2}{\varpi} < \frac{F'}{F} < 0 \tag{7.114}$$

in order that B_z and B_φ are real (Lüst and Schlüter, 1954; Parker, 1979, p. 68). The function $F(\varpi)$ is just the square of the field magnitude, $B^2(\varpi) = B_\varphi^2(\varpi) + B_z^2(\varpi)$, so that with $F'(\varpi) \leqslant 0$, $B(\varpi)$ is necessarily a maximum on the axis.

Figure 7.25(b) puts these facts together, including the torsion in $S_\perp$ as a consequence of nonvanishing α. The flux surface S may, or may not be part of the twisted flux bundle; it is drawn schematically without torsion in Fig. 7.25(b). The essential point is that B increases across the width of $S_\perp$, with a maximum on the axis QQ', around which the field lines in $S_\perp$ are wound. Note that QQ' is drawn parallel to PMP' in Fig. 7.25(b) only for convenience. In fact there is no unique relationship between their directions. The essential point is that the twisting of the field about some local axis is essential to initiate a field maximum in a nearby flux surface in regions far removed from any interfering boundaries.

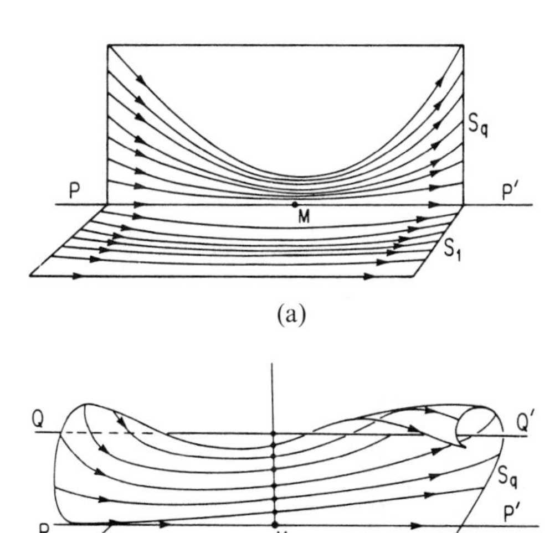

Fig. 7.25. (a) A schematic drawing of the field lines in a flux surface S in which the field magnitude B has a gentle maximum at the point M, together with the perpendicular flux surface $S_\perp$, intersecting S along the field line PMP′. (b) A schematic drawing of the spiral form of $S_\perp$, about an axis SS′.

To provide specific examples, showing the winding of the field about a straight or spiral axis, providing a maximum in field magnitude, consider the flux bundle extending uniformly from $z = 0$ to $z = L$ and wound into the twisted form

$$B_\varpi = 0,\ B_\varphi = B_2 J_1(k\varpi),\ B_z = [B_1^2 + B_2^2 J_0^2(k\omega)]^{\frac{1}{2}} \tag{7.115}$$

by the motion of the footpoints at $z = 0, L$. This equilibrium form follows from (7.113) with the generating function

$$F(\varpi) = B_1^2 + B_2^2[J_0^2(k\varpi) + J_1^2(k\varpi)] \tag{7.116}$$

or from the Grad–Shafranov equation (equation 4.11) with

$$B_z^2(A) = B_1^2 + k^2 A^2 \tag{7.117}$$

and

$$A = (B_2/k)\, J_0(k\varpi). \tag{7.118}$$

Note from equation (7.115) that $B_\varphi = 0$ at the zeros of $J_1(k\varpi)$. So cut off the solution at the first zero $\varpi = R$, where $kR = 3.83$. The field at $\varpi = R$ is in the z-direction with magnitude

$$B(R) = [B_1^2 + B_2^2 J_0^2(kR)]^{\frac{1}{2}} = [B_1^2 + 0.162 B_2^2]^{\frac{1}{2}}, \tag{7.119}$$

so that the twisted flux bundle fits smoothly to an external uniform field of magnitude $B_0 = B(R)$ in the z-direction.

This twisted flux bundle, with the maximum field on the axis ($\varpi = 0$), can be used to illustrate the field rotation effect described in §§6.1 and 7.3. For imagine that a rectangular array of such circular flux bundles is constructed in the form sketched in Fig. 7.26(a) with the uniform field $B(R)$ in the z-direction filling the interstices. The field is everywhere in equilibrium and everywhere continuous. But suppose that, instead of the array of circular flux bundles with interstices, the twisted flux bundles are packed tightly together, as sketched in Fig. 7.26(b). Their surface fields were

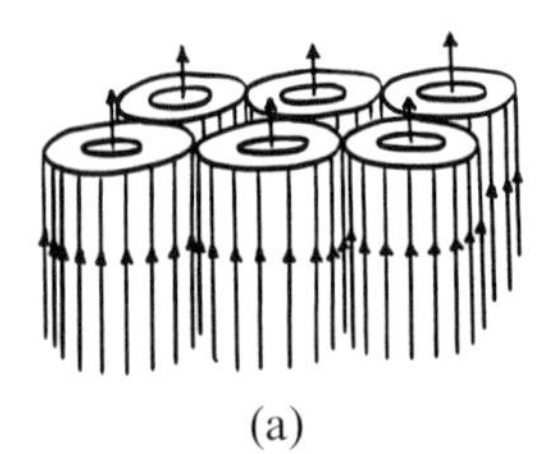

(a)

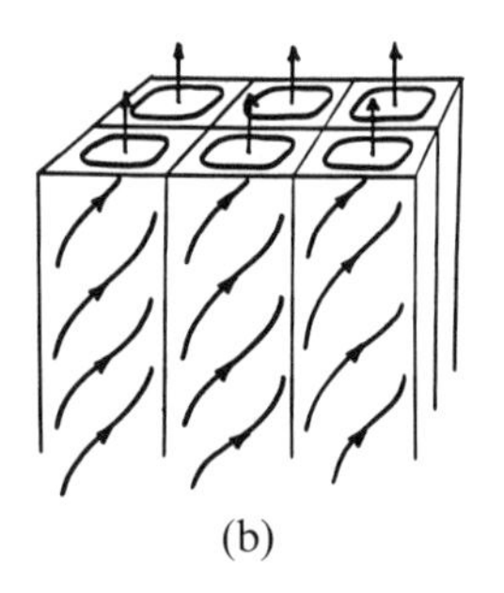

(b)

Fig. 7.26. (a) Schematic drawing of rectangular array of flux tubes with rotational symmetry and a field $(0, 0, B_z)$ at the surface $\varpi = R$, from equation (7.115). (b) Schematic drawing of the same flux tubes when squashed together without interstices, showing the refraction of the field lines at the surface.

originally all in the z-direction but upon being squashed together, the surface rotation effect produces tangential discontinuities at their surfaces or separatrices. For it must be recalled that the field immediately inside the surface, at $\varpi = R - \varepsilon$, is inclined slightly to the z-direction, with

$$B_\varphi \cong -B_2 J_1'(kR)k\varepsilon.$$

So when the twisted bundles are squashed together, the maximum B lies along the centerline of each of the four faces of the deformed tube, and the optical analogy prescribes how the field lines are refracted across each face at a finite angle in the limit of small ε, giving rise to the discontinuity described in §7.3. A more elaborate treatment of the squashing together of circular flux bundles has been published elsewhere (Parker, 1990).

For a second example of the relation of field maxima to tangential discontinuities, note again the case presented in §7.3 where a uniform slab of force-free field is compressed by a strip of enhanced pressure extending uniformly in the z-direction, producing the surface rotation effect and a tangential discontinuity at $x = 0$. The twisted flux bundle described by equation (7.115) is precisely the instrument that produces such a uniform ridge of pressure. Suppose that there is a single twisted flux bundle $\varpi < R$ matching smoothly to the uniform field $B(R)$ in the z-direction in $\varpi > R$. Then suppose that the field has the laminar force-free form

$$B_x = B(R)\sin q(y \pm b),\ B_y = 0,\ B_z = B(R)\cos q(y \pm b) \tag{7.120}$$

in $y > b$ and $y < b$, respectively, shown schematically in $y > b$ in Fig. 7.27(a), where b is any length greater than R. Note that the field is continuous across $y = \pm b$ and the magnitude is uniform with a value $B(R)$ everywhere outside $\varpi = R$. Suppose, then,

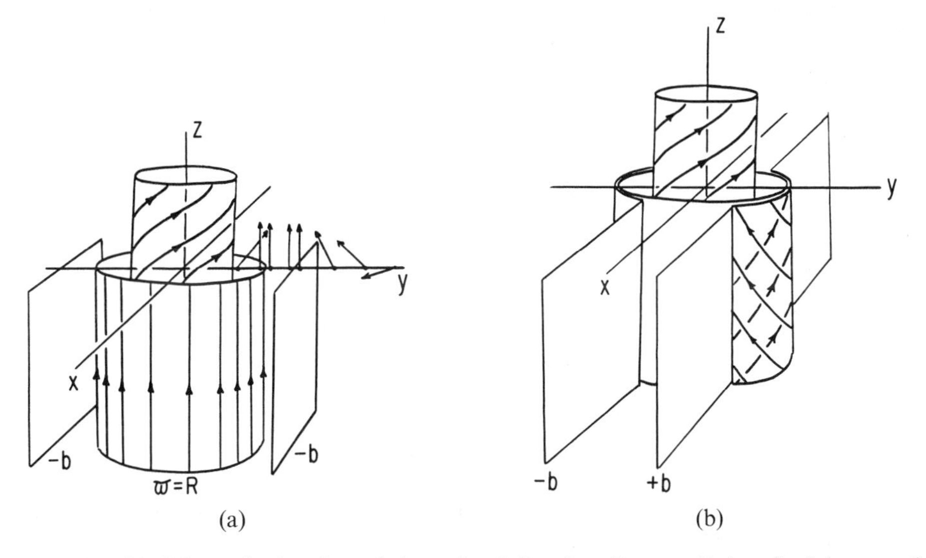

Fig. 7.27. (a) Schematic drawing of the twisted flux bundle $\varpi < R$ described by equation (7.115) and surrounded by uniform field. Beyond $y^2 = b^2$ the space is filled with the force-free field of equation (7.120). The short arrows on the y-axis indicate the field direction. (b) Schematic drawing of the same twisted flux bundle when $b < R$ so that the force-free field is indented by the bundle, which becomes somewhat flattened as a consequence.

that the uniform field in $y^2 < b^2$ is caused (by the continuous motion of the footpoints at $z = 0, L$) to flow continuously and smoothly out in both directions along the x–axis so that the thickness $2b$ of the slab of uniform field declines with the passage of time. The field remains continuous across $\varpi = R$ up to the time that b falls to the value R and the contact surfaces $y = b$ with the laminar force-free field become tangent to $\varpi = R$ at $x = 0$. Further evacuation of uniform field causes b to decrease further, so that the force-free field in $y^2 > b^2$ presses against the twisted flux bundle inside $\varpi = R$, sketched in Fig. 7.27(b). The pressure on opposite sides of the twisted bundle tends to flatten it, of course, and the pressure is a maximum along the sides at $x = 0$. Note, then, that the field immediately inside the surface of twisted bundle is refracted, according to the optical analogy, to a finite angle of inclination to the z-direction, being rotated from the yz-plane backward into the page. On the other hand, the field immediately outside the surface is refracted by the same pressure increase to the same angle of inclination out of the page. This is, of course the surface rotation effect again, described in §7.3. There is then a tangential discontinuity created at the surface of the twisted flux bundle on both sides $y = \pm b$. In fact the pressure increase extends outward through the force-free field in both directions, so that the pressure and field magnitude are a maximum at $x = 0$ for all values y, with the maximum becoming broader and weaker with increasing y^2. It follows, then, that the surface rotation effect of §7.3 occurs at $q(y - b) \cong n\pi$ where $n = 0, 1, 2$, etc. providing surfaces of tangential discontinuity of increasing characteristic width and declining strength with increasing n. These tangential discontinuities are all traceable back to the maximum B on the axis of the twisted flux bundle at $\varpi = 0$.

It is instructive to treat a more complicated example, giving up the invariance $\partial/\partial z = 0$ of the uniform twisted flux bundle and proceeding to a lumpier form of field maximum, providing localized maxima in the field magnitude, rather than the infinitely long uniform ridge of the foregoing invariant case (Parker, 1990). Now the only available continuous analytic solutions involve an ignorable coordinate, so we turn to Tsinganos (1982) who provides solutions to $\nabla \times \mathbf{B} = q\mathbf{B}$, where q is a constant, with helical symmetry rather than linear invariance. With $\varpi^2 = x^2 + y^2$ and $u = \varphi - kz$ he shows that the field is expressible in terms of the function $f(\varpi, u)$ as

$$B_\varpi = \frac{1}{q\varpi}\frac{\partial f}{\partial u}, \quad B_\varphi = \frac{qk\varpi f - \partial f/\partial\varpi}{q(1 + k^2\varpi^2)}, \quad B_z = \frac{qf + k\varpi\partial f/\partial\varpi}{q(1 + k^2\varpi^2)}, \tag{7.121}$$

where f satisfies the field equation

$$\frac{1}{\varpi^2}\frac{\partial^2 f}{\partial u^2} + \frac{1}{\varpi}\frac{\partial}{\partial\varpi}\frac{\varpi}{1 + k^2\varpi^2}\frac{\partial f}{\partial\varpi} + \frac{q^2 f}{1 + k^2\varpi^2} - \frac{2kqf}{(1 + k^2\varpi^2)^2} = 0. \tag{7.122}$$

The variables are separable and there are solutions of the form

$$f(\varpi, u) = [qZ_\nu(\lambda\varpi) - k\lambda\varpi Z'_\nu(\lambda\varpi)] \exp i\nu u,$$

where Z_ν is any solution of Bessel's equation and $\lambda^2 = q^2 - \nu^2 k^2$. The field is invariant along $u = constant$, i.e., along $\varphi = \varphi_0 + kz$.

Consider a flux rope composed of two twisted flux bundles spiraling about each other and about the z–axis. Then $\nu = 1$ and the solution bounded on the z-axis is of the form

$$f = D[(q + k)J_1(k\varpi) - k\lambda\varpi J_0(k\varpi)] \cos u. \tag{7.123}$$

where D is an arbitrary constant. The field components are

$$B_\varpi = -D\left[\frac{q+k}{q\varpi} J_1(\lambda\varpi) - \frac{\lambda k}{q} J_0(\lambda\varpi)\right] \sin u, \tag{7.124}$$

$$B_\varphi = +D\left[\frac{q+k}{q\varpi} J_1(\lambda\varpi) - \lambda J_0(\lambda\varpi)\right] \cos u, \tag{7.125}$$

$$B_z = \frac{D\lambda^2}{q} J_1(\lambda\varpi) \cos u. \tag{7.126}$$

In order to see the structure of this field, note that close to the z-axis ($\lambda\varpi \ll 1$), the components are

$$B_\varpi \cong D[\lambda(k-q)/2q] \sin u,\ B_\varphi \cong D[\lambda(k-q)/2q] \cos u \tag{7.127}$$

to lowest order, with B_z declining to zero as

$$B_z \cong D(\lambda^3/2q)\varpi \cos u. \tag{7.128}$$

The magnitude of the field is

$$B \cong D\lambda(k-q)/2q \tag{7.129}$$

to lowest order. The field points in the $u = \frac{1}{2}\pi$ direction across the z-axis.

Supposing the field to be confined within the circular cylinder $\varpi = R$, the solution must be constructed so that B_ϖ vanishes there. This can be accomplished only if q is negative. So write $q = -Q$ where $Q > 0$ and set $B_\varpi = 0$ at $\varpi = R$, with the result that

$$(Q-k)/k = -\lambda R\, J_0(\lambda R)/J_1(\lambda R), \tag{7.130}$$

and $Q = +(\lambda^2 + k^2)^{\frac{1}{2}}$. Note, then, that λ must be real if B_ϖ is finite at $\varpi = 0$ and vanishes for some value $\varpi = R > 0$. Hence $Q > k$. It follows that the left-hand side of equation (7.130) is positive, requiring that $J_0(\lambda R)$ and $J_1(\lambda R)$ have opposite signs. The region between the first zero of $J_0(kR)$ at $kR = 2.40$ and the first zero of $J_1(kR)$ at $kR = 3.83$ satisfies this condition, and the right-hand side of (7.130) varies monotonically from zero at $k\varpi = 2.40$ to $-\infty$ at $k\varpi = 3.83$. Hence there is a root of equation (7.130) somewhere in the region $2.40 < kR < 3.83$ for any choice of $Q > k$. There are roots for $kR > 3.83$ too, of course, but B_z changes sign across $k\varpi = 3.83$ so these additional roots do not apply in the present context.

It is evident from equation (7.126) that B_z changes sign across $u = \pm\frac{1}{2}\pi$. This does not affect the field magnitude and the magnetic pressure, but it is not compatible with our standard scenario in which the flux bundle is created by moving the footpoints of an initial uniform field B_0 extending in the positive z-direction between $z = 0$ and $z = L$. Negative B_z is not possible. If it is desired to bring the present example into line with the standard scenario, it is necessary only to reverse the sign of the field components $(B_\varpi, B_\varphi, B_z)$ from equations (7.124), (7.125), and (7.126) throughout $\frac{1}{2}\pi < \varphi < \frac{3}{2}\pi$. The equilibrium is unaltered, but the spiral surface $\varphi = \pm\frac{1}{2}\pi$ becomes a surface of tangential discontinuity, i.e., a current sheet. The nature of the discontinuity is evident by inspection of equations (7.124), (7.125), and (7.126), which show that B_φ and B_z both vanish at $u = \pm\frac{1}{2}\pi$, while

$$B_\varpi = \pm\left[\frac{k-Q}{Q\varpi} J_1(\lambda\varpi) - \frac{\lambda k}{Q} J_0(\lambda\varpi)\right]. \tag{7.131}$$

Thus at $u = \frac{1}{2}\pi - \varepsilon$ the upper sign applies, with the lower sign applicable at $u = \frac{1}{2}\pi + \varepsilon$, where $\varepsilon > 0$. The discontinuity is only in B_ϖ, with $\nabla \times \mathbf{B}$, or the current density, in the z-direction.

The magnetic field represents two twisted flux bundles wrapped around each other to form a rope confined to $\varpi < R$. The two bundles complete one revolution about each other in a distance $2\pi/k$ in the z-direction. The field on the surface $\varpi = R$ is

$$B_\varpi = 0,\ B_\varphi = \frac{D\lambda^2 J_1(\lambda R)}{QkR}|\cos u|,\ B_z = -\frac{D\lambda^2}{Q}J_1(\lambda R)|\cos u|.$$

The field magnitude is

$$B = \frac{D\lambda^2 J_1(\lambda R)}{Q}\left(1 + \frac{1}{k^2R^2}\right)^{\frac{1}{2}}|\cos u| \tag{7.132}$$

running in spiral strips along $\varpi = R$. The magnetic pressure is $B^2/8\pi$, varying from zero at $u = \pm\frac{1}{2}\pi$ to a maximum value

$$P_{max} = \left[\frac{D\lambda^2 J_1(\lambda R)}{Q}\right]^2\left(1 + \frac{1}{k^2R^2}\right) \tag{7.133}$$

at $u = 0, \pi$. The mean pressure $\langle P\rangle$ is $\frac{1}{2}P_{max}$, of course, upon averaging over u.

Suppose, then, that such a twisted rope is created within a rigid infinitely conducting cylinder $\varpi = R$ by suitable winding of the footpoints at $z = 0, L$ of an initially uniform field. The space outside $\varpi = R$ is filled with a uniform field B_0 in the z-direction. The rigid cylinder $\varpi = R$ is then removed. If B_0 is adjusted so that $B_0^2/8\pi = \langle P\rangle$, it is evident that the twisted field expands outward across $\varpi = R$ in the vicinity of $u = 0, \pi$ where its pressure exceeds $\langle P\rangle$ and contracts inward in the vicinity of $u = \pm\frac{1}{2}\pi$, in response to the pressure of the external field. Figure 7.28 is a sketch of the resulting outer surface of the twisted field.

It is evident that a spiral field of this nature applies a pressure varying as $\cos^2(\varphi - kz)$ to the external field surrounding it. If the external field is of laminar force-free form, as described by equation (7.120), the pressure applied to the external field consists of localized maxima as one moves in the z-direction along the external field. For instance, a laminar force-free field pressed against the right-hand side ($\varphi = \pi/2$), as in Fig. 7.27(b), experiences maxima essentially of the form $\cos^2 kz$. Hence there are patches of tangential discontinuity at regular intervals in the z-direction arising from the surface rotation effect in the force-free field. The patchiness of the discontinuity is intimately associated with the spiral character of the axis of each of the two twisted flux bundles that make up the twisted rope. It is clear that an irre-

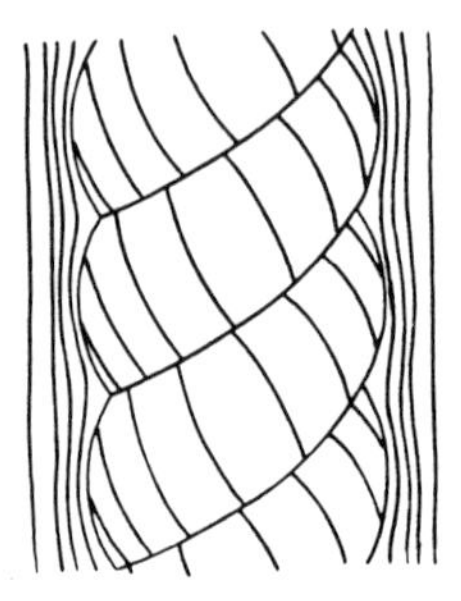

Fig. 7.28. A sketch of the twisted flux rope, described by equations (7.123)–(7.126) confined in an external magnetic field in the z-direction.

gular pattern of winding of twisted flux bundles around each other provides irregular patterns of tangential discontinuity in the surrounding force-free fields.

7.11 Displacement of Individual Flux Bundles

The development has concentrated so far on magnetic fields subject to continuous mapping of their footpoints. However, the case in which one or more elemental flux bundles is displaced discontinuously relative to the ambient field is not without relevance to a variety of circumstances in nature. For instance, the discrete fibril structure of the magnetic field at the surface of a star like the Sun guarantees that individual flux bundles are at least occasionally, if not frequently, displaced independently of the neighboring flux bundles. The same may occur in the planetary magnetosphere, driven by exposure of a particular flux bundle to the solar wind, or by some localized inflation by fast particles, etc. The net effect of a displaced flux bundle can be rapid dissipation and reconnection, with a variety of suprathermal phenomena. The suprathermal effects may inflate some portion of the flux bundle, leading to the extension or eruption of that portion into another region of field, providing further suprathermal activity, etc. This section provides a brief summary of examples (Parker, 1968, 1989a,b) of the configuration of displaced and/or inflated flux bundles in otherwise continuous fields.

A particularly simple example is provided by a displaced elemental flux bundle in the potential field.

$$B_x = +B_0 \cos kx \exp(-ky) \tag{7.134}$$

$$B_y = -B_0 \sin kx \exp(-ky) \tag{7.135}$$

described by the scalar potential

$$\Phi = -(B_0/k) \sin kx \exp(-ky) \tag{7.136}$$

and the vector potential

$$A = -(B_0/k) \cos kx \exp(-ky), \tag{7.137}$$

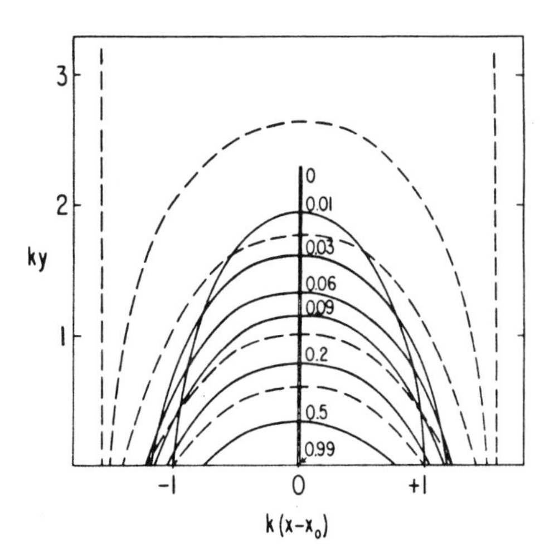

Fig. 7.29. The dashed lines represent the field lines $A = constant$ from equation (7.137). The solid lines represent the paths of an inflated flux bundle ($\beta = 10^{-2}$), with the associated value of $n(y_0)$ from equation (7.154) indicated on each curve.

representing periodic arches of field. The field lines of one arch $-\frac{1}{2}\pi < kx < +\frac{1}{2}\pi$, are indicated by the dashed curves in Fig. 7.29, obtained by setting A equal to a constant. We may imagine that the field is anchored in the plane $y = 0$.

Consider, then, the general form of the field lines in the flux surface given by the xy-plane. The eikonal equation (7.6) becomes

$$(\nabla\phi)^2 = B^2 \tag{7.138}$$

where ϕ is the scalar potential for the displaced field, and B is the strength of the ambient field $(B_x^2 + B_y^2)^{\frac{1}{2}}$. The eikonal equation is nothing more than the statement that the magnitude $|\nabla\phi|$ of the displaced field is compressed by the external field to the value B of the external field. Thus, in the present case, with the field described by equations (7.134) and (7.135),

$$(\nabla\phi)^2 = B_0^2 \exp(-2ky) \tag{7.139}$$

The effective index of refraction is just the square root of the right-hand side of this equation, with which we construct Fermat's principle and the associated Euler equations. However, we already know the form of the solution of the Euler equations because every field line of the ambient potential field, given by $A(x, y) = \textit{constant}$, satisfies the same Euler equations. Noting that the field magnitude, or index of refraction, is independent of x, it follows these field lines may be displaced by arbitrary amounts in the x-direction, and their general description is given by $\gamma = \textit{constant}$ where

$$\gamma(x, y) = \cos k(x - x_0) \exp(-ky) \tag{7.140}$$

for the field line reaching its apex $ky = -\ln\gamma$ at $x = x_0$, with footpoints on $y = 0$ at $x = x_0 \pm \cos^{-1}\gamma$. The dashed lines in Fig. 7.29 are examples of these arched field lines, of course, and they may be displaced by arbitrary amounts along the x-axis. Increasing or decreasing the separation of the footpoints at $y = 0$ increases or decreases respectively, the height of the arch, with the height becoming large without bound as the separation of the footpoints $2\cos^{-1}\gamma$ approaches π/k and γ declines to zero. The field lines are vertical ($x = \textit{constant}$) extending to $y = +\infty$ for any larger separation of the footpoints.

A somewhat similar situation obtains in the 2D dipole field

$$B_\varpi = +B_0(a/\varpi)^2 \cos\varphi \tag{7.141}$$

$$B_\varphi = +B_0(a/\varpi)^2 \sin\varphi \tag{7.142}$$

in polar coordinates $\varpi = (x^2 + y^2)^{\frac{1}{2}}$, $\tan\varphi = y/x$. This field is described by the scalar potential

$$\Phi = B_0(a^2/\varpi)\cos\varphi \tag{7.143}$$

and the vector potential

$$A = B_0(a^2/\varpi)\sin\varphi \tag{7.144}$$

with

$$B_\varpi = +\frac{1}{\varpi}\frac{\partial A}{\partial\varphi}, \quad B_\varphi = -\frac{\partial A}{\partial\varpi}$$

The field lines are given by setting A equal to a constant. In this example the field magnitude is $B = B_0 a^2/\varpi^2$ and is independent of φ. The field lines are circles of radius $R = B_0 a^2/2A$ tangent to the y-axis ($\varphi = \frac{1}{2}\pi$) at the origin. But, in fact, any circle passing through the origin satisfies the Euler equation for B proportional to $1/\varpi^2$ and independent of φ. Hence the displaced field lines are given by $\gamma =$ *constant* with

$$\gamma = \sin(\varphi - \varphi_0)/\varpi \tag{7.145}$$

for arbitrary φ_0.

In a 3D dipole magnetic field the degeneracy is absent and the problem is more complicated. The ambient field has the components

$$B_r = 2B_0(a/r)^3 \cos\theta, \tag{7.146}$$

$$B_\theta = B_0(a/r)^3 \sin\theta, \tag{7.147}$$

in spherical coordinates (r, θ, φ). This field is described by the scalar potential

$$\Phi = B_0(a^3/r^2)\cos\theta, \tag{7.148}$$

and by the vector potential

$$A = B_0(a^3/r^2)\sin\theta. \tag{7.149}$$

The field lines are given by constant $A(r,\theta)r\sin\theta$, which yields

$$r = r_0 \sin^2\theta \tag{7.150}$$

where r_0 is the radius of the apex of each field line arching over the equator at $\theta = \frac{1}{2}\pi$. The field magnitude, or index of refraction, is

$$B = B_0\left(\frac{a}{r}\right)^3 (1 + 3\cos^2\theta)^{\frac{1}{2}}, \tag{7.151}$$

and is a function of both variables r and θ, or ϖ and z. The Euler equation for the ray paths is

$$\left(\frac{r''}{r} + \frac{r'^2}{r^2} + 2\right)(1 + 3\cos^2\theta) = 3\left(\frac{r'}{r} + \frac{r'^3}{r^3}\right)\sin\theta\cos\theta \tag{7.152}$$

where the prime indicates differentiation with respect to θ. It is readily shown that the field lines given by equation (7.150) satisfy this equation. There is no ignorable coordinate and the general integration of equation (7.152) is not possible by elementary means. For comparison with the field lines of equation (7.150) that arch over the equator, consider a flux bundle displaced so that it straddles the pole at $\theta = 0$. Write $u \equiv r'/r$ and $f(\theta) = \ln(1 + 3\cos^2\theta)$, so that equation (7.152) can be expressed as

$$\frac{u' + 2(1 + u^2)}{u(1 + u^2)} = -\frac{1}{2}f'(\theta)$$

The quantity u passes through zero at the apex where $\theta = 0$, suggesting that u can be expanded in ascending powers of θ. It is then straightforward to show that

$$u \cong -2\theta - \frac{5}{3}\theta^3 + \frac{17}{60}\theta^5 + O(\theta^2)$$

for $0 \leqslant \theta \ll 1$, from which it follows that

$$r = r_0 \left[1 - \theta^2 + \frac{1}{12}\theta^4 + \frac{107}{360}\theta^6 + O(\theta^8) \right].$$

This is to be compared with equation (7.150), which can be written as

$$r = r_0 \left[1 - \vartheta^2 + \frac{1}{3}\vartheta^4 - \frac{2}{45}\vartheta^6 + O(\vartheta^6) \right]$$

in terms of the angular distance $\vartheta = \frac{1}{2}\pi - \theta$ from the apex at $\theta = \frac{1}{2}\pi$. The coefficients of θ^2 and ϑ^2 are the same in these two expressions for r, but the coefficient of ϑ^4 is larger than the coefficient of θ^4, indicating a flatter arch over the equator $\vartheta = 0$. This is to be expected from the fact that for fixed r the field peaks over the poles $\theta = 0, \pi$ and is a minimum over the equator.

Consider the consequences of a fluid pressure P in the ambient field and the fluid pressure p in the displaced flux bundle. The total pressure in the ambient field is $B^2/8\pi + P(x, y)$, which is balanced by the pressure $(\nabla\phi)^2/8\pi + p(x, y)$ in the displaced flux bundle. Hence

$$(\nabla\phi)^2 = B^2 + 8\pi(P - p), \tag{7.153}$$

in the displaced bundle. Equation (7.153) is the eikonal equation providing the effective index of refraction, equal to the square root of the right-hand side.

For the simple case that $P = 0$ and $p = p_0$, with B given by equation (7.139), the effective index of refraction can be written

$$n(y) = [\exp(-2ky) - \exp(-2ky_1)]^{\frac{1}{2}} \tag{7.154}$$

where $\beta \equiv \exp(-2ky_1)$ is the ratio of the gas pressure p_0 to the magnetic pressure $B_0^2/8\pi$ at $y = 0$. Note that $n(y)$ is real only for $y < y_1$, for beyond y_1 the fixed fluid pressure p_0 exceeds the magnetic pressure $(B_0^2/8\pi)\exp - 2ky$ and the ambient magnetic field cannot confine it. Since $\partial n/\partial x = 0$, write Fermat's principle as

$$\delta \int dy(1 + x'^2)^{\frac{1}{2}} n(y) = 0$$

where $x' = dx/dy$. Then x is an ignorable coordinate and the Euler equation is immediately integrable to

$$x' = \frac{\pm C}{[n^2(y) - C^2]^{\frac{1}{2}}}$$

where C is the integration constant. Note that $x' = \infty$ at the apex (x_0, y_0) of a field line, at a height y_0. Hence $C = n(y_0)$, and

$$x - x_0 = \pm C \int_y^{y_0} \frac{dy}{[\exp(-2ky) - \exp(-2ky_0)]^{\frac{1}{2}}}. \tag{7.155}$$

Carrying out the indicated integration leads to

$$k(x - x_0) = \pm\{1 - \exp[-2k(y_1 - y_0)\}^{\frac{1}{2}} \tan^{-1}[\exp 2k(y_0 - y) - 1]^{\frac{1}{2}},$$

conveniently rewritten as

$$\cos^2 Q(x - x_0) = \exp(-2k(y_0 - y)], \tag{7.156}$$

where the wave number Q is a function of y_0 and y_1,

$$Q = \frac{k}{\{1 - \exp[-2k(y_1 - y_0)]\}^{\frac{1}{2}}}. \tag{7.157}$$

Note, that Q becomes large without bound as y_0 increases toward y_1, becoming imaginery for $y_0 > y_1$, where $n(y)$ is imaginery.

The footpoints of the field at $y = 0$ lie at $x_0 \pm (x_f - x_0)$, and are separated by $2(x_f - x_0)$, where

$$\cos Q(x_f - x_0) = \exp(-ky_0) \tag{7.158}$$

There is always a root $Q(x_1 - x_0) = v$ to this equation for $y_0 > 0$ because the right-hand side is positive and less than one. It follows that for a given y_0, $x_f - x_0$ is inversely proportional to Q,

$$\begin{aligned} x_f - x_0 &= v/Q \\ &= (v/k)\{1 - \exp[-2k(y_1 - y_0)]\}^{\frac{1}{2}} \end{aligned} \tag{7.159}$$

and goes to zero as y_0 increases to y_1. For $k(y_1 - y_0) \ll 1$ the relation reduces to

$$k(x_f - x_0) \cong \frac{v}{[2k(y_1 - y_0)]^{\frac{1}{2}}} \tag{7.160}$$

Conversely, a flux bundle with fixed footpoint positions extends to larger and larger heights y_0 as the inflation $\beta = \exp(-2ky_1)$ increases and y_1 decreases. This follows directly from equations (7.157) and (7.158) with Q tending to increases with decreasing y_1. But $\cos^2 Q(x_f - x_0)$ decreases with increasing Q and fixed $x_f - x_0$ only to the point that $Q(x_f - x_0)$ reaches $\frac{1}{2}\pi$, where the cosine vanishes as y_0 increases to $+\infty$. In fact the solution becomes unphysical even before the decreasing y_1 meets the increasing y_0. In the limit that the footpoints are close together, $x_f - x_0 \ll 1/k$ (providing a low lying flat arch in the absence of inflation, $y_1 = +\infty$), the arch runs off to $ky = +\infty$ when ky_1 declines to $ky_0 + (2/pi^2)k^2(x_f - x_0)^2$. The plasma β at this point has a value

$$\beta = \exp[-2ky_0 - (4/\pi^2)\, k^2(x_f - x_0)^2],$$

which can be as large as one only in the limit of an infinitesimal arch, with both y_0 and $x_f = x_0$ infinitesimal.

The solid curves in Fig. 6.29 show the path of the inflated flux bundle for $\beta = 10^{-2}$ and $ky_1 = 2.303$. The index of refraction $n(y_0)$ at the apex is indicated on each curve. Note the variation in the separation of the footpoints at $y = 0$, starting with the tiny arch when $y_0 = 0.05$ and $n(y_0) = 0.99$. The path arches higher (declining $n(y_0)$) as the footpoints are more widely separated up to $n(y_0) \cong 0.06$ beyond which the separation of the footpoints diminishes for higher apices, in order that the effective buoyancy of the inflated flux tube in the pressure gradient of the ambient field can be held in check by the tension in the flux bundle. The maximum height to which the inflated flux bundle can be confined is $y_0 = y_1$, of course, equal to $ky_0 = 2.303$ in the present case that $\beta = 10^{-2}$. Note that the equilibrium paths are unstable for an apex higher than the value for the maximum separation of the footpoints. This may be seen from the fact that in this regime the higher the apex, the more the inflation $p_0 - P_0$ dominates the magnetic pressure in the neighborhood of the apex. It is for this reason that the footpoint separation must be decreased to maintain the apex of the arch against the buoyancy created by the ambient pressure gradient. Further details of this and other examples can be found in the literature (Parker, 1981a).

The essential physical point is that the displacement of a flux bundle can lead to its eruption to infinity if the footpoints become too widely separated for the tension in the bundle to overcome the force of the pressure gradient of the ambient magnetic field. In such a case, static equilibrium finds the two legs of the erupted arch extending vertically (Parker, 1979). The same can occur if the gas pressure within the displaced bundle exceeds some critical value while the footpoints remain fixed.

In the context of the present writing, primary interest in displaced flux bundles centers on their rapid dissipation (Parker, 1981b) of the displaced bundle and the associated suprathermal effects in various astronomical settings, e.g., the magnetosphere of Earth, an active region on the Sun or other star, and the magnetic – cosmic ray halo of a galaxy.

8

Topology of Tangential Discontinuities

8.1 Conservation of Discontinuities

We turn now to the topology of those surfaces of tangential discontinuity that are required by the topology of the magnetic field in static equilibrium (Parker, 1989, 1990, 1991). The surfaces are current sheets and the surface current density J, as well as the jump ΔB in the opposing field components, can be used as a direct measure of the strength of the discontinuity in the magnetic field.

Consider, then, the surface current $\mathbf{J}$ associated with the magnetic discontinuity at the surface S between two fields $\mathbf{B}_1$ and $\mathbf{B}_2$. The fields differ in direction by the angle θ where they press together at S. The field intensities B_1 and B_2 are unequal if they contain different fluid pressures p_1 and p_2, respectively. The surface S is a flux surface of both fields, and Fig. 8.1 is a sketch of the geometry. The fields $\mathbf{B}_1$ and $\mathbf{B}_2$ are shown at a distance ε on each side of S, making angles θ_1 and θ_2 with the line GG' which is perpendicular to the small rectangle CDEF with sides 2ε and $2h$. Then $\theta = \theta_1 + \theta_2$ for any orientation of CDEF. If J_n is the component of the surface current density in S perpendicular to CDEF, it follows from Stokes' theorem applied to Ampere's law (see equation 2.32), that

$$c(2hB_2 \sin\theta_2 + 2h\, B_1 \sin\theta_1) = 4\pi 2h\, J_n \tag{8.1}$$

in the limit of small ε. The left-hand side represents the line integral of $\mathbf{B}$ around the contour CDEF and the right-hand side represents the total current through the surface CDEF. The orientation of CF and DE is arbitrary. To establish the orientation of $\mathbf{J}$ relative to $\mathbf{B}_1$ and $\mathbf{B}_2$, rotate the rectangle CDEF about the vertical (in

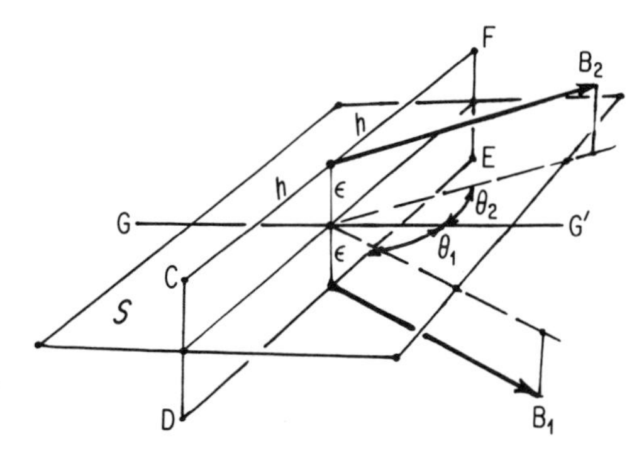

Fig. 8.1. A diagram of the two fields $\mathbf{B}_1$ and $\mathbf{B}_2$ on opposite sides of the surface S of tangential discontinuity.

Fig. 8.1) so as to maximize J_n, when the normal to CDEF points in the direction of **J**. The angle θ is given, so write equation (8.1) as a function of θ_2,

$$4\pi J_n = cB_2[\sin\theta_2 + \nu\sin(\theta - \theta_2)],$$

where $B_1 = \nu B_2$. Setting $dJ_n/d\theta_2$ equal to zero yields

$$\cos\theta_2 - \nu\cos(\theta - \theta_2) = 0.$$

It follows that

$$\sin\theta_2 = \frac{1 - \nu\cos\theta}{(1 - 2\nu\cos\theta + \nu^2)^{\frac{1}{2}}},$$

$$\cos\theta_2 = \frac{\nu\sin\theta}{(1 - 2\nu\cos\theta + \nu^2)^{\frac{1}{2}}}.$$

The surface current density is equal to the maximum value of J_n,

$$J = \frac{c}{4\pi}B_2(1 - 2\nu\cos\theta + \nu^2)^{\frac{1}{2}} \tag{8.2}$$

in the direction θ_2 relative to $\mathbf{B}_2$.

The discontinuity in the field component parallel to DE is $\Delta B = B_1\sin\theta_1 + B_2\sin\theta_2$, which has the value $4\pi J/c$, of course.

The essential point is that the electric current direction lies somewhere between the directions of the fields on either side and so is parallel to neither field. Thus the field lines and the current lines go their separate ways. They pass over the surface of discontinuity into different regions of field and connect into the boundaries $z = 0$ and $z = L$ at different locations along the discontinuity. Along the way they pass through different patterns of intermixing of the field lines (determined by the choice of the stream function $\psi(x, y, kzt)$ of the fluid motion, through equation (3.7)) so the field lines on opposite sides, as well as the surface current, may have different topologies. It follows, then, that a surface of tangential discontinuity in the field **B** is generally a surface of discontinuity in the fluid pressure p if $\nabla p \neq 0$, or a surface of discontinuity in the torsion α if $\nabla p = 0$ so that the field is force-free.

Now if $\nabla p = 0$, so that the field is force-free, described by equation (3.3), then the magnetic pressure is continuous across S. Hence $B_1 = B_2$ and $\nu = 1$, from which it follows that $\theta_1 = \theta_2 = \frac{1}{2}\theta$. The direction of the surface current lies midway between the direction of the fields $\mathbf{B}_1$ and $\mathbf{B}_2$ on either side, with a surface current density

$$J = \frac{c}{2\pi}B_2\sin\frac{1}{2}\theta. \tag{8.3}$$

On the other hand, the field lines of $\mathbf{B}_1$ and $\mathbf{B}_2$ may connect to different regions of a boundary, e.g., the photosphere or coronal transition layer of a star. Hence, there is no a priori reason to expect the same fluid pressures in the fields. Since $p + B^2/8\pi$ must be continuous across the surface of discontinuity, it follows that in general

$$\nu^2 = 1 + 8\pi(p_2 - p_1)/B_2^2, \tag{8.4}$$

$$= [1 + 8\pi(p_1 - p_2)/B_1^2]^{-1}. \tag{8.5}$$

Consider the overall requirement for conservation of electric current. If the fluid pressures p_1 and p_2 have no gradient, or if the pressure gradient satisfies the gravitational barometric law

$$\nabla p + \rho \nabla \Phi = 0, \tag{8.6}$$

where Φ is the gravitational potential and ρ is the fluid density, then the fields $\mathbf{B}_1$ and $\mathbf{B}_2$ are force-free, and their magnetostatic equilibrium is described by equation (3.3). The current densities $\mathbf{j}_1$ and $\mathbf{j}_2$ flow precisely along the field, with equation (3.3) combined with Ampere's law to give

$$\mathbf{j} = (c/4\pi)\alpha\mathbf{B}$$

through the space. It follows from Ampere's law that $\nabla \cdot \mathbf{j} = 0$, which also follows from the present relation, given equation (3.5) that α is rigorously constant along **B**. The essential point is simply that the total flux of electric current is conserved along each flux bundle, just as the magnetic flux is conserved. It follows that the volume current density **j** does not communicate with the surface current density **J** in the surfaces of discontinuity, for to do so would be to fail to conserve the flux of electric current **j** everywhere along the field **B**. The total current $(\mathbf{j} + \mathbf{J})$ is conserved, of course, from which it follows that the surface current density **J** is conserved along the surfaces of tangential discontinuity. Hence the total strength of the tangential discontinuities (integrated over the width of the surfaces) is conserved along the field from one boundary $z = 0$ to the other, at $z = L$. The tangential discontinuities may intersect and reform as they extend along the field, but they do not end anywhere in the volume $0 < z < L$. A tangential discontinuity intersecting $z = 0$ necessarily connects all the way to $z = L$. Specific examples are treated in §8.4.

In the same spirit, it should be noted that the field lines that border a current sheet at any one location pass their entire length (from $z = 0$ to $z = L$) along a current sheet. This follows from the fact that there is no way that the field lines on either side of a current sheet can veer away from the current sheet, without other field lines entering into the region between them and the current sheet, creating a surface of discontinuity along the deflected field lines (Parker, 1986a, Fig. 1).

The tangential discontinuities lie along certain flux surfaces of the magnetic field, so their topology is not unrelated to the topology of the field. However, the surfaces of discontinuity form an irregular honeycomb under the standard conditions and it is the transverse structure of the honeycomb as much as the extension along the field that is of interest.

8.2 Fluid Pressure

The presence of nonvanishing ∇p on either side of S complicates the situation, unless S is an isobaric surface. To see how this works out note from equation (3.2) that p is uniform along each field line (in the absence of a gravitational field), and the component of $\nabla \times \mathbf{B}$ perpendicular to **B**, denoted by $(\nabla \times \mathbf{B})_\perp$, is given by

$$(\nabla \times \mathbf{B})_\perp = 4\pi \mathbf{B} \times \nabla p / B^2,$$

which follows upon forming the vector product of **B** with equation (3.1). There are two directions perpendicular to **B**, of course. We choose one as the direction perpendicular to the surface S of the tangential discontinuity, denoted by the subscript n. The other direction lies in S and is denoted by the subscript s. Thus

$$(\nabla \times \mathbf{B})_\perp = (\nabla \times \mathbf{B})_n + (\nabla \times \mathbf{B})_s, \tag{8.7}$$

with

$$(\nabla \times \mathbf{B})_n = 4\pi \mathbf{B} \times (\nabla p)_s / B^2, \tag{8.8}$$

$$(\nabla \times \mathbf{B})_s = 4\pi \mathbf{B} \times (\nabla p)_n / B^2. \tag{8.9}$$

These two equations, (8.8) and (8.9), apply separately to $\mathbf{B}_1$ and p_1 on one side of S and to $\mathbf{B}_2$ and p_2 on the other side. Supposing that $(\nabla p_1)_s \neq 0$ in the field $\mathbf{B}_1$, say, it follows that the pressure p_1 varies from one field line of $\mathbf{B}_1$ to the next over the surface S, providing a fluid pressure discontinuity across S that varies with distance across S in the direction perpendicular to $\mathbf{B}_1$ at S. The fluid pressure in $\mathbf{B}_2$ cannot be adjusted to avoid a pressure discontinuity because $\mathbf{B}_2$ has a direction different from $\mathbf{B}_1$, and p_2 can be made to vary only in the direction perpendicular to $\mathbf{B}_2$. Figure 8.2 is a sketch of the field lines of $\mathbf{B}_1$ and $\mathbf{B}_2$ on opposite sides of S. The field lines are also the contours of constant fluid pressure, indicated on each field line in the figure. The net result is a fluid pressure discontinuity across S that varies at the rate $\mathbf{e} \cdot [(\nabla p_1)_s - (\nabla p_2)_s]$ in the arbitrary direction $\mathbf{e}$ in S.

In this situation, where $(\nabla p)_s$ is not zero, it follows from equation (8.8) that there is a nonvanishing current density

$$4\pi \mathbf{j}_n = c(\nabla \times \mathbf{B})_n \tag{8.10}$$

flowing into S from either side. Conservation of current requires that

$$\begin{aligned} \nabla_s \cdot \mathbf{J} &= \mathbf{j}_{1n} - \mathbf{j}_{2n} \\ &= c\left[\frac{\mathbf{B}_1 \times (\nabla p_1)_s}{B_1^2} - \frac{\mathbf{B}_2 \times (\nabla p_2)_s}{B_2^2}\right], \end{aligned} \tag{8.11}$$

where $\nabla_s \cdot \mathbf{J}$ refers to the 2D divergence of the surface current density in S. In such cases the surface current, i.e., the strength of the tangential discontinuity is not conserved. Special cases can be contrived in which a current sheet diminishes to zero strength along the direction of the current, the surface current $\mathbf{J}$ gradually diverted as $\mathbf{j}_n$ into the volume on either side of the sheet. But the special arrangements required to accomplish this disappearance means that the effect is not expected in nature. The current sheets, if not precisely conserved, are nonetheless expected to extend from $z = 0$ to $z = L$.

It goes without saying that the application of a discontinuous fluid pressure $p(x, y)$ at the boundaries $z = 0, L$ provides a surface current even if $\mathbf{B}_1$ and $\mathbf{B}_2$ would otherwise join smoothly and continuously across the surface S. If one or both of $(\nabla p_1)_s$ and $(\nabla p_2)_s$ is nonvanishing, the strength of the discontinuity varies in the s-direction. On the other hand, it should be noted that a surface of tangential discontinuity required

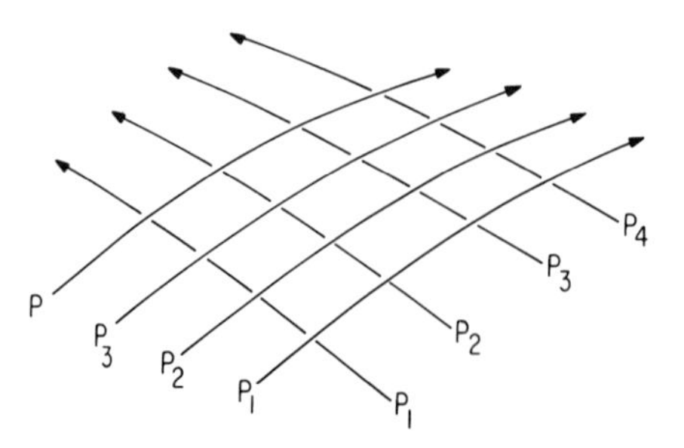

Fig. 8.2. A schematic representation of the field lines of $\mathbf{B}_1$ and $\mathbf{B}_2$ on opposite sides of the surface S of tangential discontinuity in a case in which the fluid pressure varies from one line to the next ($(\nabla p)_s \neq 0$) in both fields, indicated by the progressive change in pressure p_1, p_2, p_3, p_4 from one position to the next.

for force-free magnetostatic equilibrium by the topology of the field generally cannot be eliminated by the manipulation of the fluid pressure p at the boundaries.

Finally, note that if $(\nabla p_1)_s = (\nabla p_2)_s = 0$, then $\mathbf{j}_{1n} = \mathbf{j}_{2n} = 0$ and $\nabla \cdot \mathbf{J} = 0$, so that the strengths of the tangential discontinuities are conserved as with force-free fields. However, if $(\nabla p_1)_s = (\nabla p_2)_s = 0$, then S is an isobaric surface, in which the optical analogy applies to the complete field $\mathbf{B}$ and not merely to the projection $\mathbf{B}_s$ of $\mathbf{B}$ onto the flux surfaces of $\nabla \times \mathbf{B}$ or $\mathbf{j}$. This occurs only in special circumstances, of course, in which the applied fluid pressure (at $z = 0, L$) is suitably coordinated with the motions that provide the nonsymmetric topology of the field.

This is an appropriate place to comment on the complexity of the fluid pressure field throughout $0 < z < L$ if the fluid pressure $p(x, y)$ applied at $z = 0, L$ is not uniform. For if the applied pressure should be different at opposite ends of any flux bundle, the fluid is set into motion, streaming along the flux bundle from the high pressure end to the low pressure end. It is obvious that if equal fluid pressures are to be applied at both ends of every flux bundle, the initial inhomogeneous pressure field $p(x, y)$ must be carried along bodily in the fluid, i.e., $dp/dt = 0$ at $z = L$ while keeping $p(x, y)$ fixed at $z = 0$. Failing this special requirement, as when the pressure patterns $p(x, y, 0) = p(x, y, L)$ remain fixed in time, the fluid is accelerated along each flux bundle and only a stationary dynamical state can be achieved in place of magnetostatic equilibrium. The stationary state is a more complicated situation (cf. Tsinganos, 1981, 1982a–d, for some important special cases) which is not dealt with here.

We proceed with the force-free field ($\nabla p = 0$) leaving $\nabla p \neq 0$ for the future. The surfaces of tangential discontinuity are conserved and extend all the way from $z = 0$ to $z = L$.

8.3 The Primitive Discontinuity

Consider the local form of a force-free magnetic field $\mathbf{B}$, described by equation (3.3). Choose an arbitrary point P in the field $\mathbf{B}$ wherever $\alpha \neq 0$ and construct any unit vector $\mathbf{n}$ perpendicular to $\mathbf{B}$ at the point P, where $\mathbf{B}$ is denoted by $\mathbf{B}_p$. Then as one moves an infinitesimal distance away from P in the direction $\mathbf{n}$, the field direction is found to rotate by a comparable infinitesimal amount about the direction $\mathbf{n}$ for almost all choices of the direction $\mathbf{n}$. Except in the improbable case that P lies on an axis of symmetry of $\mathbf{B}$, the rate of rotation of the field varies with the choice of direction of $\mathbf{n}$. There is one direction — call it $\mathbf{n}_m$ — in which the rate of rotation is maximum, and the perpendicular direction $m_0 = m_m \times \mathbf{B}$ in which the rate of rotation vanishes, as sketched in Fig. 8.3. Construct the local orthogonal right-hand Cartesian coordinate system xyz with origin at the point P and with the x-axis oriented along $\mathbf{B}_p$. Point the positive y-axis in the direction of m_0 and the positive z-axis in the direction of m_m, as shown in Fig. 8.3. Then to first order in (x, y, z),

$$B_x \cong B_p + B_p\, q_{xx} x,$$

$$B_y \cong B_p(q_{yy}\, y + q_{yz}\, z),$$

$$B_z \cong B_p\, q_{zz}\, z,$$

where the q_{ij} are constants. The vanishing divergence of $\mathbf{B}$ requires that $q_{xx} + q_{yy} + q_{zz} = 0$. The equilibrium equation (3.3) reduces to $q_{yz} \equiv -\alpha_p$ where α_p is

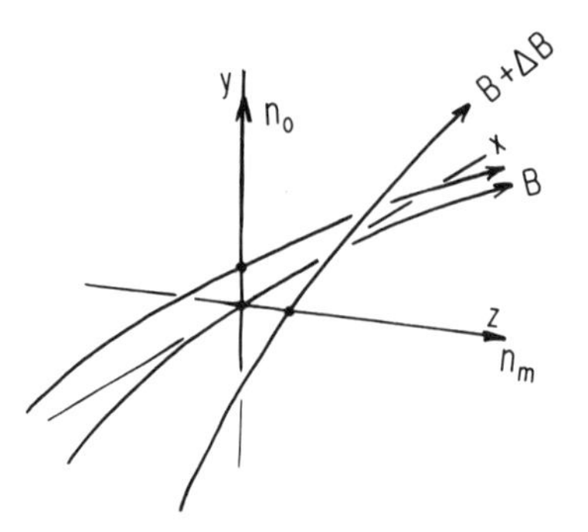

Fig. 8.3. The local Cartesian coordinate system with origin at the point P in a force-free field **B**.

the value of the torsion coefficient α on the field line through the point P. Thus q_{xx}, q_{yy}, and q_{zz} are not involved directly in the local balance of stress, and can be put equal to zero for the present purposes, so that the field magnitude may be considered uniform in the vicinity of P. The rotation of the field direction along m_m in the z-direction is simulated by the basic field,

$$B_x = +B_p \cos qz,\ B_y = -B_p \sin qz,\ B_z = 0, \tag{8.12}$$

where $q = \alpha_p = -q_{yz}$. This is the primitive force-free field treated in §§5.4.1 and 5.4.2. In that instance, a slab $(-h < z < +h)$ of such field was anchored in the distant rigid circular boundary $\varpi = (x^2 + y^2)^{\frac{1}{2}} = R$, and the slab was squeezed in the near vicinity of the origin $(\varpi < a = O(h))$. The enhanced pressure and the associated increased field magnitude produced two spiral surfaces of tangential discontinuity.

It is instructive to go back to that formal example to examine the topological properties of the two surfaces of discontinuity in more detail. With reference to the optical analogy, the enhanced pressure excludes the field in each flux surface $z = constant$ from the central region $r < a$ as sketched in Fig. 7.13 for $z = 0$. The central cross-hatched region DF symbolizes the area of enhanced pressure applied to the surfaces $z = \pm h$ of the layer of field, reducing the thickness of the slab to zero in $r < a$ by exclusion of the field. It is assumed for convenience that $a, h \ll R$. Note that Fig. 7.13 is only a schematic representation of the field, with no attempt to show a or h small compared to the outer radius $\varpi = R$ where the field is anchored. The diamond-shaped gap in the flux surface is indicated by CDEF. Figure 8.4 is a schematic drawing of many such flux surfaces $z = constant$ at $\varpi = R$ superimposed over the range $\lambda_1 < z < \lambda_2$. The boundaries of the gaps in each successive flux surface of increasing z are TMNU, SMNV, RMNW, QMNX, PMNY, etc. The surfaces of tangential discontinuity are formed where flux surfaces (at different levels of z at $\varpi = R$) come into contact through the gaps in the intervening flux surfaces. For instance, the field lines of the flux surface RMNW come into contact with the field lines of the flux surface TMNU at the two points marked with a circled plus sign because of the gap, e.g., SMNV, in all the intervening flux surfaces.

It is easy to see from the right-hand rule that the electric current is directed radially outward on the right-hand side, toward PQRST in Fig. 8.4, indicated by the short arrow, and radially inward on the left-hand side, toward UVWXY in Fig. 8.4, indicated by the short arrow. The surface current density declines outward inversely with radial distance ϖ, with the outward current in the surface $qz + \varphi = 0$ and the inward current in the surface $qz + \varphi = \pi$. The variation of the current density, i.e., the jump in the transverse field component B_φ across the discontinuity, is readily

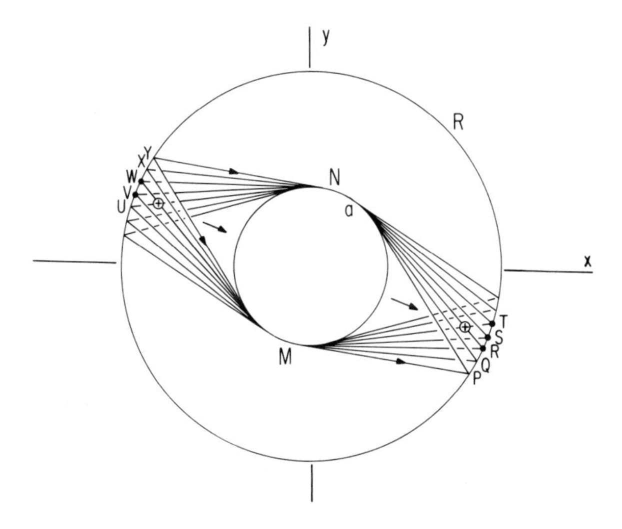

Fig. 8.4. Superposition of the boundaries of the gaps in successive flux surfaces. The two short arrows indicate the field directions in the two spiral surfaces of discontinuity. The circled plus signs indicate where the field lines RN and NU come into contact with TM and MW.

established by considering two field lines, e.g., NQ and MT, intersecting at an angle 2ϑ at a radial distance ϖ. Then $\sin\vartheta = a/\varpi$ and the transverse field components on the two lines are $\pm B_0 \sin\vartheta = \pm B_0 a/\varpi$ on opposite sides of the discontinuity.

Close inspection of Fig. 8.4 reveals that there are no field lines that lie only above or only below the spiral surfaces as the lines pass across the region $\varpi = R$. Consider, for instance, the line QN on the right-hand side, which lies on the upper side of the surface of discontinuity. This line extends into NX, which may be seen to pass under YM and the surface of discontinuity on the left-hand side. Similarly QM beneath the surface on the right becomes MX, which lies above WN and the surface of discontinuity on the left. Figure 8.5 is a sketch showing the lines of force SNV and SMV relative to the two inclined spiral surfaces of discontinuity, with SNV passing above on the right and below on the left, while SMV passes below on the right and above on the left.

An obvious point is that the field lines on one side of a surface of discontinuity connect into footpoints that are generally widely separated from the footpoints of the field on the other side of the discontinuity. Hence in almost all cases the torsion coefficient α is different on the two sides of the discontinuity, so that a discontinuity in the field implies a discontinuity in α (Parker, 1986b). Conversely, as pointed out in

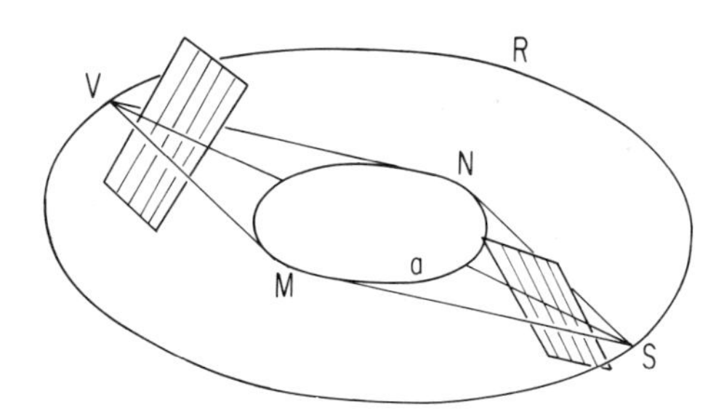

Fig. 8.5. A perspective drawing showing the two field lines SMV and SNV in a flux surface ($z = constant$ at $\varpi = R$) together with the inclined spiral surfaces of discontinuity $qz + \varphi = 0$ and $qz + \varphi = \pi$, indicated by the two inclined leaves. The essential feature is that each field line passes above one surface and below the other.

§4.9 a discontinuity in α implies a discontinuity in the field, because a discontinuity in α can arise only if the field lines pass through different regions of fluid motion.

It was proved in §7.6 (see Fig. 7.15) that the two field lines connected to a close pair of footpoints on opposite sides of a surface of discontinuity at $z = 0$ and at $z = L$ pass off that surface onto other branches of the surface so that the pair of lines become widely separated throughout the interior of $0 < z < L$. On the other hand, two field lines that pass close by each other on opposite sides of a surface of discontinuity connect to footpoints that are widely separated from each other at $z = 0, L$. The present example, of the field lines close to the spiral surfaces of discontinuity, is simply another illustration of the extension of field lines from one surface to another and of the distant rooting in the boundaries of field lines that cross over each other on opposite sides of the surface of discontinuity in the interior.

This is in contrast with the surface of tangential discontinuity created solely by discontinuous displacement of the footpoints of the field. Thus, for instance, in the context of the initial uniform field $\mathbf{B} = \mathbf{e}_z B$ extending from $z = 0$ to $z = L$, the application of the fluid velocity $v(z)$ in the y-direction of the form

$$v(z) = \omega z \begin{cases} +1 & \text{for } x > 0 \\ -1 & \text{for } x > 0 \end{cases} \tag{8.13}$$

for a time τ produces the magnetic field

$$B_y = B\omega\tau \begin{cases} +1 & \text{for } x > 0 \\ -1 & \text{for } x < 0, \end{cases} \tag{8.14}$$

with the discontinuity $\Delta B_y = 2B\omega\tau$ in the flux surface $x = 0$. This discontinuous field is in magnetostatic equilibrium, and it is obvious that the field lines at $x = +\varepsilon(\varepsilon \to 0)$ extend all the way from $z = 0$ to $z = L$, on the same side of the discontinuity $x = 0$. The field lines at $x = -\varepsilon$ also extend all the way on the same side of $x = 0$. This is possible only because the discontinuity is not produced by the Maxwell stresses in connection with the topology of the field. Hence the discontinuity is free to take on other properties, e.g., the individual field line lying on one side of a single surface of discontinuity all the way from $z = 0$ to $z = L$.

The related point, which will come up in Chapter 10 where we treat the active resistive dissipation at surfaces of discontinuity, is that the field discontinuity produced solely by discontinuous motion of the footpoints, without topological involvement, remains in magnetostatic equilibrium as resistive dissipation progresses. There may be local resistive instabilities but there is no tendency for rapid reconnection, because the topology does not demand of the Maxwell stresses that they restore the discontinuity destroyed by the resistivity. As a result the resistive diffusion is passive, described by equation (2.40). Hence the characteristic thickness of the current sheet is the diffusion scale $(4\eta t)^{\frac{1}{2}}$, which increases so that the dissipation declines to low levels with the passage of time. This is to be contrasted with the tangential discontinuity required by the field topology, which is actively maintained by the Maxwell stresses at some more or less fixed characteristic thickness as resistive dissipation progresses. This dynamical state arises precisely because the field lines extend transversely onto other branches of the surface of discontinuity. That is to say, the discontinuity required by the field topology is subject to continuing dynamical rapid reconnection so that it plays an important role in astrophysical plasmas.

8.4 Topology Around Neutral Points

Consider the three dimensional topology of the surfaces of tangential discontinuity around an "X-type neutral point." In fact in three dimensions an X-type neutral point P is a neutral point only in the transverse field components (see discussion in Parker, 1979, §14.8). The 3D field configuration is sketched in Fig. 8.6 with the projection of that field onto the perpendicular plane indicated by the dashed lines. Imagine then, that the X-type neutral point shown in Fig. 8.7 is compressed in one transverse direction or the other, so that the X-type neutral point becomes two Y-type neutral points, shown in detail in Fig. 6.5. Some simple examples of two Y-type neutral points created in this way are provided in §§6.5–6.8. The essential point for the present discussion is that the field lines on opposite sides of each separatrix are deformed out of alignment because their topology is different. That is to say, two lines facing each other across a separatrix extend away into different regions of space. Therefore, when the field is deformed, they respond differently and, in almost all cases, do not continue to be precisely parallel. Hence there is a tangential discontinuity formed at each separatrix when the field is squeezed so as to convert an X-type neutral point into two Y-type neutral points. Figure 8.7 is a 3D sketch of the separatrices (surfaces) associated with the two Y-type neutral points shown in Fig. 6.5. The separatrices AEFB and DEFC in Fig. 8.7 are surfaces of tangential discontinuity, with the strongest discontinuity across EF where the transverse components reverse sign. The other leaves of the surface of discontinuity, indicated by AE, FB, DE, FC are weaker discontinuities because they arise only from the difference in transverse components of the same sign.

To see the topology of the field and the surfaces of discontinuity on a larger scale, recall that EF represents the gap created in the field by the locally enhanced

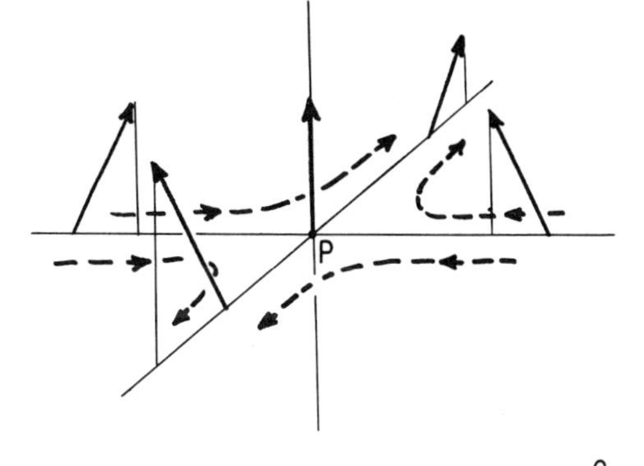

Fig. 8.6. A diagram of the magnetic field in the neighborhood of what is generally called an X-type neutral point at a point P in a 3D field. The dashed lines represent the field lines of the transverse component (in the plane perpendicular to **B** at the point P) around the neutral point.

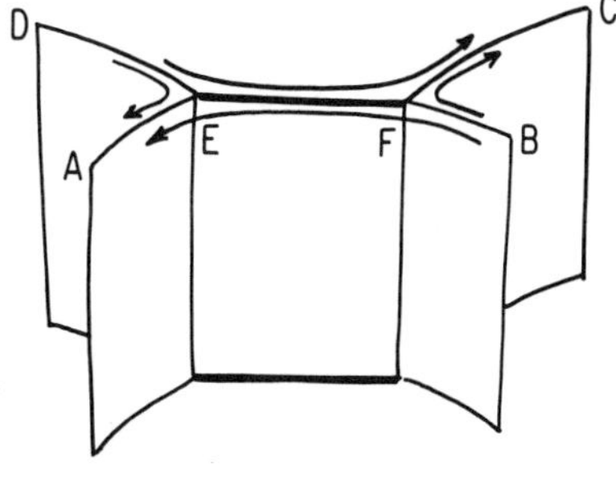

Fig. 8.7. A schematic drawing of the separatrices and surfaces of tangential discontinuity AEFB and DEFC associated with the two Y-type neutral points of Fig. 6.5. The direction of the transverse field is indicated by the arrows.

pressure applied to the region. That is to say, the X-type neutral point was converted into two Y-type neutral points by the opening of the gap. Figure 8.8, then, is a sketch of a larger view of the same configuration, showing the limited extension of the gap along the field, in response to the local pressure maximum, i.e., the local maximum in the effective index of refraction B, according to the optical analogy. The surfaces AEFB and DEFC in Fig. 8.8 are the same surfaces as appear in Fig. 7.4. Applying the optical analogy to the initial field configuration of the X-type neutral point, it is clear that the field lines on opposite sides of EF have the general form indicated by the two curved arrows in Fig. 8.8. Figure 8.9 sketches the field lines on the leaves AE and FB, indicating how they are refracted around the gap EF. This point was treated formally in §7.6 where it was shown that field lines that are parted to form a tangential discontinuity always pass off onto other leaves (e.g., AE

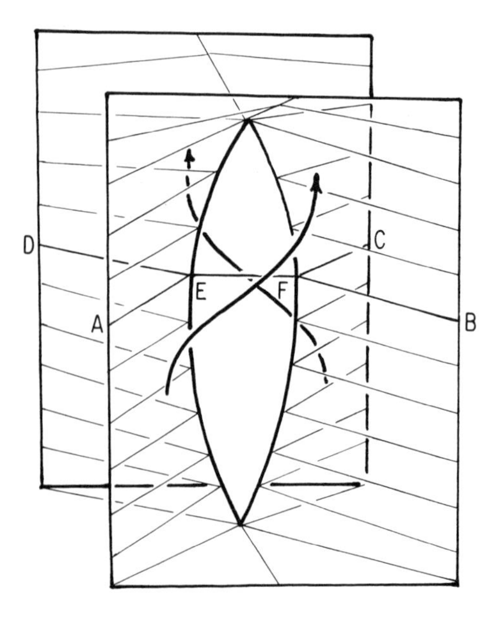

Fig. 8.8. A larger view of the field and its surfaces of discontinuity in Fig. 8.8. The two curved arrows indicate the general field pattern on opposite sides of the principal discontinuity EF located in the gap indicated by the oval. The other lines have no significance except to indicate the surfaces of discontinuity.

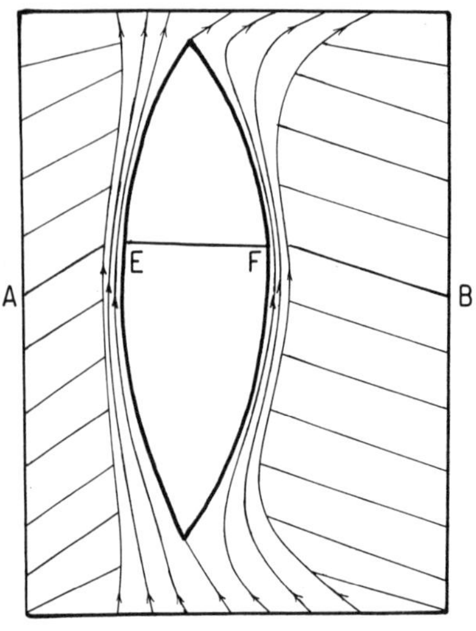

Fig. 8.9. A schematic drawing of the field lines on the near side of the separatrices AE and FB shown in Figs 8.8 and 8.9.

and FB), for not to do so would contradict the magnetostatic equations of stress balance.

A formal example of the creation of the foregoing surfaces of tangential discontinuity was provided in §6.12. The example begins with a rectangular array of flux bundles, with the transverse field components described by the vector potential $A = (B_1/k)\sin kx \sin ky$ (where $kL \gg 1$) and the longitudinal component $B_z(A) = (B_0^2 + 2k^2A^2)^{\frac{1}{2}}$, satisfying the Grad–Shafranov equation (6.127). The field is sketched in Fig. 8.10, showing the rectangular array of twisted flux bundles fitted together so as to provide a continuous magnetic field. Each flux bundle was created by a precisely defined rectangular array of alternate vortex motions extending from $z = 0$ to $z = L$, with each vortex confined within $m\pi < kx < (m+1)kx$ and $n\pi < ky < (n+1)ky$ where m and n are integers. It was pointed out that in the real physical world the mathematically precise matching of the strengths of each vortex is not possible. A simple example was provided in which the flux bundles with positive helicity were more strongly twisted than those with negative helicity. The result is an expansion (see §6.12) of the right-hand (positive helicity) twisted flux bundles, at the expense of the left-hand flux bundles. The result is sketched in Fig. 6.29, where the right-hand bundles press their way between the vertices of the left-hand flux bundles to come in contact with each other to form strong tangential discontinuities. This involves the conversion of the single X-type neutral point at each corner of the cells of continuous field $A = (B_1/k)\sin kx \sin ky$ into two Y-type vertices. Note, too, that the expansion of the right-hand flux bundles increases the obliquity of the spiral field at the surface of the bundle, while the contraction of the left-hand bundles decreases the obliquity. Consequently the fields of neighboring flux bundles are not parallel where they meet at the separatrices and the entire system of cell boundaries represents surfaces of discontinuity, as already noted in §6.12. The purpose in recapitulating these results here is to establish the basis for the manner in which the surfaces of discontinuity branch and deform as they approach the boundaries, $z = 0, L$.

Consider the general form of the surfaces of discontinuity in the imperfectly balanced rectangular array just described. If, for instance, the original vortex pattern responsible for the twisted flux bundles were precisely adjusted so as to be congruent with the final discontinuous field pattern, then the field would extend uniformly ($\partial/\partial z = 0$) from $z = 0$ to $z = L$, and so would the surfaces of discontinuity. The fluid motions, of course, would necessarily have the same discontinuities as the final

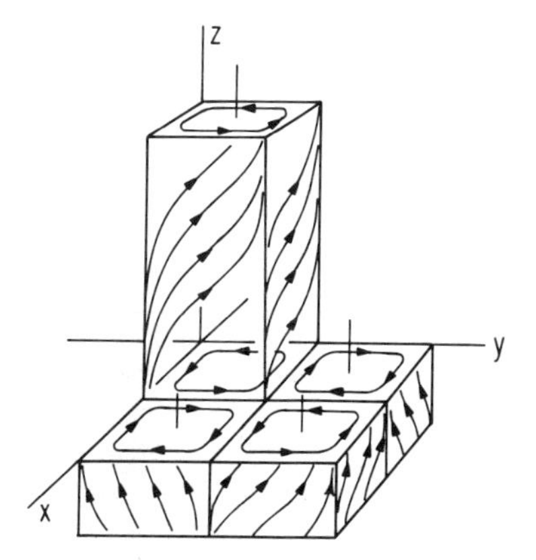

Fig. 8.10. A sketch of the close packed rectangular array of alternate twisted flux bundles providing the continuous field $A = (B_1/k)\sin kx \sin ky$.

field. More interesting is the case, already described, that the vortex motions are continuous and are confined to the basic cells $m\pi < kx < (m+1)\pi$, $n\pi < ky < (n+1)\pi$ with the motions in $0 < kx,\ ky < \pi$ stronger than in $-\pi < kx < 0,\ 0 < ky < +\pi$, and similarly throughout the x, y plane on a checkerboard pattern. In that case the field lines of each flux bundle, whether expanded or contracted, connect into the basic cells of equal size $(\pi/k \times \pi/k)$ at the boundaries $z = 0$ and $z = L$. That means that the contracted flux bundles diverge somewhat as they approach the boundaries and the expanded flux bundles converge somewhat, sketched in Fig. 8.11(a) and (b), respectively. The expansion or contraction of the flux bundles declines asymptotically to zero over a characteristic distance $1/k$ inward from each boundary. The principal discontinuities caused by the gap in the flux surfaces of the original $(\sin kx \sin ky)$ field lie on the bevel on the corners of the expanded flux bundles, which tapers down to zero width (in a characteristic distance $1/k$) at each end of each expanded bundle, as may be seen in Fig. 8.11(b). The topology of the current sheet, or surface of discontinuity, at the "southwest" corner of the expanded flux bundle is shown in Fig. 8.11(c) (Parker, 1990). Note the buttressed structure of the current sheet at its base, much like the trunks of many species of tree in the tropical rain forest.

The field lines passing by on each side of the gap, sketched in Fig. 8.9, are the field lines of the two contracted flux bundles that fit on opposite sides where the beveled corners of expanded tubes meet.

Turning to another example, note §6.11, treating the twisting and expansion of initially separate flux bundles until they come into contact to form current sheets, sketched in Fig. 6.26. Again the expanded flux bundles taper down to a radius $\varpi = b$ at each end where they intersect the boundaries $z = 0, L$. The principal current sheets lie along the surfaces of contact between neighboring bundles, but there is a weak discontinuity or current sheet, everywhere around each bundle. This may be seen from the fact that the field of the individual flux bundle is continuous so long as the bundle radius R is less than a, where $2a$ is the distance between the

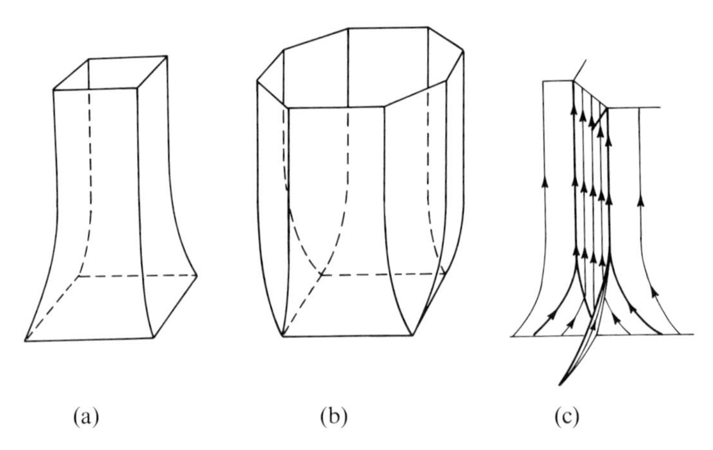

Fig. 8.11. A sketch of the twisted flux bundles of Fig. 8.10. as they approach the boundary $z = 0$, with (a) showing a compressed bundle with negative helicity with less twisting, (b) an expanded bundle with positive helicity with more twisting, and (c) the principal current sheet or tangential discontinuity at the "southwest" corner of the expanded bundle. The lines with arrows indicate the current pattern.

axes of the bundles. The azimuthal field rises linearly from zero to a maximum at $\varpi \cong R - \varepsilon$ and then drops rapidly but smoothly to zero at R. When R exceeds a, the flux bundles press against their neighbors providing an enhanced pressure along the narrow strip of contact. That strip of enhanced pressure does not extend all the way to $z = 0, L$, of course, because the bundle radius tapers down to $\varpi = b < a$ in a characteristic distance a at each end. The optical analogy shows how the successive layers of field in each twisted bundle are refracted by the strip of high pressure. Figure 8.12 (Parker, 1990) is a sketch of the successive layers starting with the outermost layer, for which $B_\varphi = 0$, and progressing inward a very small distance $O(\varepsilon)$. The oval represents the region of enhanced pressure, which, as already noted, does not extend all the way to each boundary. The essential point is that the gaps in one layer or another wrap relatively far around each flux bundle if the bundle is strongly twisted ($Qa \gg 1$). Hence there is a tangential discontinuity almost everywhere around the flux bundle. The gaps in each flux layer extend all the way to the boundaries $z = 0, L$, arriving at the boundaries at successive azimuthal positions around the flux bundle. Hence the surfaces of discontinuity extend all the way to the boundary all the way around the bundle. We may imagine a hexagonal array (see Fig. 6.26a and b) of separate flaring surfaces extending inward from each boundary and meeting to form a hexagonal honeycomb throughout the interior of $a < z < L - a$.

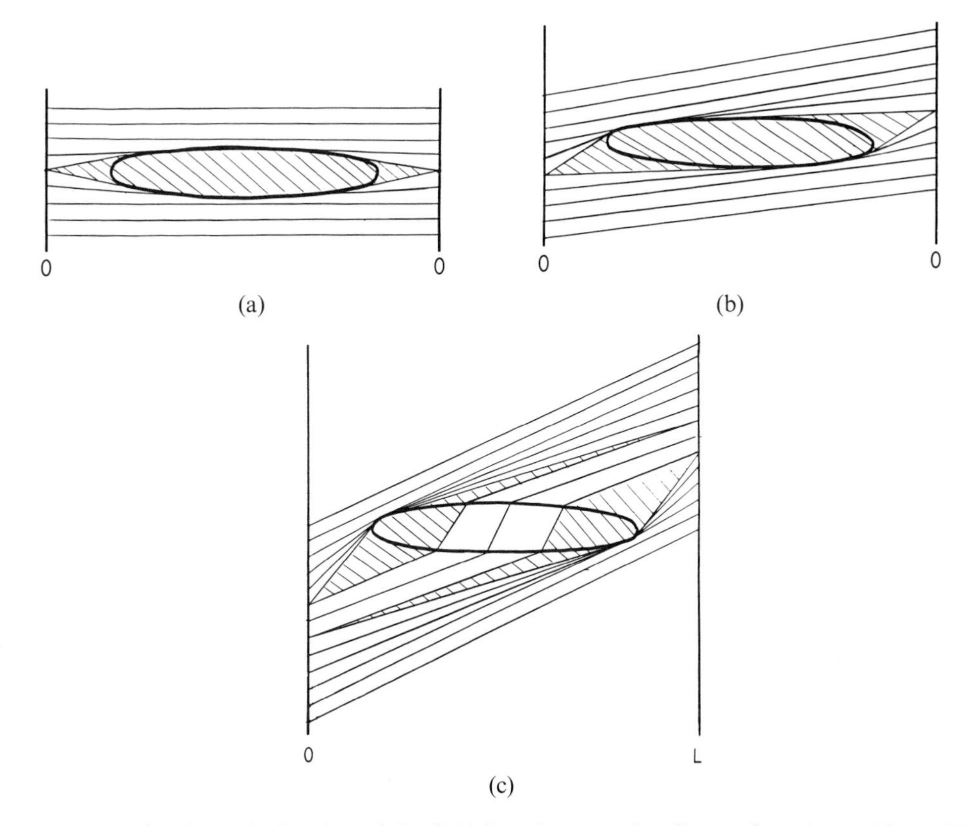

Fig. 8.12. A schematic drawing of the field lines in successive flux surfaces inward from (a) the surface of the flux bundle to (c) at some distance $O(\varepsilon)$ from the surface.

8.5 General Form of Discontinuities

The surfaces of tangential discontinuity described in §8.3 form a regular honeycomb, with tapering or flaring of the cell walls toward the boundaries $z = 0, L$. Such regularity is not expected in nature and the next step is to develop some idea of the topology under more typical circumstances. This opens up a host of different geometrical and topological forms, so that we are obliged to pursue one or two prominent special cases from which we infer the general properties of a random array of flux bundles.

To start with a particularly simple case, suppose that a single flux bundle is displaced to an oblique path across $0 < z < L$. Denote the radius of the initial flux bundle by $\varpi = a \ll L$. The upper end, at $z = L$, is then displaced in the y-direction a large distance $h = O(L)$ so that the flux bundle extends obliquely through the ambient field $\mathbf{e}_z B$ with a inclination $\theta = \tan^{-1} h/L$, as sketched in Fig. 8.13. The flux $\Phi = \pi a^2 B$ in the displaced bundle is flattened by the field pressure on each side, sketched in Fig. 8.14 where the current direction is shown by the heavy lines, the ambient field is indicated by the light lines, and the field in the misaligned flux bundle is shown by the short arrows (Parker, 1981a,b; 1990). This case is unlike the previous examples in that the currents close locally, rather than extending all the way from $z = 0$ to $z = L$. In fact we may think of the tangential discontinuities here

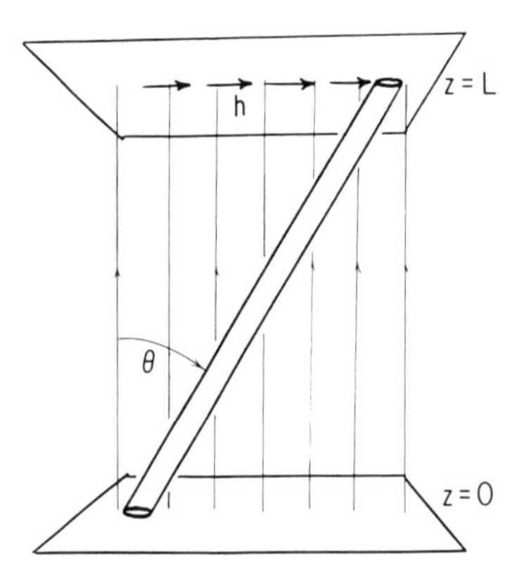

Fig. 8.13 A sketch of the displaced flux bundle, inclined at an angle $\theta = \tan^{-1} h/L$ to the ambient field.

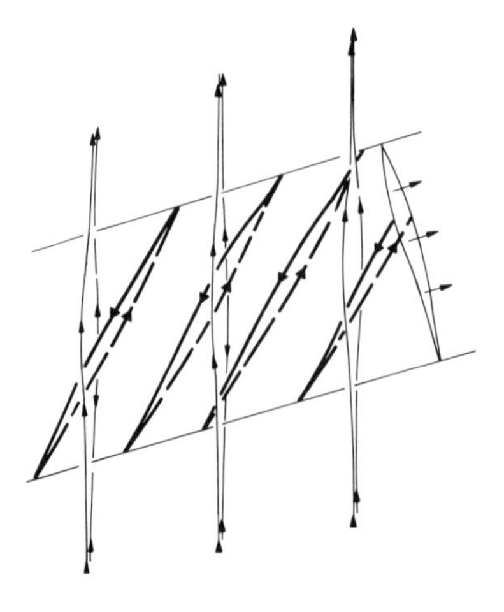

Fig. 8.14. A diagram of the electric current (heavy lines) and field (light lines) around a small flux bundle extending obliquely across an otherwise uniform field.

as two sheets which intersect along their edges, with the current flowing from one onto the other, not unlike the intersection of surfaces sketched in Fig. 8.11(c). This intersection of current sheets is a common feature of the magnetic field with random winding and intermixing, which we take up next.

Consider, then, the irregular array of current sheets or surfaces of tangential discontinuity that arises when the fluid motion, described by the stream function $\psi(x, y, kzt)$ in equations (1.3) and (1.4), represents a sequence of n random localized uncorrelated mixing patterns along the lines of force from $z = 0$ to $z = L$. Following the mixing of the field by the fluid motion ψ the field is continuous everywhere when ψ and its derivatives are continuous. The surfaces of tangential discontinuity form when the field is subsequently allowed to relax to the minimum available energy state while holding the footpoints fixed at the boundaries $z = 0$ and $z = L$.

It was shown in §3.5 that in the limit of large n the field takes on the potential form $\mathbf{B} = -\nabla\phi$ with $\nabla^2\phi = 0$ throughout the regions of continuous field between the surfaces of discontinuity. Each mixing pattern has a correlation length $\ell_\parallel$ in the z-direction along the mean field and a correlation length $\ell_\perp$ in the transverse xy-direction. Each mixing pattern extends for a distance of the order of $L/n(>\ell_\parallel)$ along the field. If the field is strongly wound, so that the transverse components are comparable to the mean field B, then $\ell_\parallel$ is of the same order as $\ell_\perp$. However, if the winding is weaker so that the transverse field components are much weaker than the mean field, the winding pattern may be stretched out in the z-direction so that $\ell_\parallel \gg \ell_\perp$.

Now tangential discontinuities form along the topological separatrices in each winding pattern, in the same way that they formed at the separatrices in the foregoing examples. It is immediately evident, then, that the surfaces of discontinuity created in any one winding pattern generally do not project precisely into the surfaces of discontinuity produced in any of the other $n - 1$ (independent) winding patterns. Denote by S_j the network of surfaces produced in the jth winding pattern in the neighborhood of $z = z_j \cong (j/n)L$. The surfaces S_j are created with separations of the order of $\ell_\perp$. Each surface S_j produced in the jth winding pattern fans out beyond the ends of the jth region, with the strength of the discontinuity decreasing inversely with distance in order to conserve the current J (see discussion in §8.1) while the width of the sheet increases in direct proportion to the distance, illustrated in the primitive spiral surfaces of discontinuity treated in §§5.41, 5.42 and 8.2. The essential point is that as any surface of discontinuity S_j produced in the jth winding pattern extends away from the location z_j of the jth pattern, it spreads out over increasingly large distances in the transverse direction and mixes into more and more uncorrelated patterns along the way. The result is that S_j intersects surfaces $S_{j\pm1}$, $S_{j\pm2}$, etc. produced in successive uncorrelated winding patterns. However, two surfaces of tangential discontinuity do not pass through each other where they intersect. The stronger generally cut off the weaker. That is to say, the edges of a surface of discontinuity are determined by the intersection with a surface of stronger discontinuity (Parker, 1986a, 1990).

The general rule is that the strong discontinuities produced in any location z_j generally absorb the intersecting weaker discontinuities that extend into the region from elsewhere. To see this in more detail, consider Fig. 8.15(a), which is a sketch of a surface of tangential discontinuity ABCD with the field lines shown on both sides of the surface and the current direction shown by the single arrow. Now imagine

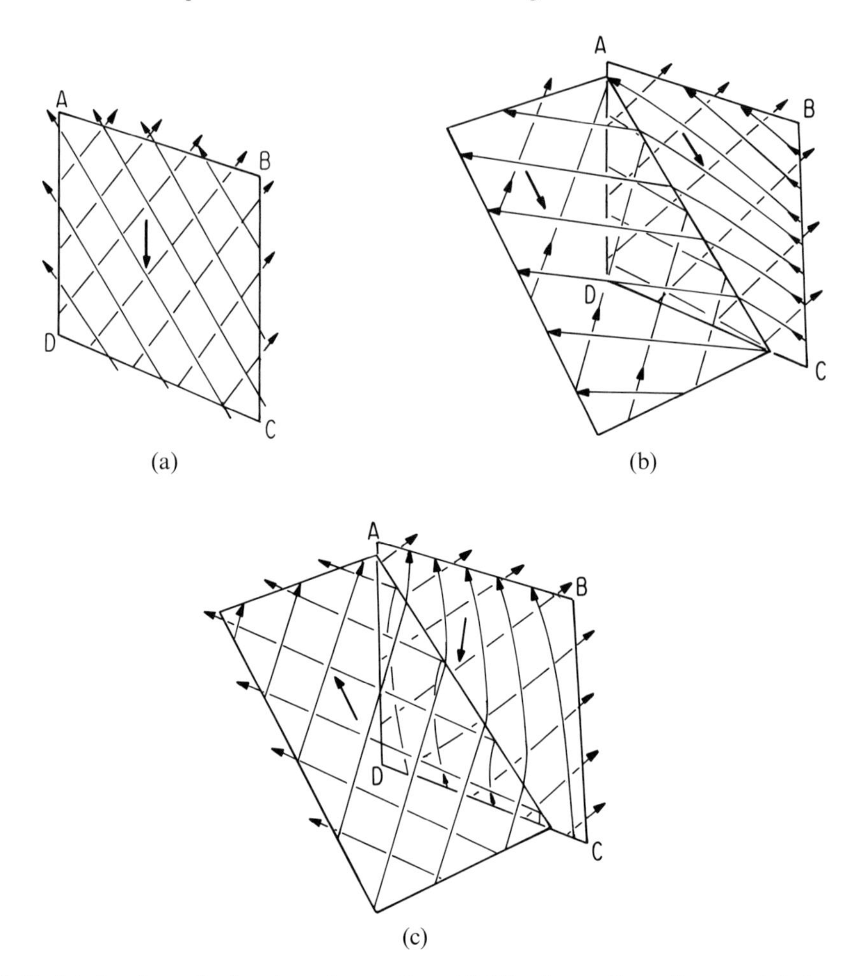

Fig. 8.15. (a) A schematic drawing of a portion ABCD of a surface of tangential discontinuity, showing the field lines on both the near and far sides. The short arrow indicates the direction of the torsion or electric current in the discontinuity. (b) The field configuration and torsion when ABCD is intersected by another weaker surface of discontinuity with torsion or current in the same direction, and (c) with torsion or current in the opposite direction.

that another weaker tangential discontinuity is created with its surface more or less perpendicular to the plane of the paper in a region located outward from the plane of the paper. This discontinuity represents a shear plane which intersects the surface ABCD and tries to extend itself across ABCD. But the shear plane is aligned with the field lines shown on the near side of ABCD. Hence, the field on the back side of ABCD lies across, rather than parallel to, the shear plane, so that the shear plane cannot penetrate through the discontinuity ABCD. What happens instead is sketched in the schematic Fig. 8.15(b) if the currents in the two discontinuities are more or less parallel. The schematic Fig. 8.15(c) indicates the configuration of field and current when the currents are more or less antiparallel. The essential point is that the stronger discontinuities terminates the weaker one. If a weaker discontinuity is subjected to a stronger discontinuity we would expect a general redistribution and reorientation of the field so that the weaker discontinuity is broken up by the stronger. But any quantitative statements in this respect are beyond the scope of the present work.

In summary, the strong discontinuities produced in any locality z_j terminate the weaker discontinuities extending into the region z_j from neighboring localities $z_{j\pm1}$, $z_{j\pm2}$, etc. Thus, we expect that the random mixing of flux by the successive n uncorrelated winding patterns in the fluid motion ψ provides an irregular honeycomb of surfaces of tangential discontinuity on a transverse scale of the order of $\ell_\perp$ in the limit of large n. What is more, the average strengths of the discontinuities are of the order of $B\ell_\parallel/\ell_\perp$ in the limit of large n because, in the absence of any net rotation of the flux bundles between $z = 0$ and $z = L$, the net torsion and the net current are zero. Hence there are as many discontinuities with positive torsion or current as there are with negative torsion or current. The net result is a general cancellation of torsion or current in the discontinuities where they meet, so that the mean strength is independent of n. The surfaces form an irregular honeycomb and we expect the scale of the honeycomb to be $O(\ell_\perp)$ and the general geometrical structure of the honeycomb to be independent of n.

It would appear that the statistical distribution of strength, if not the mean strength, of the local surfaces of discontinuity S_j, may depend upon n, because, as already noted, each discontinuity S_j absorbs the successively weaker discontinuities $S_{j\pm1}$, $S_{j\pm2}$, etc. Whether this may occasionally accumulate into a particularly strong surface of discontinuity is an interesting question, with direct implications for the structure of the filamentary X-ray emitting regions in the solar corona.

8.6 Field Structure Around Intersecting Discontinuities

The intersection of surfaces of tangential discontinuity is a general occurrence in any succession of patterns of intermixing along the field. Generally speaking, the stronger of two intersecting discontinuities terminates and absorbs the weaker, as described in the previous section. The result is three sectors around each line of intersection, rather than the four that obtain around the intersection of two planes. The present section explores the continuous field in the three sectors defined by intersecting surfaces of discontinuity (Parker, 1990). The interesting feature of the intersection is the equality of the three apex angles of the sectors, regardless of the relative strengths of the two intersecting tangential discontinuities. The intersection of a weak tangential discontinuity with a strong tangential discontinuity causes the stronger discontinuity to form a ridge with an apex angle of $2\pi/3$ on the side facing the weak discontinuity. The weak discontinuity curves around so that it bisects the angle formed by the crest of the ridge. The profile is sketched in Fig. 8.16, in which the heavy line represents the strong discontinuity and the light line the weak discontinuity. Thus, in the large, far from the intersection, the sector angles are approximately β, π, and $\pi - \beta$, but close to the intersection the sector angles are all $2\pi/3$, with the weak fields forcing the stronger transverse fields of the stronger discontinuity to give way. This result can be understood from the manner in which the transverse fields approach zero in the apex of each sector. But that can be established only by careful attention to the physical conditions in the field.

There is a variety of solutions of the Grad–Shafranov equations in corners,

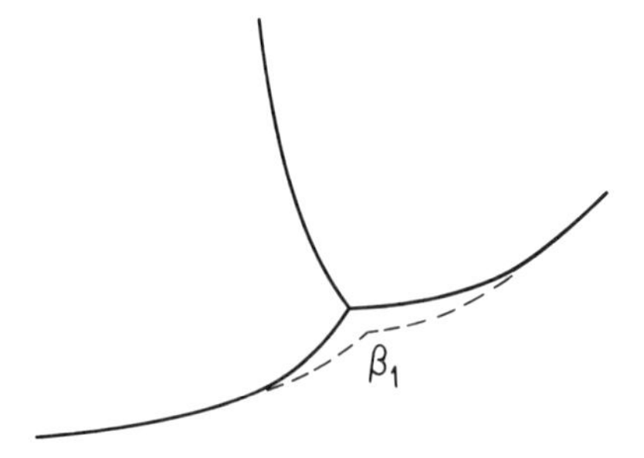

Fig. 8.16. The profile of the junction between a weak discontinuity (light line) and a strong discontinuity (heavy line), showing the 120° angle between each surface at the line of intersection. The dashed line indicates the hypothetical form in which the sector angle for the stronger field exceeds 120°.

cusps, etc. available in the literature (cf. Vainshtein, 1990; Parker, 1990; Strachan and Priest, 1991; Vekstein and Priest, 1992, 1993), and among them we may expect to find an appropriate solution for the field in the vicinity of the intersection of two surfaces of discontinuity, dividing the region around the line of intersections into three sectors. However, we must be clear as to the essential physics if we are to distinguish the correct solution from those that are inapplicable.

In view of the nonlinearity of the force-free field equation (1.5) we attempt only a near solution, applicable in some limiting small radius $r < \varepsilon$ around the vertex of each sector. This limiting solution is sufficient to exhibit the crucial properties of the field in the vertices and to determine the basic symmetry of the vertices. The essential physical properties of the field near the vertex of each sector are three-fold. First of all, the torsion coefficient α is generally nonvanishing, because α is rigorously constant along each field line, most lines extending through successive regions wherein the field is twisted and wound in arbitrary forms by the fluid motion described by $\psi(x, y, kzt)$ in equations (1.3), (1.4), and (3.7)–(3.11). A further consequence of the general nonvanishing torsion is that field lines spiral around the intersecting boundaries and across the vertex of each sector. Finally, the total magnetic pressure is continuous across the boundaries or separatrices between the sectors. These three points are essential for defining the acceptable solutions and excluding the many other mathematically interesting, but physically inapplicable, solutions currently available.

In view of the complexity of the set of all solutions, it is useful to construct a paradigm for the field. The essential point is that the field in each sector represents a twisted flux bundle with characteristic longitudinal scale $\ell_{\parallel}$. If the flux bundle were separated from its neighbors and placed in isolation in a uniform external pressure P_0, it would assume a circular cross-section (Vainshtein and Parker, 1986; Linardatos, 1993). Conversely, one may think of the field in each sector as a twisted flux bundle of circular cross-section that has been forcibly jammed into the irregular prism formed by the surfaces of tangential discontinuity.

So as a first step, consider the general nature of a long straight twisted flux bundle, with an eye to the arbitrary form of its internal twisting. The essential point is that such a flux bundle is rotationally symmetric and can be described in terms of an arbitrary generating function $f(\varpi)$, where ϖ is radial distance from the axis of symmetry. The field is $[0, B_\varphi(\varpi), B_z(\varpi)]$ and the equilibrium form (Lüst and Schlüter, 1954; Parker, 1979) is

$$B_\varphi = \pm\left[-\frac{1}{2}\varpi f'(\varpi)\right]^{\frac{1}{2}}, \tag{8.15}$$

$$B_z = \pm\left[f(\varpi) + \frac{1}{2}\varpi f'(\varpi)\right]^{\frac{1}{2}}, \tag{8.16}$$

where

$$f(\varpi) = B_\varphi^2(\varpi) + B_z^2(\varpi).$$

In order that B_φ and B_z are real, the arbitrary function $f(\varpi)$ declines monotonically with increasing ϖ, but not faster than $1/\varpi^2$. This field satisfies the force-free condition described by equation (3.3), and it is easy to show that the torsion coefficient is

$$\alpha = \frac{3f' + \varpi f''}{4(f + \frac{1}{2}\varpi f')^{\frac{1}{2}}(-\frac{1}{2}\varpi f')^{\frac{1}{2}}}. \tag{8.17}$$

Note, then, that α vanishes only in the limiting case that $f(\varpi) \propto 1/\varpi^2$, providing the potential field $B_\varphi \propto 1/\varpi$. The field lines are given by the two parameter family of curves

$$\varpi = \lambda, \tag{8.18}$$

$$\varphi = \theta + \left\{\frac{-f'(\lambda)}{2\lambda[f(\lambda) + \frac{1}{2}f'(\lambda)]}\right\}^{\frac{1}{2}} z, \tag{8.19}$$

where θ is the azimuthal angle of the field line at $z = 0$. The field lines spiral endlessly around the bundle except in the limiting case that $f(\varpi) = \mathit{constant}$, providing only a uniform field $B_\varphi = 0$, $B_z = B_0$.

A simple and adequate example follows for

$$f(\varpi) = B_0^2(1 - k^2\varpi^2)/(1 - k^2R^2)$$

inside the radius $\varpi = R$, where kR is chosen to be less that $1/2^{\frac{1}{2}}$ in order that the torsion coefficient is bounded and B_z neither vanishes nor changes sign. The quantity $1 - k^2R^2$ is introduced in the denominator as a normalizing factor, so that the magnetic pressure at the surface $\varpi = R$ is $B_0^2/8\pi$. With this choice of $f(\varpi)$, it follows that

$$B_\varphi = B_0 k\varpi/(1 - k^2R^2)^{\frac{1}{2}},$$

$$B_z = B_0(1 - 2k^2\varpi^2)^{\frac{1}{2}}/(1 - k^2R^2)^{\frac{1}{2}},$$

and the torsion coefficient is

$$\alpha = +2k/(1 - 2k^2\varpi^2)^{\frac{1}{2}}$$

across the radius of the bundle. The field lines are given by

$$\varpi = \lambda,\ \varphi = \theta + kz/(1 - 2k^2\lambda^2)^{\frac{1}{2}}.$$

The azimuthal field B_φ is derived from the vector potential

$$A = -\frac{1}{2}B_0\, k\varpi^2/(1 - k^2R^2)^{\frac{1}{2}},$$

with the result that

$$B_z(A) = \frac{+B_0}{(1 - k^2R^2)^{\frac{1}{2}}}\left[1 + \frac{4k(1 - k^2R^2)^{\frac{1}{2}}A}{B_0}\right]^{\frac{1}{2}},$$

and

$$\alpha(A) = \frac{+2k}{[1 + 4k(1 - k^2R^2)^{\frac{1}{2}}A/B_0]}.$$

The essential point is that the torsion coefficient is continuous, bounded, and nonvanishing throughout the interior, all the way out to the surface. At the same time the field lines spiral around the bundle throughout the interior and around the surface as well. As already noted, the torsion coefficient is nonvanishing and the field lines spiral around the flux bundle for every choice of $f(\varpi)$ except the limiting forms $f(\varpi) = 1/\varpi^2$ and $f(\varpi) = constant$, in which case one or the other of the two conditions is avoided. But either of these special cases is unlikely in nature.

The next step is to box in the flux bundle with three or more infinitely conducting plane boundaries all parallel to the z-direction. The intersection of the planes with the xy-plane forms an irregular polygon, and the twisted flux bundle is compressed within that polygon. The vertices of the polygon each provide an example of the vertex of the individual sector at the line of intersection of the surfaces of tangential discontinuity. The essential points are that the confinement of the flux bundle alters the torsion coefficient somewhat, but leaves it nonvanishing, while the field lines, whose topology is conserved by the infinite electrical conductivity, continue to spiral around the bundle, crossing the vertices in the process. With these points in mind, consider the solution of the magnetostatic equations close to the vertex of an individual sector. The problem proves to be more complicated than one might expect.

8.6.1 Magnetostatic Field in a Vertex

To proceed systematically, denote by r the distance measured radially outward from the line of intersection of the surfaces of tangential discontinuity. Note, then, that there is no characteristic scale in the limit of small r except r itself. The scale of variation of the field along the line of intersection is some finite length $\ell_\parallel$, so that $\ell_\parallel/r$ becomes large without bound in the limit of small r. It follows that the variation of field along the line of intersection is negligible compared to the transverse variation of the field with r close to the line of intersection. Consider, then, the local Cartesian coordinate system (ξ, η, ζ) with the origin at some point O on the line of intersection and with the ζ-axis tangent to the line of intersection at O. Rotate the coordinate system about the ζ-axis so that the ξ axis is tangent to one of the three surfaces of discontinuity in the limit of small r. Then $r = (\xi^2 + \eta^2)^{\frac{1}{2}}$, and $\partial/\partial\zeta$ is small $O(r/\ell_\parallel)$ compared to $\partial/\partial\xi$ and $\partial/\partial\eta$. In the limit of small r the Grad–Shafranov equations (4.10)–(4.12) are applicable, with

$$B_\xi = +\partial A/\partial\eta,\ B_\eta = -\partial A/\partial\xi,\ B_\zeta = B_\zeta(A) \tag{8.20}$$

and

$$\frac{\partial^2 A}{\partial\xi^2} + \frac{\partial^2 A}{\partial\eta^2} + B'_\zeta(A)B_\zeta(A) = 0. \tag{8.21}$$

for a force-free field ($\nabla p = 0$). The torsion coefficient is

$$\alpha = \alpha(A) = B'_{\zeta}(A).$$

Note from equation (1.7) that α is rigorously constant along each field line, regardless of where the line extends at large r.

Assume that the surfaces of discontinuity have a well-defined direction in the limit of vanishing r so that they may be represented locally by planes. Denote the vertex angle of one of the sectors by β, with the sector boundaries $\varphi = 0, \beta$, where φ represents azimuth around the ζ-axis, so that $\xi = r\cos\varphi$, $\eta = r\sin\varphi$. Then $B_\varphi = 0$ on $\varphi = 0, \beta$, and the boundaries are given by $A(r, \varphi) = \textit{constant}$. There is no loss of generality in using the gauge in which $A = 0$ on the boundaries. Consider the solution of the Grad–Shafranov equations in $0 < \varphi < \beta$.

The development of the force-free field in the sectors around the intersection of surfaces of tangential discontinuity is based on the assumption that $B_\zeta(A)$ is an analytic function of its argument. Then $\alpha(A) = B'_z(A)$ can be expanded in ascending powers of A,

$$\alpha(A) = \alpha(0) + \alpha'(0)\,A + \frac{1}{2}\alpha''(0)\,A^2 + \ldots, \tag{8.22}$$

with a finite radius of convergence. It follows that

$$B_\zeta(A) = B_0 + \alpha(0)A + \frac{1}{2}\alpha'(0)A^2 + \frac{1}{3!}\alpha''(0)A^3 + \ldots \tag{8.23}$$

upon integration, where the integration constant $B_0 = B_\zeta(0)$ represents the generally nonvanishing field component along the line of intersection. Since A vanishes in the limit of small r, it is sufficient to keep only first-order terms, with equation (8.21) reducing to the inhomogeneous linear form

$$\frac{1}{r}\frac{\partial}{\partial r}r\frac{\partial A}{\partial r} + \frac{1}{r^2}\frac{\partial^2 A}{\partial \varphi^2} + k^2 A + \alpha(0)B_0 = O(A^2), \tag{8.24}$$

where

$$k^2 \equiv \alpha(0)^2 + \alpha'(0)B_0. \tag{8.25}$$

The development is carried through for the case that $\alpha'(0)$ is not so negative that $k^2 < 0$. It is a simple matter to show that the near solution is the same when $k^2 < 0$ as when $k^2 > 0$, the only difference being the employment of the modified Bessel function $I_\nu(kr)$ in place of $J_\nu(kr)$ in the preliminary expressions.

The particular solution to the inhomogeneous equation is

$$A = -\frac{1}{4}\alpha(0)B_0\,r^2.$$

Combining this with the solution $J_2(kr)\cos(2\varphi - \beta)$ to the homogeneous equation, note that the form

$$A = \frac{2\alpha(0)B_0}{k^2}\left[\frac{J_2(kr)\cos(2\varphi - \beta)}{\cos\beta} - \frac{1}{8}k^2 r^2\right]$$

is a solution to equation (8.24). In the limit of small kr and A, this reduces to

$$A \cong \frac{1}{4}\alpha(0)B_0\left[\frac{\cos(2\varphi - \beta)}{\cos\beta} - 1\right]r^2, \tag{8.26}$$

neglecting terms $0(r^4)$. This reduced expression for A satisfies the boundary condition that $A = 0$ on $\varphi = 0, \beta$ to the order considered. It follows that

$$\begin{aligned} B_r &= +\frac{1}{r}\frac{\partial A}{\partial \varphi}, \\ &\cong -\frac{\alpha(0) B_0\, r \sin(2\varphi - \beta)}{2\cos\beta}, \end{aligned} \tag{8.27}$$

$$\begin{aligned} B_\varphi &= -\frac{\partial A}{\partial r}, \\ &\cong -\frac{1}{2}\alpha(0) B_0 \left[\frac{\cos(2\varphi - \beta)}{\cos\beta} - 1\right] r \end{aligned} \tag{8.28}$$

for small r. A field line in the surface $\varphi = 0$ of the sector is described by $d\varphi = 0$ and

$$\frac{dr}{d\zeta} = \frac{B_r}{B_\zeta}. \tag{8.29}$$

In the limit of small r this can be integrated to give

$$r = \lambda \exp\left[\frac{1}{2}\tan\beta\alpha(0)\zeta\right] \tag{8.30}$$

for the field line crossing $\zeta = 0$ at $r = \lambda$. It is evident that for any nonvanishing λ, the field line approaches the vertex ($r = 0$) only asymptotically, in the limit of large negative z (assuming $\tan\beta\alpha(0) > 0$). Hence, this solution fails to represent the physical situation at hand, wherein the field lines cross over the vertex. To remedy this shortcoming supplement the field given by equation (8.26) with the additional term

$$A_1 = \frac{2^{\pi/\beta}\beta\Gamma(1 + \pi/\beta) B_1}{\pi k} J_{\pi/\beta}(kr)\sin(\pi\varphi/\beta) \tag{8.31}$$

representing a solution to the homogeneous equation and tailored to vanish on $\varphi = 0$ and $\varphi = \beta$. The constant B_1 is arbitrary at this point. In the limit of small kr,

$$A_1 \cong \frac{\beta B_1}{\pi} k^{\pi/\beta - 1} r^{\pi/\beta} \sin\left(\frac{\pi\varphi}{\beta}\right)[1 + O(k^2 r^2)]. \tag{8.32}$$

This limiting expression satisfies Laplace's equation, so to this order it contributes nothing to $\nabla \times \mathbf{B}$ and nothing to the torsion coefficient. Now if $\beta > \frac{1}{2}\pi$, it follows that $\pi/\beta < 2$ so that in the limit of small kr the dominant contribution to the total vector potential is A_1 rather than the expression given by equation (8.26). Hence, in the limit of small kr,

$$B_r \cong +B_1 (kr)^{\pi/\beta - 1}\cos\frac{\pi\varphi}{\beta}, \tag{8.33}$$

$$B_\varphi \cong -B_1 (kr)^{\pi/\beta - 1}\sin\frac{\pi\varphi}{\beta}, \tag{8.34}$$

while B_ζ is still given by equation (8.23) as B_0 to lowest order. The field lines are described now by

$$\frac{dr}{d\zeta} = \frac{B_1}{B_0}(kr)^{\pi/\beta - 1}$$

in the surface $\varphi = 0$ in the limit of small kr, with B_ζ exactly equal to B_0. Integration yields

$$(kr)^{2-\pi/\beta} = (2 - \pi/\beta)\frac{B_1}{B_0}k\zeta,$$

or

$$kr = \left[\frac{(2 - \pi/\beta)B_1}{B_0}k(\zeta - \zeta_0)\right]^{\beta/(2\beta-\pi)} \tag{8.35}$$

for the field line reaching $r = 0$ at $\zeta = \zeta_0$. It follows, then, that the field lines spiral across the vertex at $r = 0$ provided only that $\beta > \frac{1}{2}\pi$. It follows that the combined solution

$$A \cong \frac{1}{4}\alpha(0)B_0\left[\frac{\cos(2\varphi - \beta)}{\cos\beta} - 1\right]r^2 + \frac{\beta B_1}{\pi k}(kr)^{\pi/\beta}\sin\frac{\pi\varphi}{\beta} \tag{8.36}$$

satisfies the two basic conditions, that $\alpha(0)$ is nonvanishing and the field lines spiral across the vertex of the sector at $r = 0$. The quantity B_1/B_0 determines the scale over which the field lines approach the vertex, which depends, of course, on the spiral angle at large r, beyond the range of validity of the present limiting solution. Sufficiently close to the vertex, the field is dominated by the potential field given by equations (8.32)–(8.34). The potential field is not associated with the torsion $\alpha(0)$ represented by the equations (8.26)–(8.28). The potential field is the response of the force-free field to the boundaries of the vertex.

The field components given by the vector potential (8.36) are readily seen to be

$$B_r = -\frac{1}{2}\alpha_1(0)B_0 r\frac{\sin(2\varphi - \beta)}{\cos\beta} + B_1(k_1 r)^{\pi/\beta-1}\cos\frac{\pi\varphi}{\beta}, \tag{8.37}$$

$$B_\varphi = -\frac{1}{2}\alpha_1(0)B_0 r\left[\frac{\cos(2\varphi - \beta)}{\cos\beta} - 1\right] - B_1(k_1 r)^{\pi/\beta-1}\sin\frac{\pi\varphi}{\beta} \tag{8.38}$$

where the subscript 1 has been attached to $\alpha(0)$ and to k to indicate the values appropriate to the sector (O, β).

Thus far we have not succeeded in constructing a physically acceptable solution for $\beta < \frac{1}{2}\pi$. That case is of no concern for the present problem, for which it will be found that $\beta = \frac{2}{3}\pi$, but it is nonetheless a real question because the flux bundle deliberately enclosed by rigid planes set at arbitrary angles may certainly involve one or more acute vertices.

It is important to note that smaller vertex angles can be fitted if the torsion is not required to be analytic at the vertex. A recent paper by Vekstein and Priest (1993) is interesting in this respect. They treat an intersection of separatrices involving a cusp, with the separatrices given by $\varphi = \pi$ and $\varphi = \pm Kr^\beta$ for small r where $O < \beta < \frac{1}{2}$. The negative x-axis forms a separatrix which splits at the origin to form the two separatrices $y \cong Kx^{\beta+1}$ for $x > 0$. They construct solutions within the cusp that satisfy pressure balance across the separatrices, with

$$B_\zeta^2(A) = B_0^2 + (3 + 2/\beta)\varepsilon A^{2\beta/(2+3\beta)}$$

inside the cusp. The field components are bounded as r declines to zero, with $A \sim r^{1+3\beta/2}$, $B_r \sim r^{\beta/2}$, $B_\varphi \sim r^{3\beta/2}$ and $B_\zeta \sim B_0$ immediately inside the surfaces of the cusp. The field line crossing the vertex at $\zeta = \zeta_0$ has the form $\zeta - \zeta_0 \sim r^{1-\beta/2}$, coming tangent to the vertex at $\zeta = \zeta_0$. The torsion $B'_\zeta(A)$ varies as $A^{-(2+\beta)/(2+3\beta)} \sim r^{-(1+\frac{1}{2}\beta)}$, to which the current density j_ζ is also proportional. The essential point is that the torsion increases without bound as r declines to zero. That is to say, there is magnetic flux on which the torsion is arbitrarily large, all the way form $z = 0$ to $z = L$. Such unlimited torsion does not arise from the limited continuous mapping of the footpoints of the field at the endplates $z = 0, L$. So these solutions do not apply in the context of the present theoretical development.

8.6.2 Compatibility of Sectors

Consider how solutions of equation (8.36) in each of the three azimuthal sectors $(0, \beta)$, (β, γ), and $(\gamma, 2\pi)$ around the line of intersection (the ζ-axis) can be fitted together. The boundary condition is continuity of magnetic pressure. Equation (8.36) provides the field in the sector $(0, \beta)$. For the sector (β, γ) write $A = A_2$ where

$$A_2 = \frac{1}{4}\alpha_2(0)B_0\left[\frac{\cos(2\varphi - \beta - \gamma)}{\cos(\gamma - \beta)} - 1\right]r^2 + \frac{(\gamma - \beta)B_2}{\pi k_2}(k_2 r)^{\pi/(\gamma-\beta)} \sin[\pi(\varphi - \beta)/(\gamma - \beta)]. \tag{8.39}$$

The field components are

$$B_r = -\frac{1}{2}\alpha_2(0)B_0\, r\frac{\sin(2\varphi - \beta - \gamma)}{\cos(\gamma - \beta)} + B_2(k_2 r)^{\pi/(\gamma-\beta)-1}\cos[\pi(\varphi - \beta)/(\gamma - \beta)], \tag{8.40}$$

$$B_\varphi = -\frac{1}{2}\alpha_2(0)B_0\, r\left[\frac{\cos(2\varphi - \beta - \gamma)}{\cos(\gamma - \beta)} - 1\right] - B_2(k_2 r)^{\pi/(\gamma-\beta)-1}\sin[\pi(\varphi - \beta)/(\gamma - \beta)]. \tag{8.41}$$

In $(\gamma, 2\pi)$ the field is given by $A = A_3$ where

$$A_3 = \frac{1}{4}\alpha_3(0)B_0\left[\frac{\cos(2\varphi - \gamma)}{\cos\gamma} - 1\right]r^2 + \frac{(2\pi - \gamma)B_3}{\pi k_3}(k_3 r)^{\pi/(2\pi-\gamma)}\sin[\pi(\varphi - \gamma)/(2\pi - \gamma)]. \tag{8.42}$$

Then

$$B_r = -\frac{1}{2}\alpha_3(0)B_0\, r\frac{\sin(2\varphi - \gamma)}{\cos\gamma} + B_3(k_3 r)^{\pi/(2\pi-\gamma)-1}\cos[\pi(\varphi - \gamma)/(2\pi - \gamma)], \tag{8.43}$$

$$B_\varphi = -\frac{1}{2}\alpha_3(0)B_0 r\left[\frac{\cos(2\varphi - \gamma)}{\cos\gamma} - 1\right] - B_3(k_3 r)^{\pi/(2\pi-\gamma)-1}\sin[\pi(\varphi - \gamma)/(2\pi - \gamma)]. \tag{8.44}$$

The magnetic pressure is continuous across the boundaries $\varphi = 0, \beta, \gamma$, which restricts the various constants in the expressions for the fields in the individual sectors. Note, then, that with $A = 0$ on the boundaries, the ζ-component follows from equation (8.23) as B_0. Since $B_\varphi = 0$ on the boundaries, the magnetic pressure is just $(B_0^2 + B_r^2)/8\pi$, from which it follows that to lowest order B_0^2 must be continuous across each boundary and to $O(r)$ the transverse field B_r must also be continuous.

Equations (8.37), (8.40), and (8.43) provide the B_r. It is evident by inspection that the sector widths must all be the same with

$$\beta = \gamma - \beta = 2\pi - \gamma \tag{8.45}$$

if the powers of r in equation (8.37), (8.40), and (8.43) are to match across the boundaries. Hence $\gamma = 2\beta$ and $\beta = 2\pi/3$. It follows that the lowest order terms in B_r are all $O[(kr)^{\frac{1}{2}}]$. Squaring B_r in each case and equating the results on opposite sides of $\varphi = 0, 2\pi/3, 4\pi/3$, we obtain the three relations

$$+\alpha_1 B_1 B_0 k_1^{\frac{1}{2}} r^{\frac{3}{2}} \tan\beta + B_1^2 k_1 r \cong +\alpha_2 B_2 B_0 k_2^{\frac{1}{2}} r^{\frac{3}{2}} \tan(\gamma - \beta) + B_2^2 (k_2 r),$$

$$+\alpha_2 B_2 B_0 k_2^{\frac{1}{2}} r^{\frac{3}{2}} \tan\beta + B_2^2 k_2 r \cong -\alpha_3 B_3 B_0 k_3^{\frac{1}{2}} r^{\frac{3}{2}} \tan 2\beta + B_3^2 (k_3 r),$$

$$+\alpha_1 B_1 B_0 k_1^{\frac{1}{2}} r^{\frac{3}{2}} \tan\beta + B_1^2 k_1 r \cong -\alpha_3 B_3 B_0 k_3^{\frac{1}{2}} r^{\frac{3}{2}} \tan 2\beta + B_3^2 k_3 r,$$

respectively, where terms $O(r^2)$ have been neglected. Equating coefficients of r yields the two relations

$$B_1^2 k_1 = B_2^2 k_2 = B_3^2 k_3. \tag{8.46}$$

Equating coefficients of $r^{\frac{1}{2}}$ yields

$$\alpha_1 B_1 k_1^{\frac{1}{2}} = \alpha_2 B_2 k_2^{\frac{1}{2}} = \alpha_3 B_3 k_3^{\frac{1}{2}}, \tag{8.47}$$

upon noting that $\tan 4\pi/3 = -\tan 2\pi/3 = 3^{\frac{1}{2}}$. The wave numbers k_1, k_2, and k_3 are the positive square roots of equation (8.25) for the corresponding $\alpha_i(0)$ and $\alpha_i'(0)$.

In the circumstance that B_1, B_2, and B_3 all have the same sign, that sign can just as well be taken to be positive. Then equation (8.46) requires that

$$B_1 k_1^{\frac{1}{2}} = B_2 k_2^{\frac{1}{2}} = B_3 k_3^{\frac{1}{2}}. \tag{8.48}$$

It follows from equation (8.47) that

$$\alpha_1(0) = \alpha_2(0) = \alpha_2(0), \tag{8.49}$$

It is obvious from equation (8.25), then, that any differences in the wave numbers k_1, k_2, and k_3 arise from differences in the $\alpha_i'(0)$, with the B_i adjusted accordingly to satisfy equation (8.48).

The ζ-component B_0 is assumed to be positive here, in keeping with the general scenario of a field extending from $z = 0$ to $z = L$. However the transverse field A can be reversed in any sector, arbitrarily changing the signs of any one, or all, of the B_i and of the torsion $\alpha_i(0)$, without affecting the balance of forces across the boundaries. In general, then, equations (8.48) and (8.49) should be written in terms of the magnitudes of the $B_i k_i^{\frac{1}{2}}$ and the $\alpha_i(0)$, etc. Note that the signs of B_1 and α_1, for instance, cannot be changed independently of each other because it is the torsion α_1 that determines the direction B_1 with which the field lines spiral across the vertex of the sector. Specifically, it is evident by inspection of equation (8.38) for B_φ that α_1 and $B_1 k_1^{\frac{1}{2}}$ must have the same sign, because the two terms are compatible, representing the same direction of spiralling only if the term $O(r)$ has the same sign as the term $O(r^{\frac{1}{2}}) = O(r^{\pi/\beta - 1})$. Finally, if the B_i and the α_i all have the same sign, there is a discontinuity in both $O(r^{\frac{1}{2}})$ and $O(r)$ across each sector boundary. This applies to the situation sketched in Fig. 8.15(b). But if one of the pairs B_i, α_i has a sign different from the other two pairs then there is a discontinuity across only one sector boundary. This choice of signs applies to the scenario illustrated in Fig. 8.15(c). Then to $O(r^{\frac{1}{2}})$ and $O(r)$ the field is continuous across two of the three boundaries in the limit of small kr, even though the field is discontinuous across all three boundaries or separatrices at large r. It follows that only where the signs of each pair B_i, α_i are the same on opposite sides of a sector boundary does the tangential discontinuity survive to small r. If the signs of B_i and α_i reverse across a separatrix, then the fields are brought into continuity by the boundary deformation to the order of r considered here, even though the fields are discontinuous across the same separatrix at large kr.

Now with $\beta = 2\pi/3$, the field lines at the surface $\varphi = 0$ of the sector $(0, \beta)$ follow from equation (8.35) as

$$kr = \left(\frac{B_1}{2B_0}\right)^2 [k(\zeta - \zeta_0)]^2.$$

The field line is half a parabola, approaching tangentially to the vertex of the sector at $\zeta = \zeta_0$, crossing over the vertex, and departing on a similar tangential half parabola on the far side. Figure 8.17 is a sketch of the path across the vertex.

The field in the sector $(0, 2\pi/3)$ follows from equations (8.33) and (8.34) to lowest order, and from equations (8.37) and (8.38) to the next higher order, with

$$B_r \cong +B_1(k_1 r)^{\frac{1}{2}} \cos\frac{3}{2}\varphi + \alpha_1(0) B_0\, r \sin[2(\varphi - \pi/3)],$$

$$B_\varphi \cong -B_1(k_1 r)^{\frac{1}{2}} \sin\frac{3}{2}\varphi + \frac{1}{2}\alpha_1(0) B_0 r\{1 + 2\cos[2(\varphi - \pi/3)]\}$$

in the limit of small kr. The field in the other two sectors has the same form, with the same coefficients $B_1 k_1^{\frac{1}{2}}$ and $\alpha_1(0)$. The projection of the field lines onto the $\zeta = 0$ plane provides the family of curves given by $A = constant$, which to lowest order is the general hyperbolic form

$$r^{\frac{3}{2}} \sin\frac{3}{2}\varphi = constant$$

within $0 < \varphi < 2\pi/3$.

Consider how it is that each of the three sectors has the same apex angle $\beta = 2\pi/3$, regardless of the relative strengths of the two intersecting tangential

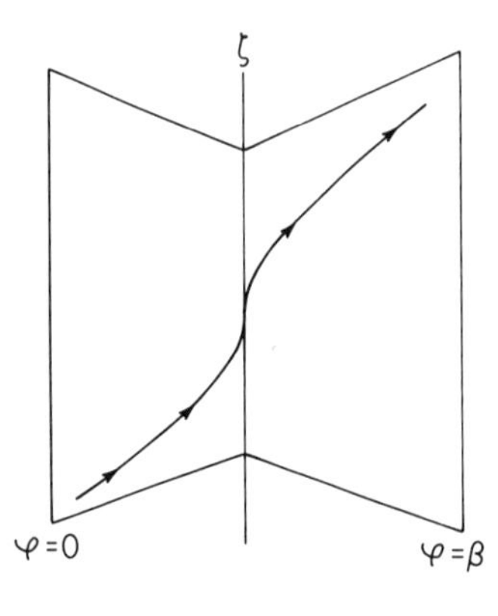

Fig. 8.17. A sketch of a field line passing across the vertex of a sector of width $\beta = 2\pi/3$, described by equation (8.34).

discontinuities. That is to say, how can a strong discontinuity, i.e., a sector with strong transverse fields, be forced into an apex angle $2\pi/3$ by the relatively weak transverse fields of the other surface of discontinuity. The answer is, first of all, that the equality of the sector angles is only in the limit of small kr. The relation does not hold for $kr = O(1)$. To understand the result as the limiting case as kr declines to zero, recall that the transverse field in any sector of angular width β falls to zero as $(kr)^{\pi/\beta - 1}$ in that limit. Hence a large β means a stronger field at any given value of kr. Thus, for instance, the fields in two sectors of width β_1 and β_2 would be in the ratio $(kr)^{\pi/\beta_1 - \pi/\beta_2}$ at any fixed value of r. Suppose, then, that the stronger transverse field lies in the sector with the larger apex angle β_1, indicated by the dashed line in Fig. 8.16. This is what one might have expected from the greater "stiffness" of the stronger field. However, with $\beta_1 > \beta_2$ the ratio of the stronger to the weaker field becomes large without bound as kr diminishes to zero. The result can only be that the pressure of the stronger field dominates the weaker field as kr declines to zero, pushing the apex toward the region of weaker field and diminishing β_1 in the process. It follows that β_1 diminishes to within some infinitesimal difference from β_2, whereupon the ratio of the fields approaches one in the limit of small kr. The result is the displacement of the dashed boundary in Fig. 8.16 into the solid boundary for which $\beta_1 = 2\pi/3$.

It is evident from this consideration that the sector boundaries curve continuously as they approach the vertex, as sketched in Fig. 8.16. Their radius of curvature $R(r)$ is large compared to r in the limit of small r, i.e., $R(r)/r$ increases without bound. Hence the plane boundaries $\varphi = 0, \beta$, etc. employed in the foregoing development of the limiting form of the fields can be presumed valid.

It is interesting to note from equation (8.49) that the limiting values $\alpha_1(0), \alpha_2(0), \alpha_3(0)$ of the torsion coefficients are all equal. For it must be remembered that these values of α project unchanged along the field lines that pass along both sides of the surfaces of discontinuity, wherever those field lines may ultimately extend. Consider, then, the elemental flux bundles on opposite sides of a single isolated surface of tangential discontinuity. The torsion associated with such bundles generally varies continuously along both sides of the surface of discontinuity, with no general connection or correlation between the torsion α on opposite sides of the surface. The variation is a consequence of the different environments of $\psi(x, y, kzt)$ into which the lines extend elsewhere in the space $0 < z < L$. In formal terms, the boundary conditions require only a balance of magnetic pressure $B^2/8\pi$ across every point on the surface, which is achieved without the same torsion in most cases. However, when the topology of the field provides intersecting surfaces of discontinuity, the foregoing calculations introduce the restrictive relation of equation (8.49) between the torsion coefficients on field lines at the surface of adjacent sectors. Namely, the torsion coefficients are the same on the three surface field lines that pass by each other at each point O on the line of intersection of the surfaces of discontinuity, sketched in Fig. 8.18. It should be noted, of course, that $\alpha(0)$ may vary with the position of the point O on the line of intersection of the surfaces of discontinuity. The present development ignores that variation, with finite characteristic scale $\ell_{\parallel}$, working only in the limit of small kr. The unanswered question seems to be how are the field lines in each of the three sectors displaced along the line of intersection in such a way as to cross over the vertex of their respective sectors in the company of field lines in the other two sectors with the same torsion

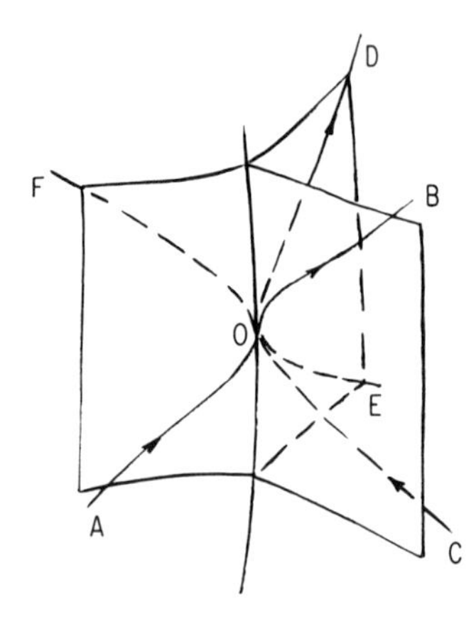

Fig. 8.18. A schematic drawing of three field lines AOB, COD, and EOF at the common surface of adjacent sectors, crossing over each other at the point O on the line of intersection of the surfaces of discontinuity.

coefficient $\alpha(0)$. The present theory cannot answer this question and, indeed, the conclusion of equation (8.49) is so remarkable that one may well ask whether there is another degree of freedom that has been overlooked in the present limited development — an additional free parameter that might get around the restriction of equation (8.49). The development, it will be recalled was based on the assumption that $B_z(A)$ is analytic at $A = 0$. The effect of poles and branch points at $A = 0$ has been examined elsewhere (Parker, 1990, Appendix C) but the solutions provide either infinite torsion $\alpha(0)$ or the field lines fail to cross over the vertex of the sector. Whatever the final disposition of the problem, it is clear that the relation between the local limiting conditions and the torsion introduced elsewhere along the individual field lines, or elemental flux bundles, must be established.

8.7 Discontinuities Around Displaced Flux Bundles

The creation of tangential discontinuities by the displacement of a single flux bundle in a magnetic field is a useful illustration of the intermixing of field lines in such places as the bipolar magnetic field above the convective surface of a star like the Sun, and perhaps in the magnetospheres of planets. The path of an elemental flux bundle (without torsion) in flux surfaces of an ambient magnetic field follows from the optical analogy (Parker, 1981a). The displacement of the flux bundle from the direction of the ambient field suggests the possibility that the fluid pressure within the flux bundle may be different from the ambient fluid pressure which must be taken into account as well.

Figure 8.13 provides a simple example of a flux bundle displaced from its original position in a uniform field in the z-direction. Figure 8.14 is a sketch of the local structure of the displaced or misaligned, small flux bundle showing that the bundle is flattened (since it contains no internal torsion) by the ambient magnetic field. In this case, the current sheets form a simple closed sheath on the surface of the bundle. However, if there is some slight torsion in the displaced bundle, so that the thickness h of the displaced flattened bundle is a finite fraction h/ℓ of the scale ℓ of the ambient field, there is at least a weak tangential discontinuity extending from the flux bundle away along the flux surface of the ambient field in which the bundle lies. The reason is that the unequal displacement of the generally unsymmetrical fields on either side of the displaced bundle causes the external fields to be slightly misaligned where they meet at the flux surface. A more important effect in many cases arises from the finite total tension $\mathcal{L} = SB^2/4\pi$ in the displaced bundle where S is the cross-sectional area so that the total magnetic flux in the bundle is $\Phi = SB$. The tension $\mathcal{L}$ displaces the ambient field in unsymmetrical fashion (unless the displaced bundle lies in some symmetrical position). The point is that the ambient field on the concave side of the curving flux bundle is pulled slightly out of alignment with the field on the convex side. Figure 8.19 is a schematic drawing of the deformation of a region of field by a displaced flux bundle that extends part way around the region. It is evident that the deformed region of field has been pulled out of precise alignment with the field on the other side of the flux surface defined by the displaced elemental flux bundle. Figure 8.19 shows the ambient flux that is deflected and misaligned by the tension in the displaced flux bundle passing around it. Not shown, of course, is the undeflected ambient fields flanking the displaced flux on either side. The flux surface defined by the displaced bundle becomes a surface of tangential discontinuity, wrapping around the surface of the displaced bundle. In this case, the strength, or angle of misalignment, θ of the discontinuity can be estimated in terms of the tension $\mathcal{L}$ in the displaced flux bundle. The transverse force exerted by $\mathcal{L}$ displaces and misaligns the ambient field up to the point where the restoring force in the ambient field balances $\mathcal{L}$. Denote by θ the angular deflection of either end of the region of dimension ℓ of deformed field. The restoring force in the deflected ambient field is then of the order of the total stress $\ell^2 B^2/4\pi$ multiplied by $2\tan\theta$ in the bundle of deflected field. This restoring force is equal to the component $\mathcal{L}_\perp$ of the tension in the displaced flux bundle , where $\mathcal{L}_\perp = \mathcal{L}\sin\psi$ for a flux bundle inclined at an angle ψ to the mean field direction. Hence,

$$\ell^2 \frac{B^2}{4\pi} 2\tan\theta \cong \mathcal{L}\sin\psi.$$

in order of magnitude. But the field within the flux bundle is of the same magnitude as the ambient field. Then for small deflection,

$$\theta \cong (S/2\ell^2)\psi$$

in order of magnitude.

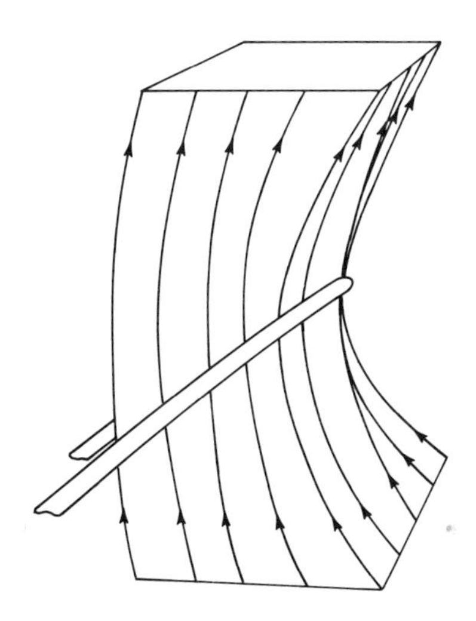

Fig. 8.19. A schematic drawing of the deformation of a region of field by a displaced flux bundle that partially encircles it.

It is evident that a simple bipolar magnetic field above the surface of the Sun may contain a complicated internal network of surfaces of discontinuity when several small internal flux bundles are strongly and unsymmetrically displaced from their normal positions by the photospheric convection. Note that the displaced flux bundle defines a flux surface in the ambient field. The width of the flux surface increases more or less linearly with time as the footpoints of the displaced bundle are transported at a small speed u along a meandering path among the footpoints of the ambient field. It is the meandering flux surface defined by the meandering displaced flux bundle that defines the surface of tangential discontinuity.

It should be noted that the convective displacement of the footpoints of an individual flux bundle, e.g., shown in Fig. 8.13, is presumably a commonly occurring phenomenon in the fibril magnetic fields of stars. The phenomenon may also occur in the magnetic fields of planets and probably galactic halos as well. It should be recognized that misaligned flux bundles may arise from other causes, such as a localized region of plasma turbulence and anomalous resistivity. This phenomenon is taken up in §10.4, where it is shown to create displaced field and discontinuities essentially along the same lines as described in the foregoing.

9

Fluid Motions

9.1 Collision of Separate Regions

The formation of tangential discontinuities in a slowly convolving magnetic field produces a variety of different, and often high speed, motions of the electrically conducting fluid in which the field is embedded. The formation of discontinuities arises through the squashing together of two initially separate regions of field, producing a gap in the intervening flux surfaces and converting an X-type neutral point into two Y-type neutral points, as described in the foregoing chapters. A continuing deformation of the magnetic field causes the gap in the flux surfaces to move across the flux surfaces, resulting in high speed motion of thin sheets of field and fluid across the gap, mentioned briefly in §7.7. The ultimate resistive dissipation and rapid reconnection of magnetic field at the surfaces of discontinuity, with or without the high speed motion of thin sheets, involves fluid motions of a complicated form, on which there is an extensive literature reviewed in Chapter 10. The present chapter treats only those motions that do not require resistivity for their existence.

9.1.1 Inviscid Fluid

We begin with the simple question of the fluid motions associated with the approach and contact of two separate topological regions of field, as they squash together to form a surface of tangential discontinuity. We treat the problem in its most elementary form, representing the two regions of field by their boundaries which for simplicity are taken to be rigid prior to contact. The field and fluid between the surfaces are represented by a single inviscid incompressible fluid. Figure 9.1(a) is a sketch of two such surfaces which are mutually concave toward each other and symmetric about the straight line that lies midway between them. It is readily seen that, when driven together by a constant force, the decline of CC′ and DD′ to zero is exponential, $\exp(-t/\tau)$ because the efflux of fluid is proportional to the declining separation. The mutual approach of two parallel plane surfaces is equivalent to the approach of the two mutually concave surfaces, again making contact only in the asymptotic limit of large time. But convex surfaces are a different matter altogether. They make contact in a finite time. They represent the dominant effect, as may be seen from the fact that if the mutual approach of C and C' were to precede the approach of D and D' in Fig. 9.1(a), the net effect is the approach of two convex surfaces, making contact in a finite time when driven by a finite force.

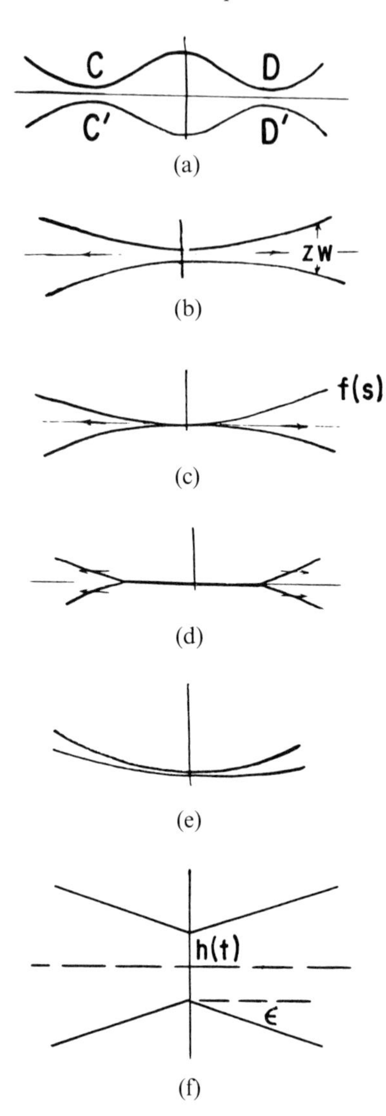

Fig. 9.1. (a) A schematic drawing of two surfaces CD and C′D′ that are mutually concave over a finite length. (b) Two mutually convex surfaces, showing the channel width $w(s, t)$, made up of $h(t)$ and $f(s)$. (c) Contact between the two convex surfaces of (b) when h falls to zero. (d) The two surfaces in (c) squashed together to make contact over a broad front, pushing the intervening fluid out each end. (e) A concave and a convex surface in contact, the channel width increasing in both directions from the point of contact. (f) Two approaching vertices.

Figure 9.1(b) is a sketch of two mutually convex surfaces. The convex surfaces expel the ideal inviscid incompressible fluid from between them, making contact, illustrated in Fig. 9.1(c), in a finite period time. If the two convex regions are squashed together beyond their first contact, the fluid is pushed away as the area of contact broadens, sketched in Fig. 9.1(d), representing two Y-type neutral points. The hydrodynamics in this stage poses no particularly interesting features, so we treat only the hydrodynamics of the initial approach, up to the instant of contact sketched in Fig. 9.1(c).

The critical aspect of the profile of approaching surfaces is the convexity in the near neighborhood of the point of contact. Note that convexity need only be relative, e.g., shown in Fig. 9.1(e). The requirement is simply that the fluid channel broaden with distance from the point of contact. We use the simple parabolic channel width for purposes of illustration.

Figure 1.4(a) shows an X-type neutral point, which, when squashed, provides the two Y-type neutral points sketched in Figs 1.4(b) and 9.1(d). The essential point is that the two Y-type neutral points, with the tangential discontinuity between them, may be approached from either the convex configuration of Fig. 9.1(b) and (c) or the X-type neutral point of Fig. 1.4(a). As we shall see, either approach takes place in a finite time.

The crucial hydrodynamics of two approaching surfaces lies in the neighborhood of the point of contact as the separation declines to zero. In that near neighborhood the fluid is confined to a very thin channel of width $2w(s,t)$, where s is distance measured along the direction of flow. It follows that $\partial w/\partial s = 0$ at the potential point of contact (for symmetric surfaces), and, in view of the thin channel, it is obvious that $|\partial w/\partial s| \ll 1$ and $|w\partial^2 w/\partial s^2| \ll 1$. The fluid motion $v(s,t)$ along the channel is essentially 1D in the neighborhood of the point of contact. It is sufficient to use the 1D Euler equation,

$$\frac{\partial v}{\partial t} + v\frac{\partial v}{\partial s} = -\frac{1}{\rho}\frac{\partial p}{\partial s}, \tag{9.1}$$

where the pressure $p(s,t)$ is applied to the fluid by the walls of the channel. Conservation of fluid requires that

$$\frac{\partial w}{\partial t} + \frac{\partial}{\partial s}wv = 0 \tag{9.2}$$

for an incompressible fluid of uniform density ρ. The mathematical procedure is now to specify the manner in which the two surfaces approach, i.e., prescribe $w(s,t)$, deducing the associated $v(s,t)$ from equation (9.2) and the pressure $p(s,t)$ from equation (9.1). A convenient form for w is

$$w(s,t) = h(t) + f(s),$$

representing a channel with fixed wall profile $f(s)$, while the separation $h(t)$ declines to zero. Equation (9.2) becomes

$$h'(t) + \frac{\partial}{\partial s}v(h+f) = 0. \tag{9.3}$$

Let s be measured from the position of minimum width of the channel, so that $f(s)$ is symmetrical $(f(s) = f(-s))$ about $s = 0$ and $v(0,t) = 0$. It follows upon integration that

$$v(s,t) = -\frac{h'(t)s}{h(t) + f(s)}. \tag{9.4}$$

Equation (9.1) can now be integrated over s to yield the pressure

$$\begin{aligned} p(s,t) = p(0,t) + \rho\Bigg\{ h''(t)\int_0^s \frac{d\xi\xi}{h(t)+f(\xi)} \\ - 2h'(t)^2\int_0^s \frac{d\xi\xi}{[h(t)+f(\xi)]^2} + h'(t)^2\int_0^s \frac{d\xi\xi^2 f'(\xi)}{[h(t)+f(\xi)]^3}\Bigg\}. \end{aligned} \tag{9.5}$$

Note, then, that $p(s,t) - p(0,t)$ is an even function of s, in view of the fact that $f(s)$ is symmetric about $s = 0$. Three forms for $f(s)$ come to mind. For the mutually concave surface shown in Fig. 9.1(a), let

$$f(s) = q^3(s^2 - a^2)^2,$$

where C and D lie at $s = \pm a$ and q is an inverse length determining the amplitude of the waviness, with $f(0) = q^3 a^4$. For the flat surface put $f = 0$. For a convex surface, shown in Fig. 9.1(b), write $f(s) = qs^2$, where again the inverse length q determines the sharpness of the profile.

Take the simplest case first, with $f = 0$. Then

$$p(s, t) - p(0, t) = \frac{1}{2}\rho s^2 \left[\frac{h''(t)}{h(t)} - 2\frac{h'(t)^2}{h(t)^2}\right]. \tag{9.6}$$

Then if $h(t) = h_0 \exp(-\omega t)$, the required pressure is

$$p(s, t) - p(0, t) = -\frac{1}{2}\rho\omega^2 s^2. \tag{9.7}$$

The pressure is constant in time and declines quadratically from a maximum at the center $s = 0$. The two surfaces approach each other in the limit of large t. Suppose, then, that $h(t) = a(-\omega t)^\alpha$ with $\alpha > 0$ so that $h(t)$ declines to zero in a finite time, as t increases to zero. It follows that

$$p(s, t) = p(0, t) = -\frac{1}{2}\alpha(\alpha + 1)\omega^2 s^2/(-\omega t)^2, \tag{9.8}$$

from which it is clear that the pressure difference $p(s, t) - p(0, t)$ necessary to drive the flow increases without bound as the width declines to zero at $t = 0$. It follows that the two surfaces, driven only by finite pressures, do not come into contact in a finite time. Note, then, that if $h(t) = a/(\omega t)^\alpha$, the pressure differential is

$$p(s, t) - p(0, t) = -\frac{1}{2}\alpha(\alpha - 1)\rho s^2/t^2, \tag{9.9}$$

declining to zero in the limit of large t and vanishing width $h(t)$. As a final example, suppose that $h(t)$ declines to zero as a $\exp(-\omega^2 t^2)$. The necessary pressure differential is

$$p(s, t) - p(0, t) = -\omega^2 \rho s^2 (1 + 2\omega^2 t^2), \tag{9.10}$$

increasing without bound as t becomes large. It is evident that with only a finite pressure differential available, the asymptotic approach can be no more rapid than exponential, $\exp(-\omega t)$. Anything faster requires pressures that increase without bound. A slower approach, e.g., $1/t^\alpha$, involves pressure differences that decline to zero with increasing t.

Now for the mutually concave boundaries, sketched in Fig. 9.1(a), it is evident that they cannot approach each other more rapidly than the flat surfaces, because, again, for a fixed width $2a$ a finite amount of fluid must be expelled through the narrowing part of the channel at $s = \pm a$. A steady applied pressure $p(s)$ leads to a steady outflow, so that equation (9.1) can be integrated to give Bernoulli's law,

$$p(s) = p(0) - \frac{1}{2}\rho v^2(s). \tag{9.11}$$

Conservation of fluid in $-a < s < +a$ requires that the outflow $v(a)$ past $s = +a$, where the width is $h(t)$, be equal to the rate of loss of fluid volume $ah'(t)$,

$$h(t)v(a) = -ah'(t). \tag{9.12}$$

Hence, with Bernoulli's law

$$\frac{h'(t)}{h(t)} = -\frac{2^{\frac{1}{2}}}{a\rho^{\frac{1}{2}}}[p(0) - p(a)]^{\frac{1}{2}}. \tag{9.13}$$

This relation can be integrated to give

$$h(t) = h(0)\exp\left\{-\frac{2^{\frac{1}{2}}t}{a\rho^{\frac{1}{2}}}[p(0) - p(a)]^{\frac{1}{2}}\right\}, \tag{9.14}$$

with the steady outflow velocity

$$v(a) = \frac{2^{\frac{1}{2}}}{\rho^{\frac{1}{2}}}[p(0) - p(a)]^{\frac{1}{2}}. \tag{9.15}$$

Thus, like the flat surfaces, the concave surfaces approach each other only asymptotically in the limit of large t under a bounded steady pressure.

The interesting case, then, is the relatively convex surface, with the channel width increasing as $f(s) = qs^2$ from the point of contact at $s = 0$. In this case equation (9.5) yields the pressure differential

$$p(s,t) - p(0,t) = \frac{\rho}{2q}\left[h'' \ln\left(\frac{h + qs^2}{h}\right) - \frac{h'^2 qs^2(2h + qs^2)}{h(h + qs^2)^2}\right]. \tag{9.16}$$

For contact at time $t = 0$ let $h(t) = a(-\omega t)^\alpha$ for $t \leqslant 0$, with the result that

$$p(s,t) - p(0,t) = \frac{\alpha\rho a\omega^2}{2q}(-\omega t)^{\alpha-2}\left\{(\alpha - 1)\ln(1 + W) - \frac{\alpha W(1 + 2W)}{(1 + W)^2}\right\}, \tag{9.17}$$

where W is the similarity variable $qs^2/a(-\omega t)^\alpha$.

The factor $(-\omega t)^{\alpha-2}$ in front of the braces provides a pressure differential $p(s,t) - p(0,t)$ that falls to zero for any $\alpha > 2$ as the surfaces come into contact. However, the point $s = t = 0$ in space time is singular because W can go to zero or infinity. In particular, the quantity in braces on the right-hand side of equation (9.17) passes through a maximum that increases without bound as ωt increases to zero in the limit of small s. So the limit needs careful examination. The second term in braces is not the problem. To show this write $\tau = \omega t$ and $\chi = qs^2/a$ in the second term. The result is proportional to

$$F_2 = \tau^{\alpha-2}\chi(\chi + 2\tau^\alpha)/(\chi + \tau^\alpha)^2.$$

Differentiating F_2 with respect to τ and setting the result equal to zero leads to the value of τ,

$$\tau^\alpha/\chi = \frac{3}{8}(\alpha - 2)^{\frac{1}{2}}\left[(\alpha - 2)^{\frac{1}{2}} \pm \left(\alpha - \frac{2}{9}\right)^{\frac{1}{2}}\right],$$

at which F_2 is a maximum. Since τ^α/χ is positive, only the upper sign applies, and τ^α/χ is real and positive for any $\alpha > 2$ at the extremum in F_2. Hence the extreme value of F_2 is finite and therefore not particularly interesting.

The first term on the right-hand side of equation (9.17) is proportional to

$$F_1 = \tau^{\alpha-2}\ln(1 + \chi/\tau^\alpha).$$

Differentiating with respect to τ and setting the result equal to zero yields

$$[(1 + W)/W]\ln(1 + W) = \alpha/(\alpha - 2)\tau^2,$$

in which W is again $qs^2/a(-\omega t)^\alpha$. It is apparent that the value of W for which F_1 is an extremum grows large as τ declines to zero and the right-hand side of this relation increases without bound. It is readily shown that in the limit of small τ,

$$W \cong \exp Q(\tau)[\![1 - [1 + Q(\tau)]\exp[-Q(\tau)] + O\{\exp[-2Q(\tau)]\}]\!],$$

where the quantity $Q(\tau)$ is $\alpha/(\alpha - 2)\tau^2$. It follows that the extreme value of the first term is $F_1 = F_m$, where

$$F_m \cong [\alpha\tau^{\alpha-4}/(\alpha - 2)]\{1 - \exp[-\alpha/(\alpha - 2)\tau^2] + \ldots\}.$$

Thus the extreme value of the pressure differential is bounded for all $s < a$ as τ declines to zero if and only if $\alpha \geqslant 4$. So a fixed pressure differential provides contact of the two convex surfaces in a finite time with $\alpha = 4$. A diminishing pressure gives $\alpha > 4$. It follows that a bounded pressure causes the two rigid convex surfaces to make a gentle contact, with a velocity falling rapidly to zero as they come close. Figure 9.2 is a plot of the pressure differential (9.17) for $\alpha = 3,4,5$, and 6 in units of $\rho a\omega^2/2q$ as a function of $(q/a)^{\frac{1}{2}}s$ at time $\omega t = -1$ and as a function of $-\omega t$ at $(q/a)^{\frac{1}{2}}s = 1$ for $\alpha = 4$ and 6.

As a final point, note that, as pointed out earlier, if the surface rigid surface $C'D'$ in Fig. 9.1(a) were canted slightly so that C and C' made contact before D and D', then the contact of C and C' would occur in a finite time, along the lines just described for two relatively convex surfaces. The subsequent approach of D and D' would still be only asymptotic, of course, as when C and C', and D and D', approach each other simultaneously. Note, too, that if the surfaces are not rigid, then a finite pressure differential concentrated in the neighborhood of CC' and DD' would bring those convex parts into contact in a finite time, with the intervening fluid displaced into the concave regions.

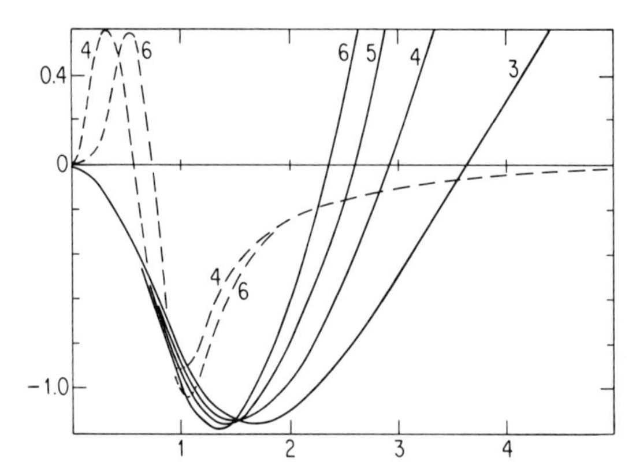

Fig. 9.2. The solid curves provide the pressure profile given by the quantity in braces on the right hand side of equation (9.17) at the time $\omega t = -1$, for the values of $\alpha = 3,4,5$, and 6 indicated on each curve. The horizontal axis represents position s in units of $(a/q)^{\frac{1}{2}}$. The dashed curves provide the time variation of the pressure in units of $\alpha\rho a\omega^2/2q$ at the position $s = (a/q)^{\frac{1}{2}}$, with $-\omega t$ plotted along the horizontal axis, for the values $\alpha = 4$ and 6 indicated on the two curves.

To get some idea of the time in which two surfaces come into contact in astronomical circumstances, consider the relation between ω and the pressure differential at $W = 1$ at the characteristic time $\omega t = -1$ and the characteristic position $s = a$ with $qa = 1$. Equation (9.17) becomes

$$\begin{aligned}\delta p &\equiv p(a, -1/\omega) - p(0, -1/\omega) \\ &= -\frac{1}{2}\rho a^2\omega^2\alpha[(\alpha - 1)\ln 2 - 3\alpha/4].\end{aligned}$$

With $\alpha = 5$ this reduces to

$$\omega^2 = +0.409|\delta p|/\rho a^2.$$

A typical situation in the active X-ray corona of the Sun might involve a field $B = 10^2$ gauss deformed so that the local rms transverse component $B_\perp$ is 25 gauss, with a pressure $B_\perp^2/8\pi \cong 25\,\text{dynes/cm}^2$. Some fraction of $B_\perp^2/8\pi$ causes two regions of field to press together, so that $\Delta p \cong 2\,\text{dynes/cm}^2$, say. The number density $N = 10^{10}/\text{cm}^3$ yields $\rho \cong 2 \times 10^{-14}\,\text{gm/cm}^3$. The characteristic granule scale $a = 5 \times 10^7$ cm then yields $\omega \cong 0.13$/sec. The characteristic time $1/\omega$ is 7 sec, with a characteristic velocity $a\omega = 6.5 \times 10^6$ cm/sec. This characteristic velocity is small compared to both the speed of sound (2×10^7 cm/sec) and the high speed jets observed in the transition zone, as well as the characteristic Alfven speed of 5×10^7 cm/sec in $B_\perp$. The essential point is that the establishment of contact of different topological domains of field may be so fast that the slow winding and interweaving of the field by the photospheric convection of the footpoints does not take the field far from magnetostatic equilibrium. We have not found any contrary situations in the presence of normal slow convective deformation of a strong field in a tenuous atmosphere. So distinct regions of field may be considered to be squashed together to form a surface of tangential discontinuity as soon as the discontinuity is called for by the stress balance in the evolving topology of the field. The evolution of the field is in quasi-static equilibrium in every detail, right down to the infinitely sharp tangential discontinuities that would obtain in an infinitely conducting fluid. For it should be remembered that once contact is established, shown in Fig. 9.1(c), the contact is easily extended in both directions from the initial point of contact shown in Fig. 9.1(d), to form a broad surface of contact.

9.1.2 Viscous Fluid

The hydrodynamics of thin sheets of fluid plays a central role in the formation and dissipation of the tangential discontinuities in a slowly evolving field. The foregoing calculations have exhibited the role of thin sheets or channels of inviscid fluid in the mechanics of contact between initially separate domains of magnetic field. Thin sheets come up again in §9.2 where the contacting domains drift relative to the surrounding field, and again as part of rapid reconnection, treated in Chapter 10. So the question naturally arises as to the role of viscosity, in the thin sheet of rapidly moving fluid. As we shall see, nonslip boundary conditions in the channel of width $2w(s, t)$ put viscosity in a dominant position, with contact between the boundaries only in the asymptotic limit of large time. On the other hand, a more realistic boundary condition, allowing slippage of fluid at the magnetic surfaces that form the channel boundaries, provides for contact in a finite time, as with the inviscid fluid in the foregoing section. Further investigation shows that in most astronomical settings, the final extreme effect of viscosity is masked by resistive diffusion. But we take the investigation one step at a time, beginning with viscosity and nonslip boundary conditions.

Consider an incompressible viscous fluid. The Euler equation (9.1) is replaced by the approximate Navier–Stokes equation,

$$\frac{\partial v}{\partial t} + v\frac{\partial v}{\partial s} \cong -\frac{1}{\rho}\frac{\partial p}{\partial s} + \frac{\mu}{\rho}\frac{\partial^2 v}{\partial x^2}, \tag{9.18}$$

in the limit of a narrow channel, where x is the coordinate measured across the channel with its origin on the midline of the channel, along which s measures distance. The coefficient of viscosity is designated by μ. Take the channel boundaries $x = \pm w(s, t)$ to be rigid with nonslip boundary conditions, so as to maximize the effect of the viscosity. Assuming that the flow in the thin channel is laminar, the viscous effects appear first in the boundary layers of thickness $(4\mu t/\rho)^{\frac{1}{2}}$ at $x = \pm w$, and we are dealing with a nonlinear partial differential equation. To make the problem tractable, consider the case where the channel is so thin that the flow is dominated by viscosity. This occurs when the characteristic time T of variation of the flow is longer than the time required for the viscous boundary layer to extend from the wall to the midline of the channel, i.e., if we write again that $w(s, t) = f(s) + h(t)$, this occurs when $(4\mu T/\rho)^{\frac{1}{2}} > w(s, T) = f(s) + h(T)$. This condition is satisfied at $s = 0$, where $f(s)$ vanishes, as the surfaces approach contact, $h(t) = 0$.

In the viscous limit equation (9.18) approximates to

$$\frac{\partial p}{\partial s} \cong \mu\frac{\partial^2 v}{\partial x^2}. \tag{9.19}$$

Then $\partial p/\partial x = 0$ and the velocity profile across the channel is parabolic, vanishing at the walls so that

$$v(s, t) = \frac{3}{2}\langle v\rangle[1 - x^2/(h + f)^2],$$

where $\langle v\rangle$ is the mean flow velocity across the channel. Substituting this form into equation (9.19), it follows that

$$\frac{\partial p}{\partial s} = -\frac{3\mu\langle v\rangle}{(h + f)^2}. \tag{9.20}$$

Equation (9.4) for conservation of fluid applies to the mean flow velocity along the channel, with the result that

$$\frac{\partial p}{\partial s} = +\frac{3\mu h'(t)s}{(h + f)^3}.$$

Hence

$$p(s, t) - p(0, t) = +3\mu h'(t)\int_0^s \frac{d\xi\xi}{[h(t) + f(\xi)]^3}. \tag{9.21}$$

For parabolic convex surfaces, $f(s) = qs^2$, the result is

$$p(s, t) - p(0, t) = +\frac{3\mu h'(t)s^2[2h(t) + qs^2]}{4h^2(t)[h(t) + qs^2]^2}. \tag{9.22}$$

Then if $h(t) = a(-\omega t)^\alpha$ for $t \leqslant 0$, so that the walls come into contact (at $s = 0$) at time $t = 0$, it follows that

$$p(s,t) - p(0,t) = -\frac{3\alpha\mu\omega s^2[2a(-\omega t)^\alpha + qs^2]}{4a(-\omega t)^{\alpha+1}[a(-\omega t)^\alpha + qs^2]^2}, \tag{9.23}$$

$$= -\frac{3\alpha\mu\omega W(2+W)}{4a^2q(-\omega t)^{\alpha+1}(1+W)^2},$$

where again $W = qs^2/a(-\omega t)^\alpha$. It follows, then, that the pressure gradient described by equation (9.20) is concentrated in a small neighborhood of the origin as ωt increases to zero. The expression (9.23) reduces to

$$p(s,t) - p(0,t) \cong -\frac{3\alpha\omega\mu}{2(-\omega t)^{2\alpha+1}}\left(\frac{s}{a}\right)^2[1 - 32W + W^2 + \ldots],$$

in the neighborhood of the contact point $s = 0 (W \ll 1)$, whereas far away $(W \gg 1)$

$$p(s,t) - p(0,t) \cong -\frac{3\alpha\omega\mu}{4aq(-\omega t)^{\alpha+1}}\left[1 - \frac{1}{W^2} + \frac{2}{W^3} - \frac{3}{W^4} + \ldots\right],$$

which is independent of s in the limit of large W. In effect, then, the characteristic pressure differential Δp is

$$\Delta p = +3\alpha\omega\mu/4aq(-\omega t)^{\alpha+1}. \tag{9.24}$$

between $s = 0$ and $s = +\infty$, with most of Δp within a declining distance of $(h(t)/q)^{\frac{1}{2}}$ around $s = 0$. This pressure differential diverges as t approaches zero, for any $\alpha > 0$. Hence, no fixed finite pressure differential can bring the rigid convex surfaces into contact in a finite time in the presence of viscosity and nonslip boundary conditions at the channel walls.

On the other hand, suppose that $h(t)$ declines only asymptotically to zero, as $h(t) = a/(\omega t)^\alpha$, in the limit of large t. The result can be written as

$$p(s,t) - p(0,t) = -\frac{3\alpha\mu\omega(\omega t)^{\alpha-1}(1+2\Upsilon)}{4aq(1+\Upsilon)^2},$$

where Υ is the similarity variable $\Upsilon = a/qs^2(\omega t)^\alpha$. Note that for any fixed value of s, however small, Υ declines monotonically to zero as ωt increases without bound, and

$$p(s,t) - p(0,t) \sim -\frac{3\alpha\mu\omega}{4aq}(\omega t)^{\alpha-1}, \tag{9.25}$$

to lowest order. Thus, if the approach is slow, $0 < \alpha < 1$, the pressure differential is bounded, declining to zero as the walls approach each other. If $\alpha = 1$, the differential approaches asymptotically to a finite value, and if the approach is so fast that $\alpha > 1$, the pressure differential increases without bound. That is to say, a fixed pressure difference can drive the two convex surfaces together no faster than $h(t) \propto 1/t$.

As a matter of curiosity, consider the approach of two similar opposing symmetric vertices whose contact would form an X-type neutral point, sketched in

Fig. 9.1(f). In this case, $f(s) = \varepsilon s$ where the vertex angle θ is equal to $\cot^{-1} 2\varepsilon$. The formalism is valid for $\varepsilon \ll 1$. The result is

$$p(s,t) - p(0,t) = \frac{3\mu h'(t) s^2}{2h(t)[h(t) + \varepsilon s]^2}.$$

Thus, for $\varepsilon s \ll 1$,

$$p(s,t) - p(0,t) = \frac{3\mu h'(t)}{2\varepsilon^2 h(t)}.$$

For $h(t) = a(-\omega t)^{\alpha}$ it follows that

$$p(s,t) - p(0,t) = -\frac{3\alpha\omega\mu}{2\varepsilon^2 t}. \tag{9.26}$$

One might have thought that two vertices, however blunt ($\varepsilon \ll 1$), might avoid the limiting effects of viscosity because the channel between them forms two pie-shaped sectors rather than two cusps upon contact. But the channel width declines to zero in the neighborhood of $s = 0$ at $t = 0$, so viscosity has the last word. Only in the limit of $\varepsilon = O(1)$ is the contact possible in a finite time.

In summary, the surfaces driven together by a fixed finite pressure differential are prevented by the viscosity from coming into contact in a finite time. This is an immediate consequence of treating the boundaries as rigid walls with nonslip boundary conditions. In fact the channel walls in the present context are separatrices of different topological regions of field. The viscosity of the fluid in the channel between the regions is free to drag along the fluid in the two magnetic regions beyond the channel walls. The fluid is free to move parallel to the fields beyond the walls, even if it does not move freely perpendicular to the field. In most cases in the 3D world there is a direction parallel to the field along which the fluid may escape. The case where this does not occur is taken up in §9.1.3. The net effect is a viscous drag at both walls of the channel, but the fluid is not constrained to a nonslip condition at the boundaries.

A simple representation of the effect of the nonslip condition at the walls $x = \pm(h+f)$ is to apply the nonslip condition to a more distant surface, e.g., $x = \pm(c+h+f)$, where c is a suitable length. In that case equation (9.20) is replaced by

$$\frac{\partial p}{\partial s} = -\frac{3\mu\langle v\rangle}{(c+h+f)^2}, \tag{9.27}$$

and in place of equation (9.21),

$$p(s,t) - p(0,t) = 3\mu h'(t) \int_0^s \frac{d\xi\xi}{[c+h(t)+f(\xi)]^2[h(t)+f(\xi)]}. \tag{9.28}$$

For the convex surface $f(s) = qs^2$, the result is

$$p(s,t) - p(0,t) = \frac{3\mu h'(t)}{2qc^2}\left[\ln\frac{(h+qs^2)(c+h)}{h(c+h+qs^2)} - \frac{cqs^2}{(c+h)(c+h+qs^2)}\right]. \tag{9.29}$$

Only the logarithmic term is singular as h and qs^2 decline to zero, in which limit

$$p(s,t) - p(0,t) \cong \frac{3\mu h'(t)}{2qc^2}\left\{\ln\left[1+\frac{c}{h(t)}\right] - \ln\left[1+\frac{c}{h(t)+qs^2}\right]\right\}. \tag{9.30}$$

With $h(t) = a(-\omega t)^{\alpha}$, and $h'(t) = -\alpha a\omega(-\omega t)^{\alpha-1}$, it is obvious that the pressure differential falls to zero for all s in the limit of vanishing ωt provided only that $\alpha > 1$. So with $c > 0$, i.e., without the nonslip condition, the viscosity does not prevent a bounded pressure difference from pushing the two magnetic regions into contact in a finite period of time. But note now that if $\alpha > 1$ allows contact of the walls in a flow limited by viscosity and driven by a finite pressure differential, it was shown from equation (9.17) that the inertia of the fluid, neglected in the viscous flow, restricts the approach to the much slower case that $\alpha \geqslant 4$. So taking all things together, it is evidently the inertia of the fluid that is the final controlling factor (providing the largest value of α) rather than the viscosity. This follows from the vanishing of the denominator $[h(t) + qs^2]^2$ in equation (9.4) for $v(s, t)$ in the neighborhood of $s = 0$ as $h(t)$ declines to zero.

9.1.3 Viscosity and Resistivity

There is now another aspect of viscosity that merits close attention, and that is the fact that the contact of two surfaces is prevented by viscosity with nonslip boundary conditions only in the final approach to vanishing separation. That is to say, viscosity has no really qualitative effect except in the final approach of $h(t)$ to zero. But resistive diffusion η effectively prevents the separation from becoming smaller than something of the order of the characteristics diffusion length $(4\eta t)^{\frac{1}{2}}$. Hence, even in the extreme case of nonslip boundary conditions, treated in §9.1.2, the final blocking of the approach by viscosity, as $h(t)$ declines to zero, does not occur because the effective channel width does not fall below the resistive diffusion length. In other terms, resistivity alone provides the characteristic diffusion velocity $\eta/h(t)$, and, when that velocity becomes as large as the rate of approach $h'(t)$, the layer of fluid becomes no thinner, with resistive diffusion dominating the expulsion of viscous fluid. Thus the viscous effects do not diverge even when $v > \eta$. This is where dynamical rapid reconnection takes over from the ideal outflow of fluid, and it occurs at the time when

$$(-\omega t)^{2\alpha-1} = \eta/\alpha\omega a^2, \tag{9.31}$$

when $h(t)$ has the form $a(-\omega t)^{\alpha}$ assumed here. The right-hand side is essentially the reciprocal of the magnetic Reynolds number. At this point in time, it follows from equation (9.24) that

$$\Delta p = \frac{3\alpha\omega\mu}{4aq}\left(\frac{\alpha\omega a^2}{\eta}\right)^{\frac{\alpha+1}{2\alpha-1}}, \tag{9.32}$$

so that

$$\omega = \left(\frac{4q\Delta p}{3\mu}\right)^{\frac{2\alpha-1}{3\alpha}} \frac{\eta^{(\alpha+1)/3\alpha)}}{\alpha a^{1/\alpha}}. \tag{9.33}$$

This relation provides a measure of the reciprocal time ω in which h decreases from its characteristic initial value a at time $t = -1/\omega$ to the small value at which resistivity becomes the dominant diffusive effect. The next step, then, is to obtain some idea of the appropriate values of η and μ under various astronomical circumstances.

The resistive diffusion coefficient $\eta = c^2/4\pi\sigma$ is given by

$$\eta \cong 0.5 \times 10^{13}/T_E^3 2\,\mathrm{cm}^2/\mathrm{sec} \tag{9.34}$$

(Cowling, 1953, 1957; Spitzer, 1956; Chapman and Cowling 1958) in ionized hydrogen at an electron temperature T_E. The resistive diffusion arises from the Coulomb scattering of electrons from ions so it is essentially unaffected by the presence of magnetic fields. It is insensitive to density because the electron conduction velocity varies inversely with the particle density while the collision rate varies directly with the particle density, the two effects largely cancelling each other.

The viscosity μ and the kinematic viscosity $\nu = \mu/\rho$ of ionized hydrogen are given by

$$\mu \cong 10^{-16} T_I^{5/2} \text{ gm/cm sec,} \tag{9.35}$$

and

$$\nu \cong 0.6 \times 10^8 T_I^{5/2}/N \text{ cm}^2/\text{sec,} \tag{9.36}$$

respectively (Chapman, 1954) in terms of the ion temperature T_I in the absence of magnetic field B. The effective viscosity for momentum transport parallel to a magnetic field B (i.e., $\nabla \times v \perp \mathbf{B}$) is essentially unaffected by B, whereas the viscous momentum transport across B (i.e., $\nabla \times v \| \mathbf{B}$) is reduced by a factor characterized by $(R_{cI}/\lambda_I)^2$ when the factor is less than one, where R_{cI} is the thermal ion cyclotron radius and λ_I is the ion mean free path. The essential point is that in the strong magnetic fields in tenuous ionized gases that make up stellar coronas, planetary magnetospheres, galactic halo, etc. the ratio R_{cI}/γ_I is small compared to one and the viscosity for momentum transport across the magnetic field is greatly reduced compared to the momentum transport along the field. Both the tension in the magnetic field and the viscosity (see equation 9.35) oppose any shearing along the field, whereas a hydrodynamic shear across the flux surfaces of a field is not opposed by the Maxwell stresses and involves only a greatly reduced viscosity. The shear in the flow in the thin channel between two approaching field domains is principally perpendicular to B, so the present question involves the reduced viscosity. Note that

$$\begin{aligned} R_{cI} &= M w_I c/B, \\ &= (MkT_I)^{\frac{1}{2}} c/B, \\ &\cong T_I^{\frac{1}{2}}/B \text{ cm,} \end{aligned} \tag{9.37}$$

where the characteristic hydrogen ion thermal velocity w_I has been set equal to $(kT_I/M)^{\frac{1}{2}}$. The characteristic ion mean free path λ_I is (Spitzer, 1956)

$$\lambda_I \cong 0.5 \times 10^4 T_I^2/N \text{ cm} \tag{9.38}$$

in the range of temperatures appropriate to the aforementioned astronomical settings. It follows that

$$R_{cI}/\lambda_I \cong 2 \times 10^{-4} N/BT_I^{\frac{3}{2}}, \tag{9.39}$$

and the appropriate reduced viscosity is given approximately by

$$\mu \cong 4 \times 10^{-24} N^2/B^2 T_I^{\frac{1}{2}} \text{ gm/cm sec,} \tag{9.40}$$

$$\nu \cong 2.5 N/B^2 T_I^{\frac{1}{2}} \text{ cm}^2/\text{sec.} \tag{9.41}$$

Note that this effective viscosity diminishes slightly with increasing temperature, because the mean free path λ_I increases so rapidly.

It follows from equations (9.34) and (9.36) that

$$v/\eta \cong 10^{-5} T_I^{5/2} T_E^{3/2}/N \tag{9.42}$$

in weak fields ($R_{cI} > \lambda_I$), and it follows from equations (9.34) and (9.41) that

$$\frac{v}{\eta} \cong \frac{5 \times 10^{-13} N T_E^{\frac{3}{\pi}}}{B^2 T_I^{\frac{1}{2}}} \tag{9.43}$$

in strong fields ($R_I < \lambda_I$).

For present purposes put $T_E = T_I \equiv T$, so that equation (9.43) reduces to

$$\frac{v}{\eta} \cong \frac{5 \times 10^{-13} NT}{B^2} \tag{9.44}$$

$$= 0.7 \times 10^2 \frac{p}{B^2/8\pi} \tag{9.45}$$

for strong fields, where p is the gas pressure $2NkT$. It follows that $v > \eta$ wherever p is not considerably smaller than $B^2/8\pi$, e.g., in stellar interiors. On the other hand, in magnetically controlled regions, were $p \ll B^2/8\pi$, we have $\eta > v$ and it is the resistivity that determines the minimum channel width between field domains.

To treat some specific cases, note that in a stellar chromosphere, where typical conditions are $N \cong 10^{12}$ atoms/cm^3 and $T \cong 10^4$ K, it follows that $v/\eta \cong 10^{-1}$ in the absence of magnetic field, so the resistive diffusion is the limiting effect. A magnetic field of 10^2 gauss, typical of a plage (Title, et al. 1992), yields $R_{cI}/\lambda_I \cong 2$, so there is no significant suppression of the viscosity by the magnetic field.

On the other hand, in an active X-ray corona $N \cong 10^{10}$ atoms/cm^3 and $T \cong 10^6$ K, providing about the same gas pressure p as the chromosphere the result is $v/\eta \cong 10^9$ in the absence of magnetic field, and $v/\eta \cong 0.5$ with 100 gauss, for which $R_{cI}/\lambda_I \cong 2 \times 10^{-5}$ and $8\pi p/B^2 = 0.7 \times 10^{-2}$. In a nonemitting coronal region (a coronal hole) the density is $N \cong 10^8$ atoms/cm^3, and the temperature is about the same, so that $v/\eta \cong 10^{11}$ in the absence of magnetic field and $v/\eta \cong 0.5$ in the presence of 10 gauss, for which $R_{cI}/\lambda_I \cong 2 \times 10^{-6}$ and $8\pi p/B^2$ is again about 0.7×10^{-2}. A temperature of $2 - 3 \times 10^6$ K might be more appropriate, with the result that v/η in a stellar corona is a little greater than one even in the presence of typical magnetic fields.

In a galactic halo, where $N \cong 10^{-3}$ atoms/cm^3, $T \cong 10^7$ K, and $B \cong 10^{-6}$ gauss, we have $p \cong 3 \times 10^{-12}$ dynes/cm^2 and $B^2/8\pi \cong 4 \times 10^{-14}$ dynes/cm^2, so that $v/\eta \cong 50$. In the geomagnetic field, where the particle pressure p is the fraction ε of the pressure of the confining field, the result is $v/\eta = 70\varepsilon$, ignoring the fact that the particle distributions are far from the Maxwellian distribution employed in the computation of η and v and ignoring the fact that T_E and T_I are quite different.

It appears, then, that v/η may be comparable to or larger than one in a variety of circumstances, in stellar, galactic, and planetary coronas. To illustrate the consequences of η with a specific example, consider the same situation in an active X-ray corona that was worked out at the end of §9.1.1 for the inviscid fluid. Let $\Delta p = 2$ dynes/cm^2, $\alpha = 5$, $a = 5 \times 10^7$ cm, $q = 1/a$, $\rho = 1.66 \times 10^{-14}$ gm/cm^3 ($N = 10^{10}$ atoms/cm^3). Put $T = 3 \times 10^6$ K, obtaining $\eta \cong 10^3$ cm^2/sec from equation

(9.34) and $\mu \cong 2 \times 10^{-11}$ gm/cm sec from equation (9.41) with the magnetic field $B = 10^2$ gauss. Then equation (9.33) yields $\omega = 0.3$/sec. This is of the same order as the estimate at the end of §9.1.1 for the inviscid fluid, limited by inertia of the fluid as the approach goes all the way to contact. The result shows that resistivity η avoids the last slow asymptotic viscous approach to contact. That is to say, the resistivity limits the effect of the viscosity to modest levels, even though equation (9.45) gives ν as being comparable to η. The role of viscosity in the high speed sheets of field and fluid, treated in the next section, is discussed in §9.4.

9.2 High Speed Sheets

It was pointed out in §7.7 (Parker, 1990) that the slow drift of a quasi-static magnetic field across the width of the gap in the flux surfaces leads to high speed motion of a thin sheet of field and fluid across the gap. This situation arises in fields whose strong deformation and internal interweaving continue to evolve slowly with time after forming tangential discontinuities. It occurs where two topological lobes of field press together to provide a local maximum in the field magnitude. The high speed of field and fluid appears when the lobes move slowly relative to the intervening field. The only way for the intervening field to get past the moving lobes is to force its way between the lobes. This occurs only after the moving lobes have stretched the intervening field to such a degree that the tension in the field is able to force the field into the enhanced pressure in the gap between the opposing lobes. Once in the region of high pressure, the Maxwell stresses accelerate the field to a velocity characterized by the Alfven speed. It is important to note that the sheet of field and fluid being forced through the gap may block rapid reconnection across the discontinuity, by keeping the opposing lobes slightly apart. The high speed sheet may in some cases be so robust as to have dynamical consequences of its own, on which we will have more to say later.

A gap in the flux surfaces of the intervening field, caused by a localized pressure maximum, is illustrated in Fig. 7.6, with the field lines in the flux surfaces parting to pass around opposite sides of the gap. The function $f(c)$, defined by equation (7.45), is a direct measure of the net enhanced pressure in the neighborhood of $x = 0$ summed along the field line crossing the y-axis at $y = c$. It is also a direct measure of the local enhancement of the field magnitude, equivalent to the index of refraction. The simple resonance functional form $f(y) = 1/k(1 + k^2y^2)$ provides the relation of equation (7.89) between the footpoints $(y = a)$ of a field line at $x = \pm L_1$ and the position $y = c$ at which the line crosses the y-axis where the pressure maximum is located. Equation (7.90) and Fig. 7.18 provide da/dc.

Now, as pointed out in §7.7, so long as $2\varepsilon kL$ is small enough that c is a continuous single valued function of a, (so that $da/dc > 0$ everywhere) there is no gap in the flux surfaces and the field is continuous everywhere in the region. In this case the opposing field lobes creating the pressure maximum do not come into contact and the intervening field is free to flow slowly and smoothly between the opposing lobes. Thus if the footpoints of the intervening field (at $x = \pm a$) have a slow uniform motion $da/dt = u$ in the positive y-direction, the field lines flow smoothly across the region with the continuous nonuniform speed $dc/dt = u\,dc/da$ at $x = 0$, where da/dc is plotted from equation (7.90) in Fig. 7.19.

When $2\varepsilon kL$ is large enough to produce a gap in the flux surface (i.e., when the condition of equation (7.52) that $\varepsilon Lf'' > -1$ is fulfilled so that $dc/da < 0$) the field lines must somehow get across the gap as their footpoints move steadily in the positive y-direction. The gap exists because there are no equilibrium paths across the gap. Hence the field lines cross the gap in a dynamical state. The tension in the field accelerates a thin sheet of field and fluid out into the gap from the boundary of the gap (representing the last equilibrium position). The acceleration continues to the far side of the gap, which is the next equilibrium path to be encountered. The equilibrium at the far side of the gap arises where the pressure gradient and the tension in the field lines can again balance each other, so that the acceleration of the field and fluid falls to zero. The flux bundles coming across the gap at high speed collide with the quasi-static field and fluid that lie beyond the boundary and they may overshoot, crashing through the boundary and going beyond, to be decelerated by the field tension which now pulls backward on them. It is evident, then, that the motion becomes dynamically complex upon arriving at the far side of the gap, and one can only guess what chaotic effects may arise, depending upon the thickness and the mass of the fluid and field coming across the gap, the viscous drag on the thin sheet of field and fluid, the dynamical Kelvin–Helmholtz instability, the resistive tearing instabilities, and the chaotic impact of the high speed sheet with the quasi-static field and fluid beyond the boundary of the gap. The present writing concentrates only on the origin of the phenomenon, without attempting to deduce the detailed consequences.

Note that the gap is rendered unsymmetrical by the slow drift u, illustrated in Fig. 7.20 and described formally in §7.7. Denote the near boundary of the gap by $c = c_1$, reached when the footpoints at $x = \pm L_1$ lie at $y = a$. Then c_1 follows from equation (7.53) as $\varepsilon Lf''(c_1) = -1$ and a follows from equations (7.51) and (7.52) as

$$a = c_1 + \varepsilon Lf'(c_1). \tag{9.46}$$

The passage of the field and fluid across the gap is swift, so that the slow footpoint motion u can be neglected during transit. In that case the farther boundary at $c = c_2 > c_1$ follows as the second root of equation (9.46) for the same a. For the example $f(c) = 1/k(1 + k^2c^2)$ employed in §7.7 it is readily shown that for $2\varepsilon kL = 4$, the first boundary lies at $kc_1 = -0.438$ with $ka \cong +0.795$. The farther boundary lies at $kc_2 \cong 1.42$, with the same ka.

Consider, then, the dynamics of the rapid transport of field and fluid across the gap, from c_1 to c_2 with the position a fixed. The optical analogy is not applicable in the form employed in §7.7 because the field is not in quasi-static equilibrium. However, it is convenient to use the same geometry and notation as was employed in §7.7, so as preparation for the dynamical case, consider the magnetic field line extending from $x = -\lambda, y = a$ to $x = +\lambda, y = a$ across the pressure maximum concentrated along the y-axis, where the field line crosses at $y = c$. For convenience we put $L_1 = L_2 = \lambda$, so that $L = \frac{1}{2}\lambda$, and we restrict a and c to small values $O(\varepsilon\lambda)$. The deflection of the field by the pressure maximum is small $O(\varepsilon)$ so that the perturbation to the otherwise uniform field $\mathbf{e}_x B$ is small. The magnetic field is conveniently written as

$$B_x = B_0(1 + \varepsilon b_x),\ B_y = B_0\varepsilon b_y,$$

so that the perturbation (b_x, b_y) is dimensionless. Consider a thin sheet of such field $z = \pm h(x, y)$ subject to the specified external pressure

$$P = (B_0^2/8\pi)[1 + 2\varepsilon b_x + O(\varepsilon^2)], \tag{9.47}$$

which deviates from the uniform value $B_0^2/8\pi$ only in some small neighborhood $O(a)$ or $O(c)$ of the origin. To this order, then, b_x is a direct measure of the externally applied pressure and of the effective index of refraction for the optical analogy for static equilibrium. We are interested in the dynamical nonequilibrium of the field crossing the gap, for which the equations of motion are

$$\rho\frac{dv_x}{dt} = \frac{B_y}{4\pi}\left(\frac{\partial B_y}{\partial x} - \frac{\partial B_x}{\partial y}\right), \tag{9.48}$$

$$\rho\frac{dv_y}{dt} = \frac{B_x}{4\pi}\left(\frac{\partial B_x}{\partial y} - \frac{\partial B_y}{\partial x}\right). \tag{9.49}$$

The right-hand side of equation (9.48) is small $O(\varepsilon)$, so v_x can be neglected. Equation (9.49) can be written

$$\rho\frac{dv_y}{dt} = \varepsilon\frac{B_0^2}{4\pi}\left(\frac{\partial b_x}{\partial y} - \frac{\partial b_y}{\partial x}\right) + O(\varepsilon^2). \tag{9.50}$$

The small variation $O(\varepsilon)$ of the pressure produces a correspondingly small variation in ρ, which is neglected here, of course.

Note that for static equilibrium

$$\frac{\partial b_x}{\partial y} = \frac{\partial b_y}{\partial x}, \tag{9.51}$$

and it is easy to show that this leads to the conditions described in §7.4. For if the applied pressure is uniform everywhere except in a neighborhood with scale $O(\varepsilon\lambda)$ at the origin, it follows that b_x and b_y are uniform except in the small neighborhood $O(\varepsilon\lambda)$. Then except in the small neighborhood $O(\varepsilon\lambda)$ of $x = 0$, an individual field line has a slope of

$$\frac{dy}{dx} = \varepsilon b_y(x, y)$$

which must be equal to $(a - c)/\lambda$. Hence, except in the small neighborhood $O(\varepsilon\lambda)$, b_y has the uniform value

$$\varepsilon b_y(x, y) \cong (a - c)/\lambda \tag{9.52}$$

for the field line extending from $y = a$ to $y = c$. We can use this result to relate b_x and $a - c$, using equation (9.51). Toward this end, note that integration along $y = c$ from $x = 0$ to $x = \lambda$ yields

$$\int_0^\lambda dx\frac{\partial b_y}{\partial x} = b_y(\lambda, c), \tag{9.53}$$

where $b_y(0, c)$ is zero from symmetry about the y-axis. Now suppose that the localized pressure maximum applied in the small neighborhood of the origin is given by the simple form

$$b_x(x, y) = X(x)Y(y). \tag{9.54}$$

which is nonvanishing only in some neighborhood $O(\varepsilon\lambda)$ of the origin. Integration along $y = c$ yields

$$\int_0^\lambda dx \frac{\partial b_x}{\partial y} = Y'(c)\int_0^\lambda dx X(x), \\ = \varepsilon\lambda Y'(c)I, \tag{9.55}$$

where

$$\varepsilon\lambda I \equiv \int_0^\lambda dx X(x). \tag{9.56}$$

The factor $\varepsilon\lambda$ is in recognition of the fact that $X(x)$ and $Y(y)$ are both $O(1)$ in the neighborhood $O(\varepsilon\lambda)$ of the origin. Now for static equilibrium it follows from equation (9.51) that

$$\int_0^\lambda dx \left(\frac{\partial b_x}{\partial y} - \frac{\partial b_y}{\partial x} \right) = 0,$$

where the integration is along $y = c$. With the aid of equations (9.53) and (9.55) this condition yields

$$b_y(\lambda, c) = \varepsilon\lambda Y'(c)I. \tag{9.57}$$

It follows from equation (9.52) that

$$a = c + \varepsilon^2\lambda^2 IY'(c). \tag{9.58}$$

This is just equation (7.46), with $\varepsilon Lf'(c)$ replaced by $\varepsilon^2\lambda^2 IY'(c)$, where the ε's are defined differently in the two cases. Then

$$\frac{da}{dc} = 1 + \varepsilon^2\lambda^2 IY''(c).$$

The boundaries of the gap lie at the roots c_1 and c_2 of

$$\varepsilon^2\lambda^2 IY''(c) = -1, \tag{9.59}$$

where da/dc falls to zero, the gap occupying the region where $da/dc < 0$. Consider the simple form

$$Y(c) = -c^{2n}/\ell^{2n}, \tag{9.60}$$

where ℓ is a scale $O(\varepsilon\lambda)$, which is conveniently written as $\ell = \beta\varepsilon\lambda$ where β is a numerical factor $O(1)$. It follows that the near boundary lies at

$$c_1 = -\beta\varepsilon\lambda \left[\frac{\beta^2}{2n(2n-1)I} \right]^{1/2(n-1)}. \tag{9.61}$$

Equation (9.58) becomes

$$a = c\left[1 - \frac{2nI}{\beta^2} \left(\frac{c}{\beta\varepsilon\lambda} \right)^{2n-2} \right], \tag{9.62}$$

so that the footpoint location of the near edge $c = c_1$ is $a = a_1$, where

$$a_1 = c_1 2(n-1)/(2n-1). \tag{9.63}$$

The simple case $n = 1$ is degenerate, with (9.62) reducing to

$$a = c(1 - 2I/\beta^2).$$

Then if $\beta^2 > 2$, da/dc is positive for all c, and if $\beta^2 < 2$, da/dc is negative for all c, i.e., there are no stable paths across the region. It is convenient, therefore, to work with the next simplest case $n = 2$, for which

$$a = c\left[1 - \frac{4I}{\beta^2}\left(\frac{c}{\beta\varepsilon\lambda}\right)^2\right]. \tag{9.64}$$

Then with a, given by equation (9.63) as

$$a_1 = \frac{2}{3}c_1, \tag{9.65}$$

it follows from equation (9.64) that

$$c_1 = -\beta^2\varepsilon\lambda/2 \times 3^{\frac{1}{2}}I^{\frac{1}{2}}. \tag{9.66}$$

The far side of the boundary at $c = c_2$, follows as the positive root c_2 of equation (9.64) with $a = a_1$. The result is

$$(\chi - 1)\left(\chi + \frac{1}{2}\right)^2 = 0,$$

where

$$\chi = 3^{\frac{1}{2}}I^{\frac{1}{2}}c/\beta^2\varepsilon\lambda. \tag{9.67}$$

The double root $\chi = -\frac{1}{2}$ gives c_1 again, while the positive root gives

$$\begin{aligned} c_2 &= \beta^2\varepsilon\lambda/3^{\frac{1}{2}}I^{\frac{1}{2}}, \\ &= -2c_1. \end{aligned} \tag{9.68}$$

The essential point is that the path $a = a_1, c = c_2$ is in equilibrium, with stable equilibrium ($da/dc > 0$) for all paths above that level ($a > a_1, c > c_2$). It follows that the field and fluid is accelerated by the tension in the field from c_1, with monotonically increasing speed up to c_2, beyond which the tension decelerates the field and fluid. In fact, there is a quasi-static accumulation of equilibrium field beginning at c_2. Hence, the accelerated field and fluid collide with the boundary c_2 of that equilibrium field, undergoing an abrupt, and probably chaotic, deceleration.

9.3 Dynamical Model

It is instructive to examine a simple dynamical model of the high speed sheet of fluid accelerating across the gap described in the foregoing section. The simplest model follows directly from equation (9.50) with the idealization that the field lines maintain the simple linear form of static equilibrium during their acceleration. That is to say, for small inclination $(a - c) \ll \lambda$, the field lines connect in a straight line from $y = c$ at $x = 0$ to $y = a$ at $x = \lambda$. Then integrate equation (9.50) along the field lines from $(0, c)$ to (λ, a), using equations (9.52) and (9.54) for b_y and b_x, respectively. The velocity v_y is constrained to

$$v_y = (1 - x/\lambda)dc/dt \tag{9.69}$$

in $0 \leqslant x \leqslant \lambda$, and the integral of the left-hand side of equation (9.50) becomes $\frac{1}{2}\rho\lambda d^2c/dt^2$. Note, then, that b_y has the uniform value given by equation (9.52), except that b_y drops rapidly to zero as x declines from $O(\varepsilon\lambda)$ to zero. Hence, going back to equation (9.53)

$$\begin{aligned} \varepsilon \int_0^\lambda dx \frac{\partial b_y}{\partial x} &\cong \varepsilon b_y(\lambda, a)[1 + 0(\varepsilon)], \\ &= (a - c)/\lambda. \end{aligned} \tag{9.70}$$

With b_x specified in the form of equation (9.54), the integral of $\partial b_x/\partial y$ becomes

$$\int_0^\lambda dx \frac{\partial b_x}{\partial y} = \int_0^\lambda dx X(x) Y'(y)$$

along the field line

$$y = c - (c - a)x/\lambda.$$

Then expanding $Y(y)$ in ascending power of x,

$$Y'(y) \cong Y'(c) - (c - a)(x/\lambda)Y''(c) + \ldots$$

to first order in x, with the result that

$$\int_0^\lambda dx \frac{\partial b_x}{\partial y} \cong \varepsilon\lambda I Y'(c) - \varepsilon^2\lambda(c - a)JY''(c), \tag{9.71}$$

where I is the integral defined by equation (9.56) and the quantity J is defined by

$$\varepsilon^2\lambda^2 J \equiv \int_0^\lambda dx x X(x). \tag{9.72}$$

The factor $\varepsilon^2\lambda^2$ multiplying J is in recognition of the fact that $X(x)$ is $O(1)$ and nonvanishing only in a small neighborhood $O(\varepsilon\lambda)$ of $x = 0$. It follows that the second term on the right-hand side of equation (9.71) is small $O(\varepsilon)$ compared to the first term and, therefore, can be neglected, so that the elementary result of equation (9.55) applies again.

Collecting the terms, the integral of equation (9.50) along the field line yields the equation of motion

$$\frac{d^2\chi}{dt^2} = \frac{2\varepsilon C^2}{3\lambda^2}(1 - \chi)(1 + 2\chi)^2, \tag{9.73}$$

where χ is the dimensionless displacement again, given by equation (9.67) and equal to $c/2(-c_1)$.

The equation of motion is readily integrated to give the energy equation

$$\left(\frac{d\chi}{dt}\right)^2 = \frac{4\varepsilon C^2}{3\lambda^2}\left[\left(\frac{3}{2} - \chi\right)\left(\chi + \frac{1}{2}\right)^3 + \chi_0\right], \tag{9.74}$$

where $\chi_0 + 3/16$ is the constant of integration, chosen so that

$$\left(\frac{d\chi}{dt}\right)^2 = \frac{4\varepsilon C^2}{3\lambda^2}\chi_0 \tag{9.75}$$

at time $t = 0$ when $\chi = -\frac{1}{2}$ and $c = c_1$. It is evident by inspection of equation (9.73) that the acceleration vanishes at $\chi = +1$, which corresponds to $c = c_2 = -2c$, at the far side of the static gap, given by equation (9.68). Note then that, if the initial velocity is taken to be zero ($\chi_0 = 0$), the velocity falls to zero at $\chi = 3/2$ as a consequence of the factor $(3/2 - \chi)$ on the right-hand side of equation (9.74), where $c = -3c_1$, with $-c_1$ given by equation (9.66). It was pointed out earlier that the motion is halted in the vicinity of the farther boundary $c_2 = -2c_1$, because the field and fluid collide with the field accumulated in static equilibrium immediately beyond c_2.

Now equation (9.74) reduces to the quadrature

$$\frac{2C}{\lambda}\left(\frac{\varepsilon}{3}\right)^{\frac{1}{2}} t = \pm \int_{-\frac{1}{2}}^{\chi} \frac{dw}{\left[\left(\frac{3}{2} - w\right)\left(w + \frac{1}{2}\right)^3 + \chi_0\right]^{\frac{1}{2}}}. \tag{9.76}$$

The motion is expressible in terms of elliptic functions. Unfortunately the time elapsed from $\chi = -\frac{1}{2}$ depends critically on χ_0 because of the factor $(w + \frac{1}{2})^{3/2}$ in the denominator, with the elapsed time increasing without bound as χ_0 declines to zero. The problem is that χ_0 does not follow from the conditions stated so far. The footpoint motion da/dt is set equal to $+u$, but that leaves $dc/dt = u dc/da$ undetermined because da/dc falls to zero as the field approaches the lower boundary of the gap. Hence dc/da, deduced from the simple quasi-static equilibrium model of equation (9.64), increases without bound, thereby violating the quasi-static assumption so that the quasi-static condition for $c < c_1$, provides no valid approximation for dc/dt at $c = c_1$.

Consider, then, the velocity with which the field arrives at the farther boundary, where it reaches into the quasi-static field and fluid beyond that boundary. That boundary lies at $\chi = +1$, as already noted, for which equation (9.74) with $\chi_0 = 0$ yields the velocity as

$$\left(\frac{d\chi}{dt}\right)_m = \frac{3C}{\lambda}\left(\frac{\varepsilon}{2}\right)^{\frac{1}{2}},$$

reducing to

$$\left(\frac{dc}{dt}\right)_m = \left(\frac{3\varepsilon^3}{2I}\right)^{\frac{1}{2}} \beta^2 C \tag{9.77}$$

at $c = c_2$. The subscript m is attached in recognition of the fact that $d\chi/dt$ reaches a maximum at $\chi = +1$. Note that εC represents the characteristic Alfven speed in the transverse field component $B_0 \varepsilon b_y$. It follows that the field and fluid reach a speed of the order of $\varepsilon^{1/2}$ times the transverse Alfven speed, becoming comparable to the Alfven speed when $\varepsilon = 0(1)$ instead of $\varepsilon \ll 1$.

9.4 Hydrodynamic Model of High Speed Sheet

The simple dynamical model of the foregoing section illustrates the basic relations between the dimensions of the gap, the Alfven speed C, and speed $O(\varepsilon^{\frac{3}{2}} C)$ with which the field and fluid crash into the far side of the gap, all for $\varepsilon \ll 1$. But the model is not formulated rigorously and it is of some interest to solve the dynamical equations in a more systematic and conventional manner. Consider, then, the thin layer of fluid and magnetic flux confined within the boundaries

$$z = \pm h(x, y) \tag{9.78}$$

by the fluid and fields on either side (Parker, 1990). The thickness h is small $O(\alpha)$ compared to the width $O(\lambda)$ of the region in the x direction (where $\alpha \ll 1$), with unlimited extent in the y-direction. Hence $(\partial h/\partial x)^2$ and $(\partial h/\partial y)^2$ are both small $O(\alpha^2)$. The fluid is assumed to be cold with density ρ. The magnetic field is in the x-direction with the uniform value B_0 at $y = \pm\infty$, where the density is ρ_0 and the thickness is h_0. Conservation of magnetic flux throughout the layer is described by

$$\frac{\partial}{\partial x} B_x h + \frac{\partial}{\partial y} B_y h = 0, \tag{9.79}$$

neglecting terms second order in α compared to one. Conservation of fluid can be written in a similar form

$$\frac{\partial}{\partial x} \rho v_x h + \frac{\partial}{\partial y} \rho v_y h = 0 \tag{9.80}$$

to the same order. The pressure $P(x, y)$ is applied to the surfaces $z = \pm h(x, y)$ and is balanced by the internal magnetic pressure, so that

$$8\pi P(x, y) = B_x^2(x, y) + B_y^2(x, y), \tag{9.81}$$

neglecting terms second order in α. The pressure has the uniform value $B^2/8\pi$, except in the neighborhood of the origin, where there is a local maximum.

The x and y-components of the momentum equations are

$$v_x \frac{\partial v_x}{\partial x} + v_y \frac{\partial v_x}{\partial y} = \frac{B_y}{4\pi\rho}\left(\frac{\partial B_x}{\partial y} - \frac{\partial B_y}{\partial x}\right), \tag{9.82}$$

$$v_x \frac{\partial v_y}{\partial x} + v_y \frac{\partial v_y}{\partial y} = \frac{B_x}{4\pi\rho}\left(\frac{\partial B_y}{\partial x} - \frac{\partial B_x}{\partial y}\right), \tag{9.83}$$

for stationary conditions. The fluid motion provides the two components (at $z = 0$)

$$v_x \frac{\partial B_x}{\partial x} + v_y \frac{\partial B_x}{\partial y} + B_x\left(\frac{\partial v_y}{\partial y} + \gamma\right) = B_y \frac{\partial v_x}{\partial y}, \tag{9.84}$$

$$v_x \frac{\partial B_y}{\partial x} + v_y \frac{\partial B_y}{\partial y} + B_y\left(\frac{\partial v_x}{\partial x} + \gamma\right) = B_x \frac{\partial v_y}{\partial x}, \tag{9.85}$$

of the magnetic induction equation, where the fluid velocity components are $[v_x(x, y), v_y\,(x, y), z\gamma(x, y)]$ in the thin layer. Thus, in place of equation (9.80) we also have

$$\frac{\partial}{\partial x} \rho v_x + \frac{\partial}{\partial y} \rho v_x + \gamma\rho = 0. \tag{9.86}$$

Simultaneous solution of these equations (9.79)–(9.85) is a formidable task, so in this first investigation we constrain the fluid motion to be in the y-direction. The insertion of many closely spaced nonconducting parallel planes $y = constant$ accomplishes the task. The inclusion of v_x should be explored eventually, because with a sufficiently strong maximum in the applied pressure $P(x, y)$, so that B_y is comparable to B_x, the motion v_x along the field is strong enough to form standing shocks in the cold gas. We ignore this aspect of the problem because the primary concern at the moment is to establish the flow in the y-direction across the local pressure maximum. The nature of that flow is not greatly altered by the present artificial constraint that $v_x = 0$. Solutions with a weak concentrated pressure maximum, so that $B_y^2 \ll B^2, B_x^2$ and v_x is naturally small, are available in the literature (Parker, 1990) for the interested reader.

With $v_x = 0$, equation (9.82) becomes superfluous. Equations (9.83)–(9.85) reduce to

$$v\frac{\partial v}{\partial y} = \frac{B_x}{4\pi\rho}\left(\frac{\partial B_y}{\partial x} - \frac{\partial B_x}{\partial y}\right), \tag{9.87}$$

$$v\frac{\partial B_x}{\partial y} + B_x\left(\frac{\partial v}{\partial y} + \gamma\right) = 0, \tag{9.88}$$

$$v\frac{\partial B_y}{\partial y} + \gamma B_y - B_x\frac{\partial v_y}{\partial x} = 0, \tag{9.89}$$

where the subscript y has been dropped from v_y. Equation (9.80) can be integrated over y to give

$$\rho h v = D(x),$$

where $D(x)$ is an arbitrary function of x, representing the total flux of matter per unit length in the x-direction. This is taken to have the uniform value $\rho_0 h_0 u$ at $y = \pm\infty$. Hence

$$\rho h v = \rho_0 h_0 u. \tag{9.90}$$

Next eliminate $\partial v/\partial y + \gamma$ between equations (9.86) and (9.88), obtaining

$$\frac{1}{B_x}\frac{\partial B_x}{\partial y} = \frac{1}{\rho}\frac{\partial p}{\partial y},$$

so that

$$B_x(x, y) = B_0\rho(x, y)/\rho_0.$$

Now write v in units of the Alfven speed $C = B_0/(4\pi\rho_0)^{\frac{1}{2}}$ at infinity, with $V \equiv v/C$. Express B_x and B_y in units of B_0, so that $b_x = B_x/B_0, b_y = B_y/B_0$. Then

$$b_x = \rho/\rho_0, \tag{9.91}$$

while the momentum equation (9.87) becomes

$$V\frac{\partial V}{\partial y} = \frac{\partial b_y}{\partial x} - \frac{\partial b_x}{\partial y} \tag{9.92}$$

upon eliminating b_x with the aid of equation (9.87). Eliminate γ between equations (9.88) and (9.89), with the result that

$$F\frac{\partial V}{\partial y} + \frac{\partial V}{\partial x} = V\frac{\partial F}{\partial y}, \tag{9.93}$$

where F is the ratio b_y/b_x. Equation (9.92) becomes

$$V\frac{\partial V}{\partial y} = F\frac{\partial b_x}{\partial x} - \frac{\partial b_x}{\partial y} + b_x\frac{\partial F}{\partial x}. \tag{9.94}$$

Equation (9.81) becomes

$$8\pi P = B_0^2(1 + F^2)b_x^2. \tag{9.95}$$

Equations (9.93)–(9.95) provide three relations for the three quantities b_x, F, and V, for arbitrary $P(x, y)$.

It is evident by inspection that equations (9.93) and (9.94) are linear in F and in b_x but nonlinear in V. Hence the mathematical solution of the three equations is most easily effected by solving for F and b_x in terms of some specified $V(x, y)$, deducing the applied $P(x, y)$ from equation (9.95). Thus it follows from equation (9.93) that

$$F(x, y) = V(x, y)\left[J(x) - \int_0^y dy' \frac{\partial}{\partial x}\left(\frac{1}{V(x, y')}\right)\right], \tag{9.96}$$

where $J(x)$ is an arbitrary function of x. The characteristic equations for equation (9.94) can be written

$$\frac{dx}{dy} = -F, \tag{9.97}$$

$$\frac{db_x}{dy} = b_x\frac{\partial F}{\partial x} - V\frac{\partial V}{\partial y}, \tag{9.98}$$

with the solution

$$\begin{aligned} b_x(x, y) = &\left\{H(\Lambda) - \int_0^y dy' V(x, y')(\partial V/\partial y')\exp\left[-\int_0^{y'} dy'' \partial F(x, y'')/\partial x\right]\right\} \\ &\times \exp\left[+\int_0^y dy' \partial F(x, y')/\partial x\right], \end{aligned} \tag{9.99}$$

where the integration is along a characteristic curve $\Lambda = \Lambda(x, y)$, obtained by integrating equation (9.97). Note that the individual characteristic curve $\Lambda = constant$ is orthogonal to the field lines, which are described by $dy/dx = +F$. Finally, the fluid density follows from equation (9.91) as directly proportional to b_x. The thickness of the fluid layer h/h_0 follows from equation (9.90) as $(u/C)/Vb_x$.

These equations take on a particularly simple form for the similarity solution

$$V(x, y) = f(\xi)/g'(\eta), \tag{9.100}$$

where $\xi \equiv x/a, \eta \equiv y/a$ in terms of the characteristic length a and where $f(x)$ and $g'(y)$ are arbitrary functions at this point. It follows from equation (9.96) that

$$F(x, y) = f(\xi)J(x)/g'(\eta) + f'(\xi)g(\eta)/f(\xi)g'(\eta). \tag{9.101}$$

In the simple case that $J(x) = 0$, the lines of force are given by the one parameter family of curves $g(\eta) = \lambda f(\eta)$. The orthogonal lines, providing the characteristics required by equation (9.97), are

$$\int d\xi f(\xi)/f'(\xi) = \Lambda + \int d\eta g(\eta)/g'(\eta). \tag{9.102}$$

To treat a single example, suppose that

$$f(\xi) = Q\exp(-q^2\xi^2), g(\eta) = \tan^{-1}\eta - \frac{1}{4}\pi. \tag{9.103}$$

The velocity follows from equation (9.100) as

$$V(x,y) = Q(1+\eta^2)\exp(-q^2\xi^2). \tag{9.104}$$

With $J(x) = 0$, it follows from equation (9.101) that

$$F(x,y) = 2q^2\xi(1+\eta^2)\left(\frac{1}{4}\pi - \tan^{-1}\eta\right). \tag{9.105}$$

The field lines are

$$\eta = \tan\left[\frac{1}{4}\pi + \Lambda\exp(-q^2\xi^2)\right] \tag{9.106}$$

and are plotted in Fig. 9.3. The characteristic curves are given by equation (9.102) as

$$\xi = \frac{\Lambda\exp[2q^2I(\eta)]}{(1+\eta^2)^{2q^2/3}}, \tag{9.107}$$

where

$$I(\eta) = \eta\left(1+\frac{1}{3}\eta^2\right)\left(\tan^{-1}\eta - \frac{1}{4}\pi\right) - \frac{1}{6}\eta^2. \tag{9.108}$$

The field b_x follows from equation (9.99) as

$$\begin{aligned} b_x(x,y) = {} & (1+\eta^2)^{2q^2/3}\exp[-2q^2I(\eta)] \\ & \times\left\{H(\Lambda) - 2Q^2\int_0^{\eta} d\eta'\eta'(1+\eta'^2)^{1-2q^2/3}\exp[2q^2I(\eta')]\right. \\ & \left.\times\exp\left[-\frac{2q^2\Lambda^2\exp 4q^2I(\eta')}{(1+\eta'^2)^{4q^2/3}}\right]\right\}, \end{aligned} \tag{9.109}$$

where equation (9.107) has been used to express ξ in terms of η' and Λ is evaluated from equation (9.107). The applied pressure $P(x,y)$ follows from equation (9.95) with F given by equation (9.105) and b_x by equation (9.109).

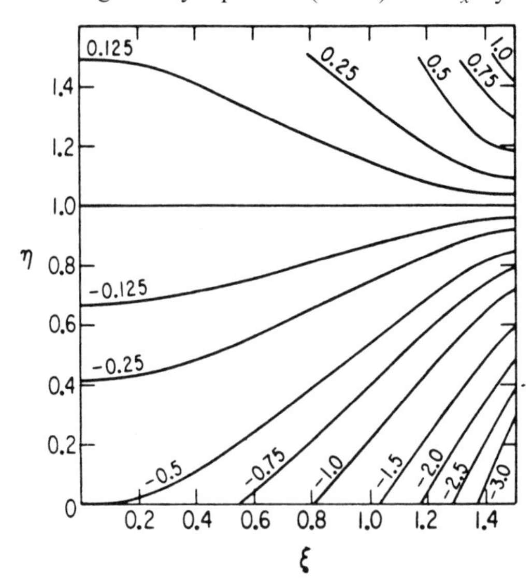

Fig. 9.3. A plot of the field lines given by equation (9.106) for the value of $4\Lambda/\pi$ indicated on each curve.

To put the overall picture in mind note from equations (9.95) and (9.105) that the applied pressure profile along the y-axis ($\Lambda = 0$), where $b_y = F = 0$, is just b_x^2, plotted in Fig. 9.4 for $q^2 = 3/2$ and $H(0) = 2Q^2$. The velocity of the field along the y-axis has the form $(1 + \eta^2)$ and is a minimum at $\eta = 0$. It follows from equation (9.100) that Q represents the velocity with which the field approaches the acceleration region, beginning at $\eta = 0$. The velocity increases rapidly as η increases beyond $\eta = 0$. The velocity of the footpoints of the field at $\xi = \pm\xi_1$ is smaller by $\exp(-q^2\xi_1^2)$. The example loses physical interest beyond $\eta \cong 1.3$, where the applied pressure falls back to the ambient value.

The thickness h of the layer of accelerated fluid follows from equations (9.90) and (9.91) as proportional to $1/Vb_x$, plotted as the unlabeled curve near the bottom of Fig. 9.4. The quantities $\exp 3I$, $\eta \exp$, $3I$, and

$$K \equiv \int_0^\eta d\eta' \eta' \exp 3I(\eta') \tag{9.110}$$

are also plotted in Fig. 9.4.

Other illustrative examples are available in the literature (Parker, 1990), all showing the obvious acceleration and thinning of the sheet of fluid once it crosses into the gap where it is accelerated. The illustration worked out in §9.2 is more informative, because the geometry of the gap appears explicitly in that development. However, the next question has to do with the stability of the steady flow of field and fluid across the gap, which is more conventionally treated with a formal stationary solution to the magnetohydrodynamic equations, e.g., the example provided here. Note that the gap can be formally reconstructed in the present model for the static case $v = 0$, starting with equations (9.87)–(9.89). For it follows from equation (9.86) that $\gamma = 0$, so that equations (9.88) and (9.89) are trivially satisfied. Equation (9.87) reduces to $\partial b_x/\partial y = \partial b_y/\partial x$ and $b_x = -\partial\phi/\partial\xi$, $b_y = -\partial\phi/\partial\eta$. Then $8\pi P = (\nabla\phi)^2$, and we have the optical analogy again, with the boundaries of the gap given by the solutions to Euler's equations.

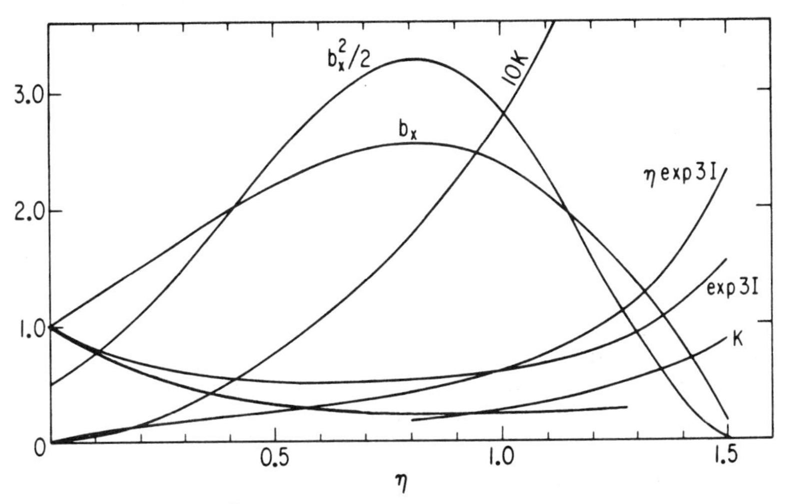

Fig. 9.4. The profile of b_x, $\frac{1}{2}b_x^2$, and the fluid sheet thickness $1/Vb_x$ (unlabeled curve) along the x-axis ($\Lambda = 0$) from equations (9.104) and (9.109) for $H(0) = 0$ and $q^2 = 3/2$. Also shown are plots of $\exp 3I(\eta)$, $\eta \exp 3I(\eta)$ and the integral $K(\eta)$ of $\eta \exp 3I(\eta)$.

As already mentioned, the stability of the high speed sheet of fluid and field sliding across the gap is an immediate question. Unfortunately the boundary conditions are not uniquely defined without specifying the details of the field on either side of the sheet, with its often complex topology. There is some complicated mode structure even in simplified idealized cases if we take into account that the fields are not parallel on opposite sides of the sheet. Then there arises the role of resistivity and viscosity, particularly when the sheet is thin. These dissipative effects were examined in some detail in §§9.1.2 and 9.1.3 for precisely the reason that they come into the picture over and over again. We will meet the same question again when we take up rapid reconnection in Chapter 10, because the slow motion of the reconnecting field lobes through the background fluid may produce a high speed sheet of fluid that is sufficiently thick as to keep the two lobes apart and thereby block the reconnection. The sometimes rapid deformation of the magnetic fields of emerging active regions at the surface of the Sun, conducive to large flares, is a prime situation for producing high speed sheets of fluid sliding along the topological separatrices between the opposing lobes of field. Such high speed sheets may act as a valve to shut off the reconnection, opening only at special times when the deformation momentarily subsides. This is all speculation, of course, because the high speed sheet has not been incorporated into dynamical reconnecting models at this point in time. It adds one more twist to the already complicated dynamics of the rapid magnetic dissipation that produces the solar and stellar flare. A similar phenomenon arises from the appearance of anomalous resistivity or an electric double layer in force-free fields, treated in §2.4.3. This form of the effect appears to be particularly relevant to the active terrestrial aurora, playing a role perhaps in the rapid changing of shapes and forms of individual auroral curtains.

The central role played by high speed sheets in regulating flare phenomena is readily illustrated with some simple examples appropriate for an active region on the Sun. Imagine, for instance, that two opposing magnetic lobes with $B = 500$ gauss and characteristic scale $\ell = 5 \times 10^8$ cm in a potential flare site are moving together at a velocity $u = 1$ km/sec through the surrounding magnetic field. The characteristic Alfven speed is $C = 10^4$ km/sec in a typical active coronal density of 10^{10} atoms/cm^3. Suppose, then, that the high speed sheet achieves a speed $v = 10^{-2}C \cong 10^2$ km/sec. Magnetic flux is swept up at a rate of the order of $u\ell B$. This same flux passes through the high speed sheet of thickness h at the rate vhB, assuming comparable field intensity in the gap. Hence on the average,

$$vh = u\ell,$$

with the result that h is of the order of 50 km. Neither viscosity nor resistivity has much effect over this thickness in the modest time $\ell/v \cong 50$ sec of passage along the thin sheet.

Another example, appropriate to the nanoflare in the active X-ray corona, is a general motion $u = 0.5$ km/sec with $\ell = 10^8$ cm, so that if v is 10^2 km again, the thickness of the sheet is 5 km, with a transit time $\ell/v = 10$ sec. So again there is no significant viscous or resistive diffusion during transit through the thin sheet. Only when u becomes so small that h falls to 10^4 cm do we expect resistive diffusion to play a strong role. For instance, with $\eta = 10^4$ cm/sec, appropriate for 10^6 K, the characteristic resistive diffusion length $(4\eta t)^{\frac{1}{2}}$ is 10^3 cm in 50 sec. This would begin to be important for $h \cong 10^4$ cm, which would be achieved if u were as small as 2 m/sec in the flare site.

These simple estimates, for all their arbitrary character, make the point that the rapid reconnection between opposing lobes of magnetic field is strongly suppressed by steady motion of the lobes relative to the ambient magnetic field. It would appear that, in the simplest case, the high speed sheet of fluid blocks the reconnection unless the thickness h of the sheet is so small that resistive diffusion arises. On the other hand, it must be realized that the dynamics of the high speed sheet with, or without, viscosity and resistivity is complicated. The sheet is surely unstable in many circumstances, with the possibility of anomalous resistivity, and the possibility of some complex intermittent dynamical reconnection. Clearly the subject needs extensive theoretical investigation. It would be helpful if observations could provide some guidance, but unfortunately the transparency of the tenuous hot coronal gases makes detection of the thin sheet impossible, with detection of associated effects dubious, and certainly ambiguous.

9.5 Displaced Flux Bundles

The foregoing sections have provided a cursory view of the high speed fluid motions that arise between contiguous topological lobes of magnetic field when the topology of the magnetic field is subject to continuing deformation (Parker, 1990). The present section takes up the related motions that arise when an individual elemental flux bundle is displaced relative to the rest of the field (Parker, 1981a,b). The anomalous motions observed at the surface of the Sun (cf. Title, Tarbell, and Topka, 1987; Title, et al. 1989; Title, et al. 1992) suggest the possibility that a magnetic fibril may be substantially displaced relative to its neighbors, thereby displacing the entire flux bundle from that fibril in the corona above (Parker, 1981a). The flux bundle becomes misaligned relative to the surrounding ambient field. The misalignment causes it to be flattened into a sheet by the pressure of the field, sketched in Fig. 9.5. If the displaced flux bundle is twisted, it resists the flattening to some degree, having an eccentric oval cross-section as indicated in Fig. 9.5. It appears that rapid reconnection dissipates the misaligned flux bundle in a short time (see discussion and examples in Parker, 1981b). An elemental flux bundle without twisting is flattened into a sheet of exceedingly small thickness and in any real situation would merge relatively quickly into the ambient field. A quantitative treatment of the precise cross-section of a misaligned flux bundle is difficult even in the case of infinite conductivity because the form of the boundary surface between the bundle and the ambient field is not known ahead of time. It would be interesting, as a first step toward dynamical reconnection, to solve the moveable boundary problem with a flux bundle of infinitesimal total flux threading its way in the z-direction between the field lines of an ambient field that is otherwise uniform

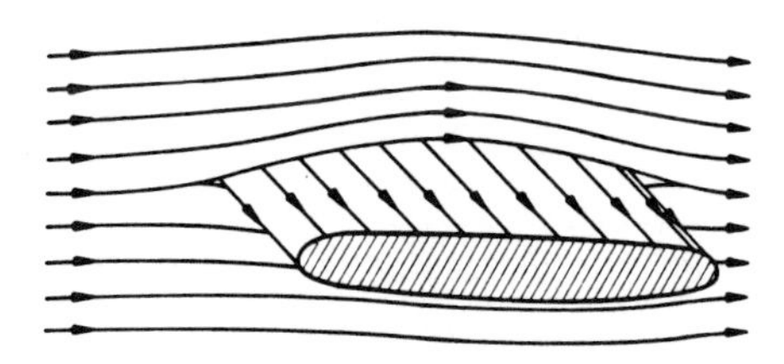

Fig. 9.5. A sketch of a displaced, misaligned flux bundle squeezed by the ambient field on either side.

with magnitude B_0 in the x-direction. The elemental flux bundle has footpoints $x = y = 0$ at the boundaries $z = \pm L$, and flattens out in between, as sketched in Fig. 9.6. The cross-section of the bundle at $z = 0$ has a characteristic thickness h and width w, as indicated in the figure. The tension $B^2/4\pi$ in the distended field lines in the flux bundle limits the width w of the bundle while the external field would flatten the bundle to vanishing thickness h and unlimited width w. The force balance equation can be written

$$\frac{\partial}{\partial x}\frac{B^2}{8\pi} = +\frac{B^2}{4\pi R}, \tag{9.111}$$

where R is the radius of curvature of the bundle field. The key point is that the ambient field is enhanced slightly where it passes around the thickness h of the flattened bundle, by an amount ΔB of the order of $B_0 h^2/w^2$. Hence the left-hand side of the above force balance equation is of the order of $(B_0^2/8\pi)(h^2/w^2)/w$, upon replacing $\partial/\partial x$ by $1/w$. The characteristic radius of curvature is related to w as $R \cong L^2/w$, in order of magnitude. Hence the right-hand side is of the order of $(B_0^2/8\pi)w/L^2$. The result is $w^2 \cong hL$. If the diameter of the elemental flux bundle at $z = \pm L$ has the small value a, the conservation of flux requires conservation of cross-section, $a^2 \cong hw$, in order of magnitude. It follows that

$$h = a(a/L)^{\frac{1}{3}}, w = a(L/a)^{\frac{1}{3}}. \tag{9.112}$$

It must be appreciated that these expressions provide only characteristic values of course, and a formal analytical solution would be highly desirable, in view of the fact that the flattened cross section of the bundle may have a slender cusp at each end.

As an illustrative example, consider a displaced flux bundle of length $L = 10^4$ km in a bipolar action region on the Sun. If $a = 10^2$ km, it follows that the displaced bundle is flattened by the ambient field to a thickness h of the order of 20 km and a width w of the order of 500 km. The characteristic resistive diffusion time across h is $h^2/4\eta$ or 4×10^9 sec $\cong$ 10 years for $\eta = 10^3$ cm^2/sec. Thus diffusion across the characteristic thickness h is insignificant in the period of hours or days in which the bundle is displaced. But it must be remembered that there is necessarily a tangential discontinuity at the surface of the bundle, across which rapid reconnection quickly devours the bundle. For even at the minimum reconnection rate $C/N_L^{\frac{1}{2}}$, where C is the Alfven speed and $N_L = wC/\eta$ is the characteristic Lundquist number,

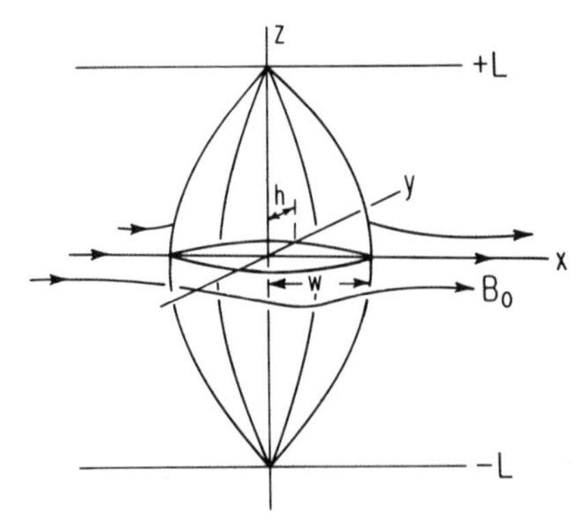

Fig. 9.6. Schematic diagram of an elemental flux bundle with footpoints at $z = \pm L$, flattened by the uniform ambient field B_0 in the x-direction to a small thickness $2h$ and large width $2w$ in the midplane $z = 0$.

the result for the bipolar field of an active region (with $C = 2 \times 10^8$ cm/sec, $\eta = 10^3$ cm/sec, and $w = 5 \times 10^7$ cm) is $N_L = 10^{13}$, with $C/N_L^{\frac{1}{2}} \cong 70$ cm/sec. This cuts across the thickness h in 3×10^4 sec or about 10 hours.

It is evident, then, that the life of a displaced flux bundle is complicated. So we begin with the idealized case in which there is no resistive diffusion, to obtain some idea of the motions associated with displaced flux bundles. It follows from equations (9.112) that the width w of a displaced flux bundle declines to zero in the limit of small radius a at the footpoints. Hence, the concept of an elemental flux bundle, whose path is represented by a mathematical curve in space, is a self-consistent concept in the limit of small a. We treat the case that the ambient field is force free. The displaced elemental flux bundle provides a curve that defines a unique flux surface in the force-free field, so that the optical analogy is available to compute the path of the bundle in the specified flux surface (Parker, 1981a).

A simple example (Parker, 1981a) follows from the elemental flux bundle in the vertical plane ($z =$ *constant*) of the periodic magnetic arcade described by equations (7.134) and (7.135). Equation (7.140) provides the path of a flux bundle with footpoints at $y = 0$ set at equal distances $(1/k)\cos^{-1}\gamma (0 \leqslant \gamma \leqslant 1)$ on either side of x_0. Figure 7.29 shows the arched paths of the equilibrium flux bundle. In the simple case that both footpoints are displaced at the same rate dx_0/dt, the entire flux bundle moves with that same velocity, and the height $y = h$ of the apex of the bundle, at $kh = -\ln\gamma$, remains fixed. On the other hand, suppose that the footpoint separation $(2/k)\cos^{-1}\gamma$ increases at the small rate $2u$, starting from zero (when $\gamma = \frac{1}{2}\pi$) at time $t = 0$. It follows that $\gamma = \cos kut$ and

$$\frac{d\gamma}{dt} = -ku(1 - \gamma^2)^{\frac{1}{2}}.$$

The apex rises at the rate

$$\begin{aligned} \frac{dh}{dt} &= -\frac{1}{k\gamma}\frac{d\gamma}{dt}, \\ &= +\frac{u}{\gamma}(1 - \gamma^2)^{\frac{1}{2}}. \end{aligned} \tag{9.113}$$

Thus, γ declines with increasing t, falling to zero at $kut = \frac{1}{2}\pi$ when the footpoint separation has increased to π/k and the apex has risen to $h = \infty$. The essential point is that for any small but nonvanishing rate u of separation of the footpoints, the apex of the stable equilibrium path of the flux bundle moves out to infinity in a finite time $\pi/2ku$. The actual moving quasi-equilibrium flux bundle lies essentially along the rising equilibrium path so long as dh/dt is small compared to the local sound velocity and Alfven velocity. When dh/dt becomes comparable to the sound and Alfven velocities, the bundle can no longer keep up, of course, and the essential point is again that very small motion u of the footpoints of the field leads to high speed, i.e., near sonic velocities, out in the field. An infinite length of time is required for the flux bundle to reach static equilibrium again, when after a time $\pi/2uk$ the footpoint separation exceeds π/k. For then the equilibrium path extends vertically from each footpoint to $y = +\infty$. Any further motion of the footpoints causes the vertical equilibrium path to shift horizontally at the same rate as the motion of the footpoints. Any horizontal displacement of the footpoints propagates outward to $y = +\infty$ as an Alfven wave in the vertical equilibrium flux bundles. The wave is coupled to the ambient field and fluid, so that the rate of propagation is only some fraction of the characteristic Alfven speed, of course.

A similar result obtains for an elemental flux bundle displaced in a two dimensional dipole field, described by equations (7.141) and (7.142). The equilibrium flux bundle may arch up from the base of the field at $\varpi = a$ at any angular position. The path of the bundle is given by equation (7.145), conveniently rewritten as

$$\varpi = a \sin(\varphi - \varphi_0)/\sin(\varphi_1 - \varphi_0) \tag{9.114}$$

for footpoints at the radius a separated by $\pi - 2(\varphi_1 - \varphi_0)$ and lying at $\varphi - \varphi_1$ and $\pi + 2\varphi_0 - \varphi_1$. The apex lies at $\varpi = r$ where $\varphi - \varphi_0 = \frac{1}{2}\pi$ and

$$r = a/\sin(\varphi_1 - \varphi_0). \tag{9.115}$$

The distance r to the apex increases without bound as the angular separation $\pi - 2(\varphi_1 - \varphi_0)$ of the footpoints increases to π, i.e., as $\varphi_1 - \varphi_0$ declines to zero. Therefore, as in the previous example, an arbitrarily small rate of increase of the footpoint separation leads to a runaway condition at the apex of the flux bundle. Quantitatively, suppose that the footpoint separation $\pi - 2(\varphi_1 - \varphi_0)$ increases at the small rate w, so that $d(\varphi_1 - \varphi_0)/dt = -w$. It follows that the apex of the equilibrium flux bundle rises at the rate

$$\begin{aligned}\frac{dr}{dt} &= \frac{2aw\cos(\varphi_1 - \varphi_0)}{\sin^2(\varphi_1 - \varphi_0)}, \\ &= 2aw\cos(\varphi_1 - \varphi_0)r^2/a^2,\end{aligned}$$

becoming large without bound as $\varphi_1 - \varphi_0$ declines to zero and r/a increases to infinity.

Again, the flux bundle keeps up with the equilibrium path change when dr/dt is small compared to the Alfven speed, but cannot keep up when dr/dt equals or exceeds the Alfven speed. The footpoint separation exceeds π in a finite time, and thereafter the flux bundle is in a dynamical state, chasing out to infinity at some velocity comparable to the Alfven speed.

Other examples of the paths of displaced elemental flux bundles in various ambient magnetic fields are available in the literature (Parker, 1981a), including the cases of a flux bundle arching symmetrically over the equator and symmetrically over the pole of a 3D dipole. Flux bundles in isothermal and polytropic atmospheres are described by Parker (1979, pp. 136–151). The path of an elemental flux bundle with an internal gas pressure is provided by equation (7.156) (Parker, 1981a). The result is qualitatively the same in each case, with the equilibrium path extending to infinity when the footpoint separation exceeds some critical value. The essential point is that the apex of a flux bundle ultimately rises at speed of the order of the Alfven speed no matter how slowly the separation of the footpoints is increased.

It is not obvious that the simple concept of a runaway flux bundle is the physical basis for such phenomena as the solar coronal mass ejection (cf. the recent descriptions by Kahler, 1991), for it must be remembered that the displaced flux bundle runs away only as far as it is squeezed by the ambient field, and the available energy is limited to the magnetic field of the bundle and the comparable energy in the deformation of the ambient field by the bundle. The limitations of the displaced flux bundle are displayed by the simple example in which the magnetic arcade, described

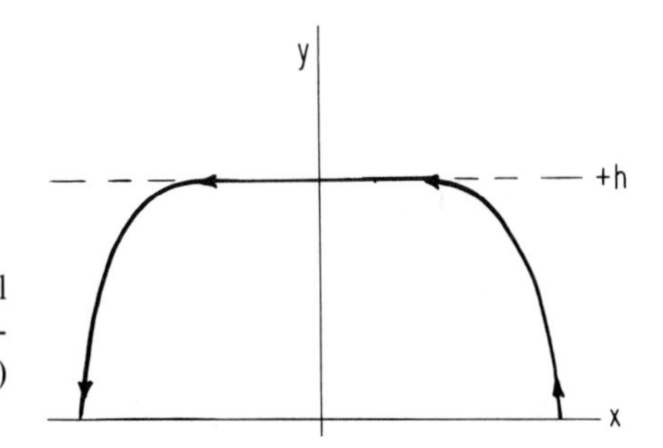

Fig. 9.7. A sketch of the path of an elemental flux bundle with widely separated ($>\pi/k$) footpoints in the ambient field of equations (7.134) and (7.135) in $y < h$ and uniform in $y > h$.

by equations (7.134)–(7.137), is inflated by a uniform gas pressure, which dominates the field above some level $y = h$ so that the field lines are essentially all vertical in $y > h$ and the field magnitude is uniform. In that case the equilibrium paths in $y > h$ are straight lines, while the paths in $0 < y < h$ are given by equation (7.140), with the form shown in Fig. 7.29. The net result is that the apex of the arched equilibrium path rises, with increasing separation of the footpoints at $y = 0$, until the apex reaches $y = h$. Further increase of the footpoint separation serves only to separate the two halves of the arched path, connected at their apices, where they come tangent to $y = h$, by a horizontal line, sketched in Fig. 9.7. In this case the motion of the flux bundle consists only of the horizontal displacement of the two halves of the arch in direct association with the motion of the footpoints.

10
Effects of Resistivity

10.1 Diffusion of a Current Sheet

Theoretical development of the spontaneous formation of tangential discontinuities in the foregoing chapters has been based on the ideal state of vanishing electrical resistivity. The purpose has been to establish and understand the basic theorem of magnetostatics, that the lowest available energy state of all but the most symmetric field topologies involves the formation of surfaces of tangential discontinuity (current sheets). We come now to the effect of a small but nonvanishing resistivity. It is perhaps simplest in principle to think of switching on the resistivity only after the final magnetostatic equilibrium has been achieved in the presence of zero resistivity. Or we may consider the small resistivity as always present but without sensible effect until the thickness $d(t)$ of the developing current sheets falls to some small value after a time t such that $d(t)$ is of the order of $(\eta t)^{\frac{1}{2}}$. After that the resistive diffusion prevents $d(t)$ from further decline and the situation is complicated by the onset of the resistive tearing instability. There is no magnetostatic equilibrium for nonvanishing d, so there is continuing motion of the fluid and field, driven by the Maxwell stresses M_{ij} in the field, striving to reduce d to the ultimate zero necessary for static equilibrium. For that reason the enhanced dissipation arising from the smallness of d (in comparison to the general large-scale ℓ of the global magnetic field) has been called topological dissipation (Parker, 1972), because it is brought about by the global topology of the field. This is, of course, just another way of describing the well-known phenomenon of rapid reconnection (Parker, 1957a, 1963; Sweet, 1958a,b; Vasyliunas, 1975), wherein the magnetic dissipation and the associated line cutting and reconnection progress at a high rate because the magnetic stresses continually drive $\nabla \times \mathbf{B}$ toward unbounded increase in certain surfaces within the field (Parker, 1979, pp. 383–391, 1981a,b, 1983a,b, 1992).

Now it is still an open question whether a finite or an infinite time is required for the Maxwell stresses to provide true mathematical discontinuities in the ideal case of infinite electrical conductivity in complicated field topologies, i.e., whether complete magnetostatic equilibrium is only an asymptotic state (see §§9.1–9.1.3). The introduction of a small resistivity renders the question moot, by limiting the potential tangential discontinuity to a shear layer of small but finite thickness d, achieved in a finite time.

The first task in understanding the role of resistivity is to show the essential nature of the Maxwell stress in maintaining relatively sharp — if not mathematically discontinuous — changes in field direction. This point is central to understanding the physics of the magnetic dissipation that provides such phenomena as the active X-ray corona and the flare on stars like the Sun (Parker, 1981a,b, 1983a,b, 1993a). For the fact is that the remarkable fibril structure of the magnetic field at the photosphere also provides discontinuities (current sheets) in the bipolar fields above. Yet such discontinuities do not result in significant dissipation of magnetic free energy unless they are also required for magnetostatic equilibrium and are, therefore, maintained by the Maxwell stresses in the overall field topology.

To demonstrate this distinction consider three cases (in this and the following two sections) of deformation of an initially uniform magnetic field B_0 extending in the z-direction through an infinitely conducting fluid (Parker, 1993a). In the first instance consider the simple plane shearing motion described by equation (8.13), deforming the uniform field by introducing the velocity shear $\mathbf{v} = \mathbf{e}_y \Omega(x) z$. The result is the transverse field component

$$B(x, \tau) = B_0 \Omega(x) \tau \tag{10.1}$$

in the y-direction after a time τ. Suppose that $\Omega(x)$ is the step function

$$\Omega(x) = \begin{cases} +\omega & \text{for } x > 0 \\ -\omega & \text{for } x < 0, \end{cases} \tag{10.2}$$

so that there is a vortex sheet at $x = 0$ and an electric current density,

$$j(x) = \frac{c}{2\pi} B_0 \omega \tau \delta(x),$$

in the z-direction, where $\delta(x)$ is a Dirac δ function. That is to say, there is a current sheet with surface current density $J = cB_0\omega\tau/2\pi$ as a consequence of the discontinuous jump of $2B_0\omega\tau$ in B across $x = 0$. The field is uniform with magnitude $B_0(1 + \omega^2\tau^2)$ on both sides of $x = 0$, so the field is in static equilibrium.

In the absence of any fluid motions, the subsequent addition of a small resistive diffusion η leads to the decay of B described by

$$\frac{\partial B}{\partial t} = \eta \nabla^2 B \tag{10.3}$$

during which time the field remains in static equilibrium if the fluid is incompressible. Otherwise it would be necessary to use the nonlinear equilibrium diffusion equation for a field in a compressible fluid (Parker, 1979, pp. 90–96). The solution to equation (10.3) for the initial $B = \pm B_0\omega\tau$ is

$$B = B_0 \omega \tau \operatorname{erf}[x/d(t)] \tag{10.4}$$

in terms of the characteristic diffusion length $d(t) = (4\eta t)^{\frac{1}{2}}$ and the standard error function

$$\operatorname{erf}(u) \equiv \frac{2}{\pi^{\frac{1}{2}}} \int_0^u ds \exp(-s^2).$$

The solution is readily verified by differentiation and substitution in to equation (10.3). The current density $j(x, y)$ is

$$\begin{aligned} j &= \frac{c}{4\pi} \frac{\partial B}{\partial x} \\ &= \frac{cB_0\omega\tau}{2\pi^{\frac{3}{2}} d(t)} \exp(-\zeta^2), \end{aligned} \tag{10.5}$$

where $\zeta = x/d(t)$. The gaussian profile is plotted in Fig. 10.1. The resistive dissipation rate per unit volume is $j^2/\sigma = 4\pi\eta j^2/c^2$. Hence the rate per unit area in the yz-plane is

$$\begin{aligned} D(t) &= (4\pi\eta/c^2)\int_{-\infty}^{+\infty} dx j^2(x,t), \\ &= \frac{(B_0\omega\tau)^2\eta^{\frac{1}{2}}}{2\pi^{\frac{3}{2}}t^{\frac{1}{2}}}, \end{aligned} \tag{10.6}$$

declining as $t^{-\frac{1}{2}}$ with the passage of time. The total energy dissipated after a time t is

$$\begin{aligned} \mathcal{W}(t) &= \int_0^t dt' D(t'), \\ &= \left(\frac{4}{\pi^{\frac{1}{2}}}\right)\frac{(B_0\omega\tau)^2}{8\pi} d(t), \end{aligned} \tag{10.7}$$

where $\pm B_0\omega\tau$ represents the initial opposing transverse field components B. The essential point is that the magnetic free energy $(B_0\omega\tau)^2/8\pi$ is dissipated over a characteristic width $d(t)$ after a time t. Therefore, the fraction of the magnetic free energy dissipated in a field of total width $2h$ in the x-direction is of the order of $d(t)/h$. This ratio is generally very small in the time available in most astronomical settings. The solar X-ray corona, for instance, has a resistive diffusion coefficient $\eta \cong 10^3\ \text{cm}^2/\text{sec}$ in fields which probably have internal scales h as small as the characteristic granule size of 500 km. The characteristic magnetic dissipation time, in which $d(t) = h$, proves to be 6×10^{11} sec or 2×10^4 years. On the other hand, in a period of 1 day, say 10^5 sec, it follows that $d(t) = 2 \times 10^4$ cm and $d(t)/h = 4 \times 10^{-4}$, converting a negligible fraction of the magnetic energy into heat.

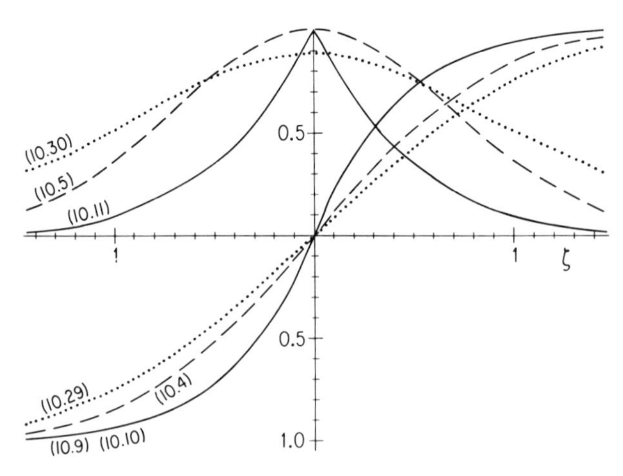

Fig. 10.1. The two dashed curves represent the error function, from equation (10.4), and the gaussian in equation (10.5). The two solid curves provide the profile of B, in the form of the quantity in square brackets on the right-hand sides of equations (10.9) and (10.10), and the field gradient and current density profile, in the form of the quantity in square brackets on the right-hand side of equation (10.11). The two dotted curves represent the profiles of v_x and v_y as a function of y, from equations (10.29) and (10.30).

Anomalous resistivity might enhance the dissipation rate where the electron conduction velocity exceeds the ion thermal velocity of some 200 km/sec ($T \cong 2 \times 10^6$ K), but this condition obtains in a typical transverse field $B_0\omega\tau \cong 25$ gauss only when $d(t) < 6 \times 10^2$ cm, during the first 90 sec. For the rest of the time, no anomalous resistivity is expected.

It is evident that the tangential discontinuity created by discontinuous fluid motion, but not required for magnetostatic equilibrium, provides only passive dissipation, and the rate is too slow to consume more than a tiny fraction of the available magnetic free energy.

10.2 Dissipation with Continuing Shear

The calculations of the previous section illustrate the negligible dissipation provided by the passive diffusion at an initial fixed tangential discontinuity or current sheet. Consider, then, what happens if the discontinuous fluid motion, described by equation (10.2), and producing the discontinuous transverse field **B** in the y-direction described by equation (10.1), continues during the period of resistive dissipation. Then in place of the passive diffusion equation (10.3) the evolution of B is described by

$$\frac{\partial B}{\partial t} - \eta \frac{\partial^2 B}{\partial x^2} = B_0 \Omega(x). \tag{10.8}$$

The solution can be constructed using the Green's function for equation (10.3), with

$$G(x, x' : t, t') = \frac{1}{[4\pi\eta(t-t')]^{\frac{1}{2}}} \exp\left[-\frac{(x-x')^2}{4\eta(t-t')}\right].$$

Then if $\Omega(x)$ is the step function described by equation (10.2), it follows that

$$B(x,t) = \frac{B_0\omega}{\pi^{\frac{1}{2}}} \int_0^t dt' \left\{ \int_0^\infty dx' G(x, x' : t, t') - \int_{-\infty}^0 dx' G(x, x' : t, t') \right\}.$$

Write $w \equiv (x' - x)/[4\eta(t-t')]^{\frac{1}{2}}$ so that for $x > 0$ the expression reduces to

$$\begin{aligned} B(x,t) &= \frac{2B_0\omega}{\pi^{\frac{1}{2}}} \int_0^t dt' \int_0^\xi dw \exp(-w^2), \\ &= B_0\omega \int_0^t dt' \operatorname{erf} \xi, \end{aligned}$$

where $\xi \equiv x/[4\eta(t-t')]^{\frac{1}{2}}$. Then $t - t' = x^2/4\eta\xi^2$ so that for fixed $x, dt' = (x^2/2\eta\xi^3)d\xi$ and

$$B(x,t) = \frac{B_0 x^2}{2\eta} \int_\zeta^\infty \frac{d\xi}{\xi^3} \operatorname{erf} \xi$$

for $x > 0$ and $\zeta \equiv x/d(t)$, with $d(t) = (4\eta t)^{\frac{1}{2}}$. The integral is readily evaluated through integration twice by parts, yielding finally

$$B(x,t) = B_0\omega t[(1 + 2\zeta^2) \operatorname{erf} \zeta + (2\zeta/\pi^{\frac{1}{2}}) \exp(-\zeta^2) - 2\zeta^2] \tag{10.9}$$

for $x > 0$. It is obvious, from the fact that the source term $B_0\Omega(x)$ on the right-hand side of equation (10.8) is an odd function of x, that the ensuing $B(x, t)$ is an odd function of x. Hence in $x < 0$, where $\zeta < 0$, it follows that

$$B(x, t) = B_0\omega t[(1 + 2\zeta^2)\,\mathrm{erf}\,\zeta + (2\zeta/\pi^{\frac{1}{2}})\exp(-\zeta^2) + 2\zeta^2]. \tag{10.10}$$

The form of $B(x, t)$ for $x > 0$ is

$$B(x, t) \cong \frac{4B\omega t}{\pi^{\frac{1}{2}}}\zeta\left[1 - \frac{\pi^{\frac{1}{2}}}{2}\zeta + \frac{1}{3}\zeta^2 - \frac{1}{30}\zeta^4 + \frac{1}{210}\zeta^6 + \ldots\right]$$

where $\zeta \ll 1$, and

$$B(x, t) \sim B\omega t\left[1 - \frac{\exp(-\zeta^2)}{\pi^{\frac{1}{2}}\zeta^3}\left(1 - \frac{3}{\zeta^2} + \ldots\right)\right]$$

for $\zeta \gg 1$. The profile of the field, in the form of the square brackets on the right-hand sides of equations (10.9) and (10.10), is plotted as a function of ζ as one of the solid curves in Fig. 10.1. Note that B deviates from $B\omega t$ only within a characteristic distance of the order of $d(t)$ of $x = 0$.

The current density is proportional to

$$\frac{\partial B}{\partial x} = \frac{4B_0\omega t}{\pi^{\frac{1}{2}}d(t)}[\exp(-\zeta^2) - \pi^{\frac{1}{2}}\zeta\,\mathrm{erf\,c}\zeta] \tag{10.11}$$

for $x > 0$ and is an even function of x, of course. The complementary error function $\mathrm{erf\,c}\zeta$ denotes $1 - \mathrm{erf}\,\zeta$. The quantity in square brackets on the right-hand side of equation (10.11) is plotted as a function of ζ in Fig. 10.1, and it is readily shown that

$$\frac{\partial B_y}{\partial x} \cong \frac{4B\omega t}{\pi^{\frac{1}{2}}d(t)}\left[1 - \pi^{\frac{1}{2}}\zeta + \zeta^2 - \frac{1}{6}\zeta^4 + \frac{1}{30}\zeta^6 + \ldots\right]$$

for $0 < \zeta \ll 1$ and

$$\frac{\partial B_y}{\partial x} \sim \frac{4B\omega t}{\pi^{\frac{1}{2}}d(t)}\frac{\exp(-\zeta^2)}{\zeta^2}\left(1 - \frac{3}{2\zeta^2} + \frac{15}{4\zeta^4} + \ldots\right)$$

for $\zeta \gg 1$. Note then that $\partial B_y/\partial x$ vanishes except in a region of scale $d(t)$ around $x = 0$, where it has a sharp corner at $\zeta = 0$, indicating a singularity in $\partial^3 B/\partial x^3$ of the form $\delta(x)$. Hence, the odd function $\partial^2 B/\partial x^2$ has a finite jump across $\zeta = 0$, arising from the step in the source function on the right-hand side of equation (10.8), of course. The dissipation depends only on $(\partial B/\partial x)^2$, and $\partial B/\partial t$ depends only on $\eta\partial^2/\partial x^2$, both of which are finite as x declines to zero from either side.

The dissipation per unit area of the yz-plane is

$$\begin{aligned} D(t) &= \frac{4\pi\eta}{c^2}\int_{-\infty}^{\infty} dxj^2 \\ &= \frac{8\eta(B_0\omega t)^2}{\pi^2 d(t)}\int_0^{\infty} d\zeta[\exp(-\zeta^2) - \pi^{\frac{1}{2}}\zeta\,\mathrm{erf\,c}\zeta]^2. \end{aligned} \tag{10.12}$$

It is readily shown by integration by parts that

$$\int_0^{\infty} d\zeta\zeta\,\mathrm{erf\,c}\zeta\exp(-\zeta^2) = (2^{\frac{1}{2}} - 1)/2^{\frac{3}{2}},$$

and

$$\int_0^\infty d\zeta\zeta^2 \operatorname{erf} c^2\zeta = \frac{2}{3\pi^{\frac{1}{2}}}\left(1+\frac{5}{2^{\frac{5}{2}}}\right),$$

so that

$$D(t) = \frac{2^{\frac{5}{2}}(7-2^{\frac{1}{2}})}{3\pi^{\frac{3}{2}}}\frac{\eta(B_0\omega t)^2}{d(t)}, \tag{10.13}$$

$$= \frac{2^{\frac{3}{2}}(7-2^{\frac{1}{2}})}{3\pi^{\frac{3}{2}}}\eta^{\frac{1}{2}}B_0^2\omega^2 t^{\frac{3}{2}}. \tag{10.14}$$

The total dissipation per unit area $\mathcal{W}$ over the time t is, then, the time integral of $D(t)$, yielding

$$\begin{aligned}\mathcal{W}(t) &= \frac{2^{\frac{3}{2}}(7-2^{\frac{1}{2}})}{15\pi^{\frac{3}{2}}}(B_0\omega t)^2 d(t),\\ &= 4.754 d(t)(B_0\omega t)^2/8\pi.\end{aligned} \tag{10.15}$$

The transverse field B at $\zeta^2 \gg 1$, far removed from the dissipation, is $B_0\omega t$ and its energy density is $(B_0\omega t)^2/8\pi$. It follows, then, that the dissipation in the neighborhood of $x = 0$ is equivalent to destroying a region of field with total thickness 4.754 $d(t)$. Comparing this result with equation (10.7), where the equivalent width is 2.257 $d(t)$, it is evident that for the same asymptotic field strength at large ζ^2, the resistivity η destroys almost twice as much magnetic free energy when the shear is uniform in time, compared to when the shear ceases before the resistivity is introduced. It follows, therefore that with the small resistive diffusion coefficient η in the hot plasmas in stellar atomospheres and interiors, the passive diffusion treated here provides no significant dissipation of magnetic free energy. The destruction of magnetic free energy is limited to scales $d(t)$ of the order of $(4\eta t)^{\frac{1}{2}}$, which consumes only a thin layer of magnetic field at the location of the discontinuity, as described at the end of §10.1.

10.3 Dynamical Dissipation

In contrast with the passive diffusion treated in the foregoing sections, consider the resistive dissipation in a magnetic field whose topology is sufficiently complex as to require tangential discontinuities for magnetostatic equilibrium. The magnetostatic equilibrium represents the lowest available energy state of the field, so any deviation from it is opposed by the Maxwell stresses, which continually strive to maintain the tangential discontinuities essential for the minimum energy. The presence of resistivity prevents the Maxwell stresses from accomplishing the true discontinuity, of course, so theirs is the task of Sisyphus. But in this case the task is not without interest, for it provides the phenomenon of rapid reconnection in an otherwise essentially permanently connected field topology. The Maxwell stresses keep the thickness of the current sheet to some value ε, rather than the characteristic growing diffusion thickness $d(t) = (4\eta t)^{\frac{1}{2}}$. The thickness ε is generally not more than $\ell/N_L^{\frac{1}{2}}$, and may be much less, in a field with overall characteristic scale ℓ, characteristic

Alfven speed C, and Lundquist number $N_L = \ell C/\eta$. Therefore, complete reconnection of the field across the characteristic scale ℓ occurs in a time of the order of $\ell\varepsilon/\eta$, which is of the order of the Alfven transit time ℓ/C multiplied by $N_L^{\frac{1}{2}}$ or less. This is to be compared to the passive diffusion time $(\ell/C)N_L$. Since N_L is typically 10^{10}–10^{20} in astronomical settings, the difference is a factor of 10^5–10^{10}. Indeed, in special circumstances with the onset of the resistive tearing instability ε may be maintained locally as small as $\ell(\ln N_L)/N_L$, and the time for complete reconnection may be as small as $(\ell/C)\ln N_L$, in order of magnitude. Since $\ln N_L = 50$, the field reconnection and the associated release of magnetic free energy may occur in times that are measured by the Alfven transit time ℓ/C. This is an important possibility when we come to the theory of the solar and stellar flare, and the activity of the magnetic halo of the Galaxy, etc.

The entire phenomenon of resistive diffusion in current sheets that are maintained thin by the Maxwell stresses is called *rapid reconneciton* or *neutral point reconnection* to distinguish it from the passive reconnection over the time $(\ell/C)N_L$. The theory of rapid reconnection is the dynamical extension of the theory of magnetostatic discontinuities required by the introduction of a small resistivity η into the fluid.

Consider, then, the jump $\Delta B = |\mathbf{B}_2 - \mathbf{B}_1|$ in the magnetic field across a quasi-static current sheet of characteristic half thickness ε, where $\mathbf{B}_1$, and $\mathbf{B}_2$ represent the magnetic fields on opposite sides of the current sheet. It follows from Ampere's law that the surface current density J in the current sheet is

$$J = c\Delta B/4\pi$$

so that the characteristic current density is $j \cong J/\varepsilon$ within the sheet. The ohmic dissipation per unit volume is $j^2/\sigma \cong 4\pi\eta j^2/c^2$. Therefore the total dissipation rate $\mathcal{W}$ per unit area of the current sheet is of the order of

$$\mathcal{W} \cong \frac{(\Delta B)^2}{4\pi} \frac{\eta}{\varepsilon} \text{ ergs/cm}^2. \tag{10.16}$$

The free magnetic energy density $(\Delta B)^2/4\pi$ is consumed at the characteristic diffusion speed η/ε cm/sec. The rate increases without bound in the limit of small ε, so the basic problem is to determine the balance between the continuing push by the Maxwell stresses to diminish ε in opposition to the resistive diffusion to increase ε.

As is evident in the sequel, the dynamical balance between the Maxwell stresses, the associated fluid motion, and the resistive diffusion is simple in general concept and remarkably ambiguous in quantitative detail. The theory has been developed over three decades by a number of theoreticians who have run down many of the byways and variations of the process. Their collective efforts have shown that the balance of forces depends critically upon the boundary conditions, and the time evolution depends critically on the initial conditions and on the possibility of anomalous resistivity. There are contradictions and puzzles between the analytical models and the numerical models. In particular, one would consider starting the inquiry with some form of steady-state rapid reconnection. But it can be shown now that there are no stationary states in some cases, while numerical simulations show that the final state may be one of vigorous turbulence and intermittency, as for instance in the intermittent explosive coalescence of the magnetic islands continually formed by the resistive tearing instability.

The reconnection rate is described in terms of the local boundary conditions in the succeeding sections of this chapter. But the theory has not been developed to the point of relating the reconnection rates within an interwoven field to the distant boundary conditions and field topology. Deep in the interior of an interwoven field, the distant boundary conditions are expected to have little, if any, effect. Longcope and Sudan (1992) provide the only attempt at relating the onset of local rapid reconnection to the large-scale topology of the field. They constructed a simple mathematical model of an interwoven field, extending through a fluid from $z = 0$ to $z = L$, truncating the complete representation of the field after the third mode. The system of equations is reduced to tractable form in this way so that they were able to track the evolution of the field by numerical methods as the field was deformed by the slow continuous motion of the footpoints at $z = 0$ and $z = L$. They demonstrate the remarkable fact that with only three modes the system already exhibits the slow build-up and sudden release through local rapid reconnection expected in the exact system described by the infinitely many modes. It shows how fundamental is the formation of current concentrations with increasing winding and interweaving of the field.

This interesting state of affairs means, nonetheless, that no quantitative statements on reconnection rates can be made from the dynamical theory of reconnection. Rather we are forced to work backwards from the magnetic dissipation required to sustain the observed active magnetic phenomenon. As we shall see, the requirements are modest and, therefore, entirely plausible, but without the more satisfying aspect of the purely deductive theory. The following sections form a brief outline and commentary on the dynamical theory of rapid reconnection in its contemporary form.

10.3.1 Basic Concepts

The fascination with the properties of an X-type magnetic neutral point can be traced back to early concerns on the nature of the solar flare. Giovanelli (1947, 1948) spoke of electrical discharges along neutral lines in the magnetic field. Dungey (1953, 1958) fixed attention on the X-type neutral point and the surrounding field, but little further progress was made because the dynamical role of the surrounding field was not recognized. It remained for Sweet (1958a,b, 1969) to put forth the relevant physical environment for rapid reconnection, employing the context of the collision of two magnetic lobes, forming a current sheet where the lobes come into contact. Sweet had in mind the current sheet formed where the bipolar magnetic fields of two solar active regions press together, but of course the precise astronomical setting is immaterial so far as the basic theory of the reconnection process is concerned.

In the language of the present writing, the contact of two distinct topological lobes of magnetic field forms a surface of tangential discontinuity or electric current sheet. Priest and Raadu (1975), Hu and Low (1982), and Low and Hu (1983) provide formal expressions for the magnetic field and for the magnetic free energy under these circumstances in two dimensions. Figure 10.2 is a schematic drawing of the transverse components $\pm B_1$ of the field in the vicinity of the discontinuity. The current sheet is represented as having a finite thickness 2ε, as a consequence of a small but nonvanishing resistive diffusion coefficient η. The width of the current

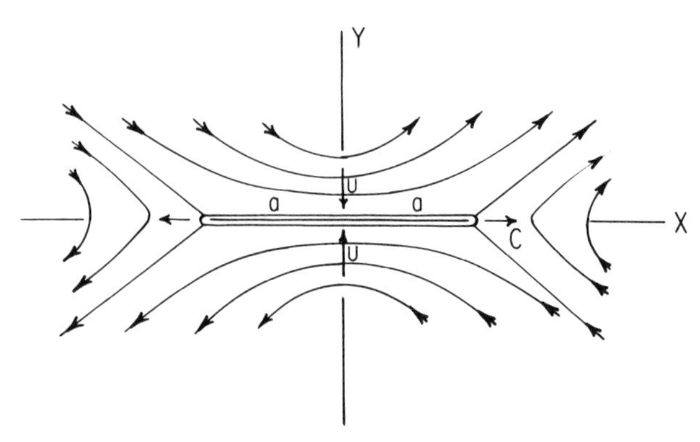

Fig. 10.2. A schematic drawing of the field and current sheet that form where two different topological lobes of field are pushed together. The half thickness ε and half width a of the current sheet are shown, together with the inflow velocity u and the outflow velocity of the order of C.

sheet is denoted by $2a$. The opposite magnetic fields, each of magnitude B_1, press inward against each side of the current sheet with the pressure $B_1^2/8\pi$. Within the current sheet the field varies more or less linearly from $-B_1$ on the lower surface ($y \cong -\varepsilon$) of the current sheet to zero on the centerline ($y = 0$), to $+B_1$ on the upper surface ($y \cong +\varepsilon$). Under quasi-static conditions the total pressure is uniform across the thickness of the current sheet, so that

$$p(x, y) + B^2(x, y)/8\pi \cong p_0 + B_1^2/8\pi \tag{10.17}$$

throughout $-\varepsilon < y < +\varepsilon$ along the y-axis $(-a < x < +a)$. Here $p(x, y)$ is the fluid pressure with $p = p_0$ in B_1 outside the current sheet, and $B(x, y)$ is the magnetic field, with $B(x, \pm\varepsilon) = \pm B_1$. Symmetry decrees that the field vanish on the centerline of the current sheet $B(x, 0) = 0$, with the result that the fluid pressure is elevated by $B_1^2/8\pi$ above the ambient B_0,

$$p(x, 0) - p_0 \cong B_1^2/8\pi. \tag{10.18}$$

Note, then, that the externally applied magnetic pressure falls essentially to zero in the neighborhood of the two Y-type neutral points at each side of the current sheet (at $x = \pm a$). Quasi-static equilibrium requires that $p \cong p_0$ at each end. Hence, along the centerline there is a pressure differential of $B_1^2/8\pi$ between $x = 0$ and $x = \pm a$, expelling the fluid out each end of the sheet with a velocity of the order of the characteristic Alfven speed $C = B_1/(4\pi\rho)^{\frac{1}{2}}$. Assuming an incompressible fluid, this result follows directly from Bernoulli's law applied to the flow along the x-axis,

$$p + \frac{1}{2}\rho v^2 = p_0 + B_1^2/8\pi, \tag{10.19}$$

given that $v = 0$ at $x = 0$ and $p = p_0$ at $x = a$. Now suppose that with the dissipation and ejection of field in the current sheet $-\varepsilon < y < +\varepsilon$, the field is convected slowly and steadily into the current sheet from each side with a velocity u, indicated by the short arrows in Fig. 10.2. The resistivity dissipates the inflowing field into heat within the current sheet of half thickness ε and half width a, where the principal field gradient $\partial B_x/\partial y$ is of the order of B_1/ε. The characteristic resistive diffusion velocity is η/ε, so that under steady conditions, in which the field is dissipated as rapidly as it is convected in at a speed u, we have

$$u \cong \eta/\varepsilon, \tag{10.20}$$

in order of magnitude. This simple result follows directly from the hydromagnetic induction equation $\partial(uB)/\partial y = \eta\partial^2 B/\partial y^2$ and the relation $\partial B/\partial y = O(B_1/\varepsilon)$. The resistive destruction of the field leaves the fluid over much of the thickness 2ε free to be expelled out the ends of the current sheet in response to the pressure difference $B_1^2/8\pi$ between the middle and ends of the sheet. The rate ua at which fluid is carried into each quadrant of the current sheet must be equalled by the rate $C\varepsilon$ of expulsion if there is to be a steady-state, so that the set of equations is completed with

$$ua = C\varepsilon \tag{10.21}$$

for an incompressible fluid. It follows, then, that

$$\varepsilon = a/N_L^{\frac{1}{2}}, \quad u = C/N_L^{\frac{1}{2}}, \tag{10.22}$$

where N_L is the Lundquist number aC/η for the system (Parker, 1957a). The characteristic dissipation time is $(a/C)N_L^{\frac{1}{2}}$, where a/C is the characteristic Alfven crossing time. The active expulsion of fluid by the magnetic pressure $B_1^2/8\pi$ limits the thickness ε of the current sheet to $a/N_L^{\frac{1}{2}}$, whereas with passive diffusion, described in §§10.1 and 10.2, the current sheet thickness would grow to be comparable to a before the magnetic free energy was dissipated. Another way to express the effect is to note that the characteristic dissipation time a/u is $(a/C)N_L^{\frac{1}{2}}$ in terms of the Alfven crossing time a/C, whereas the passive dissipation time is $a^2/\eta = (a/C)N_L$. The characteristic time for the dynamical dissipation driven by $B_1^2/8\pi$, is faster by the factor $N_L^{\frac{1}{2}}$ than the passive dissipation time. With $N_L \sim 10^{10} - 10^{20}$, the gain is by a factor, 10^5–10^{10}, as already noted.

Note for future reference the fundamental role played by the fluid pressure along the centerline of the current sheet. It is the fluid pressure that prevents the two opposite fields from pressing together, i.e., from reducing ε to very small values and providing an explosive dissipation and reconnection rate. The opposite fields can approach each other only as rapidly as the incompressible fluid can be expelled out the ends of the current sheet.

In the 3D world a field component $B_z = B_0$ perpendicular to the xy-plane should be included, along with the transverse field $\pm B_1$. With $\partial/\partial z = 0$ the 2D flow (v, u) has no affect on B_0 and vice versa, so that the result described by equation (10.22) is unaffected by the addition of B_0. However, variations in the third dimension $(\partial/\partial z \neq 0)$ have important effects and are discussed briefly in §10.3.7.

Consider, then, how the 2D field shown in Fig. 10.2 fits into the context of the 3D tangential discontinuity formed by gaps in the flux surfaces of a magnetostatic force-free field, sketched in Fig. 7.4. The 3D form of a tangential discontinuity is sketched in Fig. 10.3, in terms of the gap formed in the region of maximum magnetic pressure where two separate topological lobes of field are pressed together. Typical field lines are indicated by the heavy curves, showing how the field spirals across the surface of the tangential discontinuity in opposite directions on opposite sides. The fields passing around the gap on each side are indicated by heavy curves as well. The essential point is that the 2D field of Fig. 10.2 represents a 2D section across the gap in the field shown in Fig. 10.3. The analysis of the field dynamics is

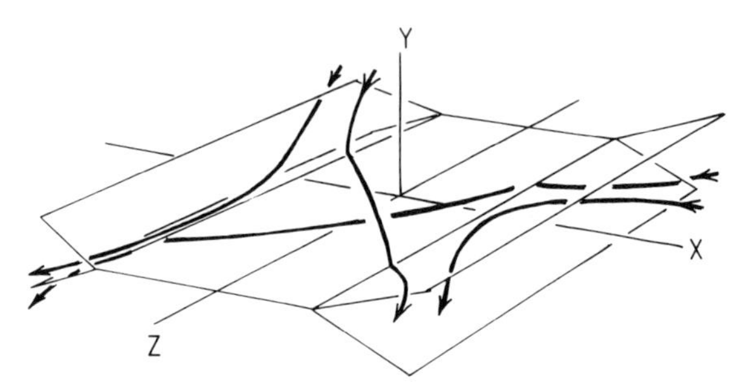

Fig. 10.3. The heavy lines are a schematic representation of the field lines above and below, and at each side of, the surface of tangential discontinuity produced by a local maximum in the magnetic pressure $B^2/8\pi$. The light lines indicate the various leaves of the surface of discontinuity.

carried out in the two transverse (x, y) directions, on the assumption that the z-component of the field is uniform in the x-direction across the width $2a$ of the gap and uniform in the z-direction as well. This idealization captures the essential physics, because the variations of B_z are small enough in many cases that they have no qualitative effect, and the pressure of a uniform B_z is simply to be added to the pressure p of the incompressible fluid employed in the basic treatment of the problem. The third dimension in not without important consequences, but its discussion in §10.3.7 is feasible only after we have examined the 2D case at some length.

10.3.2 Basic Relations

In the 2D $(\partial/\partial z = 0)$ reconnection of magnetic field embedded in an incompressible fluid it is convenient to write the transverse magnetic field as

$$B_x = +\frac{\partial A}{\partial y}, B_y = -\frac{\partial A}{\partial x}, \tag{10.23}$$

and the velocity field as

$$v_x = +\frac{\partial \Psi}{\partial y}, v_y = -\frac{\partial \Psi}{\partial x}, \tag{10.24}$$

to insure their vanishing divergences. Then the vorticity is $\nabla \times \mathbf{v} = -\mathbf{e}_z \nabla^2 \Psi$, where ∇^2 represents the 2D Laplacian operator in x and y. The vorticity equation becomes

$$\frac{\partial}{\partial t}\nabla^2 \Psi = \frac{\partial(A, \nabla^2 A)}{\partial(x, y)} - \frac{\partial(\Psi, \nabla^2 \Psi)}{\partial(x, y)}, \tag{10.25}$$

where $\partial(F, G)/\partial(x, y)$ represents the Jacobian of the transformation $x, y \to F, G$. The magnetic induction equation becomes

$$\frac{\partial A}{\partial t} = \frac{\partial(\Psi, A)}{\partial(x, y)} + \eta \nabla^2 A + cE(t), \tag{10.26}$$

where E is at most a function of time, representing the electric field in the z-direction in the fixed frame of reference induced by the fluid motions. The symmetry of these two equations is obvious, but their general solution is not. The nonlinearity is the principal problem, providing a complexity of field with an enormous sensitivity to initial conditions and boundary conditions that has been partially penetrated only by decades of effort on the part of many theoreticians.

It is easy and useful to examine the one linear aspect of the problem, viz. the state of the magnetic field in the presence of a specified fluid motion. This problem in kinetic diffusion is described by equation (10.26) in which $\Psi(x, y, t)$ is specified. It is instructive, therefore, to examine the field in the simple converging flow described by the stream function

$$\Psi = \frac{2k\eta x[1 - \exp(-k^2y^2)]}{\pi^{\frac{1}{2}}\operatorname{erf}(ky)}, \tag{10.27}$$

so that

$$\begin{aligned} v_x &= \frac{4k\eta kx\exp(-k^2y^2)}{\pi\operatorname{erf}(ky)^2}[\pi^{\frac{1}{2}}ky\operatorname{erf}(ky) + \exp(-k^2y^2) - 1], \\ &\cong k^2\eta x \quad \text{for } k^2y^2 \ll 1, \\ &\cong (4k^2\eta x/\pi^{\frac{1}{2}})ky\exp(-k^2y^2) \text{ for } k^2y^2 \gg 1, \end{aligned} \tag{10.28}$$

and

$$\begin{aligned} v_y &= -\frac{2k\eta[1 - \exp(-k^2y^2)]}{\pi^{\frac{1}{2}}\operatorname{erf}(ky)}, \\ &\cong -k^2\eta y \quad \text{for } k^2y^2 \ll 1, \\ &\cong \mp 2k\eta/\pi^{\frac{1}{2}} \quad k^2y^2 \gg 1. \end{aligned} \tag{10.29}$$

The profiles of v_x and v_y are shown by the dotted lines in Fig. 10.1, and it is readily seen that the flow represents a general inflow from $y = \pm\infty$ toward the x-axis at a speed $2k\eta/\pi^{\frac{1}{2}}$. The two opposite flows meet head-on at $y = 0$ and are deflected so as to split and stream out in both directions along the x-axis. The flow is simple in form and provides an illustration of the fluid motion driven by the Maxwell stresses and maintaining the current sheet in the theoretical development in §10.3.1. The presence of a uniform magnetic field $\pm B_1$ in the x-direction at $y = \pm\infty$, respectively, provides the field

$$B_x(x, y) = B_1\operatorname{erf}(ky), B_y = 0 \tag{10.30}$$

from the induction equation (10.26). Thus k^{-1} represents the characteristic thickness ε of the current sheet, in which the current density is

$$\begin{aligned} j_z &= -\frac{c}{4\pi}\frac{\partial B_x}{\partial y}, \\ &= -\frac{ck}{2\pi^{\frac{3}{2}}}B_1\exp(-k^2y^2). \end{aligned}$$

The total ohmic dissipation per unit area in the xz-plane is

$$\begin{aligned} \mathcal{W} &= \int_{-\infty}^{+\infty} dy j_z^2/\sigma, \\ &= \frac{\eta k B_1^2}{2^{\frac{1}{2}}\pi^{\frac{3}{2}}} \text{ ergs/cm}^2\text{ sec}. \end{aligned} \tag{10.31}$$

The rate at which magnetic field energy is convected inward from $y = \pm\infty$ to the x-axis is

$$\mathcal{E} = 2uB_1^2/8\pi,$$
$$= \mathcal{W}/2^{\frac{1}{2}}\,\text{ergs/cm}^2\,\text{sec},$$

and the rate at which the inflow does work against the magnetic pressure is the same. The total work done is the sum of these two, or $2^{\frac{1}{2}}\mathcal{W}$.

Applying this kinematical flow to a current sheet of width 2a at $y = 0$, extending from $x = -a$ to $x = +a$, set v_x equal to $C \equiv B_1/(4\pi\rho)^{\frac{1}{2}}$ at $x = a$, $y = 0$. Since equation (10.28) gives $v_x = k^2\eta a$, it follows that

$$k^2\eta a = C.$$

Hence,

$$ka = N_L^{\frac{1}{2}}. \tag{10.32}$$

It follows from equation (10.29) that the inflow velocity from $y = \pm\infty$ is

$$u = 2k\eta/\pi^{\frac{1}{2}},$$
$$= 2C/\pi^{\frac{1}{2}}N_L^{\frac{1}{2}}. \tag{10.33}$$

Equations (10.32) and (10.33) are equivalent to the results given by equations (10.22).

The inadequacy of these results for such phenomena as the solar flare is immediately evident. For if the generous values $B_1 = 500$ gauss in a gas density $N = 10^{10}$ H atoms/cm^3 are assumed, the characteristic Alfven speed is 10^9 cm/sec. A temperature of 10^6 K yields the resistive diffusion coefficient η as 5×10^3 cm^2/sec. A current sheet width $a = 500$ km, comparable to the granule radius associated with the convective deformation of the field, yields a Lundquist number $N_L = aC/\eta = 10^{13}$. The Alfven crossing time is $a/C = 5 \times 10^{-2}$ sec. The rapid reconnection velocity u is 3×10^2 cm/sec, and the characteristic reconnection time is $a/u = 1.6 \times 10^5$ sec, or about 2 days, whereas the initial intense flash phase of a flare is only 10^{-3} as long as this. The entire life of a flare is about 10^{-2} of this estimated time. Note that the passive diffusion time is $a^2/4\eta \cong 10^{11}$ sec $\cong 3 \times 10^3$ years.

The large compressibility of the gas enhances the reconnection rate to some degree (Parker, 1963). It was noted in §10.3.1 that the fluid pressure along the centerline of the current sheet plays a crucial role in regulating the reconnection rate. The pressure keeps the opposite fields $\pm B_1$ on either side of the current sheet from collapsing against each other and dissipating explosively. It is evident, then, that if the gas pressure p_0 in the ambient field B_1 is small, and if there is no B_z in the perpendicular direction, then the inflow is faster by the factor $(B_1^2/8\pi p)^{\frac{1}{7}}$ by which the density of the inflowing gas must be compressed to provide the gas pressure equal to $B_1^2/8\pi$ along the centerline. This factor can be as large as 10 or 10^2 (Parker, 1963) but it is still inadequate, and there is no reason to think that the field B_0 in the third ignorable direction is ever less than the transverse field $\pm B_1$. Hence the enhancement of compressibility is probably more like a factor of $2^{\frac{1}{2}}$ or less.

It follows that magnetic reconnection in the regime of the wide current sheet is inadequate to account for the solar flare. There is no known alternative to the idea that the flare arises from rapid magnetic reconnection in some form or other

(Parker, 1957a,b), so the observed solar flare has been the driver for the exploration and development of the theory of dynamical reconnection.

One could at this point appeal to anomalous resistivity, because the current sheet proves to be exceedingly thin in the example just provided, with $\varepsilon \cong 16\,\text{cm}$ (comparable to the ion cyclotron radius) and electron conduction velocities of the order of 10^{10} cm/sec. But historically that was not the order of events. The next step in the theory of reconnection was made by Petschek, who pointed out that the width a of the current sheet need not necessarily be identified with the characteristic scale ℓ of the magnetic inhomogeneities.

10.3.3 Limits of Reconnection

Petschek (1964; see also Petschek and Thorne, 1967) proposed that the width $2a$ of the current sheet formed by the squashing together of two magnetic lboes, each of characteristic scale ℓ, may be very much smaller than ℓ under certain circumstances. Petschek outlined a 2D scenario in which the quasi-static equilibrium outside the current sheet, assumed in the foregoing development, is replaced by slow shock surfaces extending obliquely out from each of the four corners of a narrow central current sheet. He gave plausible estimates of the self-consistent nature of the picture, although where and how the shocks terminate at large distance was not considered, the calculations being confined to a rectangular box around the reconnection region. The essential point is that instead of $a = 0(\ell)$, Petschek obtained a of the order of $\ell(\ln N_L)^2/N_L$ where now $N_L = \ell C/\eta$ and is essentially the same Lundquist number aC/η used in §§10.3.1 and 10.3.2 where a was equated to ℓ. The effective Lundquist number for the current sheet is aC/η, as before, except that a is now exceedingly small compared to ℓ. The reconnection rate follows from equations (10.22) and (10.33) as

$$\begin{aligned} u &= C(\eta/aC)^{\frac{1}{2}}, \\ &\cong C/\ln N_L, \end{aligned} \tag{10.34}$$

in order of magnitude. With $\ln N_L$ of the general order of 30–50, this provides a reconnection velocity of about 200 km/sec, adequate to cut across several thousand km of magnetic field in the characteristic time of 10^2 sec for the initial flash phase of a solar flare (Petschek, 1964; Forbes, 1991).

Subsequent analyses of the Petschek reconnection mode (Fig. 1.2) (Yeh and Axford, 1970; Syrovatsky, 1971; Yeh and Dryer, 1973; Vasyliunas, 1975; Grad, Hu, and Stevens, 1975; Sonnerup and Priest, 1975; Priest and Cowley, 1975; Parker, 1979, pp. 417–423; Priest, 1981; Matthaeus and Lamkin, 1985; Guzdar, et al. 1985; Lee, Fu, and Akasofu, 1987) have shown the complexity and the subtlety of the phenomenon, leaving unanswered the question of how the system of four slow shocks, with the tiny intense current sheet at the center developed in time and how it fits into the quasi-static large-scale fields whose mashing together initiated a current sheet in the first place.

Sonnerup (1970, 1971) and Yang and Sonnerup, (1976, 1977) proposed a precise magnetic field configuration with a point vertex at the center in place of a current sheet, with eight standing Alfven waves (i.e., slow shocks in an incompressible fluid) extending radially out to infinity, providing eight sectors (Fig. 1.3). The system was treated formally and exactly everywhere except at the central point. One imagines simply that the dimensions of the current sheet are so small that any nonvanishing η would carry out the necessary reconnection at the central point. As a result the reconnection speed u is independent of η and can be made arbitrarily large by driving the inflow with sufficient pressure relative to the outflow.

Sonnerup's model would seem to be the ultimate extreme of the ideas of Petschek. Along these same lines Grad (1978) argued that field line cutting arises in the absence of resistivity, and vortex line cutting in the absence of viscosity, as a conseuqnece of infinitely sharp kinks or corners in the lines. But we will have more to say on this later. Coppi and Friedland (1971) developed a model for reconnection that incorporates the resistive tearing instability and permits reconnection at a high rate.

The theory suggests that most any rate of reconnection between the lower limit $C/N_L^{\frac{1}{2}}$ given by equation (10.22) and the upper limit $C/\ln N_L$, where $N_L = \ell C/n$, is possible, depending upon the degree to which the initial conditions and the surrounding quasi-static fields permit achieving the Petschek regime. One might even expect to find more rapid reconnection along the lines suggested by Sonnerup if a sufficient external force to drive the reconnection were present.

10.3.4 Theoretical Developments

Consider the development of the theory of dynamical reconnection of magnetic fields across an X-type neutral point. Unless otherwise specified, the theory is restricted to the two transverse xy-directions with z ignorable, and an inviscid incompressible fluid with small but nonvanishing resistive diffusion coefficient η. The results are manifold and seemingly contradictory in some degree, so that we are obliged to proceed with circumspection.

Vasyliunas (1975) constructed solutions for reconnection in both the Petschek and Sonnerup regimes for prescribed conditions on the boundary of an enclosing rectangle around an X-type neutral point. Comparable estimates of the reconnection rate were obtained by Roberts and Priest (1975) and Soward and Priest (1977). Priest and Forbes (1986, 1989) took a somewhat different approach, beginning with the formal linearization of the momentum and induction equations in ascending powers of the small Alfvenic Mach number u/C. They expanded about the static equilibrium configuration for a uniform field in the x-direction in the presence of a uniform non-negligible fluid pressure p_0. The resulting linearized equations for the field and flow around the neutral point are amenable to solution by separation of variables and superposition of solutions. By specifying different tangential flows at the rectangular boundaries, they construct solutions in the Petschek regime and in the Sonnerup regime for the external flow. However, it is not shown how the flows might be fitted to the narrow current sheet in the Petschek flows. And, of course, in the Sonnerup regime this problem is simply shrunk to a point, as urged by Grad. It is our own view that this avoids the central physics of dynamical reconnection.

It is particularly interesting, then, to see the solutions worked out by Hassam (1990, 1991) in which the tiny central region of resistive diffusion is formally included along with the external flow of fluid and field. But there is the price to be paid that the development is restricted to a cold plasma, with vanishing pressure and unlimited compressibility. Hassam treats a field contained within the rigid circular cylinder $\varpi = a$. He linearizes about the X-type neutral point given by the vector potential $A = B_0(x^2 - y^2)/2a$, for which $\nabla^2 A = 0$. This potential field represents the minimum energy for the field, whose footpoints are fixed in the rigid infinitely conducting boundary at $\varpi = a$. The radial field component is

$$B_\varpi = -\frac{1}{\varpi}\frac{\partial A}{\partial \varphi},$$
$$= +B_0 \frac{\varpi}{a} \sin 2\varphi$$

in polar coordinates (ϖ, φ) and is presumed to be fixed at $\varpi = a$. The magnetic flux crossing $\varpi = a$ in the secton $0 < \varphi < \frac{1}{4}\pi$ connects across the x-axis to the sector $-\frac{1}{4}\pi < \varphi < 0$, while the flux crossing $\varpi = a$ in $\frac{1}{4}\pi < \varphi < \frac{1}{2}\pi$ connects across the y-axis to $\frac{1}{2}\pi < \varphi < \frac{3}{4}\pi$, etc. sketched in Fig. 10.4(a). The field line pattern is invariant to rotations of $\pm\frac{1}{2}\pi$. Hassam then chooses as the initial condition a state in which B_ϖ is the same at $\varpi = a$ but with a slightly altered topology, sketched in Fig. 10.4(b), in which a smaller sector $0 < \varphi < \varphi_1 < \frac{1}{4}\pi$ connects across the x-axis to $-\varphi_1 < \varphi < 0$ and a larger sector connects from $\varphi_1 < \varphi < \frac{1}{2}\pi$ to $\frac{1}{2}\pi < \varphi < \pi - \varphi_1$. This initial field topology can be created by adding a rotationally symmetric azimuthal field $B_\varphi(\varpi)$ to the potential field. The essential point is that with $\partial/\partial\varphi = 0$ the linearized magnetohydrodynamic equations reduce to the ordinary hypergeometric equation in the radial variable ϖ^2. Hence an analytic representation of the relaxation of the initial state to the final minimum energy is possible in terms of the familiar hypergeometric function. With this mathematical foundation Hassam shows the suprising fact that the only solutions are oscillatory, with a frequency of the order of $C/\ell \ln N_L$, decaying slowly at a rate of the order of $C/\ell(\ln N_L)^2$. This seems to be a general property of solutions in which the pressure of a third field component B_z and the pressure of the fluid are omitted. Detailed exploration of this situation by both matched asymptotic solutions and numerical modeling have been carried out by Hassam (1991, 1992), Craig and McClymont (1991), and Craig and

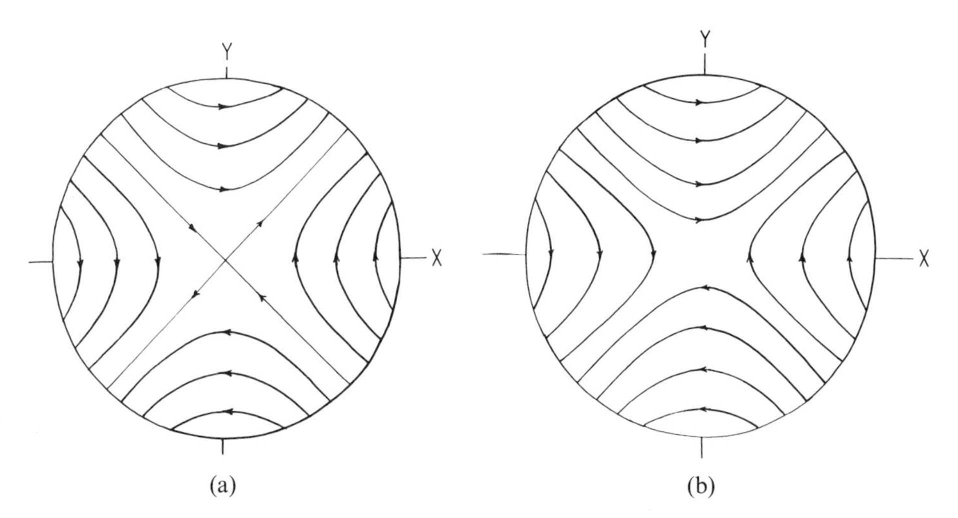

Fig. 10.4. (a) The initial quadrupole field subject to (b) the azimuthal perturbation, employed by Hassam (1991).

Watson (1992). The essential features are, first, the characteristic reconnection time of the order of $(\ell/C)(\ln N_L)^2$, not greatly different from the Petschek estimate of $(\ell/C)\ln N_L$, and, second, the curious fact that the field equations provide only oscillatory decay. The oscillatory nature of the fields may be a consequence of the enforced rotational symmetry ($m = 0$), of the perturbation, with the nonoscillatory solutions perhaps involving an $m = 2$ component as well. More recently Craig and McClymont (1993) have formulated a linear theory of reconnection in which the mechanisms of dissipation are not at all like the Petschek model but for which the characteristic reconnection time is of the order of $(\ell/C)\ln N_L$. Hassam points out that the set of solutions of the form he derives may not be complete, which leaves the question open.

The physical consequences of ignoring the plasma pressure and excluding a uniform field B_z perpendicular to (B_ϖ, B_φ) are both interesting and nontrivial, for it will be recalled from §10.3.1 that the pressure plays the fundamental role of keeping opposite transverse field components apart, thus requiring the expulsion of fluid from the current sheet in order to maintain steep field gradients as resistive diffusion progresses. Parker (1963) pointed out that the more compressible the fluid (i.e., the smaller the plasma pressure compared to the magnetic pressure) the less the volume of fluid to be expelled to maintain the steep field gradients and, consequently, the faster the reconnection (see §10.3.2). The present case of a cold plasma represents the limiting case in which no volume of fluid has to be expelled. An obvious question is whether the general calculation, starting with a nonvanishing pressure would pass to the limit obtained by setting the fluid pressure equal to zero in the first place. Now, the numerical solution of the same linearized equations with the same initial conditions and the same boundary conditions reproduces the slowly decaying oscillatory mode provided by the formal analytic solution of the equations. But the numerical solution subsequently passes smoothly over into a final nonoscillatory exponentially decaying mode that Hassam identifies as the similarity form $\exp[-(C_0^2/\eta a^2)\varpi^2 t]$, where C_0 is the characteristic Alfven speed $B_0/(4\pi\rho)^{\frac{1}{2}}$. This solution obtains when $\partial A/\partial t$ is neglected compared to $\eta\nabla^2 A$. Noting that the field magnitude B is just $B_0\varpi/a$, it follows that the Alfven speed at radial distance ϖ is $C(\varpi) = C_0\varpi/a$, so that the characteristic decay time is

$$\tau(\varpi) = \eta/C^2(\varpi).$$

at the position ϖ. This has the remarkable property that it is proportional to η. The larger the resistivity, the slower the decay. Craig and McClymont (1991, 1993) and Craig and Watson (1992) provide numerical illustrations of the oscillations, showing how the field pulses in and out of the near neighborhood of the neutral point.

In fact numerical solutions of the reconnection problem open up a whole new vista that, in spite of its limitations, shows clearly the limitations of the various analytical constructions currently available. First of all, note that numerical solutions are limited to Lundquist numbers N_L of the order of 10^3, for which $\ln N_L \cong 7$, $(\ln N_L)^2 = 49$, and $N_L^{\frac{1}{2}} = 31.6$, so that it is not always obvious how to distinguish the various regimes. But the numerical solutions have the enormous advantage that they can follow the time development of the reconnection regime. The first point to note is that no steady reconnection mode resembling the Petschek regime has been achieved in a numerical experiment without first introducing a localized region of relatively high resistivity at the ultimate reconnection point (cf. Ugai and Tsuda,

1977, 1979a,b; Ugai, 1985, Biskamp, 1986; Otto, 1990, 1991; Scholer, 1991; De Luca and Craig, 1992). Once the Petschek mode is established, with its slow shocks, the locally enhanced resistivity can be altered and a regime resembling the Petschek mode may continue if the plasma is suitably forced into the region of flow. But Ugai and Tsuda (1979b) found that the X-type neutral point was not always sustained if the resistivity were reduced to a uniform value again (Scholer, 1989, 1991). Thus the Petschek mode seems to depend for its existence on suitable nonuniform initial conditions and suitable boundary conditions. This suggests that the development of anomalous resistivity in the region of maximum current density at the neutral point may perhaps play a crucial role in initiating and maintaining the Petschek mode.

10.3.5 Tearing Instability

In the absence of an initial localized maximum in the resistivity the X-type neutral point flattens into a broad thin current sheet (Biskamp and Welter, 1989), with a relatively slow reconnection rate along the lines described in §10.3.1 and illustrated in Fig. 10.2. If the inflow of magnetic flux toward the current sheet is forced, i.e., driven by externally applied forces, the flux piles up against the current sheet causing the current sheet to grown more intense until some form of the resistive tearing instability takes over (Furth, Killeen, and Rosenbluth, 1963; Spicer, 1977, 1982; Van Hoven, 1976, 1979, 1981; Coppi, 1983). The tearing instability produces magnetic islands (i.e., plasmoids centered on O-type neutral points) with X-type neutral points between. The individual plasmoids (islands) are then expelled outward along the current sheet. Biskamp (1986) showed that the current sheet becomes unstable to the tearing mode when its width exceeds about one hundred times its thickness. Waelbroeck (1989) has shown the development of magnetic islands from the $m = 1$ kink tearing instability.

The essential point is that the tearing instability opens up a whole new scenario of reconnection and dissipation when the theory moves to three dimensions. Tetreault (1992a,b) in particular has shown how the overlapping of islands at different locations in the third dimension leads to stochastic field lines and turbulence, creating rapid dissipation beyond the basic reconnection of the foregoing sections. The purpose of the present section, then, is to outline the role of the tearing instability and the formation of islands and their consequences in two dimensions before presenting a brief description of the much more complicated interactions that arise in three dimensions in §10.3.7.

An important paper by Hahm and Kulsrud (1985) follows through the onset of the tearing instability to the formation of magnetic islands and to the final asymptotic approach to equilibrium in the presence of the islands. The calculation begins with a simple linearly varying field $B_y = B_0 x/a$ with which there is associated the uniform current density $cB_0/4\pi a$ in the z-direction. They assume a uniform magnetic field in the z-direction, and no variations in the z-direction ($\partial/\partial z = 0$) in either the initial state or in the perturbations that follow. The system is set into a dissipative state by squeezing the field locally in the x-direction, thereby expelling fluid along the field in the y-direction form the compressed region, creating a steep field gradient ($\partial/\partial x$) where the opposite fields on opposite sides of $x = 0$ press together. They identify three subsequent phases of development of the resistive tearing mode. The first phase involves the growth and increasing sharpness of the perturbation current, with the characteristic thickness declining inversely with the passage of time as the fields are driven together. This provides the resistive reconnection of the field at the developing X-type neutral points. The theory of Furth, Killeen, and Rosenbluth (1963) describes the process, and Hahm and Kulsrud (1985) provide an explicit time dependent solution for the growth of islands, valid for times of the order of $\tau_R^{\frac{1}{3}}\tau_A^{\frac{2}{3}}$, where τ_R is the characteristic resistive diffusion time $4\pi a^2/\eta$ while τ_A is the characteristic Alfven transit time a/C, with $C = B_0/(4\pi\rho)^{\frac{1}{2}}$. The final relaxation after a time $O(\tau_R)$ to an equilibrium island configuration occurs in the characteristic tearing mode time scale, $\tau_R^{\frac{3}{5}}\tau_A^{\frac{2}{5}}$. The island width δ in this phase is large compared to the resistive layer thickness. The evolution becomes nonlinear, with $\delta/a(\tau_A/\tau_R)^{\frac{4}{5}}$, and Rutherford's (1973) approach becomes appropriate as the system settles down asymptotically into its final state.

One can only imagine how this ideal evolution in the presence of fixed boundaries might behave in the absence of confining and controlling boundaries, in a large-scale magnetic field that is itself evolving slowly in time as a consequence of reconnection. One or more islands may become dominant while others are suppressed, and the final dynamical reconnection state is a matter of conjecture.

Waelbroeck (1989) and Wang and Bhattacharjee (1992) examine the final nonlinear phase in more detail, along the lines developed by Rosenbluth, Dagazian and Rutherford (1973). They find that beyond the calculations of Hahm and Kulsrud (1985) current sheets develop between islands when reconnection is forced, in the characteristic time scale $\tau_R^{\frac{1}{2}}\tau_A^{\frac{1}{2}}$, given by equation (10.22), followed by an asymptotic approach to the final state of smooth islands, first described by Rutherford (1973).

The work of Lee and Fu (1986a–c) carries on from there, examining the reconnection at multiple X-points. They show, among other things, that in such circumstances only about one-eighth of the magnetic energy released by the reconnection is converted into kinetic energy of the fluid motion. That is to say, most of the energy goes directly into heat, through resistive dissipation, contrary to the Petschek mode, where most of the magnetic energy released by the reconnection goes into accelerating the fluid. They found, too, that, contrary to the Petschek regime, the thickness of the current sheet does not vary much with increasing reconneciton rate, whereas the width increases somewhat with increasing rate. Evidently the flux pile up when the system is driven to the higher reconneciton rates forces the field more rapidly through the diffusion region where the reconnection occurs. Forbes and Priest (1987) suggest that the flux pile up may contribute to the unstable bursty character of the reconnection in many cases.

The formation of islands and multiple X-points leads eventually to unstable coalescence of neighboring islands through reconnection (Finn and Kaw, 1977). The evolution of islands has been followed by several investigators (Dickman, Morse, and Nielson, 1969; Drake and Lee, 1977; White, et al. 1977). Tajima and Sakai (1986, 1989a,b) and Sakai and de Jager (1991) emphasize the runaway explosive aspect of the coalescence, arising from the small Lundquist number of the newly developed individual islands. The coalescence of individual pairs of islands probably contributes to the bursty nature of reconnection at high overall Lundquist number. Indeed, it may be this phenomenon that helps to determine the threshold for strong bursts of reconnection and the associated current disruption in Tokamaks.

Both laboratory experience and recent numerical studies of magnetic reconnection show intermittent and bursty reconnection, attributed to a variety of effects.

Hence, one expects that such transient and chaotic reconnection may be the rule in many astronomical settings, with steady reconnection only a special, and pedagogically useful, idealization. Plasmas confined under high β conditions in the terrestrial laboratory (β = gas pressure/magnetic pressure) carry strong electric currents and are observed to undergo intermittent bursts of magnetohydrodynamic activity, evidently initiated by the onset of various internal instabilities (Rosenbluth, Dagazian, and Rutherford, 1973; Kadomtsev, 1975, 1984; Waddell, et al. 1976; Finn and Kaw, 1977; Bhattacharjee, Dewar, and Monticello, 1980; Montgomery, 1982; Ryopoulos, Bondeson, and Montgomery, 1982; Dahlburg, et al. 1986).

It is interesting in this connection to note the recent work of Jin and Ip (1991) on compressible reconnection driven by an inflow imposed at the boundaries. They find that the reconnection is always bursty. Along the same lines Matthaeus and Montgomery (1981) and Scholer (1989) show that reconnection initiated by a low level of random noise leads to the development of magnetohydrodynamic turbulence in the current sheet, with the reconnection rate controlled by the turbulence. The numerical experiments of Forbes and Priest (1982, 1987) and Matthaeus and Lamkin (1985) show a generally turbulent and bursty reconnection. Dahlburg, Antiochos, and Zang (1992) demonstrate the successive stages of development of the tearing instability into turbulence in 3D simulations. Strauss (1988) shows the development of turbulence from the tearing mode instability and the final result that the Alfven Mach number u/C of the merging velocity is proportional to $\langle(\Delta B)^2\rangle^{\frac{1}{2}}/\langle B\rangle$, where $\langle(\Delta B)^2\rangle$ is the mean square magnetic field fluctuation associated with the turbulence. The effect is that of hyperresistivity, in which the turbulent fluctuation $\langle(\Delta B)^2\rangle^{\frac{1}{2}}$ across a small-scale λ produces a local tearing instability with wave number k. The net contribution to the mean resistive diffusion coefficient is of the order of $\langle(\Delta B)^2\rangle C/k^2\langle B^2\rangle\lambda\,\mathrm{cm}^2/\mathrm{sec}$, where $k\lambda \ll 1$.

These various numerical experiments collectively show the many different states of slow and fast, steady and nonsteady modes, and regimes of neutral point reconnection in two dimensions. They show how many different initial conditions, (with free or driven flows, in magnetic fields that are free or are tied at the boundaries of the region of computation) in the presence of uniform or locally maximized resistive diffusion develop with the passage of time into various states of reconnection that are generally turbulent and sometimes intermittent. Sparks and Van Hoven (1988) provide a summary of the additional effects of thermal conduction, radiation, and resistive heating that contribute to a variety of instabilities not yet included in the numerical experiments. Forbes and Malherbe (1991) consider the formation of condensations of gas in the presence of reconnection and radiative cooling.

In summary, it appears that rapid neutral point reconnection is a complicated, highly sensitive, dynamical phenomenon providing reconnection and dissipation of magnetic fields at mean speeds u somewhere in the interval

$$N_L^{-\frac{1}{2}} < u/C < (\ell n N_L)^{-1}, \tag{10.35}$$

depending upon the initial configuration, the external conditions as reflected in the field and flow at the boundaries, the distribution of resistivity across the region, and the variations of these with the passage of time. Thus, for instance, Sakai and Washimi (1982), Sakai (1983a,b) and Sakai, Tajima, and Brunel (1984a,b) show numerical simulations of the triggering of rapid reconnection by the passage of magnetohydrodynamic waves. The combination of the many different regimes of reconnection with the sensitivity to the local environment precludes any quantitative theoretical prediction of reconnection rates in natural settings, such as the geomagnetic field, the active magnetic fields on the Sun, the magnetic halo of the Galaxy, etc. The principle suggested by the current disruptions in Tokamaks and other laboratory devices is simply that the reconnection proceeds slowly until some threshold $\Delta B/B$ in the field deformation is exceeded, whereupon there is a burst of reconnection that is soon quenched when $\Delta B/B$ is suitably reduced. This is consistent with the intermittent bursts and the turbulence exhibited by numerical simulations with Lundquist numbers up to 10^3, and may be presumed to be the character of reconnection at the extreme Lundquist numbers of 10^{10}–10^{20} appropriate for astronomical settings. Observation will have to be our guide in the astronomical setting. Perhaps one day something more definitive can be said, but the complexity suggests otherwise at the moment. Rapid reconnection is simply a powerful but capricious dynamical phenomenon.

The review papers by Priest (1985) and Scholer (1991), among others, are recommended to the reader desiring a deeper understanding of the many theoretical problems mentioned so briefly here.

10.3.6 The Complementary Approach

The theoretical development of the universal nanoflare has been carried out in this monograph beginning with an ideal infinitely conducting fluid, which leads to the basic theorem of magneotstatics that almost all magnetic field topologies form internal tangential discontinuities when allowed to relax to static equilibrium. The discontinuities then provide for rapid reconnection and dissipation of the free energy of the field when a small resistivity is introduced. In the first presentation of the basic theorem (Parker, 1972) a different formulation was employed, working with a fluid with a small resistivity rather than vanishing resistivity. In that case the field could not achieve complete equilibrium because the current sheets could not shrink to the vanishing thickness required for static equilibrium. That is to say, the current sheets were always in a state of internal rapid reconnection. Hence it was stated that any but the simplest field topologies had no equilibrium, because the field is in a state of dynamical nonequilibrium, i.e., rapid reconnection, if it has any but the simplest topologies.

The disadvantage of the earlier resistive nonequilibrium approach is that it obscures the basic theorem of magnetostatics as a cornerstone of the concept of the universal nanoflare. Hence the present writing undertakes the development with an ideal infinitely conducting fluid. But once the concepts are clear, the other approach can be revisited with advantage. In particular, recent work has shown that the slightly resistive fluid is subject to current driven instabilities that are not available in the ideal fluid. The point is that some of those resistive instabilities introduce thin current sheets that provide for the formation of magnetic islands, i.e., rapid reconnection and nanoflaring. Thus the presence of a small resistivity provides additional paths to nanoflares, besides the intrinsic tangential discontinuities that form in the field in an ideal fluid. Strauss, Bhattacharjee, and others have demonstrated the importance of this complementary approach in their work already cited.

Park, Monticello, and White (1983) studied the onset of the $m = 1$ spiral kink mode, showing the development of current sheets as the twisted portion of the magnetic flux began to kink and buckle against the surrounding field, in the nonlinear phase. The current sheets form around the regions of high pressure, according to the optical analogy, where the twisted flux bulges against the surrounding field. Otani and Strauss (1988) have picked up on this scenario, treating a twisted field with axial symmetry extending between rigid end plates. They show that the presence of a small resistivity provides an unstable ballooning mode with a growth rate proportional to the small resistive diffusion coefficient η. For large resistivity the rate is proportional to $\ell^{\frac{1}{3}}$. There are unstable modes in the ideal case $\eta = 0$ for nonaxisymmetric fields, of course, but they require a more strongly twisted field. Thus the resistivity provides for kinking and ballooning, and the ultimate formation of current sheets, under circumstances where none might otherwise occur in the ideal case $\eta = 0$. To be precise, with line tying at the end plates the $m = 1$ current driven, resistive, kink ballooning mode is the strongest instability arising when the twisting exceeds one full revolution along the length of the flux bundle. They estimate an onset time of the order of 10^3 sec under typical solar conditions, which is comparable to the preflare development inferred from observations of flares on the Sun.

Strauss and Otani (1988) go on to apply the kink ballooning mode to the formation of current sheets in the solar corona, showing that the resistive dissipation of magnetic energy is enhanced by a factor of at least $N_L^{\frac{1}{2}}$ with the onset of rapid reconnection across the current sheet. They suggest that the principal coronal heating may arise directly from the kink ballooning mode in twisted flux bundles, rather than from the general interlacing of field lines by random footpoint motions employed in the present exposition. It remains for observations with sufficient spatial resolution to determine the relative importance of rotation of the individual magnetic fibrils and the random walk of the individual magnetic fibrils.

It will be recalled that Sturrock and Uchida (1981) discussed rotation of flux bundles, but without invoking the kink ballooning mode to provide enhanced dissipation. The fact should not be overlooked that the criterion for the onset of the ballooning mode is that the twisting exceed one revolution along the entire length of the tube (between fixed end plates). Thus, for the same tube radius, the short flux bundle in a small bipolar region is more strongly twisted than the long flux bundle in a large bipolar region. The rate of energy input for a given rate of revolution of either end of flux bundle is proportional to the torque, and hence to the degree of twisting. So it is inversely porportional to the length at the time of onset. It remains to be shown that the energy input rate per unit area is approximately independent of bundle length in some final statistical equilibrium state to match the observation that the surface brightness of the X-ray corona is more or less independent of length. This is a difficult question, lying in detailed kinking of a flux bundle with many full revolutions of twisting along its length.

Bhattacharjee and Wang (1991), building on earlier work of Bhattacharjee, Dewar, and Monticello (1980) and Waelbroeck (1989), pursue the formation of current sheets through the conservation of magnetic helicity in the presence of a small resistivity (Taylor, 1974). In particular, they treat an initial layer of electric current of finite thickness extending between two regions of oppositely directed field ($y < 0, y > 0$). The resistive tearing instability produces the usual chain of magnetic islands along the separatrix ($y = 0$) between the two regions of opposite field, with X-type neutral points between each island, as described at the beginning of §10.3.5.

Suppose, then, that the region is stretched in the x-direction, increasing the length of the chain of islands. Conservation of helicity requires that the area of each island is conserved and the proportion of length to width does not change, with the result that the length of each island remains about the same. Hence the islands separate, converting each X-type neutral point into two Y-type neutral points with a current sheet between. Wang and Bhattacharjee (1992) apply the concept to the specific model studied by Hahm and Kulsrud (1985), where an initially uniform current density, for $B_x = B_0 x/a$, is pinched to form a sequence of islands, as described in §10.3.5, ultimately applying the result to the dissipation of current sheets in the solar corona.

The essential point is that there are many roads to the final conclusion that nanoflares arise universally in continuously deformed magnetic fields, providing dissipation of magnetic energy at rates enormously greater than the characteristic η/ℓ^2 for a field with overall scale ℓ.

10.3.7 The Third Dimension

It should be appreciated that the presentation so far has been based largely on the analytical and numerical developments in 2D reconnection. In nature there is always the third dimension, which can only add more degrees of freedom to the complexity of the phenomenon. The third dimension z introduces the variation of the current sheet and the longitudinal field B_z. It also permits the study of the 3D null point (at which all three field components vanish) and the topology of the field line connections between null points (Greene, 1988; Lau and Finn, 1990, 1991; Bhattacharjee and Wang, 1991).

The third dimension is an essential feature in the formation of plasmoids, which are part of the theory of the geomagnetic substorm and the dynamics of the geomagnetic tail (Birn, Hesse, and Schindler, 1989; Lau and Finn, 1991; Strauss, 1991). Plasmoids are a fascinating subject which has yet to be explored fully in regard to the dynamics of the reconnection and the propagation of the topological changes into other regions of the field. The topological changes, along with shock waves, etc., may well be the basis for the sympathetic flaring that is often observed at widely separated positions on the Sun.

As already noted, Tetreault (1990, 1991, 1992a,b) has considered the consequences of the reconnection and coalescence of magnetic islands in the third dimension along the field, leading to a new concept of the diffusion and dissipation of large-scale fields. The starting point is a current sheet that is unstable to a resistive tearing mode, forming magnetic islands. The individual island extends only a short distance in the third dimension, where it meets other islands formed farther along. The islands are not correlated, so they generally overlap each other where their ends meet. The islands reconnect rapidly where they meet, with the result that the individual field lines become stochastic, jogging randomly one way and another as they pass through successive islands. This reorganizes the field into a stochastic topology, which generally requires tangential discontinuities as part of static equilibrium, producing further tearing instability and more magnetic islands. The chaotic state of the field is treated by methods already developed to handle the hole/clump instability (Tetreault, 1983, 1988a,b, 1989), described briefly in §2.4.3. The net result is not unlike an anomalous resistivity, although it appears as a bursty, intermittent effect.

Tetreault suggests that it plays an important role in the flux transfer events between the interplanetary magnetic field and the sunward magnetopause. It is undoubtedly a major player in the reconnection of magnetic fields across X-type neutral points and across broad current sheets, described in the foregoing sections. It is an open question whether the effect can be incorporated into the existing theory in any precise quantitative manner. It joins the many other important but subtle and difficult effects already mentioned (see also Spicer, 1990).

Similon and Sudan (1989) have pointed out that the stochastic field expected in most astrophysical settings for a variety of reasons (cf. Jokipii and Parker 1968a,b, 1969a,b; Jokipii, 1973, 1975; Tetreault, 1992a,b) has the effect of scrambling any passing Alfven waves so that the waves are then dissipated by phase mixing. The waves themselves may trigger reconneciton at the current sheets, of course (Sakai and Washimi, 1982; Sakai, 1983a,b; Sakai, Tajima, and Brunel, 1984a,b) further complicating the picture and enhancing the dissipation of both the passing waves and the magnetic free energy of the ambient field. Lichtenberg (1984) showed how stochastic field lines in toroidal (Tokamak) geometry lead to excitation of the sawtooth oscillations in the $m = 1$ mode, leading to disruption of the initial field and plasma configuration.

10.4 Nonuniform Resistivity

The resistivity η is generally not uniform in either laboratory plasmas or in most astronomical settings, and subsequent resistive diffusion with any substantial variation of η with position generally introduces changes in field topology beyond the familiar reconnection at X-type neutral points. For instance, there may be enormous local variation in the effective η as a consequence of anomalous resistivity (see §2.4.3), hyperresistivity (Strauss, 1988), and the overlapping of magnetic islands and the general stochastic wandering of field lines (Tetreault, 1991, 1992a,b). It is shown in §10.5 that a local maximum in anomalous resistivity produces mutual wrapping of flux bundles that are otherwise not entangled in any way (Parker, 1993b).

Then there is the large-scale variation of the ordinary molecular resistivity associated with temperature gradients through the plasma. An example is the general increase of temperature and reduction of resistivity from the surface of a star, where $T \sim 6 \times 10^3$ K and $\eta \cong 10^9$ cm^2/sec to $T \sim 2 \times 10^6$ K and $\eta \cong 10^3$ cm^2/sec in the corona a few thousand km above the photosphere. Hence a magnetic field extending up into the corona experiences diffusion rates that vary by a factor of the order of 10^6 over a few thousand km. Diffusion that is negligible in the corona may be overwhelming in the photosphere (Parker, 1990). For it must be remembered from §8.5 that the tangential discontinuities in the corona, produced by the intermixing of the footpoints of the field at the photosphere, generally extend all the way down to the level in the photosphere where the dense fluid overpowers the field and the concept of quasi-static equilibrium disappears. However, the high resistivity at the photosphere does not permit a thin shear layer or thin current sheet. That is to say, the thick shear layers in the field in the photosphere propagate up into the corona as Alfven waves where they compete against the Maxwell stresses endeavoring to form the extremely thin intense shear layers required for magnetostatic equilibrium. Indeed, if the photospheric and chromospheric resistivity were large enough, one could imagine that the photospheric convection would not swirl the field around to provide the basic interweaving of the field in the corona above. A simple illustration of this extreme case is provided in the literature (Parker, 1983c) where it is shown how a twisted flux bundle unwinds when a thin sheet of high resistivity cuts across the flux bundle. The fact is that the resistive diffusion coefficient η is of the order of $10^9\ \mathrm{cm^2/sec}$, and this is more than adequate to guarantee that the individual flux bundles maintain their integrity during the characteristic periods of 10^3 sec that they are swirled around each other in the solar granulation.

The essential fact is that the photospheric $\eta = 10^9\ \mathrm{cm^2/sec}$ is 10^6 times larger than the resistive diffusion coefficient of $\eta \cong 10^3\ \mathrm{cm^2/sec}$ in the corona, so that the thickness $(4\eta t)^{\frac{1}{2}}$ of a current sheet in the photosphere after a given time t is 10^3 times larger than in the corona. The thickness of the photospheric current sheet propagates as an Alfven wave up into the corona, where it causes the current sheets to become so thick as to halt any significant diffusion in the relatively small coronal resistivity. The Alfvenic communication between photosphere, chromosphere, and corona is not instantaenous, of course. The dashed curve in Fig. 10.5 shows the characteristic Alfven speed as a funciton of height in a uniform vertical field B_0 of 10^2 gauss, increasing from 0.6 km/sec at the visible surface ($z = 0$) where $\rho \cong 2 \times 10^{-7}\ \mathrm{gm/cm^3}$ to about 2×10^8 cm/sec in the active corona where $N \cong 10^{10}\ \mathrm{atoms/cm^3}$ and $\rho \cong 1.6 \times 10^{-14}\ \mathrm{gm/cm^3}$. The two solid lines in Fig. 10.5 indicate the Alfven transit times upward from the visible surface and downward from a height of 1200 km in the chromosphere. Note, then, that the transit time from the photosphere to the corona is of the order of 600 sec, as a consequence mainly of the low Alfven speed below 1200 km. The Alfven transit time above 1200 km is of the order of 50 sec. The characteristic time in which the Maxwell stresses in the corona reduce the thickness of a shear layer or current sheet is of the order of the Alfven transit time in the transverse field component across the width a of the current sheet. If the transverse field $B_\perp$ is 25 gauss and the sheet width is 500 km, the characteristic transit time in $B_\perp$ is then of the order of 1 sec. So there is rapid reduction of the

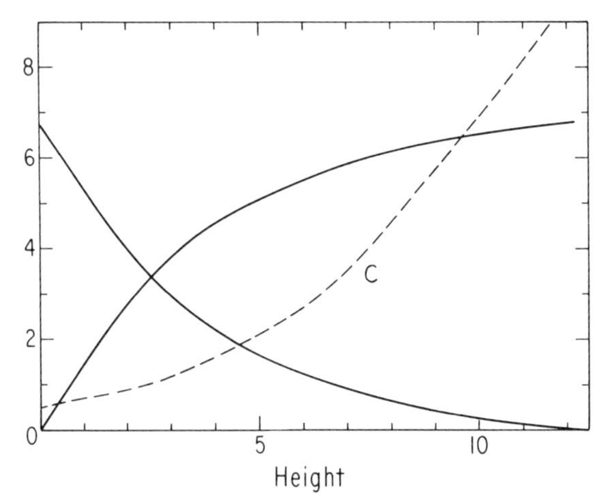

Fig. 10.5. The dashed curve is a plot of Alfven speed (km/sec) C in a field $B_0 = 10^2$ gauss as a function of height (in units of 10^7 cm) above the visible surface of the Sun. The two solid curves represent the corresponding Alfven transient times (in units of 10^2 sec) upward from the visible surface and downward from a height of 1200 km.

current sheet thickness by the Maxwell stress in the corona. But even so, the photospheric and chromospheric resistivity, being so much larger under quiescent circumstances, would severely limit resistive dissipation in the corona if the current sheets formed vertical plane or cylindrical surfaces. That this is not the case follows from the two facts that (a) anomalous resistivity is expected in the low electron densities in the corona, providing bursts of rapid reconnection in the corona, and (b) the general interweaving of the fields from one pattern to another in the corona effectively blocks the extension of the large current sheet thicknesses in the photosphere and chromosphere upward into the corona. Hence, the quantitative models based on plane or cylindrical current sheets establish the nature of the communication of current sheet thickness only in the first swirl of the field lines in the first winding pattern immediately above the chromosphere, overlooking the strong attenuation of the effects in passing through further winding patterns. The attenuation appears even in plane current sheets, as a consequence of the fanning out of the sheet with distance from the location of its creation, treated in §10.4.1.

The theoretical exploration of this complex situation proceeds a step at a time, treating the conditions observed and inferred for the Sun as the basic paradigm for stars in general. It will become apparent that the quantitative information that is available only from detailed observation of the Sun is essential for establishing any scientific opinion on the matter.

Starting at the lowest visible level, consider the fibril state of the magnetic field in the photosphere of the Sun. The individual fibrils of $1\text{–}2 \times 10^3$ gauss have diameters of the order 2×10^2 km and are generally separated by greater distances from their nearest neighbors. It is the independent motion of the individual fibrils, convected about in the intergranular lanes, that winds and interweaves the field lines in the bipolar fields of active regions on the Sun.

Now the characteristic resistive diffusion time τ_f across the fibril radius $r_f \cong 10^2$ km is $r_f^2/4\eta \cong 2.5 \times 10^4$ sec, which is 25 times longer than the estimated swirling time of about 10^3 sec for the fibrils carried in the photospheric convection. The individual fibrils are presumably individually maintained by a radial inward diffusion of fluid at a speed of the order of $4\eta/r_f \cong 4 \times 10^2$ cm/sec, with the inflowing gas cascading downward inside the fibril into the subsurface layers. It is unfortunate that the fibrils lie below the limit of resolution of ground based telescopes at the present time. It would be interesting and scientifically important to confirm these elementary inferences with direct observation. The essential point is that the individual fibrils, or flux bundles, maintain their integrity sufficiently to enforce the interweaving of the field in the corona as the fibrils are carried about in the photosphere.

The separate vertical photospheric fibrils expand to fill all the available volume at chromosperhic levels (see the model calculations provided by Kopp and Kuperus, 1968; Gabriel, 1976; Athay, 1981). So the dissipation at the current sheets between the continuous fibrils begins only in the chromosphere. On the other hand, there is every reason to believe that fibrils are continually evolving, transferring and exchanging magnetic flux with neighboring fibrils (cf. Dunn and Zirker, 1973; Simon, et al. 1988; Title, et al. 1989). So, as already noted, some current sheets are expected within the individual fibrils at the photospheric level. However, the principal resistive limitation on the thickness of coronal current sheets comes from the resistivity of the chromosphere, because the Alfven transit time to the corona is so much less, about one-tenth of the transit time from the photosphere.

The characteristic chromospheric temperature $T_{chr} \cong 7 \times 10^3$ K provides a resistive diffusion coefficient $\eta_{chr} \cong 10^7$ cm^2/sec where the number density is $N_{chr} \cong 3 \times 10^{12}$/cm^3 and the Alfven speed is $C_{chr} \cong 10^7$ cm/sec in a mean vertical field of 10^2 gauss. The surfaces of tangential discontinuity are created by the winding and interweaving of the field through a sequence of topological winding patterns in the corona where $T_{cor} \cong 2 \times 10^6$ K, $\eta_{cor} \cong 10^3$ cm^2/sec, $N_{cor} \cong 10^{10}$/cm^3 and the Alfven speed is $C_{cor} \cong 2 \times 10^8$ cm/sec in the same vertical mean field $B_0 \cong 10^2$ gauss. So it is from this chromospheric state of the field that the upward communication of thick current sheets begins. It is essential to understand the several diverse effects that make up the whole picture. We begin with a plane uniform current sheet in the simplest case of inhomogeneous resistivity.

10.4.1 Quasi-static Plane Current Sheet

Consider the idealized scenario in which a magnetic field $(0, B, B_0)$ of the form described by equations (10.1) and (10.2) extends obliquely between the rigid infinitely conducting boundaries $z = 0$ and $z = L$ at an angle $\pm\theta$ to the z-direction through an infinitely conducting fluid. With the inclination $+\theta$, where $\tan\theta = B/B_0$, in $x > 0$, and with inclination $-\theta$ in $x < 0$, there is a tangential discontinuity and current sheet at $x = 0$, between the two regions of opposite obliquity. The field configuration is indicated in Fig. 10.6 by the two oblique field lines BGK in $x > 0$ and CFJ in $x < 0$. Hold the fluid fixed and at time $t = 0$ switch on a uniform resistivity η in the thin layer $O < z < b$, while preserving the infinite conductivity in $b < z < L$. After a time t the transverse field component B spreads out in the manner described by equation (10.4), with the characteristic length $d(t) = (4\eta t)^{\frac{1}{2}}$ for the smooth transition from $+B_0\omega\tau$ in $x > 0$ to $-B_0\omega\tau$ in $x < 0$. After a sufficiently long time the transverse component approximates to zero in any interval $(-\delta < x < +\delta)$ for $4\eta t > > \delta$, and the field is in the z-direction across $0 < z < b$, indicated by the dashed lines in Fig. 10.6. Thus BGK becomes DGK and CFJ becomes AFJ. Then turn off the resistivity and release the fluid so that the field can relax again to the lowest available energy state, which is with the straight field lines

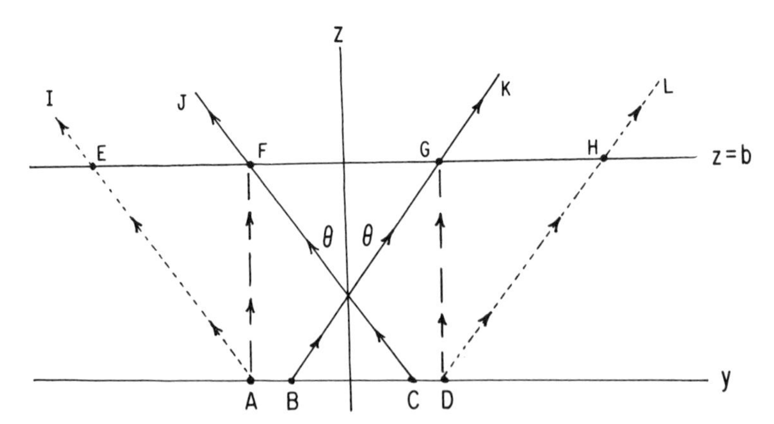

Fig. 10.6. Schematic drawing of two initial oblique field lines BGK and CFJ on opposite sides of the current sheet, going over into DGK and AFJ, respectively following diffusion, and into DHL and AEI, respectively, when magnetostatic equilibrium is again established.

DHL and AEI, indicated by the dotted lines in Fig. 10.6. With $L \gg b$ the net result is that the initial field lines BGK and CFJ (identifiable in $z > b$ where the resistivity vanishes) have each been displaced a distance $b \tan\theta$. The final field has an obliquity θ reduced by the small amount $\delta\theta = (b/2L)\sin 2\theta$ for $b \ll L$.

The matter is easily treated on a formal basis, wherein the diffusion and fluid motion occur simultaneously. To keep the calculation simple, constrain the fluid motion to the y-direction, in the amount $v(x, z, t)$. Then the field has components $(0, B(x, z, t), B_0)$ and

$$\frac{\partial B}{\partial t} = B_0 \frac{\partial v}{\partial z} + \eta \frac{\partial^2 B}{\partial x^2}, \tag{10.36}$$

with

$$\rho \frac{dv}{dt} = \frac{B_0}{4\pi} \frac{\partial B}{\partial z}. \tag{10.37}$$

Suppose, then, that $\eta = 0$ in $z \leqslant 0$ and in $b \leqslant z \leqslant L$, while η is small $(bC/\eta \gg 1)$ but nonvanishing in $0 < z < b$. Then the system is in quasi-static equilibrium evolving only very slowly as a consequence of the small diffusion in $0 < z < b$. So $\partial B/\partial t$ can be neglected compared to either of the other terms on the right-hand side of equation (10.36). The inertia ρv is small and time rate of change is even smaller, so that it follows from equation (10.38) that $\partial B/\partial z \cong 0$, i.e., $B = B(x, t)$, throughout $0 < z < L$. It follows from equation (10.36) that the fluid velocity in $0 < z < b$ is

$$v(x, z, t) = -\frac{\eta}{B_0} \frac{\partial^2 B}{\partial x^2} z, \tag{10.38}$$

given that $v(x, 0, t) = 0$.

The field lines extend essentially straight across $b < z < L$ where $\eta = 0$. In this region the fluid velocity $v = u(x, z, t)$ vanishes at $z = L$ and matches to $v(x, b, t)$ at $z = b$. Hence

$$\begin{aligned} u(x, z, t) &= v(x, b, t)(L - z)/(L - b), \\ &= -\frac{\eta b(L - z)}{B_0(L - b)} \frac{\partial^2 B(x, t)}{\partial x^2}. \end{aligned}$$

In the region where $\eta = 0$, then, there is the very slow relaxation

$$\begin{aligned} \frac{\partial B}{\partial t} &\cong B_0 \frac{\partial u}{\partial z}, \\ &= \frac{\eta b}{L - b} \frac{\partial^2 B(x, t)}{\partial x^2}, \end{aligned} \tag{10.39}$$

Since $B(x, t)$ is independent of z throughout $0 < z < L$, it follows that this result for $\partial B/\partial t$ applies to the entire region $0 < z < L$. The effective diffusion coefficient η' is readily seen to be

$$\eta' = \eta b/(L - b) < < \eta. \tag{10.40}$$

We might have anticipated that the effective diffusion coefficient η' is reduced by the fraction of the total space in which η is non zero. It follows that an initial discontinuity of $2B_1$ becomes

$$B(x, t) = B_1 \operatorname{erf}[x/(4\eta' t)^{\frac{1}{2}}]$$

with the passage of time, so that the obliquity $\theta(x, t)$ varies as

$$\tan\theta(x, t) = (B_1/B_0)\,\mathrm{erf}[x/(4\eta' t)^{\frac{1}{2}}].$$

The characteristic thickness of the current sheet is

$$d(t) = (4\eta' t)^{\frac{1}{2}}.$$

The characteristic dissipation time τ for a thickness d is

$$\tau = d^2/4\eta', \tag{10.41}$$

of course. This is also the characteristic time of decline of the obliquity θ.

Now in the tangential discontinuities formed in a wrapped and interwoven field, the current sheets are not of uniform strength. It was pointed out in §8.3 that the surface of discontinuity fans out with increasing distance from the local pressure maximum that causes the discontinuity, with the result that the surface current density in the discontinuity declines at least as fast as inverse distance. In this case there is effectively a second factor of L/b introduced into the decay time of equation (10.40), so that the effective resistive diffusion coefficient is η'' where

$$\eta'' = \eta(b/L)^2 \ll \eta' \ll \eta. \tag{10.42}$$

In fact η'' is even smaller where surfaces of tangential discontinuity from different winding patterns intersect and cut each other off. But as a first demonstration we treat a current sheet confined to a single plane surface.

Consider the surface of tangential discontinuity $\varphi + qz = 0$ in the force-free field described by equations (5.44) and (5.45)–(5.47). The field is anchored in the rigid infinitely conducting circular cylinder $\varpi = R$. A layer $-h < z < +h$ of this field is compressed in the region $\varpi < a$ around the z-axis, so that fields initially at $z = \pm(h + \varepsilon)$ come into contact through the associated gap in the flux surfaces. Their initial directions differ by $O(qh)$, producing a directional discontinuity in the vicinity of $\varpi = a$ of the same order, i.e., the discontinuity in the φ-component of the field is of the order of $B_0 qh$. We work in the limit that $a, h \ll R$ with $qa, qh = O(1)$. Thus, for $\varpi \gg a$, B_φ is of the order of $B_0 ha/\varpi$, so that

$$B_\varphi \sim B_0/q\varpi, \tag{10.43}$$

in order of magnitude. This is the result indicated by equation (5.66) at $\varpi = R$.

Now suppose that a small resistivity η is introduced in the thin layer $R - b < \varpi < R$ of thickness b at the outer boundary $\varpi = R$. The foregoing analysis of the thin layer at $z = L$ can be taken over directly, replacing the coordinate x normal to the current sheet by the new coordinate z. The old z-dimension goes over into the new ϖ, with z in equation (10.38) replaced by $R - \varpi$. Thus in place of equations (10.38) and (10.39), we have now the velocity

$$v(z, \varpi, t) = -\frac{\eta(R - \varpi)}{B_0}\frac{\partial^2 B_\varphi}{\partial z^2} \tag{10.44}$$

in $R - b < \varpi < R$, with equation (10.39) rewritten as

$$\frac{\partial B_\varphi}{\partial t} = \frac{\eta b}{R - b}\frac{\partial^2 B_\varphi}{\partial z^2}. \tag{10.45}$$

The maximum velocity occurs at $R - \varpi = b$ and can be written

$$v \cong \frac{\eta b B_\varphi}{B_0 \delta^2} \tag{10.46}$$

for a current sheet of thickness δ in place of the spiral surface of discontinuity $qz + \varphi = 0$. The characteristic dissipation time τ is of the order of the unwinding time so that $v\tau \cong R$, or

$$\tau = (\delta^2/\eta)(B_0/B_\varphi)(R/b).$$

This result is equivalent to equation (10.41), except for the factor B_0/B_φ which is of the order of (R/a), according to equation (10.42). Hence

$$\tau = (\delta^2/\eta)(R/b)^2, \tag{10.47}$$

in order of magnitude. This is equivalent to $\tau = \delta^2/\eta''$, where η'' is given by equation (10.42). The essential point here is that the fanning out of a surface of discontinuity greatly reduces the effect of resistive dissipation at some distant location. The local current sheets become increasingly insensitive to the presence of distant resistivity, and, as already noted, the intersection and cancellation of surfaces of discontinuity from different winding patterns further enhances this effect. Thus, the resistivity at some distant boundary layer does not have much effect on the thickness of the current sheets deep in an extended winding pattern.

10.4.2 Alfven Transit Time Effects

As the next step consider the limitations on current sheet thickness arising from communication by Alfven wave propagation from regions of high resistivity into regions of low resistivity (Parker, 1990). The development treats only the plane uniform current sheet, extending from a region of relatively high resistivity, corresponding to the chromosphere of a star like the Sun ($\eta_{chr} \cong 10^7\,\text{cm}^2/\text{sec}$), upward into a region of low resistivity, ($\eta_{cor} \cong 10^3\,\text{cm}^2/\text{sec}$) corresponding to the corona. The relatively thick current sheet in the chromosphere is continually transmitted to the corona in a characteristic Alfven transit time $\tau_1 \cong 50s$ (see Fig. 10.5).

It follows that for passive resistive diffusion the characteristic thickness δ_{chr} of the chromospheric current sheet is not more than the characteristic diffusion length $(4\eta_{chr}\tau_1)^{\frac{1}{2}} \cong 5 \times 10^4\,\text{cm}$ in the lifetime τ_1 in the chromosphere. At the same time the Maxwell stresses in the corona are striving to reduce the characteristic thickness δ_{cor} of the coronal current sheet. The characteristic reduction time τ_2 is just the characteristic Alfven transit time $w/C_\perp$ across the width w of the current sheet (see derivation of equation (10.22)). The transverse Alfven speed $C_\perp$ is defined in terms of the characteristic transverse component $B_\perp$ of the magnetic field, arising from the interweaving and winding of the field about the mean field B_0 in the vertical direction. Thus $C_\perp = B_\perp/(4\pi\rho)^{\frac{1}{2}}$. As a working estimate, suppose that $B_\perp = \frac{1}{4}B_0 \cong 25\,\text{gauss}$. Then $C_\perp = 5 \times 10^7\,\text{cm/sec}$ in the corona where $B_0 = 10^2\,\text{gauss}$ and the number density of the gas is $10^{10}/\text{cm}^3$. Then if $w = 4 \times 10^7\,\text{cm}$, it follows that $\tau_2 \cong 1\,\text{sec}$. If this were the only effect, we would write

$$\frac{d\delta_{cor}}{dt} \cong -\frac{\delta_{cor}}{\tau_2},$$

but with the transmission of the opposite (thickening) effect upward from the chromosphere, in the characteristic Alfven transit time τ_1, the net result is

$$\frac{d\delta_{\text{cor}}}{dt} \cong -\frac{\delta_{\text{cor}}}{\tau_2} + \frac{\delta_{\text{chr}}}{\tau_1}.$$

For steady conditions, then,

$$\begin{aligned} \delta_{\text{cor}} &= \delta_{\text{chr}}(\tau_2/\tau_1), \\ &\cong 10^3 \text{ cm}. \end{aligned} \tag{10.48}$$

Alternatively, the characteristic coronal diffusion length based on the Alfven transit time τ_1 is:

$$\begin{aligned} \delta_{\text{cor}} &\cong (4\eta_{\text{cor}}\tau_1)^{\frac{1}{2}}, \\ &\cong 0.5 \times 10^3 \text{ cm}. \end{aligned}$$

Either way, the passive diffusion in the corona is negligible because of the large value of δ_{cor}.

To look at it another way, the speed u with which field is carried into, and dissipated within, the coronal current sheet is $\eta_{\text{cor}}/\delta_{\text{cor}}$ (see equation 10.20). With $\delta_{\text{cor}} \cong 10^3$ cm, u is of the order of 1 cm/sec. A time of 4×10^7 sec (essentially 1 year) is required to cut across a flux bundle of characteristic width 4×10^7 cm. The effect of passive diffusion is negligible.

The passive diffusion rate $\eta_{\text{chr}}/\delta_{\text{chr}}$ cm/sec in the chromosphere proves to be 2×10^2 cm/sec. This is unimportant for chromospheric heating, which apparently involves something of the order of 10^9 ergs/cm^2 sec, or about 10^2 times the requirement for the active corona. It is generally believed that the principal heat input to the chromosphere is acoustic in nature (Anderson and Athay, 1989, but see the remarks in Bueno, 1991). The conclusion is simply that passive resistive dissipation in plane current sheets proceeds much more rapidly in the chromosphere than in the corona, but not fast enough to contribute significantly to the heat input in either location.

It might well be argued that the resistive dissipation in the chromosphere is not passive (as it is in the foregoing discussion) but dynamical, proceeding at a rate u at least as large as the lower bound given by equation (10.22). Then with $w = 4 \times 10^7$ cm, $N \cong 3 \times 10^{12}/\text{cm}^2$, and $B_\perp \cong 25$ gauss, it follows that $C_\perp \cong 3 \times 10^6$ cm/sec, and $N_L = 10^7$, so that we have $u_{\text{chr}} \gtrsim 10^3$ cm/sec. This cuts across a flux bundle with characteristic radius of the order of w in a time of 4×10^4 sec. The characteristic current sheet thickness follows from equation (10.22) as $\delta_{\text{chr}} \cong 10^4$ cm. Equation (10.48) then yields $\delta_{\text{cor}} \cong 0.02$, $\delta_{\text{chr}} = 2 \times 10^2$ cm so that $u = \eta_{\text{cor}}/\delta_{\text{cor}} \cong 5$ cm/sec in the corona. This is again a negligible cutting rate. It is to be compared with the lower bound of 30 cm/sec that would follow directly from equation (10.22) if the coronal magnetic field did not communicate to the chromosphere.

Note, then, that with $\delta_{\text{cor}} = 2 \times 10^2$ cm and $B_\perp \cong 25$ gauss, the electron conduction velocity in the current sheet is of the order or

$$\begin{aligned} v_c &= cB_\perp/4\pi Ne\delta_{\text{cor}}, \\ &\cong 6 \times 10^7 \text{ cm/sec}. \end{aligned} \tag{10.49}$$

(see discussion in §§2.4.3 and 2.4.3). This is nominally three times the mean ion thermal velocity and not sufficient to generate much plasma turbulence in the presence of the background electrons with temperatures comparable to the ions. But note that this calculation of v_c is a lower bound based on the upper bound on $\delta_{\rm cor}$ associated with the lower bound on the reconnection speed u. If, for instance, the communication to the chromosphere were reduced, by a simple fanning out of the plane current sheet, then the effect of the chromosperhic resistivity is reduced by a substantial geometrical factor, as in equation (10.47) (cf. equation 10.41). In that case the characteristic current sheet thickness $\delta_{\rm cor}$ may be substantially smaller, with the electron conduction velocity increasing to the electron mean thermal velocity. The onset of plasma turbulence and anomalous resistivity may then put the reconnection into a state comparable to the Petschek mode (see §10.3.3), with a reconnection rate $u10^{-2}C_\perp$. In contrast, the high density and thick current sheets in the chromosphere are not conducive to sufficiently high electron conduction velocities to provide strong anomalous resistivity, so it would appear that the chromospheric reconnection plods along at a rate close to the minimum 10^3 cm/sec estimated above. This topic is discussed at greater length in §10.4.4.

10.4.3 Upward Extension of Current Sheet Thickness

The foregoing physical arguments are readily illustrated with a formal example, showing the upward transmission of thick current sheets by Alfven waves. Use the idealized case treated in §10.2 involving the current $x = 0$ between two regions of field of opposite obliquity. The resistivity $\eta(z)$ varies with distance z along B_0, and the fluid has a velocity $\mathbf{v}(x, z, t)$ as a consequence of the Lorentz force that arises from the disruption of static equilibrium by resistive diffusion. The resistivity $\eta(z)$ is taken to decrease with increasing height z and the principal effect of the fluid motion is to communicate the thickness of the magnetic shear layer, i.e., the thickness of the current sheet, from one level of z to another. Therefore, to keep the calculation within reasonable bounds we again arbitrarily constrain the x and z components of $\mathbf{v}$ to vanish. The y-component of the fluid motion is denoted by $v(x, z, t)$. Then $\partial v/\partial y = 0$ and the flow is divergence free. The fluid density ρ is a function only of z as a consequence of a uniform gravitational acceleration g in the negative z-direction. The magnetic field has a uniform z-component B_0 and a transverse component $B(x, z, t)$ in the y-direction. The induction equation has only a y-component,

$$\frac{\partial B}{\partial t} = B_0 \frac{\partial v}{\partial z} + \eta(z) \frac{\partial^2 B}{\partial x^2} + \frac{\partial}{\partial z}\left[\eta(z) \frac{\partial B}{\partial z}\right], \tag{10.50}$$

The diffusion in the z-direction, parallel to B_0 is small compared to the diffusion in the x-direction across the current sheet, so that term can be neglected. In the steady-state, then,

$$B_0 \frac{\partial v}{\partial z} + \eta(z) \frac{\partial^2 B}{\partial x^2} = 0 \tag{10.51}$$

The shear $\partial v/\partial z$ sustains the field in the presence of the Joule dissipation. If $\eta = \eta_{\rm chr}$ in the chromosphere over a height $h_{\rm chr}$ and $\eta = \eta_{\rm cor}$ in the corona over a height $h_{\rm cor}$, then the velocity difference $\Delta v = h\partial v/\partial z$ across the chromosphere is

$$\Delta v_{\rm chr} = \frac{h\eta_{\rm chr}}{B_0} \frac{\partial^2 B_{\rm chr}}{\partial x^2} \tag{10.52}$$

The transverse field B can be expressed as the fraction s of B_0, and the thickness of the current sheet is denoted by δ, so that $\partial/\partial x$ can be approximated by δ^{-1}. The result is

$$\Delta v_{\rm chr} = \frac{s h_{\rm chr}\eta_{\rm chr}}{\delta^2_{\rm chr}} \tag{10.53}$$

with a corresponding result for the corona. If the thickness δ of the current sheet is equivalent to the diffusion length $(4\eta t)^{\frac{1}{2}}$ in some characteristic time t, then

$$\Delta v_{\rm chr} = \frac{s\, h_{\rm chr}}{4t_{\rm chr}}.$$

Presumably $t_{\rm chr}$ is of the order of the Alfven transit time over $h_{\rm chr}$, so that $h_{\rm chr} = C_{\rm chr} t_{\rm chr}$ and

$$\Delta v_{\rm chr} = \frac{1}{4} s\, C_{\rm chr}. \tag{10.54}$$

Similarly, then, for the corona

$$\Delta v_{\rm cor} = \frac{1}{4} s\, C_{\rm cor}. \tag{10.55}$$

On the other hand, suppose that the thickness δ of the current sheet is pushed to the small value $h/N_L^{\frac{1}{2}}$ for rapid reconnection, providing the lower bound on the reconnection rate given by equation (10.22), where N_L is the Lundquist number hC/η. It follows form equation (10.53) that the result is again equations (10.54) and (10.55).

It follows, then, that the shears Δv maintaining the thickness of the current sheet in the chromosphere and in the corona are in the ratio of the respective Alfven speeds. The chromospheric Alfven speed is roughly 10^{-1} of the coronal Alfven speed in the field B_0, because the chromospheric density is approximately 10^2 times larger. It follows that the velocity shear $\Delta v_{\rm chr}$ necessary to maintain the dissipation in the resistive chromosphere is approximately 10^{-1} of the shear $\Delta v_{\rm cor}$ in the corona. So the chromospheric resistivity evidently has no great effect on the thickness of the current sheets in the corona.

The momentum equation for the y-component of the velocity $v(x, z, t)$ can be written in various forms. With $\partial/\partial y = 0$ note that $(\mathbf{v}\cdot\nabla)\mathbf{v} = 0$, so that

$$\rho(z)\frac{\partial v}{\partial t} = \frac{B_0}{4\pi}\frac{\partial B}{\partial z} - \frac{\rho(z)v}{\tau} + \rho(z)v(z)\frac{\partial^2 v}{\partial x^2} \tag{10.56}$$

if we include a viscosity and a crude representation $\rho v/\tau$ of $(\mathbf{v}\cdot\nabla)\mathbf{v}$ if $\partial/\partial y \neq 0$. The equations are linear and can be solved by separation of variables or by similarity methods with $B = z^\alpha f(xz^\beta)$, etc. To take a particularly simple example, neglect viscosity and $1/\tau$, so that upon differentiating equation (10.50) with respect to t we may eliminate $\partial v/\partial t$, obtaining the wave equation

$$\frac{\partial^2 B}{\partial t^2} - \frac{\partial}{\partial z}C^2(z)\frac{\partial B}{\partial z} = \eta(z)\frac{\partial^2}{\partial x^2}\frac{\partial B}{\partial t} \tag{10.57}$$

again neglecting diffusion over the z-direction. It is convenient to introduce the dimensionless variables $\zeta = z/\ell$, $\xi = x/\ell$ in terms of some characteristic scale ℓ, writing

$$\eta = \eta_0/\zeta^\nu,\ \rho = \rho_0/\zeta^{2\mu},\ C = C_0\zeta^\mu \tag{10.58}$$

Then write

$$B(x, z, t) = F(x, z) \exp \sigma t, \tag{10.59}$$

reducing the equation to

$$n \frac{\partial^2 F}{\partial \xi^2} + \zeta^\nu \frac{\partial}{\partial \zeta} \zeta^{2\mu} \frac{\partial F}{\partial \zeta} - m\zeta^\nu F = 0 \tag{10.60}$$

where

$$m = \sigma^2 \ell^2 / C_0^2, \; n = \sigma\eta_0 / C_0^2$$

The variables are separable, yielding solutions in terms of trigonometric functions and Bessel functions. For $\mu = 1$ the similarity form is applicable, writing

$$F = \zeta^\alpha H(\chi) \tag{10.61}$$

where $\chi = \zeta\xi^{2/\nu}$ reduces the problem to an ordinary differential equation in the similarity variable χ. The further substitution $\chi^\nu = (4n/\nu^2)\psi$ reduces the differential equation to hypergeometric form. In the simplest case, let

$$1 + \nu + 2\alpha = \pm(1 + 4m)^{\frac{1}{2}} \tag{10.62}$$

with the result that

$$B = sB_0 \xi \zeta^{\alpha + \frac{1}{2}\nu} \exp \sigma t \tag{10.63}$$

where δB_0 is the strength of the transverse field at $\xi = \zeta = 1$, $t = 0$. Note, then, that σ can be replaced by $-\sigma$. The associated fluid velocity is

$$v = +s\delta C_0^2 \left(\alpha + \frac{1}{2}\nu \right) \xi \alpha_1 2\nu + 1 \exp \sigma t \tag{10.64}$$

where

$$\alpha + \frac{1}{2}\nu = \left(m + \frac{1}{4} \right)^{\frac{1}{2}} - \frac{1}{2} \tag{10.65}$$

$$\cong m - m^2 + 2m^3 - 5m^4 + \ldots \tag{10.66}$$

and is positive for all $m > 0$. With $\mu = 1$ the ratio v/C is proportional to $\zeta^{\alpha + \frac{1}{2}\nu}$ and increases with height ζ.

The current density is in the z-direction, with magnitude

$$\begin{aligned} j &= \frac{c}{4\pi} \frac{\partial B}{\partial x} \\ &= \frac{c\, s\, B_0}{4\pi\ell} \zeta^{\alpha + \frac{1}{2}\nu} \exp \sigma t \end{aligned} \tag{10.67}$$

The dissipation per unit volume is

$$\eta j^2 = \frac{\eta_0 c^2 s^2 B_0^2}{16\pi\ell^2} \zeta^{2\alpha} \exp 2\sigma t \tag{10.68}$$

increasing with height in proportion to $\zeta^{2\alpha}$. With the similarity variable $\chi = \zeta\xi^{\frac{2}{\nu}}$ the characteristic width ξ varies in proportion to $\zeta^{-\frac{1}{2}\nu}$, so that the total dissipation across ξ at a level ζ is proportional $\xi\eta j^2$ or $\zeta^{2\alpha - \frac{1}{2}\nu}$.

Noting that the growth rate σ of the fields and currents in this example is related to the parameter m, which is related to α through equation (10.65), it follows that the magnitude and sign of the exponent $2\alpha - \frac{1}{2}\nu$ depends upon the assumed growth rate. Now the Alfven speed is about 10 or 20 times larger in the corona than in the chromosphere while the temperature is a little more than 10^2 times larger. Hence η is a little more than 10^3 times smaller. So $\mu = 1$ for the Alfven speed suggests $\nu = 3$ for the resistivity, insofar as the power laws of equation (10.58) are applicable. Then

$$m = \alpha^2 + 4\alpha + 15/4$$

If $\alpha = 1$, then $m = 35/4$ and $\sigma = 2.96C_0/\ell$, representing relatively slow growth in the corona where the Alfven speed is large compared to C_0 in the chromosphere. In this case, then, the total dissipation per unit height varies as $\zeta^{\frac{1}{2}}$, i.e., proportional to $C^{\frac{1}{2}}$. So again the principal action is the corona.

Additional examples are readily constructed from similarity solutions for the stationary state with $\partial/\partial t$ and the viscosity set equal to zero in equation (10.56). Examples are provided by Parker (1990). But of more immediate interest is the cutoff of current sheet widths where the surfaces of discontinuity intersect. This is taken up in the next section.

10.4.4 Nonuniform Current Sheets

The foregoing examples of a current sheet extending along a uniform magnetic field through regions of different resistive diffusion coefficient $\eta(z)$ omit the effect of the fanning out of each current sheet with distance, described by equations (10.40)–(10.47). Thus the effect of the high resistivity of the chromosphere does not extend up into the surfaces of discontinuity fanning out from their points of origin in the corona to the modest degree suggest by the examples. In particular it must be recognized that each successive winding pattern along the field produces its own surfaces of discontinuity, which intersect and cut off the attenuated surfaces of discontinuity created elsewhere along the field, in the manner described in §8.5. It follows that the surfaces of discontinuity extending into the chromosphere are cut off within a couple of correlation lengths into the corona. So only the first winding patterns at each end $z = 0$ and $z = L$ of the coronal field are sensibly affected by chromospheric resistivity. The interior is shielded by the branching of the surfaces of discontinuity. In effect, the branching reflects the Alfven waves propagating inward from the ends.

The present analysis, including §10.4.3, indicates that the current sheets in the corona are in a state close to the lower bound on the minimum reconnection rate given by equation (10.22) much of the time. It was noted toward the end of §10.4.2 that the lower bound on the reconnection speed follows from equation (10.22) as $u \cong 30$ cm/sec for coronal conditions, with $B_\perp = 25$ gauss, $C_\perp = 5 \times 10^7$ cm/sec, a curent sheet width $w \cong 4 \times 10^7$ cm, and $\eta_{cor} = 10^3$ cm^2/sec. The maximum current sheet thickness δ_{cor} associated with this minimum reconnection velocity (see equation 10.22) is $\eta_{cor}/u \cong 30$ cm. The associated electron conduction velocity in the current sheet is given by equation (10.49) as not less than 4×10^8 cm/sec, or about half the mean electron thermal velocity in the corona with an electron temperature of the order of 2×10^6 K. Thus, it appears that the surfaces of tangential discontinuity may generally provide magnetic reconnection at some slow rate of the order of 1 m/sec or less. But this involves current sheets sufficiently thin as to reside at the brink of anomalous resistivity and the resistive tearing instability. It appears, then, that with a slowly but steadily increasing $B_\perp$, as a consequence of the continual intermixing of the photospheric footpoints of the field, there will be bursts of explosive reconnection, more or less in the Petschek mode, triggered by the onset of the tearing instability and strong plasma turbulence, along the lines described in §§10.3.4 and 10.3.5. The explosive reconnection converts magnetic free energy into jets of fluid, magnetohydrodynamic waves, and Joule heating. The jets and waves are quickly thermalized by their small-scale, so in effect the magnetic free energy goes almost entirely into heat within some modest distance of the reconnection site.

10.5 Changes in Field Topology

Resistive diffusion causes reconnection of field lines and is, therefore, capable of altering the topology of a magnetic field. Anomalous resistivity, caused by plasma turbulence, is expected to be highly irregular in space, and probably in time as well, and its large magnitude and small-scale cause qualitative topological changes on the same small-scale as the anomalous resistivity. This section demonstrates the consequences for a localized region of scale a of anomalous resistivity in a torsional force-free field. The resultant restructuring of the field topology provides wrap-around flux surfaces on a scale a, where none existed before. To be more precise, anomalous resistivity in a small region of scale a in a large-scale force-free field (i.e., with small torsion coefficient α) causes a small portion $O(\alpha a)$ of the flux to develop gradients on a scale a, or less. Hence the new field gradients are stronger than the initial torsion α. The gradients are not diminished by the anomalous diffusion but are enhanced. Hence, if the background plasma is sufficiently tenuous that α initially produces anomalous resistivity (see §§2.4.3 and 2.4.4), there is every reason to expect anomalous resistivity to occur subsequently. Thus we may expect a sequence of resistive restructuring events. Tangential discontinuities may or may not be created in the first event, but they become unavoidable as successive resistive restructuring events progressively complicate the topology.

The present writing is limited to the development of a single local intense burst of anomalous resistivity, with its rapid complication of the field topology, converting a field composed of straight parallel field lines into one in which some of the field lines wrap half a revolution around the others. Consider, then, a localized region of large anomalous resistivity appearing suddenly at some location in an otherwise force-free magnetic field $\mathbf{B}(\mathbf{r})$. The rapid diffusion and reconnection of field lines alters the field topology, thereby upsetting the initial equilibrium. The resulting combination of motion and diffusion is complicated and nonlinear, and can be treated only in simple cases (cf. Parker, 1977, 1979, pp. 90–95). Fortunately the details of the motion are of only minor interest. It is the final state of the field topology with which we are primarily concerned. Some idea of the final state can be obtained from the idealized situation in which the fluid is held fixed while the large resistivity is switched on. When the resistive diffusion has run its course, the resistivity is switched off and the fluid is released so that the system relaxes to the lowest available energy state. This final equilibrium state is then to be compared with the initial equilibrium state, completing the event. As we shall see, the field gradients around the boundary of the final state are sufficient to produce further anomalous resistivity, ushering in a second resistive event, etc.

It is appropriate to use the primitive force-free magnetic field with uniform torsion $\alpha = q$

$$B_x \cong +B_0 \cos qz, \; B_y \cong -B_0 \sin qz, \; B_z \cong 0.$$

It has a curl only as a consequence of its torsion. The other field gradients can be omitted because they are nonessential in the present topological considerations. This is the same idealized field as was employed in §5.4, and it is described in cylindrical polar coordinates by equations (5.45)–(5.47).

The resistivity of the fluid is taken to be zero and it is convenient to think of the field as anchored in a distant rigid infinitely conducting cylindrical wall $\varpi = R(qR \gg 1)$. Note that the field lines are straight and parallel in each plane $z = constant$, where the direction of the field lies at azimuth $-qz$.

Suppose, then, that the fluid is held fixed while the resistivity η in the circular cylinder $\varpi = a(a \ll R)$ becomes very large, with the resistivity outside $(\varpi > a)$ remaining zero. Then in the short characteristic time $O(a^2/\eta)$ the field in $\varpi < a$ relaxes to the potential form

$$\mathbf{B} = -\nabla\phi, \; \nabla^2\phi = 0. \tag{10.69}$$

In cylindrical polar coordinates the initial field, described by equation (5.45)–(5.47), introduces the boundary condition $B_\varpi = +B_0 \cos(\varphi + qz)$ at $\varpi = a$. The appropriate solution to Laplace's equation is, then,

$$\phi = -\frac{B_0 I_1(q\varpi)\cos(\varphi + qz)}{qI'_1(qa)}, \tag{10.70}$$

in $\varpi \leqslant a$, where $I_1(q\varpi)$ represents the modified Bessel function of first order. The field components are

$$B_\varpi = +\frac{B_0 I'_1(q\varpi)\cos(\varphi + qz)}{I'_1(qa)}, \tag{10.71}$$

$$B_\varphi = -\frac{B_0 I_1(q\varpi)\sin(\varphi + qz)}{q\varpi I'_1(qa)}, \tag{10.72}$$

$$B_z = -\frac{B_0 I_1(q\varpi)\sin(\varphi + qz)}{I'_1(qa)}. \tag{10.73}$$

We may expect the strong plasma turbulence to be confined to small volumes, so suppose that $qa \ll 1$, so that

$$I_1(q\varpi) \cong \frac{1}{2}q\varpi + \frac{1}{16}(q\varpi)^3 + \frac{1}{384}(q\varpi)^5 + \ldots$$

The radial and azimuthal components are, to lowest order,

$$B_\varpi \cong +B_0 \cos(\varphi + qz), \tag{10.74}$$

$$B_\varphi \cong -B_0 \sin(\varphi + qz), \tag{10.75}$$

which are the same as the initial field, described by equations (8.45) and (8.46). The z-component of the field was initially zero and now is

$$B_z \cong -B_0 q \varpi \sin(\varphi + qz). \tag{10.76}$$

This is small $O(qa)$, but it is sufficient to reduce $\nabla \times \mathbf{B}$ to zero from the original small value qB_0. It also has the effect of connecting the field lines from one plane $z = constant$ to another, thereby causing lines to loop halfway around other lines.

Now in $\varpi > a$ the individual field lines are straight and parallel in each plane $z = z_0$. The equation for a single line is

$$\varpi \sin(\varphi + qz_0) = \lambda, \tag{10.77}$$

where λ is the distance of closest approach to the z-axis ($\varpi = 0$) at $\varphi + qz_0 = 12\pi$. This relation also follows in $\varpi < a$ from equations (10.74) and (10.75) to lowest order in which the small z-component of the field is neglected so that $z \cong z_0$. The field lines for which $\lambda < a$ intersect the region $\varpi < a$ of high resistivity. There is an abrupt deflection by $O(qa)$, in a field line where it intersects $\varpi = a$ because within $\varpi < a$ the z coordinate is not constant. It follows from equations (10.75) and (10.76) that

$$\begin{aligned} \frac{dz}{d\varphi} &= \frac{\varpi B_z}{B_\varphi}, \\ &= q\varpi^2 \end{aligned} \tag{10.78}$$

throughout $\varpi < a$. With ϖ related to φ through equation (10.77) to lowest order, equation (10.78) can be integrated to give

$$z - z_0 = -q\lambda^2 \cot(\varphi + qz_0), \tag{10.79}$$

where the integration constant has been chosen so that $z = z_0$ at the point of closest approach ($\varpi = \lambda, \varphi + qz_0 = \frac{1}{2}\pi$) to the z-axis. The line, then, is described by equations (10.77) and (10.79) for $\varpi \leqslant a$. It follows from equation (10.77) that the line reaches $\varpi = a$ at the azimuth φ where

$$\sin(\varphi + qz_0) = \lambda/a. \tag{10.80}$$

With this value of φ it follows from equation (10.79) that at $\varpi = a$, the deviation $\Delta Z(\lambda)$ of the line from $z = z_0$ is

$$\begin{aligned} z - z_0 &= \mp q\lambda(a^2 - \lambda^2)^{\frac{1}{2}}, \\ &\equiv \mp \Delta Z(\lambda), \end{aligned} \tag{10.81}$$

to lowest order. Note, then, that $\Delta Z(\lambda)$ vanishes at both extremes ($\lambda = 0, a$) and has a maximum at $\lambda = a/2^{\frac{1}{2}}$. The upper sign applies where equation (10.80) is satisfied by

$$\varphi + qz_0 = \sin^{-1} \lambda/a, \tag{10.82}$$

and the lower sign were

$$\varphi + qz_0 = \pi - \sin^{-1} \lambda/a, \tag{10.83}$$

the principal value of $\sin^{-1}\lambda/a$ being employed in both cases. Figure 10.7 provides a sketch of the passage of the line across $\varpi < a$, showing the displacement $\mp\Delta Z$ in the z-direction.

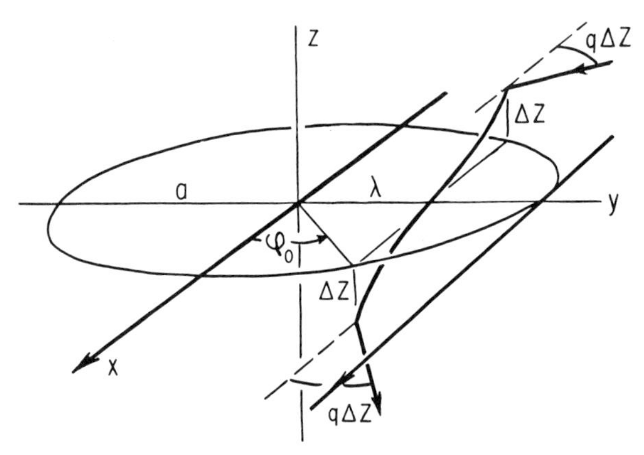

Fig. 10.7. A sketch of the field line crossing the y-axis at $y = \lambda$, $z = 0$, showing the displacement $\pm\Delta Z(\lambda)$ in the z-direction where the line reaches $\varpi = a$. Outside $\varpi = a$ the line runs parallel to the xy-plane at $z = \pm\Delta Z$ in the direction $\varphi = q\Delta Z$ and $\varphi = \pi - q\Delta Z$.

The line extends outward in both directions from its intersections with $\varpi = a$ at the levels $z = z_0 \mp \Delta Z(\lambda)$, so that equation (10.77) becomes

$$\varpi \sin[\varphi + qz_0 \mp q\Delta Z(\lambda)] = \lambda' \tag{10.84}$$

for the line in $\varpi > a$. The constant λ' is evaluated by requiring that the line is continuous across $\varpi = a$. Then since ΔZ is small $O(qa)$, it follows from equations (10.82) that

$$\lambda' \cong \lambda[1 - q^2(a^2 - \lambda^2)]. \tag{10.85}$$

For the intersection described by equation (10.83) the same result is obtained, because $\cos(\varphi + qz_0)$ is negative. In any case, the correction is second order in qa and therefore negligible.

Recall that the field lines in the initial field passing across $\varpi = a$ at the height $z = z_0$ are all parallel to the direction $\varphi = -qz_0$. Following diffusion, the lines with this direction in $\varpi < a$ extend out to $\varpi = R$ at $z = z_0 \mp \Delta Z(\lambda)$ in the direction with azimuth φ given by equation (10.84) as

$$\sin(\varphi + qz_0 \mp q\Delta Z(\lambda)] = \lambda'/R.$$

Hence

$$\varphi \cong -qz_0 \pm q\Delta Z(\lambda) + \lambda'/R. \tag{10.86}$$

The direction is deflected by $2q\Delta Z(\lambda)$ in crossing $\varpi < a$ because the line lies at $z = z_0 - \Delta Z(\lambda)$ on one side of $\varpi = a$ and at $z = z_0 + \Delta Z(\lambda)$ on the other side. The complete field line both inside and outside the cylinder $\varpi = a$ is sketched in Fig. 10.7 for $z_0 = 0$.

The nature of the restructuring can be seen in another way, from the locus at $\varpi = R$ of the footpoints of the field lines that intersect the radial line $\varphi + qz_0 = \frac{1}{2}\pi$ at $z = z_0$. In view of the spiral invariance of the field, there is no loss in generality in putting $z_0 = 0$, so that equation (10.77) reduces to $y = \lambda$. The y-coordinate at $\varpi = R$ is $R \sin\varphi$ and, with $z_0 = 0$, equation (10.86) yields the coordinates (y, z) at $\varpi = R$,

$$y \cong \lambda[1 + q^2R(a^2 - \lambda^2)^{\frac{1}{2}} + O(q^2a^2)], z \cong \mp q\lambda(a^2 - \lambda^2)^{\frac{1}{2}}$$

in terms of the distance λ of closest approach of the line to the z-axis. The y and z coordinates of the intersection of this line with $\varpi = R$ are plotted in Fig. 10.8 for the special case $qa = 0.1$, with $qR = 10, 10^2$, and 10^3. The outstanding feature is the deeply folded flux surface at $\varpi = R$ that forms a flat ribbon in the resistive cylinder. The field in each such flux surface ($z = z_0$ at $\varpi < a$) is wrapped half a revolution around the other flux surfaces. The topology of the wrapping is sketched in Fig. 10.9 in the spiral geometry appropriate for the 3D force-free field. The figure shows the flux surface for those field lines with maximum $\Delta Z(\lambda)$, for $\lambda = a/2^{\frac{1}{2}}$, as the flux surface wraps around the field in $a/2^{\frac{1}{2}} < \lambda < a$ indicated by the straight

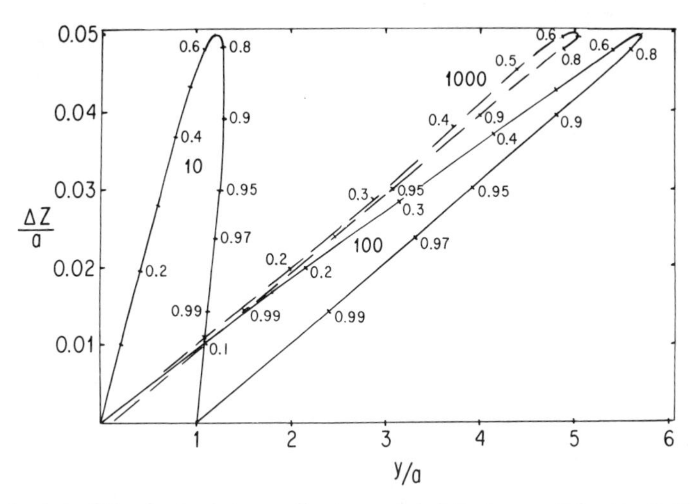

Fig. 10.8. A plot of the footpoint coordinates y and ΔZ at $\varpi = R$ for the field line shown in Fig. 10.7. for $qa = 0.1$ and the values of $\varphi R = 10, 10^2$, and 10^3 indicated on each curve. The dashed curve, for $qR = 10^3$, is compressed by a factor of ten in the y-direction. The numbers by each tic mark indicate the value of λ for that footpoint.

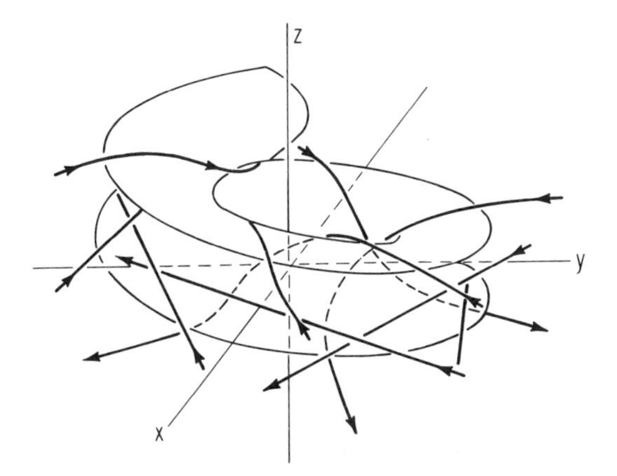

Fig. 10.9. A schematic drawing of the spiral flux surface defined by the field lines with $\lambda = a/2^{\frac{1}{2}}$. The form of the field lines in the surface is indicated by the heavy curves, showing how they wrap around the less displaced magnetic flux in $a/2^{\frac{1}{2}} < \lambda < a$. The heavy straight lines represent the undisturbed field lines passing by at $\lambda \geqslant a$.

lines. Another view is sketched in Fig. 10.10, with the field ines in the upper half representing the field in the plane $z = 0$ around which the field lines at $z = \pm\Delta Z(\lambda)$ are wrapped by half a revolution. The lower half of Fig. 10.10 sketches the same field when it has relaxed to magnetostatic equilibrium with zero resistivity so that the topology sketched in the upper half is preserved. The wrap-around flux pulls the field outward a distance $O(qaR)$ leaving a region of reduced field strength between there and the z-axis. There is the obvious possibility of tangential discontinuities forming where the reconnected flux wraps around and pulls against the field lines extending straight across, because this wrap-around creates a local maximum in the field magnitude. Another effect arises from the diminished field strength on the inward side $(O < \varpi < O(qaR))$ of the flux bundle that is pulled outward by the wrap-around flux. If the resistive cylinder extended only from $z = -h$ to $z = +h$, instead of $-\infty$ to $+\infty$, there is the possibility that the reduced field is squeezed out of the way by the undiminished force-free field immediately outside each end $z = \pm h$, producing a tangential discontinuity for sufficiently large ratios qaR/h.

In this context, the effect of a finite length $2h$ for the resistive cylinder is worked out in the literature (Parker, 1993b), as is the effect for a spherical region. The results are essentially the same as those obtained here in the simple case of a cylinder of infinite length. For instance, the effect in the equatorial plane of a spherical region of radius a is precisely half the $\Delta Z(\lambda)$ obtained for the cylinder of infinite length.

Consider, then, the magnitude of the field gradients before and after the event. Before the event the curl of the field has the uniform value qB_0 throughout the region. The effect of the event is to reconnect the field through $\varpi < a$ so that there is a transverse field compnent, relative to the initial straight field lines, in the amount $q\Delta Z(\lambda)B_0$, with a maximum value of $12qaB_0$ for $\lambda = a/2^{\frac{1}{2}}$. The winding of one flux

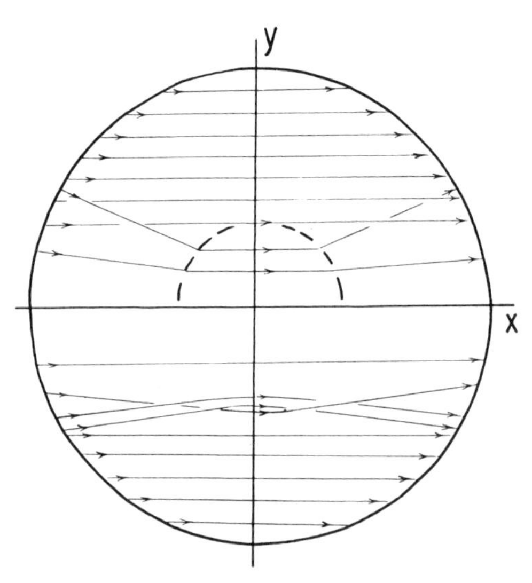

Fig. 10.10 The field lines in the upper half of the figure represent a schematic drawing of the field lines shown in the foregoing three figures, with their connection into the rigid boundary $\varpi = R$. The lower half is a schematic drawing after the field shown in the upper half is allowed to relax to magnetostatic equilibrium.

layer around another substantially enhances the gradient of this transverse component, to a characteristic scale $O(\varepsilon a)$ where $\varepsilon \leqslant 1$. Hence the local torsion is $O(q/2\varepsilon)$ which is evidently at least as large as the initial torsion q. Hence we expect anomalous resistivity to continue to arise, providing a second resistive restructuring event, etc. The field topology after several events becomes so complicated that spontaneous tangential discontinuities are unavoidable, whether they occur in the first event or not. The importance of the effect in flaring in various astronomical settings depends upon the detailed quantitative behavior of the anomalous resistivity, about which little can be said at the present time.

11

Solar X-Ray Emission

11.1 The Origin of X-Ray Astronomy

The curious optical emission spectrum of the solar corona puzzled spectroscopists for decades until it was suggested by Grotrian (1931, 1933, 1939) that the temperature is of the order of 10^6 K. With this possibility in mind Edlen (1942a,b) was able to identify nineteen of the coronal lines (Lyot, 1939) as arising from forbidden transitions of highly ionized atoms (e.g., Fe_{X}, Fe_{XIV}, Ni_{XIII}, Ca_{XII}), clearly establishing an electron temperature of the general order of 10^6 K. Radio observations corroborated this result. The coronal scale height indicated that the mean of the electron and ion temperatures is about 10^6 K and the coronal emission line widths implied an ion temperature of the same order (cf. Waldmeier, 1945). But the 10^6 K temperature was so astonishing that several speculative alternatives were explored, (see the brief summary in Billings, 1966). The emission line widths were attributed to "turbulence," which in the early 1950s was the universally accepted explanation for the fact that theoretical model atmospheres of all kinds generally predicted line widths substantially narrower than observed (largely because the models did not properly treat deviations from local thermodynamic equilibrium). The work of Billings and others finally cleared up this point for the solar corona. There is now no escaping the fact of a corona at 10^6 K, much of which is dense enough to emit a detectable flux of soft X-rays.

A temperature of 10^6 K suggests a peak in radiative output at wavelengths of the order of 30 A, i.e., soft X-rays. It is interesting to note, then, that physicists had been concerned for some years with the erratic behavior of the terrestrial ionosphere, which controls the worldwide propagation of low frequency ($\nu \gtrsim 1.5 \times 10^7$ H) radio signals. The problem had received considerable attention before and during World War II, and it was well known that the E and F layers of the ionosphere were affected by changing levels of solar activity. The F-layer is affected most by UV radiation, and the E-layer, at about 10^2 km altitude, is affected by X-rays. So the picture began to come together.

On August 5 1948 T.R. Burnight flew a photographic plate enclosed in 0.76 mm of Be in a V-2 rocket, providing an exposure time of several minutes above the terrestrial atmosphere (height $> 10^2$ km). The plate showed substantial darkening, which could only be the result of X-rays shortward of 4A. On November 18, more plates were flown, one shielded by only 0.25 mm of Be and another by 0.015 mm of Al, without detectable darkening. On December 9 a third flight was made with plates shielded by 0.0076 mm of Al and by 0.25 cm of Be. The plates shielded by the very thin Al showed darkening but those shielded by the Be did not. Burnight (1949, 1952) concluded from these observations that the Sun emits both soft and hard X-rays, with an intensity that varies enormously, evidently in association with the varying magnetic activity of the Sun.

These crude observations were the inception of X-ray astronomy. They detected the presence of X-ray sources at the Sun, demonstrating that an ordinary star like the Sun emits X-rays as a normal part of it activity, thereby confirming the theoretical implications of the high temperature for the solar corona. These facts led to the inference that many other stars emit X-rays, presumably as thermal emission from their outer atmospheres.

The highly ionized state of the solar corona was the subject of intensive theoretical study (Waldmeier, 1945; Biermann, 1947; Woolley and Allen, 1948; Miyamoto, 1949; Shklovskii, 1951; Hill, 1951; Elwert, 1952, 1954; de Jager, 1955). Observational progress came rapidly. The first X-ray photon counters were soon developed and lofted above the atmosphere (Friedman, Lichtman, and Byram, 1951; Burnight, 1952). Hinteregger (1961), Detwiler, et al. (1961), Hinteregger and Watanabe (1962), and Watanabe and Hinteregger (1962) pushed the observations down to wavelengths of 100 A, while Byram, et al. (1958) and Kreplin (1961) went after harder radiation below 100 A. Chubb, Friedman, and Kreplin (1960) and Vette and Casal (1961) extended observations into the hard X-ray region, to 0.1 A. Chubb, Friedman, and Kreplin (1960) constructed an X-ray pinhole camera, obtaining the first X-ray pictures of the Sun and finding a correlation with the location of active regions (Blake, et al. 1963). Blake, et al. (1965) combined spectroscopy with a pinhole camera in the X-ray region below 60 A and found that the harder X-rays came from active regions.

It became clear from these accumulating X-ray astronomical observations that the total X-ray emission from the Sun is of the order of 10^{27}–10^{28} ergs/sec, largely thermal in origin, and widely variable, depending upon the presence of active regions on the surface of the Sun. The brightness of the most intense steady emission is of the order of 10^7 ergs/cm^2 sec, with brightness 10 or even 10^2 times higher in occasional intense transient solar flares. Lower intensities are to be found in surrounding regions, down to the threshold of the detector. It was evident that the emission involves areas of $10^{10} \times 10^{10}$ cm and more, representing the larger bipolar magnetic active regions.

This suggested that it might be possible to detect the X-ray emission from other stars. The task requires much greater sensitivity, of course, because the nearest stars are 3×10^5 times farther away than the Sun. It turned out, fortunately, that there are X-ray sources very much brighter than the Sun, so that over the next decade many were detected (Bradt and McClintock, 1983; Pallavicini, 1989; Gioia, et al. 1990). That is to say, the positions of detectable X-ray sources were located on the celestial sphere within the observational error of the detecting instrument. Then it was a matter of inspecting the error box on the celestial sphere to identify which of several optical objects in the box was the most likely culprit. The subject advanced rapidly with the ANS X-ray satellite (Grindlay, et al. 1976), the Vela Satellites (Belian, Conner, and Evans, 1976), and the SAS-3 satellite (Lewin, et al. 1976a,b; Lewin, 1977). The general sky surveys by OSO-7 and OSO-8 led to catalogs of X-ray stars (Markert, et al. 1977, 1979), as did the extended Uhuru observations (Forman, et al. 1978) and the HEAO-2 Einstein Observatory (Marshall, et al. 1979; Giacconi, et al. 1979; Amnuel, et al. 1979).

The discoveries of spectacular extra-galactic sources, X-ray bursters, etc. are described in the literature. The brightest stellar X-ray emitters turn out to be close multiple star systems with the high temperature of their outer atmospheres caused by strong tidal interaction, or by matter from the inflated atmosphere of one member falling onto the surface of a close compact companion star (white dwarf or neutron star). A massive accretion disk around a solitary star may also produce intense X-ray emission. Solitary O and B stars have outer atmospheres with temperatures of 10^6–10^7 K, evidently arising from the vigorous chaotic state caused by the unstable levitation of the atmosphere by the radiation pressure. None of these violent mechanical mechanisms apply to the solitary late-main sequence star like the Sun. The subdued X-ray emission arises from subtler processes, and the challenge is to determine what that might be.

Consider, then, the detection of stars of more ordinary mien. Schnopper, et al. (1976) reported the detection and identification of Algol, and Topka, et al. (1979) detected and identified α-Lyrae (Vega) with an X-ray output of 3×10^{28} ergs/sec and η Bootis (a GOIV star) with 10^{29} ergs/sec. The detection and identification of the dMe dwarf flare star Proxima Centauri by Haisch, et al. (1980) and the α–Centauri system by Golub, et al. (1982), showed the variety of stellar types that emit X-rays. Giampapa, et al. (1985) made comparisons between the stellar X-ray emission and the coronal loops and flares seen on the Sun. So X-ray astronomy began with the observation of X-rays from the Sun, and the X-ray emission of all late-main sequence stars leads back to the Sun as the only star that can be resolved and closely studied with the telescope. Indeed, as will become apparent in the sequel, the Sun is none too close for observation, as the critical action is on scales of the order of 100 km (0.15″) and can be observed in visible light only with a diffraction limited telescope of at least 1 m aperture above the terrestrial atmosphere, or perhaps by a substantially larger mirror equipped with active optics at a ground-based site of unusually good seeing. Indeed the necessity to coordinate optical, UV, EUV, and X-ray observations for sufficiently long observing runs requires some, if not all, of the telescopic coverage to be above the atmosphere. It is a curious observational fact of the sociology of scientists that there is little or no enthusiasm for this fundamental study among the general astronomical community, who otherwise avidly pursue X-ray astronomy. Consequently progress has been slow, with the federal funding agencies largely avoiding the necessary support for the work.

11.1.1 Theoretical Considerations

Consider, then, the general physical principles that constrain theoretical understanding of the physics of the corona of a late-main sequence star like the Sun. The corona of the Sun is at least 200 times hotter than the temperature minimum of 4400 K in the photosphere. The second law of thermodynamics assures us that the outer atmosphere cannot be a direct consequence of heat transfer by thermal conduction or radiation from the photosphere. The heat flows the other way, from the corona down into the chromosphere (at 6–8 $\times 10^3$ K) and from there into the photosphere. There were suggestions that the highly ionized atoms are created in the high temperatures deep in the Sun and somehow transmitted to the outer atmosphere through the cool dense intervening layers without significant recombination (Menzel, 1941; Vegard, 1944). Scudder (1992a,b) has recently proposed a similar idea, where some unstated mechanism in the chromosphere produces the fast electrons and ions, which, with their long Coulomb mean free paths, evaporate upward to form the corona with random motions of the order of 200 km/sec for hydrogen ions and 10^4 km/sec for the electrons, equivalent to the inferred temperature of 10^6 K. It has been suggested that the network microflares may be the principal source of heat for the coronal holes (Martin, 1984; Porter, et al. 1987; Porter and Moore, 1988; Parker, 1991) producing jets of gas and high frequency magnetohydrodynamic waves that are quickly damped, and perhaps bursts of gas directly heated to 10^6 K or more in the intense current sheet of the microflare. Much of the microflare is at coronal levels, so this is not what Scudder has in mind. Nor is there any basis to thinking that there is sufficient energy in the microflaring (Habbal, 1992).

Turning to the basic energy source, Biermann (1946, 1948), Alfven (1947) and Schwarzschild (1948) independently noted that the heat engine represented by the subsurface convection is the only thermodynamically available source for the heat that creates the solar corona. The work done by the convective heat engine is thermodynamically available for any enterprise, without regard to entropy etc. As an example they suggested that some fraction of the work goes into producing acoustic waves and Alfven waves in the ambient magnetic field. The waves propagate both upward and downward from their place of origin in the subsurface convection. Dissipation of the waves reaching the corona is assumed to supply the heat that maintains the coronal temperature near 10^6 K. Significant wave heating in this form now appears dubious, as we shall see. But the physical principle remains, that the thermal convection beneath the surface of the Sun is the heat engine that does the work that ultimately heats the corona. The question is how that energy is transferred upward through the surface of the Sun into the corona, where it is dissipated into heat. As already mentioned, acoustic and Alfven waves have been considered at length (see also Schatzman, 1949; Piddington and Minnett, 1951; Piddington and Davies, 1953; Kulsrud, 1955; de Jager and Kuperus, 1961; Osterbrock, 1961; Uchida, 1963; Moore and Spiegel, 1964). In terms of Fourier modes (ω, k), the smallest wave number k_m that can interact effectively with a coronal region of characteristic scale L is the fundamental mode $k_m \cong \pi/L$. Thus the frequency ω must satisfy $\omega \gtrsim k_m V_s$ for acoustic waves and $\omega \gtrsim k_m C$ for Alfven waves ($V_s \ll C$ in the corona). Frequencies below $k_m C$ represent a quasi-static deformation of the magnetic field of the active coronal region. The question, then, is the characteristic frequency ω of the principal convective motions.

To put the matter differently, there is no known alternative to the idea that the corona is somehow an external manifestation of the work done by the subsurface convection. The manner in which the work is transmitted to the corona in a form that can then be converted to heat depends upon the characteristic convective frequencies relative to the dimensions of the coronal regions. This is the crucial physical question to which we now turn our attention.

11.2 Conditions in the X-Ray Corona

The early X-ray observations of the Sun from rockets showed the emission to originate in arcs or loops (cf. Vaiana, et al. 1968; Van Speybroeck, Krieger, and Vaiana, 1970). Subsequent observations at higher resolution by instruments on *Skylab* (Tousey, et al. 1973; Vaiana, et al. 1973; Underwood, et al. 1976; Reeves, et al. 1976) confirm this picture, showing clearly that the emitting loops lie along the field lines of the underlying bipolar magnetic active regions. The recent extremely high resolution X-ray pictures from rockets, using normal incidence optics (Walker, et al. 1988; Golub, et al. 1990; Golub, 1991), show the remarkably fine filamentation of the emitting loops, down to the limit of resolution of the X-ray telescopes (~500 km). The frontispiece is an X-ray picture of the Sun reproduced with the kind permission of Dr Leon Golub from one of his photographs. The corona, then, is made up of many thin filaments, arching up in the form of the capital Greek omega. The filaments lie along the field lines of local intense (50–200 gauss) bipolar magnetic fields, presumably because the gas moves freely along the field, whereas the high electrical conductivity of the gas prevents it from moving across the field, whose pressure greatly exceeds that of the gas.

It has been emphasized by observers that the inhomogeneous, and evidently filamentary, structure extends to scales below the limit of detection and resolution by contemporary instruments (Huber, et al. 1974; Feldman, 1983; Feldman, Purcell, and Dohne, 1987). Feldman and Laming (1993) emphasize again that the transition region above 3000 km, between the chromosphere and the corona, is an unresolved structure of vertical filaments of gas with temperatures of 10^5 K on the one hand and 10^6 K or more on the other. The filling factor of the hot gas increases with height to a limiting value of one a few thousand km higher. One assumes that the filamentary structure is sustained by the magnetic field.

In the active X-ray corona the gas density N is approximately 10^{10} atoms/cm^3 and the temperature T is typically 2×10^6 K. It is sufficient for present purposes to treat the gas as pure hydrogen, fully ionized, with an isotropic pressure $p = 2NkT$, equal to about 6 dynes/cm^2. The thermal energy density if 9 ergs/cm^3. The mean free path of an ion with the mean thermal velocity is of the order of 30 km, and the mean ion thermal velocity, as well as the speed of sound, are of the order of 200 km/sec.

The characteristic magnetic field is 10^2 gauss, with a pressure $B^2/8\pi \cong 400$ dynes/cm^2 and the same energy density in ergs/cm^3. The Alfven speed $C = B/(4\pi NM)^{\frac{1}{2}}$ is 2×10^3 km/sec.

The ratio β of the gas pressure to the magnetic pressure is approximately 0.02. The gas is firmly held in the field and is free to enter and leave the coronal loop only by sliding along the field, i.e., along the legs of the Ω.

Thermal conduction along the field is unimpeded by the presence of the magnetic field, whereas it is reduced by the large factor $(\omega_e\tau)^2$ across the field, where ω_e is the electron cyclotron frequency (1.6×10^9 Hertz in 10^2 gauss) and τ is the time in which the electron is strongly deflected from a straight line by coulomb interaction with the ions and other electrons. With the gas density in the brighter X-ray filaments, $N \cong 10^{10}$ atoms/cm^3 at a temperature of 2×10^6 K, it follows that $\tau \gtrsim 10^{-4}$ sec and $\omega_e\tau \sim 10^5$, so that cross-field conduction is negligible, being 10^{-10} of the conduction along the field.

In the direction along the field, the thermal conductivity κ is

$$\kappa(T) \cong 2 \times 10^{-6}\, T^{\frac{5}{2}} \text{ ergs/cm sec } K \tag{11.1}$$

(Chapman, 1954; Spitzer, 1956). The thermometric conductivity K is

$$\begin{aligned} K &= \kappa(T)/5Nk \\ &\cong 3 \times 10^9 T^{\frac{5}{2}}/N \text{ cm}^2/\text{sec} \end{aligned} \tag{11.2}$$

For $T = 2 \times 10^6$ K and $N = 10^{10}$ atoms/cm^3 the result is $K \cong 1.7 \times 10^{15}$ cm^2/sec. The characteristic thermal relaxation time t_T over a length ℓ along the coronal loop is

$$t_T = \ell^2/4K. \tag{11.3}$$

Thus the equilibration time of a small temperature perturbation over a length $\ell = 10^4$ km is 140 sec, and over 10^5 km it is 1.4×10^4 sec or 4 hours. It follows that the coronal loops, which characteristically evolve over periods of 10^5 sec, are not far from overall thermal equilibrium with their average heat source. On the other hand, a distributed heat source is necessary to account for the nearly uniform temperature distribution along an extended coronal loop ($L \cong 10^5$ km), as may be seen from the fact that the characteristic temperature gradient $|dT/dz|$ associated with the coronal energy input $I \cong 10^7$ ergs/cm^3 sec is I/κ. Then with $\kappa \cong 10^{10}$ ergs/cm sec K at 2×10^6 K it follows that $|dT/dz|$ would be 10^{-3} K/cm. This amounts to a temperature difference of 10^6 K over a relatively short length $\frac{1}{2}L \cong 10^4$ km. For longer loops, the highest temperatures appear across the apex, with somewhat lower temperature in the legs, but clearly a distributed energy source centered more or less on the apex is required to maintain the temperature down the legs within 10^6 K or so of the apex.

It appears that the gas is more or less in barometric equilibrium along the magnetic field, with a pressure scale height $\Lambda = 2kT/Mg$, where M is the mass of the hydrogen atom and g is the acceleration of gravity at the surface of the Sun, 2.7×10^4 cm^2/sec. For $T = 2 \times 10^6$ K the scale height is $\Lambda = 1.2 \times 10^{10}$ cm, indicating that even in the largest loops, with radii of 10^{10} cm, the pressure varies no more than a factor of two as a consequence of the gravitational field.

The radiative emission from an optically thin gas can be written in the form

$$\varepsilon = N^2 Q(T) \text{ ergs/cm}^3 \text{ sec}, \tag{11.4}$$

Rosner, Tucker, and Vaiana (1978, Appendix A and references therein) use a value of Q that can be approximated as

$$Q(T) = 10^{-22}(10^6/T)^{0.6}$$

over the range $10^5 \leqslant T \leqslant 10^7$. This gives $Q \cong 7 \times 10^{-23}$ at $T = 2 \times 10^6$ K. Hildner (1974) and Peres, et al. (1982) estimate $Q(T)$ to lie in the range 3×10^{-23} to 10×10^{-23} at 2×10^6 K. The principal uncertainty is the abundance of the heavier elements relative to hydrogen (Davis, et al. 1975) because a major part of the radiation from coronal gas is in the emission lines of heavy ions that are not completely stripped of their electrons. Chapman (1958) emphasized the enhancement of the rare strongly ionized atoms in the high temperature of the corona. The enhancement arises from the vertical electric field that maintains charge neutrality in the ambient hydrogen plasma in the presence of a gravitational field. In the vertical legs of a magnetic Ω-loop, the vertical temperature gradient is the dominant effect. As a case in point, a rare ion such as Ni_{XVI} in an otherwise pure hydrogen plasma tends toward an equilibrium distribution with a density proportional to T^α where α is of the order of 600. So extreme a variation is physically impossible in the presence of any substantial temperature difference, of course. There are not enough particles in the universe to reach equilibrium across a 50 percent change in the temperature if there is even one highly ionized particle on the cold side. But the point is clear. Heavy ions are enhanced relative to hydrogen in the high temperature of the corona.

The characteristic time τ in which diffusion of the rare heavy ions increases their density in the region of high temperature can be estimated from the diffusion coefficient, given by Chapman as

$$D = 2.5 \times 10^{10} T^{\frac{5}{2}}/NZ^2 \text{ cm}^2/\text{sec}. \tag{11.5}$$

For $N = 10^{10}$ hydrogen atoms/cm^3 at $T = 2 \times 10^6$ K the diffusion coefficient for $Z = 15$ ($\mathrm{Ni_{XVI}}$) is 6×10^{13} cm^3/sec. Thus the characteristic relaxation time in which the density distribution of the rare ions readjusts by a factor e over a length ℓ parallel to the magnetic field is

$$\tau = \ell^2/4D$$

This is 4×10^3 sec, or about 1 hour, for a short loop $L = 2 \times 10^4$ km and $\ell = \frac{1}{2}L$. Over a length 6×10^4 km it is 2×10^4 sec or about 5 hours, etc. It is evident, then, that the enhanced abundance of heavy elements may reach one or more factors of e in the life of a filament. It must be recalled that all coronal loops emerge through the surface of the Sun so that they first appear in the corona with a length $L \cong 10^4$ km. The principal enhancement of heavy ions occurs in this initial phase and thereafter increases little if at all. And therein lies the difficulty in estimating the radiative losses.

We proceed, then, with equation (11.4) to see where it leads with $Q = 7 \times 10^{-23}$, remembering that the results are substantially uncertain. For $T = 2 \times 10^6$ K and $N = 10^{10}$/cm^3, the result is $\varepsilon \cong 7 \times 10^{-3}$ ergs/cm^3 sec. The thermal energy density $U = 3NkT$ for fully ionized hydrogen yields a characteristic cooling time

$$\tau = U/\varepsilon \tag{11.6}$$

With $U = 9$ ergs/cm^3 this proves to be 1.3×10^3 sec, or about 20 minutes. The overall X-ray emission changes substantially over periods of 10^5 sec, but not over periods as short as 20 minutes. Only in the finest filaments, and in the transition region where temperatures are of the order of 10^5 K and N^2 is 400 times larger, does one find rapid flickering, on times of minutes, attributable mainly to the network activity (cf. Habbal and Grace, 1991; Habbal, 1992).

Note, then, that thermal conduction and radiative cooling have the same characteristic times (1.3×10^3 sec) over loop length $L = 2\ell \sim 6 \times 10^9$ cm, representing a coronal loop of moderate length. The much longer characteristic life ($\sim 10^5$ sec) of a coronal X-ray loop is evidently a result of a long-lived, i.e., slowly varying, heat source. The source is quasi-steady and the brightness of the individual coronal loop indicates where the heat source is strongest around the apex of the loop.

Now it is important to understand that the addition of a small amount of heat to the upper atmosphere of the Sun is all that is necessary to provide a corona at temperatures of the order of 10^6 K. The essential point is that the radiative emission rate ε falls off as N^2 with declining density, so that at some sufficiently high level above the surface of the Sun the density is so small that a modest heat input, e.g., 10^{-4} of the photospheric intensity of 0.6×10^{11} ergs/cm^2 sec, can overcome radiative cooling and send the temperature soaring, with a corresponding decline in N. Thermal conduction (by the highly mobile electrons), back into the cool chromospheric layers below, removes heat from the superheated tenuous gas at the top of the atmosphere, of course. The result is that the gas immediately below is heated and expands upward to enhance the density of the superheated gas, moderating the otherwise runaway temperature. Martens, Van Den Oord, and Hoyng (1985) work out an illustrative example. Equilibrium is achieved either when the density rises to where radiative cooling becomes large enough to balance the heat input (minus the downward conduction losses) or the gas expands away into space to form the solar wind. It is a curious coincidence that both limits provide for coronal temperatures in the 2×10^6 K range. The difference lies in the strength of the local magnetic field. It is observed that, where the field at the surface of the Sun has a mean value of 10 gauss or less, the field is pushed out into space by the coronal gas, leaving an open field configuration through which the coronal gas slowly expands (Parker, 1963; Pneuman, 1973). The field of 10 gauss at the photosphere may have a pressure in excess of the local coronal gas, but it must be remembered that the strength of a bipolar field declines asymptotically to zero with increasing altitude, so that the field at the apex is very much weaker and cannot resist even modest coronal gas pressures. The heat input, of about 0.5×10^6 ergs/cm^2 sec (Withbroe and Noyes, 1977; Withbroe, 1988) in these weak field regions goes primarily into maintaining the temperature of the slowly expanding gas in the vicinity of 1.5×10^6 K. Indeed, it is precisely the maintenance of the coronal temperature out to distances of many solar radii that causes the expansion and gradual acceleration of the gas to reach speeds of the order of 500 km/sec far out in space, thereby providing the solar wind. The gas density close to the Sun is limited by the expansion to about 10^8 atoms/cm^3. It follows immediately from equation (11.4) that the radiation losses at such low gas densities are only 10^{-4} of the emission from the denser X-ray corona and hence quite negligible. Such regions, then, appear dark in the X-ray telescope and are referred to as *coronal holes*.

In contrast with the coronal hole, the corona in the fields of 10^2 gauss or more, that make up a bipolar active region, is held fast by the overpowering magnetic field ($\beta \cong 0.02$, as already noted). The gas cannot escape, so the upward expansion of gas from the chromosphere continues until the density reaches about 10^{10} atoms/cm^3 and radiative cooling, together with a small loss by downward thermal conduction, balances the heat input. The temperature hovers in the range $2\text{–}3 \times 10^6$ K with a heat input of about 10^7 ergs/cm^2 sec (Withbroe and Noyes, 1977). Beaufume, Coppi, and Golub (1992) provide a quantitative model of the formation of the corona in response to a modest heat input.

In summary, it follows that a small heat source must inevitably provide a hot atmosphere at some sufficient height above the surface of a star. To understand the height at which this may occur, note that the pressure scale height $\Lambda = kT/Mg$ for neutral hydrogen is approximately 200 km at the visible surface of the Sun where $N \cong 10^{17}/\text{cm}^3$ and $T = 5600$ K. It follows that the density falls to $10^{12}/\text{cm}^3$ at a height of $11.5\Lambda \cong 2300$ km, to $10^{10}/\text{cm}^3$ at $16\Lambda = 3200$ km, to $10^8/\text{cm}^3$ at $21\Lambda = 4200$ km, and finally to interstellar densities of $1/\text{cm}^3$ at $39\Lambda = 7800$ km if the temperature remained at 5600 K. In fact the temperature declines to a minimum of about 4400 K a few hundred km above the visible surface, for which $\Lambda \cong 150$ km. But the temperature soon rises to 6000–8000 in the chromosphere where the gas becomes largely ionized and $\Lambda \cong 2kT/Mg \sim 500$ km. The essential point is that a heat source of mechanical origin manifests itself already at chromospheric levels, where it produces a slight elevation of the temperature and scale height. At about 3000km the density has fallen so low that the heat input there is able to run away with the temperature to 2×10^6 K in regions of strong field.

As a final comment, it appears from both theory and observation that the heat input to the X-ray corona is intermittent and spotty, with large fluctuation in temperature as the gas is suddenly heated by a local nanoflare and subsequently cools by radiation and conduction. The temperature, and hence the pressure scale height, vary by a factor of two or more, as will become clear in the sequel. It follows from the foregoing quasi-static considerations that the gas surges up and down along the field as the temperature varies in each elemental flux bundle, so the mean X-ray corona is anything but static on a small scale. This microsurging is an essential part of any detailed quantitative model of the X-ray corona, but the present primitive state of observational and theoretical knowledge of the corona precludes any really quantitative treatment of the phenomenon, so we continue with the concept of a mean quasi-static corona.

11.2.1 Variability in the X-Ray Corona

The foregoing discussion summarizes the average conditions in the active X-ray corona, from which it must not be concluded that the corona is only slowly varying in space and in time. Indeed, the outstanding characteristic of the corona is its extraordinary inhomogeneity, indicating the concentrated intermittent character of the heat input. The frontispiece is an X-ray photograph of the Sun, from which it can be seen that the X-ray corona is made up of filaments whose widths extend down to the limit of resolution of the X-ray telescope at about 500 km. It is not unlikely that improved resolution will turn up even smaller scales.

The finely striated filaments of X-ray emission (Golub, 1991) indicate the strong inhomogeneity of N and T across the magnetic field (Feldman, 1983). Recent detailed analysis of the emission spectrum indicates that the temperature at any one point is highly variable, spending most of the time cooling from occasional brief transient injections of heat (Sturrock, et al. 1990; Raymond, 1990; Feldman, et al. 1992; Laming and Feldman, 1992). The lower ends of the coronal loops are observed to flicker (10–30 percent) with characteristic times of 1–2 minutes in the transition zone where $T \sim 10^5$ K (Habbal, 1991; Habbal and Grace, 1991, Rabin and Dowdy, 1992) attributed to microflaring in the network activity. Brueckner and Bartoe (1983) observe bursts of high speed turbulence and jets of gas at the same levels, which also appear to be associated with microflaring in the network fields (Dere, et al. 1991; Porter and Dere, 1991; Habbal and Grace, 1991; Habbal, 1992). Cheng (1991) and Cook (1991) note the strong variations of emission in the upper chromosphere with characteristic time scales ranging from 2 to 30 sec. Habbal and Grace (1991) note the strong variability over 5 minute intervals in the EUV bright points.

Looking specifically at the small magnetic bipoles in the Ca network, i.e., at the magnetic fields swept into the boundaries of the supergranule cells, Porter, et al. (1987) find localized brightening in the lines of $C_{IV}(10^5\,K)$ throughout the magnetic network, occurring in small magnetic bipoles with lengths of $2\text{–}4 \times 10^3$ km. The brightenings (which are often called microflares) have characteristic lives of the order of 10^2 sec in the smaller bipoles, but a few of the brightenings last several minutes, and occasionally up to an hour. Some bipolar fields remain brighter than others for a time, with 10 percent of the bipoles continuously and variably bright. The larger and stronger the bipole, the more likely the rapid succession of brightenings as if there were a more or less fixed probability of brightening per unit of total magnetic energy or surface area of the Sun. The continuous brightening of the larger bipoles is then a statistical effect of the larger volume and magnetic energy. Porter, et al. (1987) suggest that the individual brightenings may be related to the events observed by Brueckner and Bartoe (1983). It is clear, then, that the heating in the smallest bipoles, which hardly extend above the transition region, is largely through small but intense bursts of energy, i.e., microflares. Habbal (1991) emphasizes that most of the observed fluctuations of this character are associated with the network activity. Very few small bipoles exist outside the converging flows forming the network. Esser (1992) and Habbal (1992) point out that the mean radiative energy output associated with the network activity comes to only the order of 10^5 ergs/cm^2 sec when averaged over the whole surface of the Sun, so it is not obvious that such activity is a major energy input to the active X-ray corona.

Larger bipoles, with lengths of 10^9 cm, contain many thin bright loops which individually turn on and off, providing a continuous and variable brightness in EUV and X-rays. The individual loops fluctuate on time scales of the order of 400 sec (Sheeley and Golub, 1979; Nolte, Solodyna, and Gerassimenko, 1979). Occasionally a large increase in brightness occurs. Thus the X-ray bright points formed by the small (10^9 cm) bipolar magnetic field behave to some degree like scaled down versions of a normal ($\sim 10^{10}$ cm) active region (Krieger, Vaiana, and Van Speybroeck, 1971; Golub, et al. 1977; Moore, et al. 1977).

The size distribution of these small bipolar regions, i.e., the X-ray bright points, is what might be expected, with the smaller regions much more numerous than the larger regions. The smaller regions again are more rapidly fluctuating and shorter lived than the larger regions (Golub, Krieger, and Vaiana, 1976 a,b; Habbal and Withbroe, 1981). As with the very small bipoles already mentioned, it appears from observations that the heat input to the bipolar regions is the result of many small reconnection events, microflares, whose statistical density is more or less independent of the size of the bipolar region. Thus the surface brightness is essentially independent of the length L of the bipolar region as already mentioned, and the brightness fluctuations are the result of the statistical fluctuations of the more or less randomly appearing reconnection events.

This condition prevails in the normal (larger) active regions ($L \cong 10^{10}$ cm) as well. Porter, Toomre, and Gebbie (1984) observed the intensity of emission lines of Si_{IV} and O_{IV} (from $T \cong 10^5$ K in the transition region) with a spatial resolution of 2×10^8 cm. They found the fluctuations in brightness in an area 2×10^8 cm square to be about the same as observed in small isolated bipoles of similar dimensions. Specifically, they observed 20–100 percent fluctuations in each 2×10^8 cm square ($3'' \times 3''$) with characteristic time scales of 20–60 sec. They remark that the fluctuations suggest heating as a result of many independent small reconnection events, and they provide a bibliography of observational information on the variability of the brightness in solar active regions. The source of the variability is not identified in their observations, except that the emission is from gas at a temperature of 10^5 K, but it is possible to make a rough estimate of the energy of the individual microflares or nanoflares responsible for the fluctuations. Using the standard figure $I = 10^7$ ergs/cm^2 sec for the radiation loss from the active X-ray corona, the emission from the area $S = 4 \times 10^{16}$ cm^2 represented by the resolution element 2×10^8 cm on a side becomes 4×10^{23} ergs/sec. Thus a 20 percent brightening over a period of

20 sec involves 1.6×10^{24} ergs, while a 100 percent brightening over 60 sec involves 2.4×10^{25} ergs. The individual reconnection events are, then, of the order of, say, 2×10^{24} ergs in this case. The stronger brightenings may involve large single reconnection sites or several smaller reconnections occurring simultaneously in the same area. These numbers will prove useful in §§11.4.5 and 11.4.7.

There is independent evidence for intense flaring on small scales in the short (1–2 sec) bursts of hard (>20 keV) X-rays detected by Lin, et al. (1984). Extrapolating the detected energy back to the Sun shows that the stronger bursts represent a total X-ray energy of the order of 10^{27} ergs emitted from the individual microflare at the Sun, with more frequent bursts at smaller energies down to the instrumental cutoff at about 10^{24} ergs. Clusters of X-ray spikes appear at random intervals of about 300 sec, suggesting a microflare continuing for 5–100 sec. It is clear that there are energetic events in great numbers, and one assumes that each spike of hard X-rays represents the onset of an individual reconnection event. The mutual triggering of reconnection events presumably accounts for the clusters of spikes that make up a microflare, on which more will be said in §11.4.7 and in the succeeding chapter.

There is another aspect of the distribution of coronal brightness that should be noted here. Saba and Strong (1991) find local line width excesses in a 9A line of Mg_{XI} (10^6 K) emitted near the tops of coronal loops to be two or more times the thermal width detected elsewhere along the field. The line width is enhanced by 45–60 km/sec toward both the red and blue, while the line center is not significantly displaced. They suggest that these line widths may be a direct signature of the heating process. As already noted, Feldman (1983) emphasizes the evidence for strong temperature variations in the form of unresolved filaments extending upward from the top of the chromosphere at a height of about 3000 km.

11.3 General Considerations

It appears that there are probably several different forms of mechanical work that supply heat to the upper atmosphere of the Sun. The chromosphere appears to be heated primarily by the dissipation of acoustic waves and downward conduction of heat from the corona (cf. Osterbrock, 1961; Anderson and Athay, 1989; see also the conference proceedings edited by Ulmschneider, Priest, and Rosner, 1991) at a rate of the order of 10^6 ergs/cm^2 sec (but see Bueno, 1991, who emphasizes the considerable uncertainty introduced by unresolved inhomogeneities). The coronal hole poses a complex problem involving heating near and far from the Sun at a total rate of about 0.5×10^6 ergs/cm^2 sec, probably by two or more different mechanisms (see discussion in Parker, 1991; Esser, 1992) none of which are unequivocally identified at the present time. The active X-ray corona consists of discreet filaments of hot dense gas confined within the intense bipolar magnetic fields of active regions and heated by intermittent distributed bursts of energy at a mean rate of about 10^7 ergs/cm^2 sec. Taking the individual burst to represent a magnetic reconnection event, it is not surprising that the mean rate of heating depends on the strength of the magnetic field. There is then the diffuse, or quiet, corona which fills the space between the intense X-ray coronal loops and all the space around the Sun except for coronal holes. The quiet corona emits a faint glow of X-rays, at the level of about 10^5 ergs/cm^2 sec. The gas is evidently confined by the magnetic field, for if it were not, its temperature of

10^6 K would cause it to expand upward, reducing the gas density and the X-ray emission to the level of a coronal hole.

The present writing is aimed at the bright X-ray coronal loops, as the principal source of X-ray emission from the Sun and other late-main sequence stars. Consider, then, the specific properties of the coronal loops that might allow unique theoretical inference of the heat source that produces them. A general summary of the observed physical characteristics of the coronal loops has been provided by Martens, Van Den Oord, and Hoyng (1985). There are two basic lines of thought for heating the outer atmosphere of a star like the Sun, already mentioned. One is the wave motion generated with $\omega > k_m V_s, k_m C_s$ in the subsurface convective zone, some of which propagates up into the corona where it dissipates its energy into heat. The other is the slow convective deformation of the magnetic field $\omega < k_m C_s$ building up magnetic free energy in the coronal magnetic field. The free energy then dissipates by some means to heat the corona. One can assert apriori that neither mechanism can be entirely absent. The problem is to determine which is the more important and whether it is adequate to account for the necessary heat input of 10^7 ergs/cm^2 sec estimated from the observed EUV and X-ray emission (Withbroe and Noyes, 1977).

The general mean physical characteristics of the X-ray coronal loops have been summarized by Rosner, Tucker, and Vaiana (1978), as well as others. The basic fact, already noted is that the X-radiation is emitted principally by thin filaments (coronal loops) of relatively hot (2×10^6 K) dense gas (10^{10} atoms/cm^3) extending along the individual flux bundles of the closed bipolar fields that protrude above the surface of the Sun. The gas filling the spaces between the X-ray loops is evidently not much cooler than the 2–3×10^6 K of the emitting gas, but is substantially less dense so that the radiative cooling, and the heat input, are a factor of 10 or more below the intense X-ray emitting filament. The emitted intensity is found to be approximately proportional to the integrated thermal energy density in the line of sight, i.e., proportional to $\int dsNkT$. Since the surface brightness is essentially independent of the scale L of the bipolar field, it follows that coronal thermal energy densities are approximately inversely proportional to the characteristic scale of the emitting region. It is also found that the energy density of the emitting gas declines inversely with the age of the emitting region (Landini, et al. 1975; Kahler, 1976). That is to say, the X-ray emitting coronal loop first appears as a thin filament of relatively high intensity and then gradually fades into the background as the length and thickness of the rising emitting filament increases with time over characteristic periods of the order of 10^5 sec. A particularly surprising aspect of the X-ray emitting loops is the narrow range of mean temperature (1–3×10^6 K) of the different loops, with scales ranging from 10^9 to 10^{10} cm and densities of 10^8–10^{10} electrons/cm^3 (Davis, et al. 1975; Withbroe 1975; Vaiana, et al. 1976). The overall coronal loop is often unchanged over periods of hours, generally waxing and waning over a few days (10^5 sec or more). Therefore, in view of the short characteristic radiative and conductive cooling times, described in §11.2, it follows that the loop is generally in an overall quasi-static equilibrium state during most, if not all, of its life. The exception is in the small-scale internal structure of the loop and in the flaring loop, wherein the plasma density, temperature, and radiative emission rise abruptly for a period of time of the order of 10^3 sec or more (Pallavicini, Serio, and Vaiana, 1977).

Rosner, Tucker, and Vaiana (1978) construct a simple hydrostatic thermal equilibrium model of an individual coronal loop, balancing various hypothetical heat input distributions against thermal conduction and radiative cooling. The resulting hydrostatic structure is stable against thermal and gravitational collapse only if the temperature is a maximum at the apex of the loop, which can be achieved only if the heat input is distributed across the apex. The loop model leads to the estimate of the local heat input function E_H of the form

$$E_H \cong 10^5 p^{\frac{7}{6}} L^{-\frac{5}{6}} \text{ergs/cm}^3 \text{ sec} \tag{11.7}$$

and a temperature

$$T \cong 1.4 \times 10^3 (pL)^{\frac{1}{3}} \text{K}, \tag{11.8}$$

where p is the gas pressure and L is the height or half-length of the loop. It follows that the energy input per unit surface area of the Sun is of the order of

$$LE_H \cong 10^5 p^{\frac{7}{6}} L^{\frac{1}{6}} \text{ergs/cm}^2 \text{ sec}. \tag{11.9}$$

The essential point is that E_H is nonvanishing over the entire upper portion of the coronal loop, and the heat input LE_H per unit surface area is essentially independent of the length of the loop. A variation in L by a factor of 10 implies a variation in the heat input of only a factor 1.4. This simple hydrostatic model of the coronal loop fits the observational facts fairly well. Golub, et al. (1980) take the model a step farther, with the idea that the heat input per unit volume, E_H, is supplied by the dissipation (by unspecified means) of the magnetic free energy provided by a steady rotation of the photospheric footpoints of the coronal loop. They used a single characteristic rotational velocity parameter, $v_\varphi = 10^4$ cm/sec and a fixed azimuthal field component B_φ provided by the rotational shear. Consequently E_H scales as the square of the longitudinal field in the loop, which they take as 40 gauss in the small bipolar field of an X-ray bright point ($L \cong 2 \times 10^4$ km), 75 gauss in a full–scale active region ($L \cong 6 \times 10^4$ km) and 20 gauss in a large-scale diffuse emitting region ($L \cong 12 \times 10^4$ km), in rough accord with the observations. These facts, that E_H is proportional to $B^2/8\pi$, that the heating extends along **B** to provide field aligned filaments, and that the heating is centered around the apex of each loop, suggest that the heat input is magnetic in origin.

11.3.1 Detailed Modeling of Coronal Loops

In view of the foregoing observational description and simple modeling of the coronal loop, this is an appropriate point to note the advanced state to which the modeling of coronal loops has developed in recent years. Krishan (1985), Chandra and Prasad (1991), Krishan (1991), and Brown and Durrant (1991) have investigated the gas pressure structure and the radiative emission across a 2D loop. Radiative diagnostics and density determinations have been pursued by Dwivedi and Gupta (1991). Thermal equilibrium and thermal instability are of central importance and the possibilities have been treated by Mok, Schnack, and Van Hoven (1991), Steele and Priest (1990, 1991a,b) and Van der Linden and Goossens (1991). The loss of mechanical equilibrium as a result of shearing and twisting coronal loops (Lothian and Hood, 1992; Robertson, Hood, and Lothian, 1992) as well as the onset of the resistive ballooning mode (Otani and Strauss, 1988; Hardie, Hood, and Allen, 1991) have been explored. Vekstein, Priest, and Steele (1991) consider the consequences of the relaxation of a complex force-free field to the minimum energy state with the same total helicity, along the lines proposed by Taylor (1974, 1986) for the relaxation of the reverse pinch and other laboratory magnetic configurations. Plasma instabilities caused by thermal conduction in a coronal loop are examined by Takakura (1990, 1991) and Ciaravella, Peres, and Serio (1991), with the idea that the resulting plasma turbulence may have interesting effects such as anomalous resistivity. Finally Brown (1991) has looked at the larger picture of the entire bipolar active region, made up of many loops. This larger picture is best summarized by noting that coronal X-ray loops connect into the Sun in regions of enhanced magnetic field, i.e., in active regions, in which the field for some reason is generally pretty close to 10^2 gauss, except near sunspots. These magnetic regions show bright chromospheric emission, and a patch of such enhanced emission is called a *plage*. The brightness is also enhanced at photospheric levels with the patches called *faculae*. Neither the plage nor the faculae is presently understood. They constitute one more mysterious magnetic effect. Recent work by Rabin (1992) indicates mean fields up to about 800 gauss in plages, composed of minute magnetic fibrils of 1200–1700 gauss with filling factors of as much as 0.5 over the 1500 km $\times$ 1500 km resolution of the magnetograph.

The conspicuous exception to the positive correlation between heating and magnetic field is the sunspot umbra ($2–3 \times 10^3$ gauss), which seems never to be the end point of a coronal X-ray loop. On the other hand, coronal loops have been observed to connect into the weaker mean fields ($\sim 10^3$ gauss) of penumbrae.

It appears, then, that magnetic fields up to local mean values of about 10^3 gauss are conducive to strong chromospheric and coronal heating. Stronger fields do not have this effect, nor do weak fields, below some value of the order of 30 gauss.

Consider, then, how it is that magnetic fields of the general order of 10^2 gauss provide a strong energy flux from the subsurface convection upward to heat the corona. There are three stages in this process. The magnetic field of the order of 10^2 gauss evidently provides a stress field against which the convection is able to work effectively. The resulting free energy, created by the work, may then be freely transmitted by the magnetic field upward into the corona. Finally the magnetic field is somehow responsible for the rapid degradation of the energy into heat in the corona. In fact, no one of these three essential steps can be missing if the magnetic field of 10^2 gauss is primarily responsible for coronal heating.

11.4 Toward a Theory of Coronal Heating

With the foregoing description of the X-ray emission from the coronal loops in the Sun, the next step is to consider the theoretical basis for the appearance of the X-ray emitting loops. What ideas can be excluded tentatively, on the basis of existing knowledge, what ideas cannot be excluded, and what future observational and theoretical information might be needed to accomplish an unambiguous identification of the principal physical cause of the solar X-ray emission? It must be kept in mind that a solid theoretical foundation for the emission of X-rays from a solitary late-main sequence star like the Sun is essential for X-ray astronomy and also for understanding and appreciating the origin of the extreme temperature variations of the upper atmosphere of the planet Earth.

As will become clear in the sequel, the many theoretical possibilities can be sorted out, leaving the spontaneous formation of tangential discontinuities as the simplest and most direct route for heating the X-ray corona. The primary energy input is the slow continuous swirling of the footpoints of the field in the subsurface convection, which has yet to be quantified by direct observations. But assuming no surprises in that direction, it would appear that the basic theorem of magnetostatics holds the key to the X-ray astronomy of solitary late main sequence stars. The full set of critical observations to put the theory on a solid foundation involves optical and UV observations with resolution of 100 km (0.13″) or better at the Sun, and comparable X-ray pictures. This is well within present technical capabilities. Indeed, they are a small part of the scientific pursuit of the remarkable variation of the total solar luminosity (cf. Zhang, et al. 1993), which is a major driver in the decade and century variations of the terrestrial climate (Eddy, 1976, 1977, 1983; Eddy, Gilliland, and Hoyt, 1982). For it is the same high resolution studies of the plage, the filament, the sunspot, and the structure and behavior of the individual magnetic fibrils in the photosphere that is required in both cases, with the X-ray emission depending mainly on the mutual swirling of neighboring fibrils.

Now, as already noted, a theory of solar X-ray emission has three essential parts. First, the theory must identify the form of the work done by the convective motions beneath the visible surface of the Sun. Second, the theory must establish how the free energy created by the work is transported up through the visible surface of the Sun into the corona, and, third, the theory must establish how the energy, in whatever form it may be, is converted into heat in the corona. The theory can be properly judged only by comparison with quantitative observational information. Note, then, the crucial fact that the brightness of the coronal X-ray loop (ergs/cm^2 sec) is approximately independent of the length L of the loops over a range of 10^9–10^{10} cm. In this connection note that the widths of the individual loops bear a rough proportionality to the length L. The model described by equation (11.9) provides for a brightness variation of $L^{\frac{1}{6}}$, or only a factor of 1.4 over the entire range. Recall that the surface luminosity of a bright coronal loop reaches approximately 10^7 ergs/cm^2 sec (Withbroe and Noyes, 1977) in a typical bipolar magnetic field of 10^2 gauss. Again, the gas density is 10^{10} atoms/cm^3, and the gas temperature is typically 2–3 $\times$ 10^6 K. Accordingly the speed of sound V_s is 2×10^7 cm/sec, while the Alfven speed C is 2×10^8 cm/sec. The sound transit times over coronal loop lengths of 10^9 and 10^{10} cm are 50 sec and 500 sec, respectively, while the Alfven transit times are 5 and 50 sec, respectively.

11.4.1 Waves in the Corona

Consider the traditional idea that the corona is caused largely by the dissipation of waves propagating up from their source in the thermal convection below the visible surface of the Sun.

To begin with the subsurface convection, the problem is to develop the theory for the production of various wave modes, e.g., acoustic waves, internal gravity waves, and slow, fast, and Alfven magnetohydrodynamic waves. This is a complex undertaking, whose historical development is contained largely in the references in the recent publications cited below. Thus, it is pointed out by Goldreich and Kumar (1988) that the acoustic emission from a turbulent fluid depends strongly on the degree of stratification (see also Moore and Spiegel, 1964) of the atmosphere and also on the manner of driving of the turbulence. Thus, for instance, freely coasting turbulence provides only the relatively weak quadrupole emission of sound waves, while convectively driven turbulence provides dipole emission (stronger by the square of the reciprocal of the Mach number). Musielak and Rosner (1987, 1988) show, among other things, that the introduction of a uniform background magnetic field does not enhance the generation of compressible wave modes. Their calculations cover a range of late-type main sequence stars. Recognizing that the magnetic fields in the Sun are in a fibril state, they treat the generation of waves in fibrils with internal fields of 1100 and 1500 gauss (Musielak, Rosner, and Ulmschneider, 1989). For solar conditions they find that 1500 gauss is so strong that it resists deformation by the turbulence and substantially decreases the magnetohydrodynamic wave emission below the level achieved with 1100 gauss.

Collins (1989a,b, 1992) employs a general method devised by Lighthill (1960) for computing the asymptotic radiation field produced by a localized fluid motion, and applies it to the production of fast, slow, and Alfven waves from specified motions in the presence of a uniform magnetic field. The net result is both a comprehensive set of scaling laws for the emitted radiation and a formalism giving the radiation in terms of the vector harmonics describing the form of the driving forces. It is interesting to note, then, that the sound speed V_s and the Alfven speed C are equal at the visible surface of the Sun ($\rho \cong 2 \times 10^{-7}$ gm/cm^3) for a magnetic field about 1200 gauss. So generally $V_s > C$ and the fast mode corresponds to acoustic waves in the convective zone beneath the surface. Collins finds that the fast mode waves are only weakly generated compared to the production of slow mode and Alfven mode waves, which are created in about equal amounts.

It appears from these general studies that in the presence of mean magnetic fields below 10^3 gauss it may be possible to provide upward propagating acoustic wave fluxes and Alfven wave fluxes in excess of 10^7 ergs/cm^2 sec. The periods of these waves are comparable to the characteristic time of the applied convective forces, i.e., comparable to the characteristic times associated with the most vigorous convective cells. The Alfven wave flux falls off rapidly for magnetic fields below 10^2 gauss because the Alfven speed falls below the convective velocity ($\sim$1 km/sec).

Consider, then, the properties of the waves that might transport the energy flux $I = 10^7$ ergs/cm^2 sec into the upper atmosphere. A wave with rms fluid velocity $\langle v^2\rangle^{\frac{1}{2}}$ propagating in a fluid of density ρ at a speed V carries an energy flux

$$I = \rho \langle v^2 \rangle V. \tag{11.10}$$

For an acoustic wave the sound speed is $V_s \cong 10^6$ cm/sec in the photosphere where $\rho \cong 2 \times 10^{-7}$ gm/cm^3 (10^{17} atoms/cm^3). The result is $\langle v^2\rangle^{\frac{1}{2}} = 0.7 \times 10^4$ cm/sec, representing a wave of small amplitude $\langle v^2\rangle^{\frac{1}{2}}/V_s \sim 0.7 \times 10^{-2}$. The sound speed varies but little with height through the chromosphere, so that in the high frequency limit $\langle v^2\rangle^{\frac{1}{2}}$ varies with height as $\rho^{-\frac{1}{2}}$. Thus $\langle v^2\rangle^{\frac{1}{2}}$ increases to the sound speed in the chromosphere where the number density has fallen to about 10^{13} atoms/cm^3. The nonlinear effects steepen the acoustic waves to shocks, which dissipate rapidly in the chromosphere. Obliquely emitted acoustic waves are also refracted around and back into the photosphere by the rapid rise in temperature at the top of the chromosphere. Detailed theoretical studies (Osterbrock, 1961; Stein, 1968; Stein and Schwartz, 1972, 1973; Stein and Leibacher, 1974; Schwartz and Stein, 1975; Hammer and

Ulmschneider, 1991) show that no significant flux of acoustic waves, gravitational acoustic waves, and fast and slow mode magnetohydrodynamic waves penetrates through the chromosphere into the corona.

That leaves Alfven waves as the only available transport into the corona. The Alfven speed C is 0.6 km/sec in a mean field of 10^2 gauss in the photosphere. Thus there is a good impedance match to the granule convective cells with rms fluid velocities $\langle w^2\rangle^{\frac{1}{2}} \sim 1$ km/sec. It is interesting to note that the necessary energy flux I requires an rms velocity $\langle v^2\rangle^{\frac{1}{2}} \sim 0.3$ km/sec in the waves, comparable in order of magnitude to C, and representing a strong wave, with $\langle v^2\rangle^{\frac{1}{2}}/C = 0.5$. The rms transverse field of the wave is $\langle b^2\rangle^{\frac{1}{2}} \cong 40$ gauss, compared to the mean field of 10^2 gauss. It is apparent that sufficient generation of such strong Alfven waves is possible only with the near coincidence of the Alfven speed and the rms turbulent convective velocity $\langle w^2\rangle^{\frac{1}{2}} \sim 1$ km/sec. The characteristic period of the turbulence is $\ell/\langle w^2\rangle^{\frac{1}{2}} \cong 300$ sec for the characteristic granule scale length $\ell \cong 300$ km. The associated wavelength λ is only a little less than ℓ, of course.

One would expect a decline in power in both stronger mean fields and weaker mean fields, which take the Alfven speed away from the rms convective velocity of about 1 km/sec. Thus, a mean field of 10^3 gauss gives $C = 6$ km/sec. There is very little power in the convective turbulence at such high phase velocity ω/k, which is why Musielak and Rosner (1988) found the power in the waves dropping off rapidly above 10^3 gauss. For weaker fields the fixed energy transport I requires wave rms velocities $\langle v^2\rangle^{\frac{1}{2}}$ increasing as $B^{-\frac{1}{2}}$. Thus the ratio $\langle v^2\rangle^{\frac{1}{2}}/C$ increases as $B^{-\frac{3}{2}}$ with declining B. But that ratio is equal to one at about 50 gauss, so that it would have to be large compared to one in the weak mean fields of 10–20 gauss to be found in quiet regions. On the other hand, an eddy cannot produce a wave with velocity amplitude, or rms velocity, in excess of C, because the eddy with characteristic velocity w, scale ℓ, and characteristic life τ (where $w \sim \ell/\tau$) can do no more than deform the mean field to a geometrical amplitude of the order of ℓ. At most this provides a wave with a characteristic time τ of variation and a mean square velocity of the order of

$$\begin{aligned}\langle v^2\rangle &\sim (\ell/\tau)^2 \\ &\sim \langle w^2\rangle\end{aligned}$$

In fact it provides rather less, because when $C < \langle w^2\rangle^{\frac{1}{2}}$ the wave cannot propagate out of the region before the next uncorrelated eddy has gotten underway, partly obliterating the field deformation of the initial eddy. So the 10^2 gauss mean field is in the neighborhood of the best value for extracting Alfven wave energy from the typical photospheric granule. This conclusion is not in conflict with the points noted earlier on the generation of Alfven waves in the individual magnetic fibril of 1100 gauss.

The observed fact that most coronal X-ray loops have one or both footpoints in regions where the mean field is of the order of 10^2 gauss, or a little more, can be understood in terms of the impedance match of the field to the fluid convection at this field strength.

11.4.2 Wave Dissipation

Consider, then, the Alfven waves in the corona, where $\rho = 1.6 \times 10^{-14}$ gm/cm^3 ($N = 10^{10}$ atoms/cm^3) and $C = 2 \times 10^8$ cm/sec ($B = 10^2$ gauss). It follows from equation (11.10) that if their energy flux is as large as the required intensity I, the rms fluid velocity associated with the waves is about 17 km/sec, or 12 km/sec in each of the two directions perpendicular to the magnetic field. No wave motion has been observed in the coronal loops, but of course there may be many uncorrelated waves along a single line of sight so that one would not detect a net oscillation. An observational upper limit on the rms wave motion has been set by Cheng, Doschek, and Feldman (1979; see also Cheng, Doschek, and Feldman, 1976; Beckers, 1976, 1977; Beckers and Schneeburger, 1977; Doschek and Feldman, 1977; Athay and White, 1978; Feldman, Doschek, and Mariska, 1979; Mariska, Feldman, and Doschek, 1980) who observed the widths of coronal emission lines. Subtracting out the expected thermal broadening they place upper limits on the rms velocity of unresolved motions. Their result is an upper limit in the line of sight of 10–20 km/sec from Si_{VIII} (at 0.9×10^6 K), 10–17 km/sec from Fe_{XI} (at 1.5×10^6 K) and 10–25 km/sec from Fe_{XII} (at 1.7×10^6 K). These are upper limits, of course, and not direct detections, but they clearly leave room for the rms velocity of 12 km/sec in the line of sight. However, it must be appreciated that if $\langle v^2 \rangle^{\frac{1}{2}}$ is restricted to 15 or 20 km/sec, then the Alfven waves must be largely dissipated in a single pass around the coronal loop, because the necessary $\langle v^2 \rangle^{\frac{1}{2}} = 12$ km/sec refers to upward propagating waves. If there are downward propagating waves as well, because the waves propagating up the opposite end of the loop only partially damp while passing over the loop, it is obvious that $\langle v^2 \rangle^{\frac{1}{2}}$ would be much larger. Very approximately, if the fraction α of the wave energy is dissipated in a single pass around a coronal loop, then the surviving fraction $(1 - \alpha)$ propagates down the opposite leg of the loop. With $I = \alpha\rho\langle v^2 \rangle C$ in place of equation (11.10), where $\langle v^2 \rangle$ is again the mean square fluid velocity in upward propagating waves, it follows that the mean square fluid velocity in the downward propagating waves at the far end of the loop is $(1 - \alpha)\langle v^2 \rangle$. The total mean square velocity at either end of the loop, assuming waves from opposite ends to be uncorrelated (i.e., no resonance), is the sum of the upward and downward propagating waves, with a total of $(2 - \alpha)\langle v^2 \rangle$. Expressing this in terms of I, the result is

$$(2 - \alpha)\langle v^2 \rangle = (2/\alpha - 1)I/\rho C$$

for the total mean square velocity at either end of the loop. The rms velocity is, then, larger by $(2/\alpha - 1)^{\frac{1}{2}}$ and unless α is at least 0.5, the observational upper limits do not permit a sufficient net wave energy flux into the corona.

In recognition of this limitation Hollweg (1984, 1986; Hollweg and Sterling, 1984; Sterling and Hollweg, 1985) pointed out that some dissipative effect comparable to eddy viscosity, must be invoked, i.e., some effect that disperses a wave in its own characteristic period. Various wave mode coupling ideas have been explored, converting Alfven waves into acoustic gravity waves (Fla, et al. 1984; Mariska and Hollweg, 1985). Strauss (1988) has investigated the effect of "hyperresistivity," involving the onset of the tearing mode instability across the field shear in the transverse scale ℓ between trains of Alfven waves. Unfortunately, the small wave amplitudes $\langle v^2 \rangle^{\frac{1}{2}}/C \sim \langle(\delta B)^2\rangle^{\frac{1}{2}}/B \sim 10^{-2}$ make all these nonlinear effects too slow to be of interest, just as plasma turbulence is not expected in the modest current densities $j = c\delta B/4\pi\ell$, where δB is the magnetic amplitude of the wave. Alfven waves of small amplitude are subject to no known instability.

Kuperus, Ionson, and Spicer (1981) turned to the possibility of wave resonance, with the fundamental mode given by $\lambda = 2L$. The idea is simply that each loop length resonates to a wave period $\tau = 2L/nC$, where $n = 1, 2, 3, \ldots$ The theory was subsequently refined to the concept of surface resonance, in which the resonance condition is satisfied only along certain surfaces in the loop for a given wave period τ. The surface resonance of an Alfven wave produces a thin layer of longitudinal oscillation at the resonance surface (cf. Lee and Roberts, 1986; Davila, 1987) into which the energy of the transverse oscillations on either side is directed. It is a remarkable phenomenon, which has received considerable attention over the years (Grossman and Tataronis, 1973; Chen and Hasegawa, 1974; Tataronis, 1975; Kappraff and Tataronis, 1977; Wentzel, 1979a,b; Sakurai and Granik, 1984; Hollweg, 1984, 1985, 1987a,b; Donnelly, Clancy, and Cramer, 1985; Einaudi and Mok, 1985, 1987; Grossman and Smith, 1988; Strauss and Lawson, 1989; An, et al. 1989; Mok and Einaudi, 1990; Hollweg, et al. 1990 and references therein). Very large oscillatory velocities are associated with the resonance, of course, although the published literature is strangely silent on the specific numbers that might apply in a coronal loop. One imagines that the amplitude of the resonant oscillation builds up to a level that is sufficient to excite nonlinear turbulence, such as a runaway Kelvin–Helmholtz instability (Hayvaerts and Priest, 1984; Browning and Priest, 1984) or hyperresistivity (Strauss, 1988) thereby fulfilling Hollweg's criterion in some way.

On the other hand, there are reasons to doubt the existence of a resonance. The footpoints of each end of a bipolar magnetic field are shuffled and intermixed to some degree by the photospheric convection. Therefore, the field lines in the bipolar field are wrapped and interlaced in the manner illustrated in Fig. 11.1. Indeed, it is to be expected that any magnetic field rooted in a convecting body such as a star or interstellar gas cloud is subject to such internal interweaving, so that the individual field lines are stochastic (Jokipii and Parker, 1968, 1969a,b; Jokipii, 1973). Similon and Sudan (1989) have shown how the wandering field lines break up an Alfven wave front, and this break-up would be expected to prevent the build-up of any strong resonance over many periods of oscillation.

11.4.3 The Wave Dilemma

As a matter of fact there is a more elementary difficulty with the wave-heating hypothesis, viz the characteristic time τ for the subphotospheric convection is too

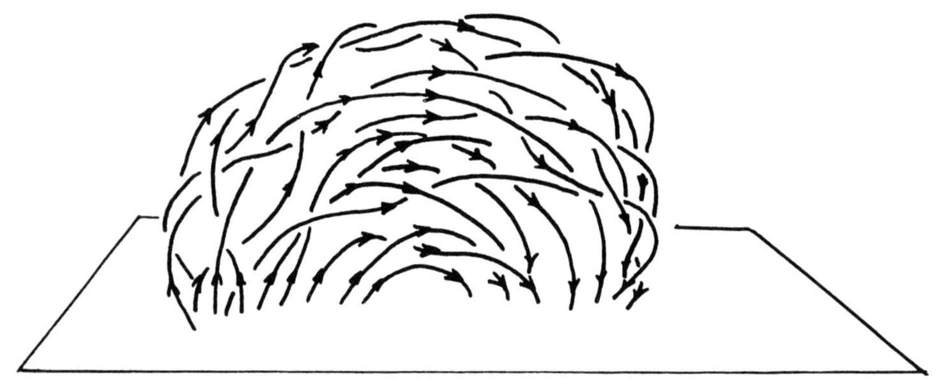

Fig. 11.1. A schematic drawing of the interlaced and woven field lines arising from the general shuffling and swirling of the footpoints of the field in the subphotospheric convection.

long, providing Alfven wavelengths that are considerably longer ($\lambda > 2L$) than most of the coronal X-ray loops they are supposed to heat. Recall that the observations show that the surface brightness is essentially independent of the loop length over the extended range 10^9 cm $< L < 10^{10}$ cm. In particular, the passage of long wavelengths ($\lambda > 2L$) causes only an overall quasi-rigid displacement of the magnetic bipole, with relatively little deformation and negligible dissipation. The difficulty is evident at once from the fact that the shortest characteristic convective turnover times are for the uppermost convective cells, viz. the granules, for which $\tau \cong 300$ sec. The principal wave production involves 5 minute oscillations, therefore, and for $C = 2 \times 10^8$ cm/sec in 10^2 gauss the characteristic wavelength $\lambda = C\tau$ is 6×10^5 km, far longer than the coronal loops. The longest wavelength λ that fits into a coronal loop of length L is the fundamental resonant mode $\lambda = 2L$, which may be as long as 2–3 $\times 10^5$ km for the longest loops. Weaker mean fields (~ 50 gauss) yield shorter waves (3×10^5 km) so that the longest X-ray loops might contain the fundamental mode. But what of the shorter coronal X-ray loops, with lengths of 10^9 cm? They are too short to contain even $\frac{1}{2}\lambda$, so the waves provide only a weak quasi-static deformation with little or no heating of the shorter bipole magnetic fields.

Another obvious possibility is the turbulent nature of the granules, which have Reynolds numbers of the order of 10^{10} or more. If the characteristic turn over time of the granule is $\tau = 300$ sec, there are surely smaller eddies with shorter characteristic times. Weak fluctuations are detected with periods of 100 sec, and the turbulent cascade must extend down to much smaller eddies, with periods of 1 sec or less. In a Kolmogoroff spectrum the characteristic time $\tau(\ell)$ associated with an eddy of characteristic scale ℓ varies as $\ell^{\frac{2}{3}}$. Since the velocity of the eddy varies as $\ell^{\frac{1}{3}}$, it follows that the kinetic energy density of eddies with scale ℓ varies as $\ell^{\frac{2}{3}}$, i.e.,

$$\frac{1}{2}\rho v^2(\ell) \sim \tau(\ell)$$

and

$$v(\ell) \sim \tau^{\frac{1}{2}}(\ell).$$

The essential point is that only those waves with $\lambda < 2L$, i.e., with $\tau < 2L/C$, are able to fit into a coronal X-ray loop of length L. The kinetic energy in the convection with characteristic periods satisfying

$$\tau \leqslant 2L/C$$

is directly proportional to τ and hence to L. The result is that the available energy density declines in proportion to the period of the wave, and one would expect the brightness of a coronal X-ray loop to vary approximately in proportion to its length L. Observations, as already noted, show that the brightness varies only as $L^{\frac{1}{6}}$ and is essentially independent of L. It is this observational fact, that the brightness is more or less independent of the length L, that cannot be achieved by dissipation of the Alfven waves expected from the subphotospheric convection. To put the matter differently, an energy input I that is independent of loop length L might be effected by postulating that the principal Alfven wave energy is contained in waves with periods of 5 sec (for which $\lambda = 10^9$ cm) or less. Just how one might expect to dissipate these waves effectively in coronal loops of different lengths remains to be established, of course. The possibility of any resonance has the unfortunate property of singling out a particular loop length at which the brightness is a conspicuous maximum. On the other hand if one postulated all wavelengths present with equal power in each period interval $\delta\tau/\tau$, the longer loops would contain all waves with $\tau < 2L/C$, experiencing substantially more wave power than the shorter loops.

In any such scenario the basic dilemma would be the question of how convection, with a characteristic turn over time of about 300 sec, could produce strong Alfven waves with periods exclusively of the order of 5 sec or less. The first stage of the theory would then be cast into deep mystery in the effort to construct the second and third stages. So instead, we accept the characteristic 300 sec and inquire where it leads the theory.

11.4.4 Quasi-static Fields

The relatively long characteristic turn-over time of the subsurface convection in general, and the $\tau \cong 300$ sec of the granules in particular, is longer than the Alfven transit time of about 100 sec along the largest coronal loop, indicating that the deformation of the bipolar magnetic fields is quasi-static rather than oscillatory. The natural and obvious effect is the relatively slow mixing and swirling of the footpoints of the field on some convective scale ℓ at velocities v of the order of magnitude of a km/sec. Thus, after a time the field lines in the largely nondissipative atmosphere of the magnetic active region become interlaced and woven in the general manner sketched in Fig. 11.1.

There is indirect evidence of the random mixing of the footpoints of the field in the observed diffusive spreading of the field of active regions over the surface of the Sun, with an effective diffusion coefficient $\mathcal{D}$ of the order of 10^{12}–10^{13} cm^2/sec (Leighton, 1966; Sheeley, Nash, and Wang, 1987; Sheeley, Wang, and DeVore, 1989; Wang, Nash, and Sheeley, 1989; Wang and Sheeley, 1990; and references therein). The combined granule and supergranule motions are presumed to be the primary cause of the global diffusion of field. The diffusion coefficient $\mathcal{D}$ is expressible in terms of a characteristic mixing length ℓ and eddy velocity v, with

$$\mathcal{D} \cong \frac{1}{3}\ell v$$

The supergranules, for which $\ell \sim 10^9$cm and $v \cong 0.3$ km/sec would appear to be the major contributor to $\mathcal{D}$, with $\mathcal{D} \cong 10^{13}$ cm^2/sec. The granules, with $\ell \sim 3 \times 10^7$ cm and $v \cong 1$ km/sec provide $\mathcal{D} \cong 10^{12}$ cm^2/sec. However, for the purposes of winding up the magnetic field lines in complicated patterns, the granules produce much more vorticity and swirling, with $v/\ell \sim 3 \times 10^{-3}$/sec (i.e., a characteristic rotational period of 300 sec) whereas the supergranules provide 3×10^{-5}/sec.

The aerodynamic drag of the granule motion on a flux bundle seems to be more than adequate to overpower the Maxwell stresses that oppose the winding. This question is taken up in §12.3 for the much stronger magnetic fields in which flares occur. The essential point here is that the magnetic fibrils cannot avoid being carried along in the motion of the granules and supergranules. The mixing and swirling of the fibrils increases the magnetic free energy in the bipolar fields in the solar atmosphere, at heights of the order of $\frac{1}{3}L$ for a bipole of length L. It remains only for observations to determine the precise rate of mixing, from which one can get some idea of the rate at which the convection does work on the magnetic field, introducing magnetic free energy into the bipolar fields. This is taken up again in §11.5.

Now it was recognized years ago that the dissipation of quasi-static magnetic fields in the solar corona can provide sufficient heat only if for some reason the electric currents are concentrated into thin sheets (Gold, 1963; Tucker, 1973, Levine, 1974, Rosner, et al. 1978; Rosner, Tucker, and Vaiana, 1978), much along the lines imagined for solar flares (Sweet, 1958a,b; Gold and Hoyle, 1960; Coppi and Friedland, 1971; Coppi, 1975). Craig, McClymont, and Underwood (1978) suggested small-scale flaring of some form as the principal heat input. The resistive dissipation is proportional to the square of the current density for a given resistivity, and the mean square current density increases with the nonuniformity of the current distribution for a given mean current density. The mean current density is determined by the overall torsion in the quasi-static field, with the mean current $\langle j \rangle$ of the order of $cB/4\pi\ell$ in a force-free field B with characteristic scale ℓ. If the current is concentrated into thin sheets with $\langle j \rangle$ occupying a fraction ε of the region, then $\langle j^2 \rangle = \langle j \rangle^2/\varepsilon$. The resistive dissipation is increased by the large factor $1/\varepsilon$. There may be a much larger enhancement of the dissipation if the peak current density, of the order of $\langle j \rangle/\varepsilon$, excites plasma turbulence and anomalous resistivity, as emphasized by Rosner, et al. (1978) (see discussion in §§2.4.3 and 2.4.4).

The essential point is that the basic theorem of magnetostatics provides the mechanism for the concentration of the mean current into thin sheets — mathematical tangential discontinuities in the ideal case of an infinitely conducting fluid (Parker, 1972, 1981, a,b, 1983a,b). The theorem asserts that any strong continuous swirling of the footpoints of the field causes the formation of tangential discontinuities in the coronal field, in the manner described in Chapters 4–8. Therefore there is rapid dissipation because the Maxwell stresses continue to sharpen the discontinuities in opposition to the broadening by resistivity, guaranteeing active dissipation and reconnection of the field.

As noted in §1.10 Van Hoven (1976, 1979, 1981) and Spicer (1976, 1977, 1982) pointed out the general occurrence of the resistive kink and resistive tearing instabilities throughout a sheared magnetic field, with the specific suggestion that they are the major source of dissipation of magnetic free energy in coronal loops. We are now in a position to reassess this possibility in light of the spontaneous formation of tangential discontinuities. The essential feature of the resistive instability is the relatively long onset time τ, of the order of

$$\tau \sim (kh)^{\frac{2}{5}}(h^2/\eta)^{\frac{3}{5}}(h/C)^{\frac{2}{5}}$$

for a wavelength $2\pi/k$ in a shear of characteristic scale h and field strength B, with $C = B/(4\pi\rho)^{\frac{1}{2}}$. In units of the characteristic resistive diffusion time h^2/η across the scale h,

$$\tau = (h^2/\eta)/N_m^{\frac{2}{5}}$$

where $N_m \equiv C/k\eta$ is the characteristic Lundquist number. Thus the onset of the resistive tearing mode and the formation of islands is relatively slow (not quite as fast as the minimum characteristic rapid reconnection time $(h^2/\eta)/N_m^{\frac{1}{2}}$).

Consider, then, the onset in a strongly deformed magnetic field of $B = 10^2$ gauss over a characteristic scale as small as $h \cong \ell \cong 300$ km. The coronal resistive diffusion coefficient $\eta \cong 2 \times 10^3\,\text{cm}^2/\text{sec}$, yields $h^2/\eta \sim 5 \times 10^{11}$ sec, and $N_m = 5 \times 10^{11}$ if $k = 2\pi/h$. Then $N_m^{\frac{2}{5}} \cong 5 \times 10^4$, and τ is of the order of 10^7 sec, too long to be of much interest.

So the primary effect is the formation of tangential discontinuities, on a characteristic time of the order of h/C, measured in seconds. It is here, then, that the resistive tearing instability enters the picture, at the rapidly steepening localized shear layers on their way to forming tangential discontinuities. When h falls to one km, h^2/η falls to 5×10^6 sec. With the same wavelength of 300 km, it follows that N_L is still 5×10^{11} and τ becomes 10^2 sec. Several seconds later h falls to 10^4 cm and τ becomes 1 sec, etc. So we expect the resistive instabilities proposed by Van Hoven and Spicer to play an important role in the dissipation of magnetic free energy through the onset of rapid reconnection at the special locations determined by the incipient tangential discontinuities.

It was noted in §§10.1–10.3 that the dissipation of magnetic field does not depend upon whether the field in the photosphere is continuous or is broken into distinct intense individual magnetic fibrils. The point is that the *topology* of the bipolar coronal field becomes sufficiently complex after a time that the Maxwell stresses force the creation of tangential discontinuities as the stresses strive to bring the field into static equilibrium. It follows that no matter what the state of the field in the photosphere, and no matter how small the resistivity of the coronal gas, the thickness of the current sheets soon becomes so small as to provide resistive dissipation.

The next step, then, is to determine the rate of dissipation of the magnetic field in the corona. It was clear from the description of rapid reconnection of field across a surface of discontinuity in Chapter 10 that the reconnection rate is nonlinear and hypersensitive to the initial conditions, boundary conditions, and the likelihood of a time varying, spatially inhomogeneous plasma turbulence and anomalous resistivity, which in turn depend sensitively on local electron velocity distributions, etc. Dissipation limits can be set, e.g., equation (10.35), for a given Lundquist number N_L, but the limits in N_L and $N_L^{\frac{1}{2}}$ vary widely for coronal values $N_L \sim 10^{12}$–10^{15}.

A variety of detailed modeling of special cases has appeared in the literature providing a vivid picture of the process in idealized cases. Thus, for instance, Beaufume, Coppi, and Golub (1992) treat the individual coronal loop as a single coherent flux bundle subject to continual twisting by the circular motions of its footpoints, along the lines suggested earlier by Sturrock and Uchida (1981). The result is a sheath of intense electric surface current. Beaufume, Coppi, and Golub (1992) explore the onset of resistive tearing modes and anomalous resistivity, leading to formation islands, etc. They emphasize the spatial inhomogeneity and intermittent character of the magnetic reconnection and dissipation of magnetic free energy into heat, consistent with the small filling factor of the hot plasma in coronal loops inferred from observation by Martens, Van Den Oord and Hoyng (1985). They treat the downward conduction of heat along the filament and the associated upward expansion of chromospheric gas to provide the enhanced coronal gas density in the loop.

Robertson, Hood, and Lothian (1992) provide models in one, two and three dimensions exhibiting the kinking instability in three dimensions that is so effective in producing discontinuities in the field (Parker, 1975). Otani and Strauss (1988) and Hardie, Hood, and Allen (1991) model the resistive ballooning modes, suggesting that the resulting current sheets play a role in coronal heating.

Bhattacharjee and Wang (1991) demonstrate the formation of current sheets as a necessary consequence of conservation of magnetic helicity in the presence of a small but finite resistivity (Walbroeck, 1989). They go on to investigate the reconnection in some detail. Wang and Bhattacharjee (1992) explore "free" magnetic reconnection, arising from an instability in the formation of the current sheet, and "forced" reconnection driven by the continuing deformation of the large-scale field by the convective motion of the footpoints. They emphasize that reconnection of both types is expected and both conserve helicity so as to produce current sheets. They work along the lines of the basic model of Hahm and Kulsrud (1985), to show in detail how the reconnection evolves, providing in the end an explicit model calculation of coronal heating.

The enormous complexity of the reconnection phenomenon and the associated nonuniqueness of present models, suggests that a less specific approach to the problem is in order. Consequently the present writing treats the reconnection and heating in generic terms, beginning with the intermittency. Empirically it is known (§§10.3.5 and 10.3.6) that there is little reconnection and dissipation of the magnetic free energy until the intensity of the current sheet, as it is driven by the Maxwell stresses toward a magnetic tangential discontinuity, exceeds some threshold, whereupon there is an explosive burst of reconnection, which substantially reduces the strength of the current sheet before quenching and restoring the trend toward magnetic discontinuity to its original quasi-steady monotonic state. We expect, then, that the continuing intermixing of the footpoints of the field in the photospheric convection produces a continual slow increase in the strength of the many surfaces of discontinuity. The increase is interspersed with, and limited by, transient local bursts of reconnection, each burst converting a portion of the local free magnetic energy into heat. The individual reconnection events are literally small flares and the term *nanoflare* is an appropriate appellation, indicating their general magnitude (see §11.4.4) relative to the largest flares of 10^{32}–10^{33} ergs. The small size and large number of nanoflares renders the individual nanoflare largely unresolvable. On the basis of this theoretical picture, the X-ray coronal loop is to be regarded as a sea of nanoflares.

11.4.5 Energy Input to Quasi-static Fields

Now, if it is not possible to deduce the detailed dynamics of the nanoflares from first principles, it is nonetheless a fact that the average rate at which the subphotospheric convection does work on the interwoven quasi-equilibrium field of the magnetic bipolar region must equal or exceed the heat input I to the coronal X-ray loops. This follows simply from the fact that the work goes directly into the magnetic free energy of the interlacing of the bipolar field. The principal escape of the accumulating magnetic free energy is through dissipation into heat by rapid reconnection, largely in the corona, where the Alfven speed is high and the plasma density is low, and where the winding and wrapping of the field lines is concentrated by the upward decline of the mean field.

To estimate the rate at which the photospheric motions do work on the field above, consider the idealized situation, sketched in Fig. 11.2, showing a uniform field B_0 extending from $z = 0$ to $z = L$ through an infinitely conducting fluid. The field is fixed at $z = L$ while the footpoint at $z = 0$ of one slender flux bundle of small radius a

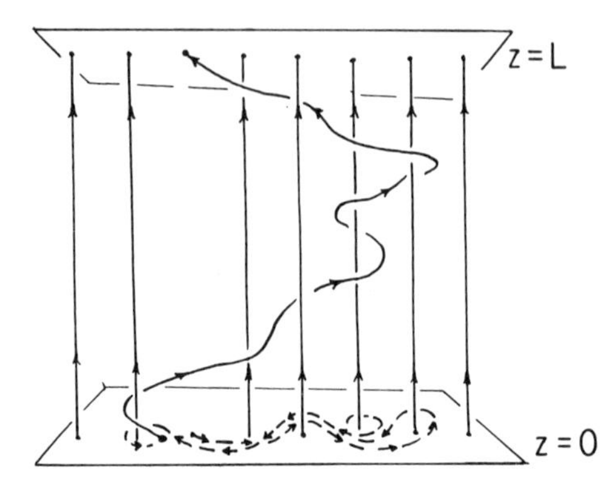

Fig. 11.2. A sketch of the idealized situation in which only one of all the flux bundles that make up a uniform magnetic field is subject to footpoint motion, causing that one bundle to trail out behind as the moving footpoint wanders at random among the other flux bundles.

is set to wandering at random among its neighbors. For simplicity the ambient field is considered as a close packed array of similar flux bundles, each of radius a.

To fix ideas suppose that the wandering consists of successive steps of length $b(>4a)$. The direction of each step is random, uncorrelated with the preceding step. That is to say, the step length b is essentially the correlation length of the random motion of the wandering footpoint. With the opposite end of the flux bundle fixed at $z = L$, it follows that the flux bundle trails obliquely out behind the wandering footstep in the manner shown in Fig. 11.2 (Parker, 1981b, 1983b). There is always the probability $p \sim a/2\pi b$ that a given step moves back over the previous step, thereby cancelling the previous step. But for $b > 4a$ this probability is less than $1/8\pi$ and can be neglected. The essential point is that the flux bundle, trailing out behind the moving footpoint, threads through and among the neighboring flux bundles along the same tortuous path traced out on $z = 0$ by the footpoint. After a time t the length s of that meandering path is vt and is composed of $n = vt/b$ independent steps, each of length b. In the idealized situation shown in Fig. 11.2, the inclination θ of the flux bundle to the z-direction is given by

$$\begin{aligned}\tan\theta &= s/L,\\ &= vt/L.\end{aligned}$$

In a more realistic case, in which the footpoints of all the flux bundles are set to wandering simultaneously, θ is increased above this estimate by the spreading out of the other flux bundles around which the test bundle winds. On the other hand, θ is somewhat diminished in that case by the tension in the test bundle, deforming the path of each neighboring bundle around which it winds. We expect the net result to be of the same order of magnitude as the estimate based on the present idealized situation.

With the inclination θ of the flux bundle, the ratio of the transverse field component $B_\perp$ to the z-component B_0 is just $\tan\theta$, so that

$$\begin{aligned}B_\perp(t) &= B_0\tan(\theta)\\ &= B_0 vt/L\end{aligned} \tag{11.11}$$

The force F per unit area exerted by the displaced flux bundle on its moving footpoint is the off diagonal component of the Maxwell stress tensor,

$$\begin{aligned}F &= B_0 B_\perp(t)/4\pi,\\ &= B_0^2 vt/4\pi L.\end{aligned} \tag{11.12}$$

Since the flux bundle above the footpoint trails backward from the forward displacement b of each independent step, the step is made in opposition to F. Hence the rate W per unit area at which work is done on the field by the convective forces at $z = 0$ is Fv, or

$$W(t) = vB_0 B_\perp(t)/4\pi, \tag{11.13}$$

$$= (B_0^2/4\pi)(v^2 t/L). \tag{11.14}$$

The magnetic free energy $\mathcal{E}$ per unit area is

$$\begin{aligned} \mathcal{E}(t) &= LB_\perp^2/8\pi \\ &= \int_0^t dt' W(t') \\ &= \frac{B_0^2 v^2 t^2}{8\pi L} \end{aligned} \tag{11.15}$$

This result has been deduced on the assumption that the motion of the footpoint is slow, with the time for each step b long compared to the Alfven transit time $L/C \cong 5$–100 sec along the loop.

Another concern is the assumption that the growing magnetic pressure $(B_0^2 + B_\perp^2)/8\pi$ of the flux bundle does not substantially alter the cross section of the bundle. As will be shown, $B_\perp^2 \ll B_0^2$, so that the pressure change is only a small correction, of no consequence for estimating the total rates.

Equation (11.14) allows a simple estimate of the rms inclination θ of the field lines to the mean field direction, and the time required to achieve this state. The rate W at which the motion v of the footpoints of the field does work on the field is, under steady state conditions, equal to the mean rate of dissipation of magnetic free energy, which is just the coronal heating rate I. Equating I to W yields

$$\tan\theta = 4\pi I/B_0^2 v \tag{11.16}$$

As a guess, suppose that the velocity v with which the footpoints of the field are mixed among each other is 0.5 km/sec, as compared to granule velocities of 1–2 km/sec at the visible surface of the Sun. With $I = 10^7$ ergs/cm^2 sec and $B_0 = 10^2$ gauss, the result is $\tan\theta \cong 0.25$ and $\theta \cong 14°$, representing a modest interweaving and inclination of the field lines. This steady state is reached in 5×10^3 sec in a bipolar field of length $L = 10^9$ cm and in 5×10^4 sec in the field of an active region with $L = 10^{10}$ cm (Parker, 1983b; Berger, 1994).

The next section examines the dissipation of magnetic free energy in more detail, arriving ultimately at these same estimates of θ for the choice $v = 0.5$ km/sec.

11.4.6 Dissipation of Quasi-static Fields

The magnetic free energy $\mathcal{E}$ in the corona increases at the slow rate W. The build-up of the free energy goes on until $B_\perp$ passes some threshold $B_{\perp c}$ for explosive reconnection. This occurs presumably (see §10.3.5) when the current sheets, formed by the stress requirements of static equilibrium, are sufficiently thin and intense as to produce a runaway resistive tearing instability, perhaps in connection with a local burst of anomalous resistivity, initiating a brief epoch of extremely rapid reconnection. The reconnection takes place at the expense of the free energy $\mathcal{E}$ in $B_\perp$, which is converted partly into heat with a non-Maxwellian particle velocity distribution, partly into jets of fluid, and in view of the burst-like character of the reconnection, partly into relatively short period magnetohydrodynamic waves ($\tau \lesssim 1$ sec). The fluid jets are turbulent and quickly thermalized. The short period magnetohydrodynamic waves are also quickly thermalized, with only the longer period Alfven waves escaping from the locale of the reconnection. So most of the reduction of $\mathcal{E}$ goes quickly into local heating.

We shall use this qualitative picture of reconnection to construct a simple representation of the dissipation of the growing tangential discontinuities and the associated magnetic free energy of the bipolar magnetic fields on the Sun. The basic unit of reconnection is referred to as a *nanoflare*, already mentioned. It remains to be determined to what degree the nanoflares are truly separate and distinct in space and time, as opposed to blending into a general background of continuing slow reconnection everywhere throughout the bipolar field. The intense inhomogeneity of the temperature in the X-ray emitting coronal gas suggests that the idealization of the nanoflare is at least a useful concept, whether it is literally correct or not. The point is that an analysis based on a quasi-steady state of reconnection throughout the field provides the same overall conclusions except that it does not account for the rapid fluctuation of temperature indicated by analyses of the coronal emission lines. Both concepts account for the intense unresolved filamentary structure of 10^5 K and 10^6 K gas in the low corona.

Assume, then, that the reconnection occurs in bursts which reduce the magnetic free energy $\mathcal{E}$ from the critical value $\mathcal{E}_c$ at the threshold $B_{\perp c}$ to some fraction $\gamma\mathcal{E}_c(\gamma < 1)$, whereupon the rapid reconnection quenches, and reconnection returns to a negligible rate, perhaps near the lower limit indicated by equation (10.35). The slow deformation and interlacing of the flux bundles continues and $\mathcal{E}$ gradually builds up again until once more $B_\perp$ reaches the threshold $B_{\perp c}$ and another nanoflare is initiated, etc. If this intermittent sawtooth relaxation oscillation of $\mathcal{E}$ is to serve as the principal coronal heat input I, then the mean value of W must equal I, with

$$\langle W \rangle \cong 10^7 \text{ ergs/cm}^2 \text{ sec.} \tag{11.17}$$

The mean value of W over the simple sawtooth relaxation oscillation between $\mathcal{E}_c$ and $\gamma\mathcal{E}_c$ is the energy difference $(1-\gamma)_c$ divided by the time T in which $\mathcal{E}$ grows from $\gamma\mathcal{E}_c$ to $\mathcal{E}_c$.

Starting with $B_\perp = 0$ at time $t = 0$, it follows from equation (11.15) that $\mathcal{E}$ reaches $\gamma\mathcal{E}_c$ at time t_γ where

$$t_\gamma = \gamma^{\frac{1}{2}} \chi^{\frac{1}{2}} L/v \tag{11.18}$$

and where χ is the dimensionless ratio

$$\chi = 8\pi\mathcal{E}_c/LB_0^2. \tag{11.19}$$

of the free energy per unit volume, $B_\perp^2/8\pi$ at the threshold $B_\perp = B_{\perp c}$ for the onset of the nanoflare, to the background energy $B_0^2/8\pi$ per unit volume. Then $B_{\perp c} = \chi^{\frac{1}{2}} B_0$ and $\chi = \tan^2 \theta_c$. The free energy reaches the critical level $\mathcal{E}_c$ after the time t_c where

$$t_c = \chi^{\frac{1}{2}} L/v. \tag{11.20}$$

With $T = t_c - t_\gamma$ it is readily shown that

$$\begin{aligned}\langle W \rangle &= \mathcal{E}_c(1-\gamma)/T, \\ &= (1+\gamma^{\frac{1}{2}})\chi^{\frac{1}{2}}\, vB_0^2/8\pi \text{ ergs/cm}^2\text{ sec.}\end{aligned} \tag{11.21}$$

It is interesting to note from this relation that the rate at which the footpoint velocity v does work on the field and creates magnetic free energy is proportional to $\chi^{\frac{1}{2}}$, proportional to the square root of the magnetic free energy necessary to trigger a rapid reconnection event. The higher this threshold, i.e., the less effective the rapid reconnection process, the larger the energy input to the corona. This is, of course, the opposite of the situation with other forms of coronal heating, where generally the coronal heating increases with the effectiveness of the dissipation process, and in fact finds serious difficulty for lack of an effective dissipation process.

To continue, then, the mean $B_\perp$ is

$$\begin{aligned}\langle B_\perp \rangle &= \frac{1}{2} B_0(v/L)(t_c + t_\gamma), \\ &= \frac{1}{2}(1+\gamma^{\frac{1}{2}})\chi^{\frac{1}{2}} B_0,\end{aligned} \tag{11.22}$$

$$= \frac{1}{2}(1+\gamma^{\frac{1}{2}})\, B_{\perp c}. \tag{11.23}$$

The mean square transverse field is

$$\langle B_\perp^2 \rangle = \frac{1}{3}(1+\gamma^{\frac{1}{2}}+\gamma)\,\chi\, B_0^2, \tag{11.24}$$

$$= \frac{1}{3}(1+\gamma^{\frac{1}{2}}+\gamma) B_{\perp c}^2. \tag{11.25}$$

Equating $\langle W \rangle$ to the heat input rate I yields

$$\chi = \left[\frac{8\pi I}{(1+\gamma^{\frac{1}{2}})v\, B_0^2}\right]^2 \tag{11.26}$$

Now we expect the nanoflare to be a localized burst of reconnection occurring at some position along the considerably larger length L of the magnetic loop. Hence, even if a major portion of the local $B_\perp^2/8\pi$ were converted into heat, the net effect on the total free energy $LB_\perp^2/8\pi$ over the full length would be small. It follows, then, that γ, which pertains to the whole length, can be approximated by $\gamma = 1$ in the above expressions. It follows that

$$\chi \cong [4\pi I/v\, B_0^2]^2. \tag{11.27}$$

This result is equivalent to equation (11.16), of course, since, with $1 - \gamma \ll 1$, the magnetic free energy and the rate W at which work is done on the field remain close to the critical value. There is, of course, the sawtooth ripple of small amplitude $(1-\gamma)\mathcal{E}_c$ on top of the mean.

With $I = 10^7$ ergs/cm^2 sec and $B_0 = 10^2$ gauss, assume again that the velocity v with which the photospheric footpoints are mixed among each other is 0.5 km/sec. The result is $\chi \cong 0.06$, yielding $\tan\theta = 0.25$ and $\theta = 14°$. The winding up time to the critical level follows from equation (11.20) as $t_c = L/4v$, yielding again 5×10^3 sec for $L = 10^9$ cm and 5×10^4 sec for $L = 10^{10}$ cm. These wind up times are somewhat less than the characteristic time of variation of bipolar magnetic fields of dimension L, indicating that the quasi-steady state assumed in the development is not inappropriate.

To pursue the matter further, it should be noted that the field does not start from a uniform state ($B_\perp = 0$). It first appears as an Ω shaped bipolar flux bundle rising up through the surface of the Sun. The arched flux bundle emerges, perhaps already in a state of strong internal winding and interweaving, becoming detectable when its length L is perhaps as small as 10^3 km. The characteristic winding up time at that length is only a matter of 10 minutes, even if the flux bundle has no initial internal interweaving and winding. As the X-ray emitting bundle rises and lengthens, the winding and interweaving may be maintained near $\mathcal{E}_c$ by the footpoint motion. It is obvious from observation that not all flux bundles in a bipolar magnetic field are intense emitters of X-rays. Evidently only a fraction of the flux bundles have their footpoints mixed and swirled around each other fast enough to supply the necessary energy input. For it must be remembered that the estimate of $\chi = 0.06$, $\theta = 14°$, is based on an assumed footpoint swirling at $v = 0.5$ km/sec and a fixed length L for the flux bundle above the surface of the Sun. If L is increasing with time, the rate of motion of the footpoints at each end is $v = \frac{1}{2}(dL/dt)\tan\theta$ in the presence of constant θ, so if $dL/dt \cong 1$ km/sec, it follows that with $\theta = 14°$ the swirling and mixing must proceed at the rate $v = \frac{1}{8}$ km/sec at each end merely to maintain a steady interweaving in the flux bundle. Obviously a swirling at $v \cong 0.5$ km/sec provides for additional winding and interweaving, increasing $\mathcal{E}/\mathcal{E}_c$ as the flux bundle rises. Indeed, one wonders if the variation in X-ray brightness of different coronal loops is not partly a consequence of their different rates of emergence and lengthening, as well as their different swirling rates. This question cannot be answered, of course, until sufficiently high resolution long-term observations have established the nature and the variability in space and time of the footpoint motions.

11.4.7 Energy of the Basic Nanoflare

Consider the magnetic free energy associated with the individual nanoflare, from the estimated threshold $\chi = \tan\theta \cong 1/4$ based on the assumption that the swirling velocity v is 0.5 km/sec. Denote by ℓ the transverse scale of the honeycomb of surfaces of tangential discontinuity created in the bipolar magnetic field by the intermixing of the footpoints of the field at the surface of the sun. This scale is essentially the scale of swirling and intermixing of the footpoints at the photosphere. With rapid reconnection occurring at a field inclination θ, the associated characteristic length along the field is $\ell\cot\theta$. Hence the characteristic volume occupied by the reconnection event is of the order of $\ell^3\cot\theta \cong 4\ell^3$. The magnetic free energy Υ in this volume is

$$\begin{aligned}\Upsilon &= \ell^3\cot\theta B_\perp^2/8\pi \\ &= \ell^3\tan\theta B_0^2/8\pi\end{aligned} \tag{11.28}$$

If $\tan\theta = \frac{1}{4}$ and $\ell \cong 300$ km, corresponding to the characteristic scale of the granule, or the separation of photospheric fibrils, then with $B_0 = 10^2$ gauss, the free energy is

3×10^{24} ergs. If the fraction $1 - \gamma$ is dissipated in a single burst of reconnection, the nanoflare energy is $(1 - \gamma)\Upsilon$ equal to 3×10^{23} ergs for $\gamma = 0.9$. This is small compared to the conventional microflare at about 10^{26} ergs so we have called it a *nanoflare* (Parker, 1988), being of the order of 10^{-9} times the energy of a great flare at 10^{32}–10^{33} ergs, just as the microflare is approximately 10^{-6} of the energy of a great flare.

It should be noted that the characteristic energy density released by the reconnection is comparable in magnitude to the thermal energy density of the ambient gas in the reconnected volume of field. Thus, if one-tenth of the field in the volume $\ell^3 \cot \theta = 10^{23}$ cm^3 reconnects, providing 3×10^{23} ergs, the energy release is about 3 erg/cm^3 over the reconnected volume, to be compared with the thermal energy density of about 10 ergs/cm^3. Thus, the mean heat input to the gas is not large, except that the input is concentrated in a thin layer, causing a substantial temperature increase in that layer.

It must be appreciated that the scale ℓ is not known at the present time and the association of ℓ with the characteristic length within a granule is a conjecture. If for instance ℓ is associated with the diameter of a granule, at 10^3 km, the nanoflare energy is 30 times larger, or 10^{25} ergs. The larger scale of the shearing motions in active regions, for which ℓ may be 10^9 cm, yields 10^{28} ergs, and if B_0 is larger, say 500 gauss, the result is 2×10^{29} ergs, etc. providing a small flare, call it a *milliflare*. On the other hand, some much smaller length ℓ may be involved, so that the nanoflare is little more than an unimportant fluctuation in a mean reconnection rate. So there is no unique universal value for the energy of the basic reconnection event. The energy probably depends upon the circumstances and only detailed observation can provide the appropriate values. The development of the ideas on coronal heating is carried on here for the small nanoflare of 3×10^{23} ergs as the current best guess for the quasi-steady coronal X-ray loop. A spectrum of energies is expected in the real situation on the Sun, and this question is taken up in §§12.2, 12.2.2, and 12.2.4.

It is possible to say something about the mean nanoflare rate based on the foregoing estimate of 3×10^{23} ergs per nanoflare. An X-ray emitting area of 10^{16} cm^2, the size of a granule, requires 10^{23} ergs/sec, i.e., 10^7 ergs/cm^2 sec. Then one new nanoflare of 3×10^{23} ergs every 3 sec would suffice. The total life of the individual nanoflare has a characteristic value of perhaps 10^2 sec, so that there would be 30 nanoflares at one stage of development or another in the area S at any one time. The observational difficulties are immediately obvious, with present resolution of ground-based telescopes usually 10^3 km, occasionally reaching 300 km under the most unusual, and hence transient, moments of seeing at the best of sites.

The observations of flickering and fluctuating visible and UV brightness of active areas on the Sun were described briefly in §11.2. The brightness fluctuations indicate individual releases of energy in the range 10^{24}–10^{27} ergs essentially the nanoflare–microflare range, and evidently are associated mainly with the network activity. These amounts exceed the single number of 3×10^{23} ergs for the basic reconnection event estimated here for the coronal X-ray loops with the arbitrary assumption that the energy release is one tenth of the characteristic magnetic free energy computed from the assumption that the characteristic scale ℓ is 300 km. As noted in §11.2, Lin, et al. (1984) detected bursts of hard X-rays in space indicating an energy output of 10^{24} ergs and up, with 10^{24} ergs representing the threshold for detection. So it would appear that there is a range of energies of, say, 10^{23} ergs and up for the basic reconnection event — the elemental nanoflare — and there is every reason to expect that more than one nanoflare may be set off at one time, to give a brightening substantially in excess of the basic elemental energy release. For there is no reason to think that the individual nanoflares occur independently. The local reconnection of field that produces a nanoflare causes a readjustment of the magnetic field in the neighboring winding patterns of the field. We expect that this often pushes other sites past the critical point for reconnection (Parker, 1987; Lu and Hamilton, 1991; Lu, et al. 1993). Hence the basic reconnection event of 3×10^{23} ergs may come in clusters of several individual bursts at a time. That is to say, the energy release may come from the stimulated, and therefore coordinated, reconnection at a number of neighboring sites. The question is taken up in §12.1 in connection with the individual flare or microflare, whose main phase may be the result of many stimulated nanoflares.

One may hope that the rapid advance of X-ray telescopes will soon resolve the basic dimensions, $\ell \cong 300$ km, of the individual nanoflare. Again, the obvious question is whether the individual basic nanoflare is in fact a distinct and well-defined pyrotechnic event, as has been assumed in the foregoing analysis, or whether the sea of nanoflares is a more continuous field, in which one nanoflare blends into, or stimulates, the next. The fluctuations observed so far suggest that the average sea contains distinguishable flurries of stimulated nanoflares.

11.4.8 Physical Properties of the Basic Nanoflare

With these general ideas in mind, consider the nanoflare defined as the basic independent reconnection event insofar as the mutual triggering of rapid reconnection events permits such identification. The essential point is that the mean magnetic field B_0 is almost always approximately 10^2 gauss so that if the photospheric intermixing and swirling of the magnetic fibrils permits the definition of a basic transverse scale ℓ, then the basic unit of reconnection is defined as occupying a domain with total magnetic free energy of the order of that given by equation (11.15). The simultaneity of such basic blocks of reconnection determines the instantaneous brightness of the overall event, and the continuing activation of fresh blocks determines the duration. Observations detect no individual flare events in most coronal X-ray loops, indicating that the interweaving of the magnetic field in such loops is usually not of such a nature as to foster reconnection events beyond a cascade of a modest number of nanoflares. The flaring loop is, of course, an exception to this state, where the interweaving and interlacing of the field lines is evidently on a larger scale, giving larger basic blocks of reconnection. There is the additional possibility that in some subtle way the topology of the interlacing is such as to provide a "domino" effect, whereby the onset of a single reconnection event triggers a growing sequence of reconnection events. The result would be a self-sustaining "fire storm" of significant duration and detectable magnitude. This question is treated at length in Chapter 12. For the present consider the basic nanoflare.

The magnetic free energy released by the basic block of reconnection provides the individual nanoflare phenomenon, spreading out primarily along the magnetic field as a shock wave and thermal pulse. The speed of sound at the characteristic hydrogen gas temperature of 2×10^6 K is of the order of 200 km/sec. The gas is ejected from the reconnection of $B_\perp$ (~ 25 gauss) at the corresponding Alfven speed ($C_\perp \cong 500$ km/sec), so the shock wave may be strong initially, but is rapidly damped by thermal conduction, etc. Thermal conduction is a consequence of the highly mobile thermal electrons, with rms velocities of about 10^4 km/sec at 2×10^6 K, and is described by equations (11.1)–(11.3).

The reconnection of $B_\perp$ provides a substantial change in the topology of the field lines involved. The topology change propagates away along the field in a form resembling an Alfven wave. The pulse is strongly scattered by the interweaving of the field lines (cf. Similon and Sudan, 1989), which is another way of saying that the reconnected flux bundles readjust their positions through the next few windings among the ambient bundles, with the readjustment strongly attenuated farther away. It is the nearby readjustment that may trigger further nanoflaring. The triggering of rapid reconnection by passing Alfven waves has been studied at length in the literature (Sakai and Washima, 1982; Tajima, Brunel, and Sakai, 1982; Sakai, 1983a,b; Sakai, Tajima, and Brunel, 1984; Matthaeus and Lamkin, 1985, 1986).

The characteristic times associated with a nanoflare can be estimated only very crudely in the context of the basic reconnection unit. Alfven transit times are short, with the perpendicular and parallel transit times being

$$\ell/C_\perp \cong \ell \cot\theta/C \cong 1\,\text{sec}.$$

Thus the onset of the nanoflare may be sudden. The subsequent reconnection, cutting across the field at a speed $u = \varepsilon C_1$, where $\varepsilon \sim 0.01 - 0.1$, provides a characteristic time

$$\ell/u \cong 0.6/\varepsilon \cong 6\text{–}60\,\text{sec}$$

This is comparable to the characteristic Alfven transit times of $L/2C \cong 5$ sec and 50 sec, respectively for loop lengths $L = 2 \times 10^9$ cm and $L = 2 \times 10^{10}$ cm.

Returning to the heat transport by thermal conduction away from the site of the nanoflare, consider a weak heat pulse $\delta T(z, t)$ described by

$$\frac{\partial \delta T}{\partial t} = K \frac{\partial^2 \delta T}{\partial z^2}$$

along a uniform magnetic field in the z-direction. The Green's function is

$$G(z, t; z', t') = \frac{1}{[4\pi K(t - t')]^{\frac{1}{2}}} \exp\left[-\frac{(z - z')^2}{4K(t - t')}\right], \qquad (11.29)$$

representing $\delta T(z, t)$ as a result of the introduction of a pulse $\delta(z - z')$ at $z = z'$ at time $t = t'$. The characteristic width of this pulse is subsequently $[4K(t - t')]^{\frac{1}{2}}$. The typical coronal values $T = 2 \times 10^6$ K and $N = 10^{10}$ atoms/cm^3 yield $K \cong 1.7 \times 10^{15}$ cm^2/sec (see equation 11.2).

The characteristic width $(4Kt)^{\frac{1}{2}}$ reaches the half length (1200 km) of the nanoflare in about 2 sec, which shows that thermal conductivity plays an important role in distribution the heat throughout the nanoflare. The characteristic width of the heat pulse reaches 10^9 cm, representing the half-length of an short X-ray bright point, in about 10^2 sec, comparable to the estimated total life of the nanoflare. Recall from equation (11.6) that the characteristic radiative cooling time is of the order of 10^3 sec, during which the thermal pulse width increases to 3×10^9 cm. Thus the heating by the nanoflare is not restricted to the immediate neighborhood ($\sim 10^8$ cm) of the nanoflare but distributes the heat over dimensions of the order of

10^9 cm during the lifetime of the nanoflare. The peak temperature of the gas in the midst of the nanoflare is determine by the competition between the thermal conduction and the rate of release of magnetic energy.

Taking a larger view, if the characteristic nanoflare converts 3×10^{23} ergs of energy into heat, then one nanoflare every 3 sec is required to supply $I = 10^7$ ergs/cm^2 sec to an area 10^8 cm square. The characteristic coronal volume associated with the surface area of 10^{16} cm, i.e., the volume of each leg of the Ω-loop is of the order of $\frac{1}{2}L$ times the surface area, where L is the length of the coronal loop. Arbitrarily putting $L = 4 \times 10^9$ cm yields a volume of 2×10^{25} cm^3. The characteristic volume of the basic reconnection event is $\ell^3 \cot \theta$ (see §11.4.4) or about 10^{23} cm^3 for $\ell = 300$ km and cot, $\theta = 4$. It follows that the characteristic loop volume is about 200 times the nanoflare volume, suggesting that on the average any given point is associated with a reconnection every 600 sec, which is comparable to the radiative cooling time as well as the time in which the thermal pulse spreads out over the length of leg of the Ω-loop.

These simple considerations, based on the order of magnitude estimates of §11.4.7 for the characteristic size and energy of the basic reconnection event, suggest an X-ray corona made up of sheets of gas with widely different and rapidly fluctuating (characteristic times of 1–10^2 sec) temperatures. It follows, therefore, that the coronal gas continually surges up and down along the field as the gas is expelled from local nanoflares and as the overall pressure scale height 2 kT/Mg varies with the dramatic changes in temperature. Thus the quasi-static mean X-ray corona is, in fact, a dynamical phenomena on a microscale of 10^2 km. As noted at the end of §11.2, the mean corona is a useful vehicle for discussing the general physical principles of coronal heating, but it must someday be upgraded to include the microsurging when sufficient observational information becomes available to define the problem (cf. Laming and Feldman, 1992). For the moment, it should be recognized that the small-scale turbulence or surging contributes to the pressure scale height of the mean corona.

It should be understood that the foregoing linearized treatment of the heat pulse from a nanoflare is highly idealized. For in fact the δT at the site of explosive reconnection is not small compared to T. The local energy input density is several times the thermal energy density of the gas. The result of the actual large δT is to provide the shock waves already mentioned, thereby substantially reducing the local gas density and the associated X-ray emission (proportional to N^2). The emission at some distance from the reconnection site may be enhanced, but unfortunately the precise fluctuation in radiative output cannot be calculated until both the steady emission measure $Q(T)$ and the transient emission measure (including the ionization relaxation dynamics) are better known. For instance, what is the precise form of the time dependence of the radiative output of a volume of gas suddenly compressed adiabatically and not necessarily irreversibly? Such complex questions do not have ready answers. The analysis of Sturrock, et al. (1990), Laming and Feldman (1992), and Feldman, et al. (1992) indicates that the emission spectrum can be understood only if the heating is in the form of short bursts so that the temperature fluctuates over a large range, say 0.5–5×10^6 K, as one might expect within the near neighborhood of a nanoflare. The rapid decline of the bursts of extreme temperature required by Feldman, et al. (1992) are presumably the result of cooling by both radiative emission and thermal conduction along the stochastic field into the surrounding space. But of course this must be checked eventually with quantitative theory and more quantitative observations. Habbal and Grace (1991) note the evidence for a wide diversity of temperature from 2×10^5 K upwards simultaneously existing within a coronal X-ray emission region.

11.4.9 Variation Along a Coronal Loop

Consider the variation of conditions along a coronal loop. It is observed that X-ray emission is brightest and hardest, i.e., the gas is hottest and most intensely heated, across the apex of an Ω-loop. It appears, therefore, that the heat input is distributed broadly across the apex of the loop (Rosner, Tucker, and Vaiana, 1978) particularly when it is recalled that thermal conduction can distribute the heat input of 10^7 ergs/cm^2 sec only over distances of the order of 10^9 cm, whereas coronal X-ray loops can be 10 times longer, as described in §11.2.

The discussion so far has largely ignored the variation of the mean field B_0 along a flux bundle, concentrating instead on the gross properties of otherwise straight field lines and uniform B_0 of the idealized model sketched in Figs 3.1 and 11.2 and used in the preceding chapters to develop the theory of the basic magnetostatic theorem. There are, however, some simple effects associated with the Ω-shaped flux bundles in coronal loops that are worth mentioning at this point in the development.

First of all, there is a small effect, proportional to the plasma β (the ratio of the gas pressure to the magnetic pressure). That is the dislocation of a flux bundle from its "normal" aligned position among the neighboring flux bundles by an anomalous internal gas pressure p. An excess β extends the coronal loop or flux bundle farther out from the surface of the Sun for a given separation of its footpoints at the photosphere. A diminished β allows the tube to retract. But since $\beta \cong 0.02$ in the active X-ray corona this effect is not expected to play a major role.

More interesting is the substantial effect that arises from the expansion of the mean field of a bipolar magnetic region with height above the surface of the Sun. Thus the magnetic field strength passes through a minimum in the neighborhood of the apex of each Ω-shaped flux bundle. The result is that the magnetostatic interweaving of the field lines is strongest in the upper part of the coronal loop, suggesting that the nanoflaring is most intense there (Parker, 1976). It is no surprise, then, that the coronal loops generally are hottest and brightest at their upper levels (Rosner, Tucker, and Vaiana, 1978; Saba and Strong, 1991) noted in §11.2.1. The effect is enhanced by whatever downward conduction losses there may be to the chromosphere, of course, cooling the legs most and the apex least.

Now the concentration of torsion, i.e., interweaving and twisting, into the region of weaker field along a flux bundle is easily seen from elementary considerations. Consider how the torsion varies along a single twisted flux bundle with circular cross-section of radius R that varies slowly along the length of the bundle. Denote the mean longitudinal component of the field by $B_\parallel$ so that the total flux $\Phi \equiv \pi R^2 B_\parallel$ is constant along the tube. The shear stress across any plane perpendicular to the axis of the tube is $B_\perp B_\parallel / 4\pi$, so that the torque $\mathcal{T}$ transmitted along the tube is of the order of the cross-sectional area πR^2 times the characteristic level arm R times the shear stress,

$$\begin{aligned} \mathcal{T} &\cong \pi R^3 B_\perp B_\parallel / 4\pi \\ &= R \Phi B_\perp / 4\pi \end{aligned} \tag{11.30}$$

In static equilibrium $\mathcal{T}$ must be constant along the tube. Hence $B_\perp$ varies as $1/R$. The spiralling of the field lines is given by $B_\perp/B_\parallel$, which is

$$B_\perp/B_\parallel \cong (4\pi^2 \mathcal{T}/\Phi^2)R. \tag{11.31}$$

The essential point is that the characteristic pitch angle ϑ of the spiralling field lines is

$$\begin{aligned} \vartheta &= \tan^{-1} B_\perp/B_\parallel \\ &= \tan^{-1}\left[(4\pi\mathcal{T}/\Phi^2)R\right] \\ &\propto R. \end{aligned} \tag{11.32}$$

These simple order-of-magnitude considerations indicate that the torsion transmitted along a force-free magnetic field increases more or less in proportion to the transverse scale of the field, i.e., inversely with $B_\perp^{\frac{1}{2}}$. Exact analytical calculations can be found in the literature (Parker, 1974, 1976, 1979, pp. 183–200). Parker (1975) pointed out that the enhanced twisting of flux bundles may lead to the kinking of the individual bundles, thereby creating tangential discontinuities and rapid reconnection to heat the corona. The relative importance of this particular process has yet to be assessed. The recent work of Robertson, Hood, and Lothian (1992) illustrates the process in detail.

The concentration of the interweaving and wrapping near the apex of the Ω-loops suggests that the tangential discontinuities are strongest (i.e., largest θ) and the nanoflares are most intense in the general neighborhood of the apex. This scenario would suggest that the winding and interlacing of the field lines, created by the subsurface convection, migrates slowly upward from the surface of the Sun as the reconnection proceeds, each individual winding gradually disappearing as it is reconnected and dissipated in the neighborhood of the apex. The characteristic upward migration velocity u is readily deduced from the requirement that the migration at the base of the field must supply the necessary magnetic free energy $I = 10^7$ ergs/cm^2 sec, where

$$I = uB_\perp^2/8\pi.$$

With $B_\perp = \frac{1}{4}B_0 = 25$ gauss, it follows that $u \cong 4$ km/sec. Considering that the characteristic Alfven speed is 2000 km/sec for $B_\parallel = 10^2$ gauss, the upward migration of magnetic free energy is clearly a quasi-static effect.

Now the interweaving of the field lines and flux bundles is not necessarily produced only by the convection (the granules) at the visible surface of the Sun. It is obvious that convection deeper in the Sun may twist and braid the flux bundles internally, as well as wrap and braid them around each other, and both the twisting and wrapping spread out along the field in an attempt to relax to the minimum energy state. Then insofar as the coronal fields are in magnetostatic equilibrium with the subsurface fields, any strong internal and external interweaving and interlacing of the (presumably) very intense subsurface fields progressively migrates up into the relatively weak fields in the corona. The external wrapping involves the relative swirling of neighboring magnetic fibrils. The internal twisting and braiding does not. An extreme example arises in the exact analytical calculations of the torsion in a simple twisted flux tube (Parker, 1974, 1976, 1979, pp. 183–200), where it is shown that the torque in the outer flux layers of an expanded segment of a twisted flux tube of infinite length falls to zero when the expansion exceeds a certain amount. But in the rest of the flux tube these same flux layers have nonvanishing torque, with the result that their torsion propagates as Alfven waves into the expanded portion of the tube, establishing a purely azimuthal sheath. The simple example, in which the pitch angle $\theta = \tan^{-1} B_\varphi / B_z$ varies with radius as ϖ/a out to a maximum $\theta_{\max} = a/R$ at the surface $\varpi = a$ of the tube, shows that expansion beyond a critical radius R_c gives rise to the azimuthal sheath, where

$$\begin{aligned} R_c &\cong 1.2024R \\ &= 1.2024a/\theta_{\max} \end{aligned}$$

To what extent this effect arises in the Sun cannot be determined at the present time, because the torsion in the individual magnetic fibrils is not known at the photosphere. The effect is noted here primarily to emphasize the possible role of the torsion and random braiding within the individual subphotospheric fibrils in addition to the mutual winding and interlacing of those fibrils.

11.5 Observational Tests

The need for advanced X-ray telescopic resolution of coronal X-ray loops was described in the foregoing section, to determine the magnitude and coordination, if any, of the basic nanoflares. The individual nanoflare may be too small to be detected directly, in which case its properties may have to be inferred only indirectly from the comparison of observed spectra and line width with theoretical models. We may hope for substantial advances in the next several years, but the problem is not an easy one on either the observational or theoretical side.

The fundamental number to be determined by observation is, of course, the total energy release in a single reconnection event, given by $(1-\gamma)\Upsilon$ with Υ from equation (11.28). We have used 3×10^{23} ergs, for want of a better number, based on the supposition that the footpoints of the field wander among each other on the characteristic scale $\ell \cong 300$ km of the granules. The direct observational determination of the basic reconnection energy release is severely complicated by the question emphasized at the end of §11.4.7, that one reconnection may often stimulate another nearby reconnection. The topological changes in the field produced by a single localized reconnection event (Parker, 1993a) increase the magnetic stress and the strength of the tangential discontinuities in nearby critical locations, so that a flurry of nanoflares may arise as a chain reaction from the first event. Without adequate observational resolution in space and time the flurry or avalanche would appear as a single event, perhaps on the magnitude of a microflare.

The most fundamental test of the idea that the corona is created by a sea of nanoflares is the direct detection and measurement of the photospheric motions of the individual magnetic fibrils and the motions, if any, within the individual fibril. For these motions represent the energy input to the system, as described in §11.4.5. The entire nanoflare theory of the origin of the X-ray corona is based on the assumption that the footpoints of the bipolar magnetic fields are swirled and intermixed at a velocity v on a characteristic scale ℓ. This involves both the swirling of neighboring magnetic fibrils around each other and the internal swirling of the footpoints within the individual fibril. Lacking direct observational determinations we have carried on the theoretical discussion for the hypothetical case that the mixing of the photospheric fibrils carries one elemental flux bundle around another at a speed of 0.5 km/sec on scales of 300 km, in times of the order of 300 sec characteristic of the photospheric granules. It remains for observations to determine the precise rate of mixing and to provide the correct values for v and ℓ. Simple considerations on aerodynamic drag (see §12.3) indicate that the magnetic stress in the small magnetic fibril cannot effectively resist the granule fluid motions. Indeed, one can expect v and ℓ to have different values, depending on the mean field strength, proximity to shear lines, etc. Observational determination of ℓ would provide an indirect result for the nanoflare energy limits, through equation (11.28).

It must be borne in mind that the swirling of one fibril around another, as well as the swirling within a fibril, may depend on the diameter of the individual fibrils. The fibrils with larger total magnetic flux would be expected to resist the aerodynamic drag of the local convection more effectively than the smaller fibrils (Parker, 1982a–d) so that the mutual wrapping and winding of the field above the surface of the Sun may perhaps be more a result of the small fibrils than of the larger ones. On the other hand, the larger fibrils are more likely to have internal swirling and winding. The presently limited resolution of ground-based telescopes permits only the larger, more steadfast, fibrils to be detected, with as much as ninety percent of the total flux in smaller unresolved fibrils. It remains yet to resolve the individual fibrils. So it appears that observations from above the atmosphere are required.

Now there are several ways in which the photospheric velocity v of the individual magnetic fibrils, and the associated interweaving of the lines of force in the field above can come about. In the simplest case, the rising flux bundles are essentially without internal interlacing of their field lines until they form a bipolar magnetic field above the visible surface of the Sun. At that point the granules swirl and intermix their footpoints so that the field lines become internally interwoven, and tangential discontinuities are the result. This is the approach taken in §11.4.5 in estimating $\langle B_\perp^2 \rangle$ and θ from the assumed v and ℓ. In this case the observed motion of the magnetic fibrils would correlate closely with the granule motions. Alternatively, the intermixing and weaving of the field lines may originate mainly within the individual fibril.

On the other hand it is entirely possible that the field lines have been strongly interwoven by the convection before the flux bundles break the surface, so that the motion of the fibrils at the visible surface is to some extent the result of the intersection of the interwoven branches of a magnetic "tree" with the surface while the tree rises up through the surface (Zwaan, 1985). In this extreme case the motions of the individual fibrils are determined as much by the more or less rigid structure of the rising tree, as by the local granule motions. The fibril motion would exhibit a component that is proportional to the rate of rise u of the tree, with a surface velocity $v = u \tan \theta$ for a fibril inclined at an angle θ to the vertical. The rate of rise u and the associated rate of lengthening dL/dt of a coronal flux bundle are observed to be a fraction of a km/sec (Harvey and Martin, 1973; Gaizauskas, et al. 1983; Brants, 1985a,b; Zwaan, 1985; Martin, Livi, and Wang, 1985; Zwaan, Brants, and Cram, 1985).

Finally, there is the possibility that the interweaving is created below the surface and propagates up through the surface as long period torsional Alfven waves into the corona (Sturrock and Uchida, 1981). The torsional waves would be damped to some unknown degree by the aerodynamic drag on the individual fibrils, of course. The torsion passing up through the surface concentrates in the apex of the bipolar field, as described in §11.4.7, where it is dissipated by the rapid reconnection driven by the large Alfven speed, $C_\perp \cong 500\,\text{km/sec}$. The torsional waves have periods characterized by $2\pi\ell/v \gtrsim 10^3$ sec, so they are quasi-static in the coronal bipolar fields. They cannot be distinguished from the effect of the rising rigid magnetic tree, except that they would occur even if the field were not rising.

We expect that all these effects are present simultaneously. Note, however, that if the magnetic free energy density $\langle B_\perp^2 \rangle$ rises or propagates upward through the photosphere with a speed u, the energy flux is $u\langle B_\perp^2 \rangle/8\pi$. The Alfven speed is 0.6 km/sec in a mean field of 10^2 gauss at the photosphere (where $\rho \cong 2 \times 10^{-7}\,\text{gm/cm}^3$) so put $u \sim 0.6\,\text{km/sec}$ and $\langle B_\perp^2 \rangle^{\frac{1}{2}} = 25\,\text{gauss}$ $(\tan\theta = 1/4)$. The energy flux is then only $10^6\,\text{ergs/cm}^2\,\text{sec}$, short by about a factor of 10. In fact the fibril state of the photospheric field increases the Alfven speed by the reciprocal of the filling factor f $(f \sim 0.05 - 0.1)$, which effect is cancelled in the mean by the energy being transported upward only over the fraction f of the solar surface. However, recall from discussion leading to equation (11.30) that $B_\perp$ is inversely proportional to the radius R of the flux bundle, while the mean longitudinal field $B_\parallel$ varies as R^{-2} and f varies as R^2. It follows that $\langle B_\perp^2 \rangle/8\pi$ varies as $1/f$ and $u\langle B_\perp^2 \rangle/8\pi$ is proportional to $1/f$. With $f \sim 10^{-1}$, the net energy flux may be as large as $10^7\,\text{ergs/cm}^2\,\text{sec}$.

This brings us to the question of whether the wave motion v is internal to the individual fibril or external, in the form of mutual circling of two or more individual fibrils. The low electrical conductivity in the photosphere appears to limit all internal modes to a basic torsional mode. Kopecky and Obridko (1968) estimate the electrical conductivity to be only about 10^{10}/sec at the temperature minimum. The characteristic resistive diffusion coefficient is, then, $\eta \cong 10^{10}\,\text{cm}^2/\text{sec}$. The characteristic diffusion time across a radius r is

$$t = r^2/4\eta.$$

For a flux bundle with radius 10^2 km, the time is $t \cong 10^3$ sec. So a simple torsional wave with a period of 10^2 sec gets by without excessive dissipation. But not if it has internal structure, so that the characteristic transverse scale is much smaller than 10^2 km.

Note, then, that there is no reason to think that the magnetic flux that makes up a single isolated photospheric fibril at one end of a bipolar magnetic field connects into a single isolated fibril at the other end. We would expect apriori that the magnetic flux of a single fibril connects into two or more fibrils at the other end of the field. Thus it is possible that the simple rotation of each fibril about its axis can provide an interweaving of the field lines that is subject to dynamical reconnection rather than just passive resistive diffusion.

The conclusion is that observations with high spatial resolution ($\lesssim 10^2$ km) are required to establish the energy input to the quasi-static bipolar magnetic fields in which the X-ray corona is created. Once observations provide a quantitative picture of the fibril swirling, and rotating, it should be possible to redo the foregoing theoretical considerations with more precise values for $v, B_\perp, \theta$, etc. It is obvious that the final construction of a quantitative model of the coupled photosphere, chromosphere and coronal X-ray region with proper inclusion of the fibril structure of the field is unavoidably an enormously complicated ad hoc undertaking.

It is to be hoped that eventually it will be possible to use observations of the Sun to construct a semi-empirical relation between the characteristic convective velocities, the mean field, and the X-ray brightness which may then be tested by comparison with the expected convective motions and X-ray luminosity of other classes of main sequence stars. Unfortunately the mean field strength and fibril filling factors for other stars will remain free parameters since there is at present no known basis for inferring them directly from observation.

An interesting theoretical question is the interlacing of the field lines and flux bundles below the level of the most active convection. Thus, to take the simplest case again, suppose that the principal swirling and intermixing of the field is introduced by the granules. It follows that the topology of the field lines leading up to the granules from below is some sort of mirror image of the winding and interweaving introduced into the field above the granules. The interweaving above is continually dissipated by the rapid reconnection rate in the tenuous corona, whereas reconnection below the surface is relatively slow. It would appear, then, that the interweaving may accumulate to a much greater degree beneath the surface of the Sun, where it is not dissipated so effectively. The net result depends upon the fibril or nonfibril state of the field far below the surface. Such questions are important in several contexts, including the nature of the solar dynamo (Parker, 1993b). There are both observational and theoretical reasons to think that the field is in a fibril state throughout much, if not all, of the convective zone (Parker, 1984, 1985, 1993b; Vainshtein, Parker, and Rosner, 1993; Goldreich, et al. 1991) but the case is certainly not yet settled.

There is the amusing but unlikely theoretical possibility, for instance, that the subsurface mirror image interweaving, accumulating during some early phase of a bipolar X-ray emitting region, could, in principle, subsequently drive the coronal heating for a time after the surface swirling by the granules is shut off. That is to say, the mirror image interweaving would propagate up into the bipolar field, and the coronal heating would continue, driven by the reversed wrapping of the field lines after the initial wrapping above the surface has been dispelled through rapid reconnection.

12
Universal Nanoflares

12.1 General Considerations

The basic theorem of magnetostatics asserts that the Maxwell stresses in all but the simplest field topologies continually strive to produce current sheets of vanishing thickness and unbounded current density, i.e., surfaces of tangential discontinuity in the magnetic field. In astronomical settings the small resistivity of the gas takes over in the final stages, as the current sheet thickness declines toward zero, introducing sites of impulsive rapid reconnection of field, i.e., nanoflares and microflares. Using the term nanoflare in its generic sense, then, we expect nanoflares to occur to some degree in all astronomical magnetic fields that are extensively deformed by convection and allowed to relax toward quasi-static equilibrium. The nanoflare produces elevated temperatures, and, in regions of sufficiently low density, extended high velocity tails in the particle velocity distributions, i.e., suprathermal particles.

The X-ray corona of the Sun is a prime example, treated in Chapter 11. The solar flare is another important case, treated in the section below. The X-ray corona of the Sun is the paradigm for the X-ray emission of all solitary late-main sequence stars, providing the direct observational guidance essential for a scientific understanding of this universal process. The same is true for the solar flare, as the paradigm for the magnetic flaring of stars in general.

In the broadest terms, the magnetic fields in almost all astronomical settings are stochastic (Jokipii and Parker, 1968, 1969a,b) and stochastic magnetic fields strive to develop internal tangential discontinuities. Anything that contributes to the stochastic nature of the field lines, e.g., convection, localized anomalous resistivity, rapid reconnection, etc. leads to nanoflares if the system is allowed to relax to static equilibrium. Therefore, tangential discontinuities, and nanoflaring to some degree, are ubiquitous in the magnetic fields rooted in interstellar gas clouds, in the gaseous disk of a galaxy, in the intergalactic gas, in a cluster of galaxies, in the interstellar wake of the heliosphere, in protostellar accretion disks, in accretion disks around black holes, etc. Indeed, the nanoflare is expected in every astronomical setting where field lines are mixed and interwoven by convection in the parent body while approaching magnetostatic equilibrium in the tenuous atmosphere outside the body.

The magnetic fields of Earth and other planets (Mercury, Jupiter, Saturn, etc.) are an inside-out version of the same process, wherein the mixing and interweaving of the field lines arises from external causes, viz. the high speed flow of the solar wind over the unstable magnetopause, and, in the case of Jupiter and Saturn, the high rotational velocity of the magnetopause relative to the solar wind. The essential point is that the wind picks up flux bundles from the magnetopause and stretches them out into the planetary magnetic tail, forming sites for tangential discontinuity within the magnetic tail and particularly where the polar field lines that form the tail depart from the undisturbed field at lower latitudes. The aurora is evidently a direct consequence of the complex current sheets created in this way.

One can go on to imagine the extended magnetic patterns and the associated nanoflaring and X-ray emission arising in the strongly interacting magnetic fields of close binary stars.

The magnetic field of the Galaxy is unusual, developing extended lobes that are inflated outward at high speeds (30–100 km/sec) by the powerful cosmic ray gas generated in supernovae, etc. Within a few hundred pc of the galactic disk the fountains of hot gas from OB associations and supernovae also contribute to the inflation of the field. The result is an extended galactic halo, dominated in the outer regions by cosmic ray gas and magnetic field. One expects that nanoflaring in the tangential discontinuities within each magnetic lobe, and at the separatrices between lobes, is a major cause of the 10^6–10^7 K temperature of the tenuous X-ray emitting galactic halo gas. The basic facts are that cosmic rays are generated at a rate of the order of 10^{40}–10^{41} ergs/sec by supernovae, etc. in the Galaxy. This input is a direct measure of the rate at which the escaping cosmic ray gas does work on the inflating magnetic lobes. It is also an indication of the order of magnitude of the heat input to the halo gas by the nanoflaring and an indication of the X-ray luminosity of the halo gas. Other spiral galaxies exhibit the same magnetic inflation phenomenon, i.e., a magnetic halo and the associated nanoflaring and X-ray emission.

Now if the nanoflare is expected on theoretical grounds to be a universal phenomenon, it must be kept in mind that the case for rapid reconnection and nanoflaring has not been established by observation in all cases of conspicuous suprathermal activity. Thus, for instance, it is not yet established by direct observation that the bipolar magnetic fields that contain the X-ray corona of the Sun are internally mixed at the rate expected from the granule motions observed at the surface of the Sun. But that is only a quantitative question. The foremost example of a qualitative nature is the mass loss, i.e., the stellar wind, from a late-main sequence star such as the Sun. The solar wind arises primarily from the thermal expansion of coronal gas at $\sim 2 \times 10^6$ K in regions in which the mean magnetic field at the surface of the Sun is so weak ($\lesssim 10$ gauss) that the gas pressure is able to push open the more extended field lines and escape into space (Parker, 1958a, 1963). The expanding gas carries the extended field lines out to "infinity." It is precisely this open-ended structure of the magnetic field that precludes nanoflares, because the swirling and intermixing of the footpoints of the field by the photospheric convection does not accumulate interweaving of the field lines, as it does in a bipolar field with both ends attached to the Sun. Instead the intermixing propagates outward along the field in the form of torsional Alfven waves of small amplitude. Such waves may dissipate at distances of $5R_\odot$ and beyond, but there is no known mechanism by which they can provide the hot expanding coronal gas near the Sun (Fla, et al. 1984; Parker, 1991).

Moore, et al. (1991) point out the interesting theoretical possibility of reflection and trapping of Alfven waves of a given frequency ω in a specified radial magnetic field if the coronal gas density is so low that the wavelength exceeds the pressure scale height of the coronal gas. One could imagine, then, that the trapped waves build up to the point that they are damped, giving up their energy to heat the coronal gas. On the other hand, the generation of a sufficient energy flux of about

0.5×10^6 ergs/cm^2 sec (Withbroe, 1988) in the form of Alfven waves in the weak field regions is problematical (cf. Parker, 1991; Collins, 1992).

The alternative, suggested by Martin (1984, 1988) and by Porter and Moore (1988), is that the primary heat source is the network activity, in the form of the microflaring observed in the small bipolar magnetic fields in and around the boundaries of the supergranules. However, more recent quantitative studies of the microflare energy (Habbal, 1992a,b) indicate only about one-tenth of the necessary energy. But there is no evident alternative (Parker, 1991, 1992a). At the present time, then, the energy source responsible for the solar wind is as much a mystery as it was 35 years ago (Parker, 1958a, 1960, 1963). It means that the mass loss from solitary main sequence stars is without a complete theoretical basis. The physical importance of the mass-loss merits a concerted effort to understand the underlying energy source.

In view of the observational inaccessibility of the essential details of magnetic fields throughout most of the astronomical universe, the present chapter begins with the occurrence of tangential discontinuities in solar flares, followed by a general survey of the nanoflares expected in other settings. It was described in §10.3.1 how observational studies of the solar flare played a seminal role in the development of the theory of rapid reconnection. Once again the observational accessibility of the solar flare, for all of the limitations, plays a leading role in the development of the theory.

12.2 The Solar Flare

The term *flare* is given to any of the intense transient suprathermal brightenings easily identifiable in H_α on the Sun. The largest such events are the great flares with a total output of the order of 10^{33} ergs over a couple of thousand seconds, at random intervals of the order of 10 years. Flaring occurs on all smaller scales down to the limit of detection in the vicinity of the nanoflare at 10^{24} ergs. There is no obvious reason to think that the phenomenon cuts off at the detection limit at about 10^{24} ergs. Observations (Dennis, 1985, 1988; Lu and Hamilton, 1991; Bromund, McTiernan, and Kane, 1994) indicate that the number of flares $\mathcal{N}(\mathcal{E})d\mathcal{E}$ per unit time in an energy interval $(\mathcal{E}, \mathcal{E} + d\mathcal{E})$ varies as $\mathcal{E}^{-1.4}$ over the flare energy range 10^{29} ergs $< \mathcal{E} < 10^{33}$ ergs, while the number $\mathcal{N}(\mathcal{P})d\mathcal{P}$ per unit time with peak power $\mathcal{P}$ in the interval $d\mathcal{P}$ varies as $\mathcal{P}^{-1.8}$. This leads to the interesting relation that $\mathcal{P}$ varies as $\mathcal{E}^{0.5}$. Hence the characteristic life $\mathcal{E}/\mathcal{P}$ also varies as $\mathcal{E}^{0.5}$.

Note, then, that the energy spectrum $\mathcal{E}^{-1.4}$ provides for finite total energy as $\mathcal{E}$ declines to zero, but diverges at the high energy end. Thus

$$\int_{\mathcal{E}_1}^{\mathcal{E}_2} d\mathcal{E}\mathcal{E}\mathcal{N}(\mathcal{E}) \propto \mathcal{E}_2^{0.6} - \mathcal{E}_1^{0.6} \tag{12.1}$$

We may expect some sort of rollover or cutoff at the lower end determined by the primitive nanoflare size estimated for the active X-ray corona in §11.4.7 at 10^{23}–10^{24} ergs. The upper limit or cutoff at $\mathcal{E}_2$ is presumably determined by the maximum size and maximum magnetic free energy of the active region in which the flare occurs. If U is the total magnetic free energy of an active region, we would expect $\mathcal{E}_2$ to be limited to some fraction 10^{-2}–10^{-1} of U. The gigantic flares, of the order of 10^{35} ergs, occurring frequently on some dM dwarf stars, attest to the monstrous scale and intensity of the magnetic field regions on those faint stars (cf.

Schaefer, 1991; Skinner, 1991; Henry and Hall, 1991; Hawley and Pettersen, 1991; Cheng and Pallavicini, 1991; Mullan, Herr, and Bhattacharjee, 1992; Hawley and Fisher, 1992 and references therein). It is a pity that the other stars are not resolved in the telescope and studied in more detail. For the present, then, the Sun is the principal laboratory for studying the flare phenomenon.

Flares have a variety of forms, and Svestka (1986) provides an extensive summary of their principal features. For present purposes it is sufficient to note (cf. Dennis, 1988) that the larger flares ($\mathcal{E} \gtrsim 10^{10}$ergs) can be grouped into three main categories A, B, and C. Type A flares are the extremely hot thermal flares, which show extended but steep X-ray emission spectra above 40 keV, while fitting a thermal spectrum below 40 keV, with an effective temperature of $3–4 \times 10^7$ K. Such flares are compact (< 5000 km) and make up 0.25 percent of the total number. Type C are gradual flares, with the X-ray emission spectrum above 30 keV becoming increasingly hard with the passage of time, from an initial energy power law exponent $\gamma \geqslant 5$ to $\gamma \leqslant 2$ some minutes after the peak of the flare. The flare may continue for 30 minutes or longer. The emission source is located at high altitudes ($\geqslant 4 \times 10^4$ km). About 18 percent of the flares are of this type.

Finally, the type B flare is the impulsive flare, making up some 82 percent of the total and characterized by the impulsive onset of intense spikes of soft X-ray, hardening somewhat through the peak of the flare. The emission is initially from low altitudes, including the footpoints of the field, and evolves into a more compact source at high altitude later on, at which time the emission spectrum approaches a single power law.

The different flare forms A, B, and C are presumably a consequence of the different magnetic configurations in which the flares occur. Thus, for instance, a flare arising from the strong shearing of a magnetic arcade provides an energy release relatively high in the solar atmosphere (above 10^4 km) and produces the *two ribbon* (Moore, et al. 1980) or *ejective* (Svestka, 1986) flare (type C). The collision of two magnetic bipoles provides the *compact* (Moore, et al. 1980) or *confined* (Svestka, 1986) flare (type A), and presumably the most common variety (type B), the impulsive flare (see also Haisch, Strong, and Rodono, 1991).

Note, then, that microflares appear more or less continually throughout an emerging active region. The more complex and time varying the magnetic pattern, the more frequent the microflares and the more frequent the full-scale ($> 10^{29}$ ergs) flares as well. The larger flares arise in magnetic regions showing the strongest large-scale deformation.

A flare appears as a luminous arch when seen on the limb, with the chromosphere obscuring the lower 3000 km so that small flares are not visible. The luminous arch may be made up of more than one discernible bright filament, and it may sometimes be in a state of eruption. On the other hand, it should be noted that flares are distinct from the more energetic coronal mass ejections and eruptive prominences, with the flare occurring a half hour or so after ejections and eruptions, if there is an associated flare at all.

A flare is conspicuous against the disk of the Sun in H_α, with only the rare *white light* flare visible in the continuum (cf. Mauas, 1990, Mauas, Machado, and Avrett, 1990). The sudden brightening in H_α is referred to as the *flash phase*, which may expand rapidly over a significant part of the magnetic active region in some cases, providing an *explosive phase*. The flare is characterized by a rapid rise of brightness, in a period of a few minutes, followed by a slow declining phase with a characteristic decay time of 10–30 minutes, often remaining detectable for 10^2 minutes or more. The fast-particle acceleration takes place during the abrupt onset of the flare, giving impulsive bursts of radio waves, hard X-rays (> 10 keV), and gamma-rays, making up the *impulsive phase* of the flare (Dennis and Schwartz, 1989). The impulsive phase if followed by the extended slow decline of the flare (see the extensive review edited by Chupp, 1990).

Here we focus attention primarily on the common type B impulsive flare, although the ideas on the stimulated onset of nanoflares presumably apply to other flare types as well.

12.2.1 Basic Nature of the Flare Release

The remarks put forth here (Parker, 1987) are based on the observational studies of Machado, et al. (1988a) of the impulsive, type B flares. Machado, et al. combined the Solar Maximum Mission X-ray pictures of the flares with ground-based vector magnetic maps of the region. As a result they were able to establish the precise location of the energy release within the configuration of the magnetic field responsible for the flare. They found that the magnetic structure of the flare usually consists of an initiating bipole plus one or more adjacent bipoles pressed against it. The flare energy release, which they observed in the form of soft X-ray emission, begins either within the initiating bipole or at the interaction site between it and the impacted bipole. In either case the flare begins at the location of the maximum nonpotential magnetic stress, i.e., at the location of the maximum magnetic free energy density, associated with the maximum electric current density. The soft X-ray emission then spreads over the two (or more) impacted bipoles simultaneously with the impulsive emission of the hard X-rays, indicating the time of peak particle acceleration (to energies above 10 keV). The footpoints of the field in which the particle acceleration is strongest can be discerned by their hard X-ray emission, indicating that they occur in the region of maximum magnetic free energy density. The simultaneity of the brightness variations in the initiating and impacted bipoles shows the strong coupling between them during the impulsive phase. It is interesting to note, then, that when an adjacent bipole has relatively little internal magnetic free energy (as indicated by the vector magnetograph) there is no significant energy release within that bipole. However, energetic particles and hot plasma are injected into it from the interaction site. An essential point is that most of the total energy emitted during the impulsive phase of a flare comes from the interior of the bipoles rather than from the interaction site between them, with most of the hard X-rays from the bipole that is brightest in soft X-rays. Machado, et al. (1988b) estimate that only about one-tenth of the total flare energy is released at the interaction site between the colliding bipoles.

Consider, then the extended decay phase that follows the impulsive phase. The decay phase involves primarily soft X-ray emission, with no indication of significant particle acceleration and hard X-rays. It is interesting to note that in some cases the soft X-ray emission spreads throughout the dimensions of the large-scale fields of the active region, including X-ray emitting loops extending from one active region to another within an activity complex. Machado, et al. (1988b) studied the spread of the soft X-ray emission (3.5–8.0 keV) from the initial flare site throughout an active region, noting that the X-ray front advances at about 10^3 km/sec. They modeled the front as a shock wave smoothed by thermal conduction. As initial conditions they assumed a number density $N = 0.5 \times 10^{10}$ atoms/cm^3 at 5×10^6 K, into which the flare injects a sudden localized burst of heat. There is then an initial thermal wave followed by the shock wave. The model exhibits the basic properties of the expansion of the flaring volume within the active region. Poletto, Gary, and Machado (1993) note the tendency for brightening at one location to be followed soon by brightening at nearby locations, even when no evident magnetic connection exists between the two locations.

Now the conventional idea of a flare is the onset of rapid explosive reconnection across an intense current sheet formed by pressing two or more magnetic bipoles together (Sweet, 1958), or by the coalescence of parallel current loops (i.e., two strongly twisted flux tubes with the same sense of twisting) pressed together along their length (Gold and Hoyle, 1960; Tajima, et al. 1987; Sakai and Ohsawa, 1987; Choudhuri, 1988), or one flux bundle drawn transversely across another (Sakai and de Jager, 1991) or some variation or combination of these. In fact, one supposes that flares with different characteristics arise from correspondingly different magnetic configuration (cf. Forbes, 1991; Forbes and Malherbe, 1991). The various levels of fast particle production are presumably sensitive to the detailed magnetic and plasma configuration (Sakai, 1990a,b). Theoretical models have been constructed for the purpose of exploring the sudden and efficient particle acceleration to high energy indicated by the observations (cf. Forbush, 1946; Ehmert, 1948; Meyer, Parker, and Simpson, 1956; Parker, 1957a; Sakai, 1990a,b; Ryan and Lee, 1991; Winglee, et al. 1991; Miller, 1991; Cliver and Kahler, 1991; Kallenrode and Wibberenz, 1991, McTiernan and Petrosian, 1991; Debrunner, Lockwood, and Ryan, 1992; Hamilton and Petrosian, 1992; Miller and Steinacher, 1992; Holman and Benka, 1992; Klecker and McGuire, 1992 and references therein). The fact that flares at the limb of the Sun appear as bright arches is an indication that the injection of heat is not usually confined entirely to a thin current sheet where two bipoles press together but fills a volume of the field as well. The detailed observations (Jakimec, et al. 1986; Machado, et al. 1988a,b; Haisch, et al. 1989; Dere, et al. 1991) emphasize this fact, showing that the reconnection at the surface between the bipoles — the interaction site — provides only 10 percent of the total flare energy with the other 90 percent of the heat appearing throughout the volume of the flare loops, as already noted.

Van Hoven (1976, 1979, 1981) and Spicer (1976, 1977, 1982) recognized the main energy release to be a volume effect many years ago, suggesting the onset of resistive kink and tearing instabilities to provide many small reconnection sites throughout the strongly deformed magnetic bipoles involved in the flare. It was an important conceptual step. They noted the difficulty presented by the slow growth of the resistive instabilities as described in the discussion in §11.4.4. Using an Alfven speed $C = 10^9$ cm/sec (for 300 gauss) and $\eta \cong 10^3$ cm^2/sec with a wavelength and characteristic scale h as small as 10^7 cm, the characteristic growth time is of the order of 10^6 sec, whereas the preflare build up lasts no more than 10^3–10^4 seconds and the flare itself flashes on in a matter of a few seconds.

Instead of the resistive tearing modes Parker (1987) suggested that the convective displacement of the footpoints of the bipolar fields, deforming the bipoles and bringing them into collision on the large-scale, also swirl the footpoints of the field on a small scale so that the field lines are wound and interwoven in the manner shown schematically in Fig. 11.1. It follows from the basic theorem of magnetostatics that the interacting bipoles are filled with many small-scale surfaces of tangential discontinuity. So when the bipoles are mashed together by the large-scale convection, there is no waiting for the slow growth of a resistive instability in a continuous shear. Instead, the resistive instabilities arise quickly in the strong shear of the tangential discontinuities. Thus we have a scenario of resistive instabilities as proposed by Van Hoven and Spicer, but set up and triggered by the indigenous tangential discontinuities. The picture is, then, that the squashing of the fields strengthens some of the discontinuities, which then flash into nanoflares throughout the volume of the deformed field. The discontinuities are strongest where the field is most strongly interwoven. The explosive reconnection at the discontinuities modifies the field topology, of course, setting off other discontinuities whose strength is further increased by the topological changes of the first reconnections. The flare continues as a fire storm of nanoflares until the magnetic free energy is depleted to a level that no more local tangential discontinuities exceed the threshold for the onset of explosive reconnection.

The topological changes created by local nanoflares are communicated largely as torsional Alfven waves along the mean field (Parker, 1983) developing an increasingly complex form as they propagate along the stochastic field lines (Similon and Sudan, 1989). Hence the onset of nanoflares arises in the more strongly interwoven flux bundles and propagates primarily along the mean field, accounting for the arched filament structure of the flare in a bipolar magnetic field (when seen in profile at the limb of the Sun).

In these terms the onset of a flare has the character of a "flame front" — a conflagration wave — that sweeps across the magnetic "tinder" that has accumulated from the small-scale swirling and intermixing of the field lines by the photospheric convection.

The flame front sweeps through any region with enough initial field line interlacing to provide sufficient tangential discontinuities near the threshold for explosive reconnection. The region behind the front is a sea of nanoflares, with each nanoflare restructuring the field lines so that there is a non-negligible probability that another nanoflare is set off somewhere nearby. A quantitative statistical model of this avalanche of nanoflares has been formulated by Lu and Hamilton (1991) and is described in the next section. The concept is compatible with the observational properties of the flare cited by Machado, et al. and others, and it provides the appropriate soft X-ray front that is sometimes observed to sweep away from the point of onset of a flare with a speed of the order of 10^3 km/sec (Machado, et al. 1988b).

As already noted, the region behind the flame front is a sea of nanoflares, as in the nonflaring X-ray corona, except that the sea is much more intense in the flare because the mean field has an energy density as much as 10^2 times larger than in the usual X-ray corona (10^3 gauss instead of 10^2 gauss).

12.2.2 *Nanoflare Size*

The flares described by Machado, et al. are typified by volumes with characteristic dimensions of 10^9 cm (10^{27} cm^3) in fields of 10^3 gauss with initial densities of about 10^{10} atoms/cm^3. Suppose that the magnetic field is internally interwoven on small scales to much the same degree $\langle B_\perp^2 \rangle^{\frac{1}{2}} \sim \frac{1}{4}\langle B \rangle$ and on much the same scale

$\ell \cong 3 \times 10^7$ cm, as was inferred in the exposition in §§11.4.5 and 11.4.7 of coronal heating in a nonflaring X-ray region. It follows that the magnetic free energy density is of the order of $\langle B_\perp^2 \rangle / 8\pi \sim 2 \times 10^3$ ergs/cm^3. Over the characteristic nanoflare volume $4\ell^3 \cong 10^{23}$ cm^3 (see §11.4.7) this gives a free energy of 2×10^{26} ergs, suggesting basic nanoflares at 10^{25} ergs within the flare. The total free energy amounts to 2×10^{30} ergs over the characteristic flare volume. Note, then, that the total characteristic magnetic energy of the field B is 4×10^{31} ergs, which can be considered as free energy only in the context of reconnection with another comparable magnetic bipole. The internal free energy of 2×10^{30} ergs plus reconnection with another bipole would appear to be adequate to provide a typical flare of 10^{30} ergs studied by Machado, et al. when two such bipoles are pressed together.

The basic questions concerning the nanoflare model of the solar flare are not unlike those concerning the active X-ray (nonflaring) corona. The fundamental question in both cases is whether the footpoints of the field are intermixed and swirled on the small scales of 10^3 km (or less) necessary to provide the internal weaving and winding of the field. For if the intermixing of the individual small magnetic fibrils at some such speed as 0.5 km/sec is present, then the bipoles are filled with many thin current sheets with widths of the order of 10^2–10^3 km, while the Maxwell stresses in the field continually strive to reduce the thicknesses of the sheets to zero to achieve magnetostatic equilibrium. The continual work done on the bipolar field by the subsurface convection accumulates as small-scale magnetic free energy and can be relieved only by small-scale reconnection across the individual current sheets or tangential discontinuities. Evidently in fields of 10^2 gauss this leads to the brightening within individual coronal loops to provide the typical nonflaring X-ray corona whose loops come and go over 10–20 hours. When the field is substantially stronger, of the order of 10^3 gauss, the epochs of brightening go much faster, providing the 10^2 times more intense and much shorter lived flaring loops, as seen on the limb. We suggest, then, that the X-ray loops and the flaring loops are similar, with the flare sporting an initial impulsive phase because it is set off by the strong deformation of colliding strong magnetic bipoles, already loaded with internal discontinuities. Whereas the quasi-static coronal X-ray loop glows more or less continuously, with only some rapid small-scale flickering, as the magnetic free energy of the small-scale discontinuities is maintained by the small-scale footpoint convection and stimulated to rapid reconnection only internally, by neighboring nanoflare reconnections. We can imagine how, in principle, a quiescent X-ray loop can be converted into a flaring loop by being squashed against another bipole, generally intensifying most of the internal tangential discontinuities in the process. If the magnetic field has the typical strength of 10^2 gauss, it will not be much of a flare, just a brightening of the loop. But if the energy density is 10^2 times higher at 10^3 gauss, the result is evidently the spectacular flare phenomenon.

Zirker and Cleveland (1993) use Berger's (1991a,b) analysis of braided magnetic fields to arrive at a nanoflare energy of 4×10^{27} ergs in a coronal field of 100 gauss. An outburst of this magnitude is in the microflare class and marginally detectable. Their large value for the energy arises partly from the assumed slow rate of braiding suggested by Berger's numerical model of footpoint mixing, so that their nanoflare rate is 0.01–0.1 the value hypothesized in the present analysis, requiring a much large energy release per nanoflare. It will be interesting to see what rate of footpoint mixing is more appropriate.

12.2.3 Preflare Deformation

We cannot determine by theoretical considerations the degree of swirling and intermixing of the footpoints of the fields of 10^3 gauss in an active region. That can be accomplished only by observations of the individual fibrils with telescopic resolution of 10^2 km or better, as described in §11.5 for the X-ray corona. But we can treat the question of the degree to which the subsurface convection is expected to drag the magnetic fibrils along with the flow. There is reason to believe that the fibril structure observed at the surface extends for some distance below the surface (Parker, 1984; Goldreich, et al. 1991). The aerodynamic drag of the convection on the magnetic fibrils has been treated in general terms (Parker, 1979, 1982a,b, 1985; Tsinganos, 1979). So consider the aerodynamic drag on a motionless magnetic fibril of radius R extending downward a distance L through convective cells of individual scale ℓ in which the characteristic mean square fluid velocity is $\langle v^2 \rangle$. The characteristic aerodynamic drag per unit length of magnetic fibril is

$$\mathcal{F} = \frac{1}{2}\rho\langle v^2\rangle RC_D$$

where C_D is the drag coefficient. The Reynolds number of the typical photospheric granule is large, so that the flow is turbulent, providing an eddy viscosity ν_e of the order of $0.1\ell\langle v^2\rangle^{\frac{1}{2}} \cong 3 \times 10^{11}$ cm^2/sec for $\ell \cong 3 \times 10^2$ km and $\langle v^2\rangle^{\frac{1}{2}} \cong 1$ km/sec. The effective large-scale Reynolds number is, then, of the order of $N_R = R\langle v^2\rangle^{\frac{1}{2}}/\nu_e \sim 3$ and the effective drag coefficient is approximately 3.

Consider, then, a vertical flux bundle of radius R extending a distance L upward across a horizontal flow v, with aerodynamic drag opposed by the magnetic tension $B^2/4\pi$. The flux bundle is deflected form the vertical by an angle θ at the upper end, so that the horizontal component of the tension is $\pi R^2(B^2/4\pi)\sin\theta$ at the upper end. Equating this to the drag, the result is

$$\tan\theta = 2C_D\rho\langle v^2\rangle L/RB^2$$

That is to say, the drag is sufficient to deform the fibril by the angle θ. As an example, consider the case that the diameter of the magnetic fibril is 10^2 km, with $L = 500$ km, $B = 10^3$ gauss, and $\langle v^2\rangle^{\frac{1}{2}} \cong 1$ km/sec at a depth of $\frac{1}{2}L$ where $\rho \cong 5 \times 10^{-7}$ gm/cm^3. The result is $\sin\theta \cong 0.3$, not greatly different from the value $\tan\theta \cong 0.25$ assumed in the foregoing estimate ($B_\perp = \frac{1}{4}B$) of the magnetic free energy. So the aerodynamic drag may be adequate to interlace the field in the manner assumed, but it remains to be shown by direct observation that the hypothesized small-scale swirling of the convection actually carries out the expected intermixing of the magnetic fibrils.

12.2.4 Flare Size Distribution

It is not possible at the present time to deduce the distribution $\mathcal{N}(\mathcal{E})$ of flare energies $\mathcal{E}$ from magnetohydrodynamics and the fluid motions beneath the surface of the Sun. As pointed out at the end of §10.3.5 and noted in §11.4.4, there is no quantitative deductive theory for the threshold for the onset of explosive reconnection at a tangential discontinuity in an interwoven magnetic field, nor is there a means for calculating the amount of reconnection before the rate lapses back to the minimum slow burn given by equation (10.22). Order of magnitude estimates are all that are available. Therefore, the recent work of Lu and Hamilton (1991) and Lu, et al. (1993) is of particular interest. They find that a simple linear avalanche model leads to a universal spectrum $\mathcal{N}(\mathcal{E})$ for the frequency of flares with total energy $\mathcal{E}$. The universal spectrum agrees well with the observed power-law spectrum $\mathcal{E}^{-1.4}$ over four factors of 10 in $\mathcal{E}$ from about 10^{29} ergs to 10^{33} ergs. The spectrum is remarkably insensitive to the detailed properties of the local nanoflare. What is more, the observed upper and lower cutoff or rollover energies $\mathcal{E}_2$ and $\mathcal{E}_1$ (see equation 12.1) provide the basic nanoflare energy and duration, from which the characteristic nanoflare dimension can be deduced along the lines of equation (11.29), noted in §12.3.

Lu and Hamilton (1991) begin with the situation that the magnetic field is filled more or less at random with many small-scale magnetic discontinuities, which suddenly give up a portion of their local magnetic free energy when their strength $\Delta B_\perp$ exceeds some threshold B_c. They point out that a power law spectrum $\mathcal{N}(\mathcal{E})$ can be obtained only if the probability of triggering an event depends only on the instantaneous local conditions. That is to say, the onset is independent of previous history and depends only on the conditions (the field gradients) at the moment. That is precisely the nature of the quasi-static tangential discontinuity of course. The discontinuities all grow (on the average) slowly with the passage of time, going into a state of explosive reconnection when $\Delta B_\perp$ surpasses some specified threshold.

The model used by Lu and Hamilton is a 3D cubical grid. They start with a uniform magnetic field so that the field $\mathbf{B}^{(i)}$ at the ith grid point is everywhere the same. A small random field $\delta\mathbf{B}^{(i)}$ is added to each point on the grid at the beginning of each step in time. The three components of each $\delta\mathbf{B}^{(i)}$ are given by random numbers between -0.2 and $+0.8$, so that the average field grows with the passage of time while the field becomes irregular. The average field gradient $d\mathbf{B}^{(i)}$ at the ith grid point is determined as the difference between $\mathbf{B}^{(i)}$ and the average of the fields $\mathbf{B}^{(j)}$ at the six nearest neighbor grid points. Then

$$d\mathbf{B}^{(i)} \equiv \mathbf{B}^{(i)} - \frac{1}{6}\sum_{j=1}^{6} \mathbf{B}^{(j)}$$

where the sum is over the six nearest neighbors. Initially $d\mathbf{B}^{(i)} = 0$, but it grows with the passage of time, of course, as the random $\delta\mathbf{B}^{(i)}$ are successively added, representing the quasi-static build-up of the field intensity and irregularity with the passage of time. Nothing happens until $|d\mathbf{B}^{(i)}|$ exceeds some predetermined threshold dB_c. When $|d\mathbf{B}^{(i)}|$ exceeds the threshold, an explosive reconnection, or nanoflare is declared, in the form of one-seventh of $d\mathbf{B}^{(i)}$ being transferred to each of the nearest neighbors, so that $\mathbf{B}^{(i)}$ becomes $\mathbf{B}^{(i)} - \frac{6}{7}dB^{(i)}$ and each of the six nearest neighboring fields $\mathbf{B}^{(nn)}$ becomes $\mathbf{B}^{(nn)} + \frac{1}{7}d\mathbf{B}^{(i)}$. It is readily shown, then, that the new $d\mathbf{B}^{(i)}$ is zero. This readjustment is likely to raise $|d\mathbf{B}^{(nn)}|$ at one of the nearer, if not nearest, neighbors above the critical value dB_c, producing another nanoflare, etc. The resulting stimulated reconnection events must eventually run their course because the small random increments $\delta\mathbf{B}^{(i)}$ added at each step are not sufficient to sustain a continuing fire storm of nanoflares. Flaring ceases at this point, giving a well defined integrated flare energy $\mathcal{E}$, peak output P, and duration T. The next flare event begins when after a time $|d\mathbf{B}^{(i)}|$ again builds up to dB_c at some grid point so as to initiate another stimulated sequence of nanoflares.

Lu and Hamilton have run long sequences of the nanoflaring to obtain a precise histogram of the numbers of flares with a given total energy $\mathcal{E}$, obtaining $\mathcal{N}(\mathcal{E}) \propto \mathcal{E}^{-1.4}$ to within very small error. The outcome, in the form of the exponent -1.4, is remarkably robust. For instance, the -1.4 is insensitive to the redistribution rules for dispersing $d\mathbf{B}^{(i)}$ after each reconnection event, provided only that the sum of the magnetic field vectors is conserved and the field is redistributed statistically isotropically. The requirement for conservation of field is in keeping with the idea that the mean field is preserved, while the isotropy of the redistribution maintains the 3D character of the grid. If there were a preferred direction, the result would be to build-up $|d\mathbf{B}^{(i)}|$ along filaments extending in the preferred direction.

Lu, et al. (1993) go into more detail using a $50 \times 50 \times 50$ lattice, to show that, in a typical active region with characteristic scale 5×10^9 cm, the characteristic scale ℓ of the individual event is approximately $\frac{1}{125}$ of 5×10^9 cm, or 400 km. Assuming that the characteristic time for the nanoflare is to ℓ/V_A, with $V_A = B_\perp/(4\pi\rho)^{\frac{1}{2}}$ in $B_\perp \cong 250$ gauss and 10^{10} atoms/cm^3, the result is $V_A \cong 5 \times 10^8$ cm/sec and the characteristic time is 1 sec. The high energy cutoff arises from the finite number of grid points and is $\mathcal{E}_2 = 10^{33}$ ergs. The peak luminosity is $\mathcal{P} = 10^{30}$ ergs/sec and the total duration is 3×10^3 sec. These results are all compatible with the observed characteristics of the largest flares. The elementary energy release (the nanoflare) then proves to be $\ell^3 B_\perp^2/8\pi$ or about 3×10^{25} ergs. This is to be compared with the 10^{25} ergs per nanoflare from simple dimensional considerations on a field $B = 10^3$ gauss and $B_\perp = \frac{1}{4} B$ in §12.2.2.

In summary, it is curious that a unique universal flare rate spectrum $\mathcal{E}^{-1.4}$ emerges from the simple assumptions invoked by Lu and Hamilton, fitting closely to the observed extended power law spectrum over four decades in $\mathcal{E}$. A unique power law spectrum seems to be a property of avalanche theories of a variety of stimulated events in which the probability of each event depends only on the local circumstances of the moment. The spectral index depends only upon the number of dimensions of the system and the symmetries of the system (Lu and Hamilton, 1991; Bak, Tang, and Wiesenfeld, 1988; Kadanoff, et al. 1989; O'Brien, Wu, and Nagel, 1991). It will be interesting to see if there is an underlying statistical principle that allows direct deduction of the exponent -1.4. Evidently we need not concern ourselves with the detailed properties of the individual nanoflare (which, in any case, lie beyond our grasp at the present time), so far as the overall statistical properties of flares are concerned. But of course such physical questions as the acceleration of fast particles and the emission of hard X-rays depend very much on the detailed quantitative properties of the nanoflare.

There is no reason at present to think that the nanoflares that provide the continuing X-ray corona are part of the flare spectrum. The basic nanoflare in the X-ray corona is estimated at 3×10^{23} ergs in §11.4.7, a factor of 30 below the nanoflare energy deduced for the flare. The nonequivalence of the flare spectrum and the coronal nanoflares has been emphasized by Hudson (1991) and Zirker and Cleveland (1993).

12.2.5 Internal Dynamics

The gas within a flaring magnetic bipole is in a violently chaotic state as a result of the spotty explosive release of energy in the sea of nanoflares. The characteristic Alfven speed of 5×10^8 cm/sec in the transverse field component $\langle B_\perp^2 \rangle^{\frac{1}{2}}$ of 250 gauss guarantees rapid reconnection and short characteristic lives for each nanoflare, of the order of a second or so. The field reconnection in each nanoflare ejects gas at the characteristic Alfven speed, which is immediately diluted and shocked in the surrounding gas to provide temperatures of the order of 10^7 K or more, associated with extreme gas velocities up to 500 km/sec or more. These motions are on small scales, and involve relatively transparent volumes of gas, so that the line of sight passes through a variety of such motions. The extraordinary emission line widths observed in flares, up to $\pm 10^3$ km/sec (Svestka, 1986) are direct evidence for the random successive stimulated explosive reconnection events that collectively make up the flare. The analysis of Lu, et al. (1993) estimates the basic reconnection event, or nanoflare, at 3×10^{25} ergs, capable of accelerating a volume of gas 10^3 km on a side with a mean density of 10^9 atoms/cm^3 to a velocity of 10^3 km/sec. The line widths exceed the thermal broadening because of the excessive small-scale gas velocities involved in the reconnection process. The high velocities are quickly thermalized by collision with the ambient gas surrounding the reconnection region, providing most of the thermal emission at low gas velocity. The broad wings of the emission lines are from the ejected gas before it has had time to be thermalized fully.

12.3 The Geomagnetic Field

The solar wind presses against the outer boundary (the magnetopause) of the geomagnetic dipole field, pushing the field in to a distance of about $10R_E$ ($R_E =$ radius of Earth, 6.4×10^8 cm) and an intensity of the order of 5×10^{-4} gauss on the sunward side. The magnetopause is subject to the Kelvin–Helmholtz instability in one form or another (Parker, 1957b; Dessler, 1962), so that the wind exerts a strong drag on the field at the magnetopause, picking up bundles of flux and stretching them in the antisolar direction, "like smoke from a chimney" (Parker, 1957b) to form a comet-shaped geomagnetic tail (Johnson, 1960; Dessler, 1964; Dessler and Juday, 1965; Axford, Petschek, and Siscoe, 1965; Ness, 1969). In fact the magnetopause does not possess a static equilibrium configuration on the scale of the ion cyclotron radius (Parker, 1967a,b; 1968a,b; Lerche, 1967). Eviatar and Wolf (1968) note that the only possible state of the magnetopause involves strong plasma turbulence with an associated intermixing of solar wind plasma and field with the geomagnetic field.

Axford and Hines (1961) in a classic paper pointed out that the strong drag of the solar wind drives a large-scale convection of the field and plasma within the magnetosphere. The essential point is that the nonconducting atmosphere of Earth permits the ionosphere, and the magnetosphere above, to move relative to the Earth (Gold, 1959), carrying the magnetospheric field along with the convection. The ionosphere possesses a large inertia and provides a viscous drag on the convection of the magnetosphere, so that the convection is generally limited to velocities of the order of 2 km/sec or less at the ionosphere (Walbridge, 1967). The magnetospheric convection is driven by the effective viscous drag of the wind on the magnetopause as well as by the accumulation of flux bundles carried bodily from the sunward magnetopause into the geomagnetic tail. This general transport of magnetic flux in the antisolar direction drives a return flow deeper in the magnetosphere, of course.

Dungey (1961) made the important point that some degree of reconnection of the northward geomagnetic field is to be expected at the sunward magnetopause when the magnetic field (the interplanetary magnetic field) carried in the solar wind has a southward component. A priori one expects a southward component about half the time. The reconnected geomagnetic field lines are swept by the wind into the geomagnetic tail. This is evidently the primary cause of the flux transfer events encountered in the magnetosheath just outside the magnetopause, in which flux bundles from the geomagnetic field are observed in the wind, on their way to becoming part of the geomagnetic tail (Cowley, 1982; Berchem and Russell, 1984). The flux bundles are detached near the sunward equator (Daly, et al. 1984) by patchy reconnection initiated by the resistive tearing instability (La Belle-Hamer, Fu, and Lee, 1988; Scholer, 1988). Thomsen, et al. (1987) find a mixture of magnetospheric and magnetosheath plasma within the individual flux bundles. The individual bundles have characteristic diameters of $1 R_E$ (Saunders, Russell, and Schopke, 1984) and surface transition layers of the order of 10^8 cm, or about 10 ion gyroradii (Rijnbeek, et al. 1987)

There are, then, two effects that transfer magnetic flux into the geomagnetic tail and drive the convection and the aurora. One is the reconnection of the geomagnetic field with any southward interplanetary field blown against the sunward magnetopause and the other is the general dynamical instability and turbulence at the magnetopause, mixing small flux bundles of geomagnetic field into the wind. There are different views of the relative importance of the two effects and on the details of either (cf. Lundin, 1988; Heikkila, 1990). There is an extensive literature on the related theory and observation of the structure of the geomagnetic tail and the associated magnetospheric convection (cf. Levy, Petschek, and Siscoe, 1964; Ness, 1969; Hultquist, 1969; Stern, 1984; Schindler and Birn, 1986; Lyons, 1992 and references therein).

12.3.1 *Aurorae*

The essential point for the present discussion is that the geomagnetic field is divided into two topological domains with a thin, topologically complex transition layer between. The inner region, between geomagnetic latitudes of about $\pm 68^\circ$ under quiet conditions, consists of relatively undisturbed closed bipolar field in the form of the traditional geomagnetic dipole. On the other hand the polar field lines above $\pm 68^\circ$ latitude are carried away to form the extended geomagnetic tail. Thus the northern half of the tail is composed of magnetic flux directed inward to the north polar region and the southern half is composed of outward directed field from the south polar region. There is a tangential discontinuity in the tail between the northern and southern halves, across which some degree of reconnection occurs (Axford, Petschek, and Siscoe, 1965).

Figure 12.1 is a sketch of the outer magnetic surface of the inner undisturbed field and a few of the outer field lines that connect into the tail. There is a thin transition between, made up of recently transported flux bundles mixed with field participating in the geomagnetic convection.

Dungey (1961) and Dessler and Juday (1965) suggested that the aurora is a consequence of the intense electric currents that flow in the thin transition layer between the polar fields that connect into the tail and the undisturbed geomagnetic field at lower latitudes. It is now well documented that the aurorae lie on the pole-

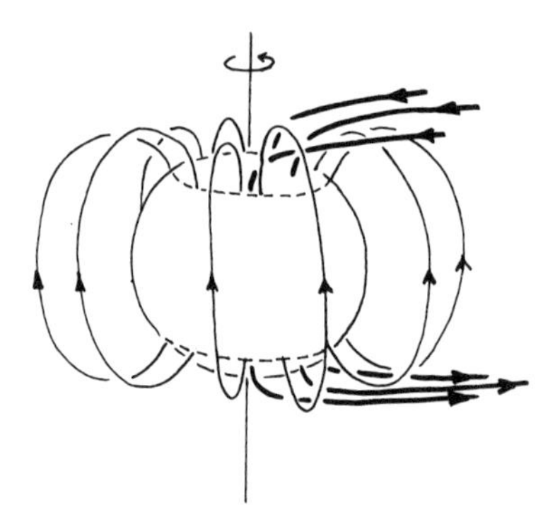

Fig. 12.1. A schematic drawing of the inner undisturbed geomagnetic field lines (light lines) and the polar field lines (heavy lines) that are carried into the tail.

ward side of the outer boundary of the undisturbed region of geomagnetic field (Davis, 1962a,b, 1967; Feldstein, 1969).

From the point of view of the basic theorem of magnetostatics, the essential point is that the flux bundles picked up in the solar wind, by whatever means, and subsequently stretched back into the tail provide an irregular field. Large flux transfer events from reconnection are mixed with the smaller flux tubes carried along the magnetopause without reconnection. This interwoven mixture of flux bundles is then caught up in the irregular convection in the geomagnetic tail. The basic theorem tells us that the Maxwell stresses strive to create intense current sheets between misaligned flux bundles, while the whole skein of magnetic flux is convected sunward from the midnight side. This is the spontaneous source of current sheets, mostly in the transition layer between the geomagnetic tail field and the undisturbed dipole field, and the aurorae are the result.

To provide some idea of the characteristic times involved in the process of making the aurora, note that the solar wind speed in the magnetosheath is typically 200–300 km/sec, traversing a distance of $1 R_E$ in about 25 sec. It follows that the transit time of a flux bundle from the sunward magnetopause to the beginning of the geomagnetic tail is of the order of 500 sec. This is comparable to the Alfven transit time from the magnetopause to the top of the ionosphere. It follows that a flux bundle picked up by the wind, for one reason or another, on the sunward side is in a dynamical state for at least the next 10^3 sec, as it is stretched back along the geomagnetic tail. Therefore, the amount of flux picked up in a period of 10^3 sec provides some measure of the thickness of the active transition layer, and vice versa.

It appears likely that the transported magnetic flux bundles have time to relax toward static equilibrium once they are embedded in the tail and carried in whatever convection patterns are present. The essential point is that the lower (ionosphere) ends of the flux bundles are limited by the ionospheric drag to velocities of about 2 km/sec or less, whereas the Alfven speed is of the order of 10^2 km/sec. Therefore, it is expected that some of the potential tangential discontinuities succeed in relaxing to thin current sheets in accordance with the basic theorem of magnetostatics. It is generally supposed that these individual thin current sheets provide the visible aurora (see Fig. 8.55 of Akasofu and Chapman, 1972). The existence of the aurora rays and curtains is, then, a direct consequence of the gross magnetohydrodynamic structure of the magnetosphere. Needless to say, when it comes to the detailed form and internal structure of an auroral curtain, with the local ionospheric effects, parallel electric fields, etc. macroscopic magnetohydrodynamics is no longer adequate and the more detailed small-scale plasma physics must be treated (cf. Lyons, 1992). But it is the large-scale magnetohydrodynamic forces in the fields outside the auroral forms that largely determine the sites for the aurora. Zhu (1994a,b) provides a quantitative treatment of the convection and principal magnetic stresses.

The observed auroral intensity, involving a total of one or more auroral curtains within the auroral oval, is usually quoted as a vertical electric current J of the order of 10^{-3} amperes/cm, or 10^5 amperes per thousand km along the oval. This surface current density is deduced through Ampere's law from the observed deflection $\Delta B/B$ of the geomagnetic field across the region, with $4\pi J = c\Delta B$. Typical values of ΔB are 10^{-3} gauss. The deflection, by an angle $\Delta B/B \cong 2 \times 10^{-3}$ radians, or 0.1°, is a direct result of the transport of magnetic flux into the geomagnetic tail. This simple estimate is complicated to some degree by the chafing across the auroral zone between the closed magnetic field lines and the tail field lines as a consequence of the rotation of the Earth about its axis. But the chafing provides detailed fluctuation, probably without influencing the net deflection and the gross features outlined here.

The quantitative correspondence between the formation of the geomagnetic tail and the observed deflection at ionospheric altitudes (100–200 km) is readily demonstrated. Consider Fig. 12.2, which is a schematic drawing of a column of magnetic flux extending up from the polar ionosphere and pulled around into the geomagnetic tail at its upper end. The single heavy line depicts a typical field line. The thickness of the flux column is designated by $h(r)$, where r is radial distance, with the ionosphere at $r = a$ and the geomagnetic tail beginning at $r = c$. The width of the sheet is $w(r)$ and conservation of magnetic flux requires that the magnetic field $B(r)$ in the column satisfies

$$h(r)w(r)B(r) = h(a)w(a)B(a). \tag{12.2}$$

The flux column is pulled in the antisolar direction by the tension of the field lines extending into the geomagnetic tail. Denote the deflection from the ambient undisturbed field direction by the angle $\theta(r)$, so that the normal field component introduced by the deflection is $b(r) = B(r)\tan\theta(r)$. The shear stress is

$$F = b(r)B(r)/4\pi \tag{12.3}$$

providing a total force Fwh across the flux bundle. In the idealized case that the surrounding field is in a normal state, the flux bundle would be in equilibrium for

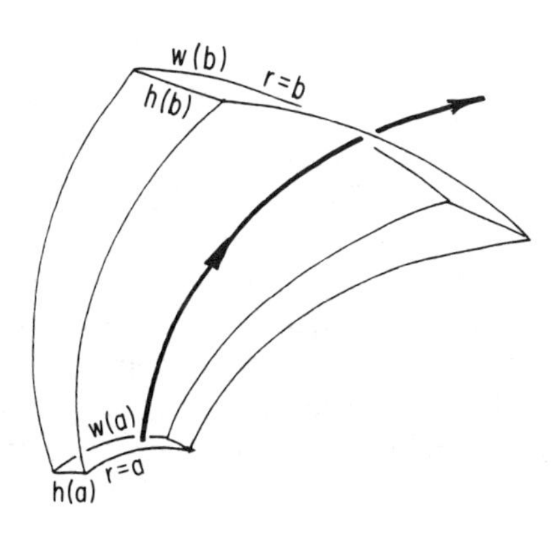

Fig. 12.2. A schematic drawing of a flux bundle of width $w(r)$ and thickness $h(r)$ extending up from the ionosphere at $r = a$. The single heavy line

$\theta = 0$. The linear displacement $r\theta(r)$ upsets the equilibrium, pulling the flux bundle up or down the local pressure gradient in the ambient field, but, in order of magnitude, the total shear stress in flux bundle in magnetostatic equilibrium is independent of r. Hence, very approximately,

$$h(r)w(r)B(r)b(r) \cong h(a)w(a)B(a)b(a) \tag{12.4}$$

from which it follows with the aid of equation (12.2) that $b(r) \cong b(a)$. That is to say, the transverse field component $b(r)$ varies little compared to $B(r)$ between the ionosphere and the near end of the magnetic tail at $r = c$. This condition indicates that the deflection θ of the field lines connecting into the tail is $b(a)/B(a)$ at the top of the ionosphere with $b(a)$ comparable in order of magnitude to the field at the near end of the tail where $b(b) \cong B(b)$.

Arbitrarily defining the near end of the geotail to lie at $r = c = 8R_E$, it follows that $b(c) \cong 10^{-3}$ gauss. Therefore, $b(a) = 10^{-3}$ gauss at the ionosphere and the deflection of the field is

$$\begin{aligned} \theta &= b(a)/B(a) \\ &\cong 1.6 \times 10^{-3} \text{ radians.} \end{aligned}$$

across an auroral current sheet. The current per unit width associated with a change $b(a)$ in the transverse field follows from Ampere's law as

$$\begin{aligned} J &= (c/4\pi)b(a) \\ &\cong 2.5 \times 10^{6} \text{ esu} \\ &\cong 10^{-3} \text{ amperes/cm} \end{aligned}$$

This magnetic deflection of 10^{-3} gauss, (associated with currents of 10^{-3} amperes/cm) is typical of the quiet-day vertical current in the auroral oval, of the order of 10^{5} amperes spread over a thousand km or so.

12.4 Diverse Settings

Consider some of the diverse settings in which the spontaneous appearance of tangential discontinuities is expected, mentioned at the beginning of this chapter. The discontinuities are more than a theoretical curiosity only when their rapid reconnection and dissipation of magnetic free energy produces nanoflaring at some significant level, of course.

One can imagine intense nanoflaring in the expected external magnetic fields of accretion disks around massive compact objects, or in protostellar nebulae such as the one that formed the Sun and the planets. There is no hard information on the strength or form of the magnetic fields, but if the magnetic field played a major role in transferring angular momentum outward through the disk, it is entirely possible that a significant fraction of the gravitational energy released in the accretion went into the magnetic field. Presumably the magnetic energy is dissipated largely in nanoflares in tangential discontinuities between the lobes and sectors of the spiral magnetic field generated in the nonuniformity rotating accretion disk. The negative gravitational energy of the Sun is approximately 10^{49} ergs. If even 1 percent of that amount went into

Fig. 12.3. A schematic drawing of the magnetic field carried out through the heliopause (inner sphere) by the solar wind and then off into the downstream interstellar wake. The outer light line enclosing the region indicates the contact surface between the wake and the interstellar wind, with the short arrows indicating the motion of the gas. The heavy spiral lines represent field lines from the southern hemisphere of the heliosphere at a time when the magnetic dipole of the Sun is oriented opposite to the dipole of Earth. The light spiral lines represent field lines from the northern hemisphere. The actual spiral field is wound much more tightly than indicated in this sketch. The dashed lines overlying the figure symbolize the local interstellar magnetic field.

nanoflares in the solar accretion disk over a period of 10^6 years, say, the suprathermal output of the disk would be about comparable to the present luminosity of 4×10^{33} ergs/sec of the Sun. However, direct quantitative observation of contemporary protostellar accretion disks is the only basis for any scientific opinion.

Another site for nanoflaring is the interstellar wake produced by the passage of the Sun and the heliosphere through the local interstellar medium. Observations of the UV radiation from interstellar neutral hydrogen and helium entering into the heliopause (where they are ionized by charge exchange with the solar wind and by solar UV) indicates that the interstellar medium is streaming past at about 9 km/sec from the direction of $\alpha = 263°$, $\delta = -23°$ (Thomas and Krassa, 1971; Bertaux and Blamont, 1971; Fahr, 1974; Weller and Meier, 1974, 1979).[1]

Figure 12.3 is a schematic drawing of two field lines in the southern hemisphere (heavy lines) emerging through the heliopause and extending away into interstellar space in the outflowing wake of the solar wind. The two field lines in the northern hemisphere are shown (light lines) spiralling away along the core of the wake, while enveloped by the magnetic field of the southern hemisphere. Now the topology of

[1]The motion of the interstellar gas relative to the local stars has been studied in some detail by observations of the interstellar absorption of Na and Ca lines of stars within about 50 pc (Bertaux, et al. 1985; Lallement, Vidal-Madjar, and Fexlet, 1986; Welsh, et al. 1991; Vallerga, et al. 1993).

the actual heliospheric magnetic fields is more irregular than indicated by this idealized sketch. But the point is that, even without any complications, surfaces of tangential discontinuity are expected throughout the interior of the wake, as well as at the separatrix between the billowing wake and the local interstellar field. Theoretical descriptions of the wake, as it extends off through interstellar space, have been given by Yu (1974), Baranov (1990), and Fahr and Fichtner (1991) and others.

The relative unimportance of nanoflaring in the interstellar wake can be seen from the fact that the magnetic energy density in the solar wind at the orbit of Earth is of the order of 4×10^{-10} ergs/cm^3 (for $B = 10^{-4}$ gauss) and the kinetic energy density of the supersonic wind is 6×10^{-9} ergs/cm^3 (for $N = 5$ H atoms/cm^3 at a speed of 400 km/sec). Beyond the orbit of Earth the magnetic field is principally azimuthal, declining outward as $1/r$, so that the magnetic energy density declines as $1/r^2$, in the same way as the density N declines. Therefore, the ratio of magnetic to kinetic energy density does not change much from the value of 10^{-1} at the orbit of Earth. The kinetic energy of the wind is largely converted into heat at the shock transition at 10^2 au, and the ratio of plasma pressure to magnetic pressure is of the order of $\beta = 10$ or more. Therefore, the downstream nanoflares cannot have a major influence on the state of the wake, since they have available less than 10^{-1} of the total energy.

The possibility of nanoflaring in the interstellar or galactic magnetic field may be of interest in a variety of circumstances, some speculative and others more substantial. First of all the tumbling, billowing, interstellar gas clouds, driven by the formation of O, B, and A stars within clouds and between clouds, must interweave the field lines of the interstellar field to a substantial degree on scales as large as $\ell = 10^2$ pc. Observations (Kaplan, 1966; Jokipii, Lerche, and Schommer, 1969) detect field fluctuations $|\delta \mathbf{B}|$ as large B itself, over scales of 10^2 pc. There are weaker fluctuations on smaller scales down to about 10^{11} cm (Lee and Jokipii, 1976; Parker, 1967c). But in any case the magnetic free energy density is of the order of $(\delta B)^2/8\pi \sim 0.4 \times 10^{-12}$ ergs/cm^3 for $\delta B = 3 \times 10^{-6}$ gauss, comparable to the kinetic energy density of the observed cloud motions of 7 km/sec in a mean gas density of 1–2 atoms/cm^3. The Alfven speed is about 7 km/sec and one expects a continuing tendency to form surfaces of tangential discontinuity in the stochastic field lines (Jokipii and Parker, 1969a,b) created by 10^{10} years of interweaving and reconnection across random field fluctuations. Therefore, it seems likely that there are regions in interstellar space in which there are significant suprathermal effects produced by the local nanoflaring.

On the other hand, it must be appreciated that the strongest suprathermal effects are from supernovae and perhaps active galactic nuclei (Jokipii and Morfill, 1985, 1987, 1990). The blast waves from supernova produce the hot (10^5–10^7 K) component of the interstellar medium and burst out through the surface of the disk in many cases. It is generally believed that the supernovae and their respective blast waves are the principal source of cosmic ray gas. Both the hot gas component and the cosmic ray gas inflate the galactic magnetic field. The cosmic rays in particular produce extended lobes of magnetic field extending outward from both sides of the gaseous disk to form a close-packed outward billowing magnetic halo (Parker, 1958b, 1965a,b,c, 1966a, 1968c, 1969; Lerche and Parker, 1966). Figure 12.4 is a sketch of the general topological structure of the halo fields. The expected internal irregularities in the inflation of the individual lobe are not shown.

Fig. 12.4. A schematic drawing of the close packed lobes of magnetic field inflated outward from both sides of the gaseous disk of the Galaxy to create the galactic halo.

12.5 Cosmic Rays and Galactic Halos

The formation of the galactic halo begins with the fact that the undisturbed horizontal magnetic field in the gaseous disk of the Galaxy is strongly inflated by the powerful cosmic ray gas at the same time that it is confined to the gaseous disk by the weight of the interstellar gas. It follows that the galactic magnetic field is strongly unstable to forming hills and valleys along the field, with a particularly short scale across the field (Parker, 1966b, 1967c, 1979). The heavy cool gas tends to slide downward along the field lines into the valleys, thereby unloading the hills so that they expand upward, and weighing down the valleys so that they sag downward. The situation is further aggravated by the cosmic ray gas, streaming freely along the field in defiance of gravity and inflating the upward bulging lobes of field. The characteristic growth time for the instability is of the order of 10^7 years, and the instability proceeds to large amplitudes (Parker, 1968d, 1979; Mouschovias, 1974) with the magnetic field billowing upward through the atmosphere (Shibata, et al. 1989a,b, 1990; Shibata, Tajima, and Matsumoto, 1990).

The fact is that the continuing generation of cosmic ray gas in the gaseous disk, at a rate of the order of 10^{40}–10^{41} ergs/sec, causes the inflation of the magnetic lobes to continue until eventually rapid reconnection between neighboring magnetic lobes cuts off the connection to the galaxy and the supply of cosmic ray gas. The rate of inflation of a magnetic lobe can be estimated from the time the average cosmic ray particle spends in the disk, while the characteristic distance to which the lobes extend follows from the measured age of the cosmic ray particles.

The basic fact is that the collisional breakage of the heavier nuclei (C,N,O,. . .,Fe) among cosmic rays indicates that cosmic rays have passed through approximately $\mathcal{M} = 5\,\text{gm/cm}^2$ of interstellar gas (Shapiro and Silberberg, 1970;

Garcia-Munoz, Mason, and Simpson, 1975a,b; Morfill, Meyer, and Lüst, 1985). Employing the simple model that the mean gas density is N atoms/cm^3 over a disk thickness of $2h$, it follows that the average path length λ traversed by a cosmic ray particle in interstellar space is

$$\lambda = \mathcal{M}/NM$$

where M is the mass of the hydrogen atom. A particle with velocity c traverses the distance λ in a time

$$\begin{aligned} t &= \lambda/c \\ &= \mathcal{M}/NMc \end{aligned}$$

Thus, if $N \cong 2/\text{cm}^3$ (Schmidt, 1957) the result is $t \cong 5 \times 10^{13}$ sec $\cong 1.6 \times 10^6$ years. This is, of course, an upper limit on the time spent in the gaseous disk, because part of $\mathcal{M}$ may have been traversed in the much denser gas in the supernova blast wave (or whatever) that produced the cosmic rays. The only escape for the cosmic rays from the gaseous disk is through inflation of the lobes of the galactic magnetic field. So if the time spent in the gaseous disk of half thickness h is designated as t, the cosmic rays stream outward across both surfaces of the disk at an average speed $u = h/t$. For $h = 10^2$ pc, this is $u = 60$ km/sec over the entire surface area. If the lobes cover only the fraction α of the surface, the speed is $u = h/\alpha t > 60$ km/sec.[2] The Alfven speed in the gaseous disk is typically 10 km/sec, whereas in a magnetic lobe it is of the order of 10^2 km/sec or more.

It is interesting to note, then, that the depletion of cosmic ray B^{10} by radioactive decay indicates that the cosmic rays in the neighborhood of the Sun have an average age of about 10^7 years (Garcia-Munoz, Mason, and Simpson, 1975a, 1977; Hagen, Fisher, and Ormes, 1977; Morfill, Meyer and Lüst, 1985). Taking this to be the characteristic age in general indicates that the cosmic rays maintain free access to the gaseous disk for 10^7 years while spending only about one-sixth of the time in the gaseous disk and five-sixth of the time outside the disk in a galactic halo of low background gas density (10^{-4}–10^{-3} atoms/cm^3). It follows that the individual magnetic lobe, through which the cosmic rays circulate freely, remains fully connected to the disk while it extends out an average distance of $6h \cong 600$ pc in 10^7 years. If all lobes were inflated at the same fixed rate, this would imply that their magnetic connection to the gaseous disk is severed when they extend to some distance of the order of 1 kpc. It follows, too, that, if the individual cosmic ray particle moves freely between the gaseous disk and the extended magnetic lobes during its 10^7 year association with the galaxy, then the field in the lobe must be strong enough to confine the cosmic ray pressure of about 0.5×10^{-12} dynes/cm^2, requiring 3×10^{-6} gauss. The reader is referred again to Fig. 12.4, indicating each extended magnetic lobe, anchored in the dense gaseous disk and rapidly inflated outward by the cosmic ray gas produced in the gaseous disk. A more detailed account of the inferred properties of the galactic halo created by cosmic rays and hot interstellar gas has been provided in the literature (Parker, 1992b) in which it is suggested that the lobes are an essential part of the galactic dynamo.

[2]This raises some interesting questions on the streaming of cosmic rays within the disk, which is believed to be limited to a few times the Alfven speed by the onset of instabilities producing Alfven waves that interact strongly with the gyrating cosmic ray particles (see Hartquist and Morfill, 1986 and references therein).

It follows from the basic theorem of magnetostatics that there are tangential discontinuities where adjacent lobes press together, sketched in Fig. 12.5(a), providing substantial reconnection of the magnetic field in a mixture of the two patterns sketched in Fig. 12.5(b) and (c). The reconnection in (b) between adjacent lobes represents a net loss of magnetic flux to the gaseous disk, leaving behind closed loops of magnetic field with their lower sides deeply embedded in the gaseous disk. The basic theorem indicates that these loops form discontinuities where they press sidewise against their neighbors (as well as discontinuities in their stochastic internal fields with flux tubes further dislocated by nonuniform inflation) indicating that reconnection proceeds on all fronts. In Fig. 12.5(c) the reconnection cuts off each magnetic lobe. The release of a magnetic lobe leaves it free to expand and accelerate rapidly away with the internal load of thermal gas and cosmic ray gas, leaving behind a much shortened magnetic lobe, which continues with the inflation.

Now, in an extended lobe of field $B \cong 3 \times 10^{-6}$ gauss, where nanoflaring drives the temperature up to 10^7 K, the gas density is estimated to be about 10^{-4} atoms/cm^3, so that the gas pressure does not exceed the pressure in the gaseous disk on which it rests. The resulting characteristic Alfven speed is of the order of 700 km/sec. If reconnection proceeds at $0.03C \cong 20$ km/sec, it cuts 200 pc, or about halfway, into each side of a lobe in 10^7 years, thereby freeing the lobe. Thus the 10^7 year cosmic ray age in the galaxy is consistent with the nature of the extended magnetic lobes.

It follows that the escaping magnetic loops filled with cosmic rays may expand to infinity, contributing to a general galactic wind (Holzer and Axford, 1970; Kopriva and Jokipii, 1983; Ko, et al. 1991; Breitschwerdt, McKenzie, and Volk, 1991).

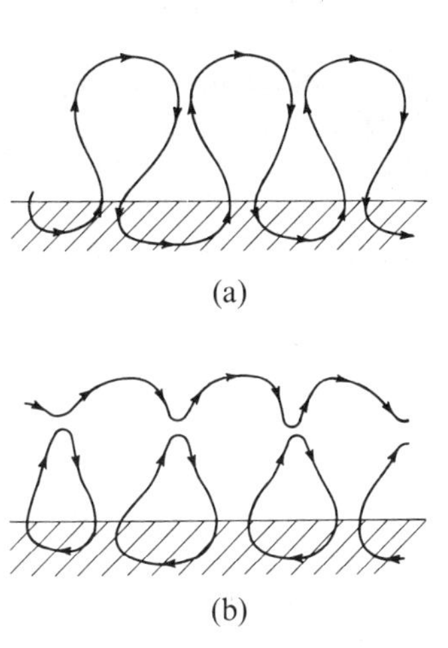

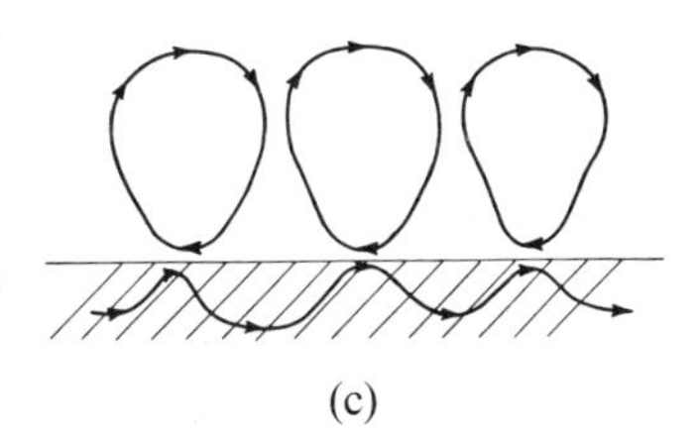

Fig. 12.5. A schematic drawing (a) of the lobes of azimuthal magnetic field extending out through one surface of the gaseous disk of the Galaxy. (b) The magnetic topology, following reconnection between adjacent magnetic lobes, so as to cause a net loss of magnetic flux. (c) Reconnection across the foot of each lobe, allowing the free escape of the lobe and its cosmic rays without loss of net magnetic flux from the Galaxy.

The rate at which the cosmic ray gas does work on the inflating lobes of field is $\langle uB^2/8\pi \rangle$ per unit area on each side of the gaseous disk. With $B = 3 \times 10^{-6}$ gauss and $u = 60$ km/sec, the rate is 2×10^{-6} ergs/cm^2 sec. Over both faces of the disk out to a radius of 10 kpc the total rate is 2×10^{40} ergs/sec, comparable to the estimated cosmic ray output in the disk, of course. This is the rate at which the magnetic field is extended and magnetic free energy is created. It follows that the rapid reconnection between and within magnetic lobes dissipates magnetic free energy at some substantial fraction of 2×10^{40} ergs/sec, the rest going into expansion of disconnected magnetic flux and cosmic rays into intergalactic space. It follows, then, that the heat input to the galactic halo is a major fraction of 10^{40} ergs/cm^3. The halo gas has a temperature of the order of 10^6–10^7 K, emitting mostly in the region of soft X-rays. It would appear, then, that the X-ray luminosity of a galactic halo, such as our own, is a direct measure of the cosmic ray production rate in the gaseous disk of the Galaxy (Parker, 1992b). It is only a very rough measure, of course, because there is substantial heat input from blast waves and fountains of hot gas from the disk (Savage, 1990) providing additional inflation and heat input as well. But it appears that the X-ray luminosity of a galactic halo well away from an active galactic nucleus is probably sustained by cosmic rays more than by other mechanisms. This predicts an X-ray luminosity for the outlying portions of the halo of the Galaxy as being of the order of 10^{40} ergs/sec. The halo spreads 4π steradians around us at a mean distance of the order of 10 kpc, of course, so that it provides an X-ray background that is difficult to distinguish from a more cosmic background. The recent paper by Kahn and Brett (1993) describes the dynamics of the near halo dominated by the escaping hot gas ejections from the disk, which provide their own interweaving and reconnection of the magnetic lobes.

Observations of the halos of distant spiral galaxies show X-ray emitting halos as bright as 10^{42} ergs/sec, with others at lower luminosities down to the observational limit around 10^{39} ergs/sec (Bregman and Glasgold, 1982; Fabbiano and Trinchieri, 1984; Forman, Jones, and Tucker, 1985; Fabbiano, 1988). We conjecture that these observable halos originate in much the same way as the halo of our own Galaxy, with the X-ray emitting temperatures produced in large part by the rapid reconnection at the current sheets formed within and between the many magnetic lobes produced by inflation by cosmic rays. In this respect the basic theorem of magnetostatics plays a central role in the heating of X-ray coronas and halos on all scales from the Sun to the Galaxy.

12.6 Observational and Experimental Studies of Nanoflaring

A conspicuous feature of the universal nanoflaring implied by the basic theorem of magnetostatics is the difficulty in direct detection and observation of the individual nanoflares and of the rate of convective winding and interlacing of the magnetic fields that produce the nanoflares. In all cases, from the solar X-ray corona to the galactic halo, the rate of winding, i.e., the basic energy input, has been estimated from other rates, such as the photospheric convective speed and scale, the age of cosmic rays inferred from their isotopic composition, etc. It remains, then, to accomplished the basic studies that would move stellar X-ray astronomy from a taxonomic study into hard physics. A similar statement applies to solar and stellar flares, and, to the degree permitted by contemporary technology, to galactic X-ray halos.

To define the problem more broadly note that the principal energy input to the solar X-ray corona is assumed in all contemporary theories to be the motions of the individual magnetic fibrils, with characteristic fibril diameters of the general order of 10^2 km (0.13 sec). The diameters are too small by a factor of three or more for telescopic resolution of the detailed motion of the fibrils, particularly at the crucial times when one fibril is close to another. The terrestrial atmosphere above the telescope limits the resolution to 0.3 sec under the very best (and, therefore, relatively rare) seeing conditions. The minimum requirement for the observations is a balloon-borne diffraction-limited 1 m telescope carried to the top of the atmosphere for several periods of 5–20 days. Alternatively one might consider what can be accomplished with suitable active optics in a ground based telescope of much larger aperture at the best observing site available, but there are serious limits to the resolution that can really be achieved in this way. The optical observations must be complemented from time to time by high resolution UV and X-ray images, from sounding rockets or spacecraft. The ideal instrument for decisive study would be a 1 m, or larger, diffraction limited telescope in an orbiting spacecraft, accompanied by suitable UV and X-ray telescopes for simultaneous observation of the suprathermal activity associated with the photospheric motions. The continuous coverage over the extended period of a year or more is essential for a definitive study. Ultimately, of course, the observations should sample conditions at all phases of the 11 year magnetic cycle.

Interest in the microscopic examination of the magnetohydrodynamic activity at the visible surface of the Sun is not limited to the magnetic fibril motions that are responsible for the solar X-ray corona and the solar flare, which, we point out again, serve as the prototype for the X-ray emission and flares on all solitary late-main sequence stars, of course. The sunspot and the plage and faculae are conspicuous phenomena whose nature and cause lies at the small scales of 100 km, and which are currently without definitive physical explanation. Is the plage a consequence of acoustical heating that is somehow enhanced by the mean fields of 50 gauss or more in which the plages occur? Or are plages heated by the dissipation of the magnetic trash or U-tubes (Spruit, Title, and Van Ballegooijen, 1987) as the fields are mashed together in the converging flows between granules (cf. Yi and Engvold, 1993)? The energy input is high, of the order of 10^9 ergs/cm^2 sec or more, in order to provide the enhanced brightness of the plage (compared to the average photospheric brightness of 6×10^{10} ergs/cm^2 sec).

Observational studies of the luminosity variation of the Sun indicate that the plage is a major player, along with faculae and the opposing sunspots, in the fluctuation in brightness viewed from Earth (Sofia, Oster, and Schatten, 1982; Chiang and Foukal, 1985; Foukal and Lean, 1986; Hudson, 1988; Spruit, 1991; Willson and Hudson, 1991). Yet so little is known about the plage that it is not established by direct observational resolution whether the plage is a hole (Spruit, 1979; Deinzer, et al. 1984; Hasan, 1985, 1988; Hasan and Sobouti, 1987) or a hillock (Schatten, et al. 1986). Evidently both features are present, the hillock represented by the enhanced chromosphere.

In fact there is a whole new world waiting to be discovered by a modern Antonie van Leeuwenhoek with a "solar microscope," as such a space-borne telescope has been called. That microscopic world has been scanned so far only with the blurred eyesight of the conventional ground based telescope. The excellent seeing occasionally available from the ground at locations in the Canary Islands, and, on special occasions at the best mountain-top sites, has shown how active is the world at the limit of resolution. The 30 cm SOUP orbiting telescope (Title, et al. 1989) has established the visibility of the separate individual, but still unresolved, magnetic fibrils and has mapped the large-scale flow of the fibrils. The SOUP resolution is approximately 300 km on the Sun. One of the many important results besides the streaming of fibrils is the observed increase of the rms fluctuating horizontal gas velocity to 0.75 km/sec on scales of 700 km or less, compared to 0.3 km/sec over dimensions of 2000 km.

The observations of Title, et al. (1992) with a resolution of about 300 km show an extraordinary activity and "jiggling" of magnetic fibrils within the plage, without resolving the details. It should be added that sunspots show remarkable diversity and motion on very small scales too, but lack a physical theory for their overall existence (Parker, 1992c). So it can be safely asserted, from the limited knowledge that already exists, that there is much to see and to learn of the microscopic world of the magnetic and hydrodynamic activity on the surface of the Sun. The implications of what is observed will be far reaching, which brings us back to the variable luminosity of the Sun.

The luminosity variations evidently involve more than surface emissivity because the energy is supplied to the surface by the convective motions. So one expects an associated small variation of the solar radius and a change in motion throughout the convective zone (Eddy and Boornazian, 1979; Spiegel and Weiss, 1980; Hudson, 1987; Spruit, 1991; Ribes, et al. 1991; Nesme-Ribes and Mangeney, 1992). There is evidence for a small variation in the rotation as a function of latitude (Eddy, Gilman, and Trotter, 1976; Ribes, et al. 1991) and in meridional circulation in the Sun (Ribes, et al. 1991; Nesme-Ribes, Ferreira, and Vince, 1993).

It was pointed out a century ago (Sporer, 1889; Maunder, 1890, 1894, 1922; Clerke, 1894) that the activity level of the Sun sank to remarkably low levels in the last half of the 17th century (approximately 1645–1715 AD). Eddy (1976, 1977) substantiated Maunder's statement and went on to show that the Sun has experienced such prolonged periods of deep inactivity during 10 of the last 70 centuries and has been raised to a state of hyperactivity during eight other centuries (Eddy, 1980, 1982, 1988). Eddy then pointed out that the available historical and geological records of climate, extending back about three millenia in the northern temperate zone, show the remarkable fact that the mean annual temperature dropped by about 1°C during the centuries of inactivity, with devastating effects on agriculture in Europe, China, and the northern plains in North America. The mean annual temperature was also found to rise a comparable amount during the centuries of hyperactivity, accompanied by prolonged devastating drought in the otherwise semi-arid regions of southwestern United States. Studies of the ocean water surface temperatures from the last 130 years show a close tracking of the level of solar activity, from the 11 year cycle to the long-term trends of $\pm 1°$C (Reid, 1991).

The observed variation of solar luminosity with the level of activity suggests the basis for the variation in the terrestrial climate. The observed variation of the luminosity of the Sun over the last 15 years has been modest, about two parts in 10^3, which is not expected to have much effect on climate (Hoffert, Frei, and Narayanan,

1988). But the deep Maunder minimum of the 17th century, and the nine earlier minima established from the ^{14}C record, are another matter. Observations of many solar type stars, along with the Sun, establish a general relation between the luminosity, the level of activity, and the length of the activity cycle (Baliunas, Soon, and Zhang, 1994). This calibration indicates that the luminosity of the Sun was fainter by 5 ± 2.5 parts in 10^3 during the Maunder minimum, and that amount is estimated to provide a decline of the order of 1°C (about three parts in 10^3) in the mean annual temperature in the northern temperate zone of Earth (Zhang, et al. 1994).

The estimate of the solar luminosity variation from the seawater studies is roughly one part in 10^2 over a century. This is about twice as large as the luminosity variation estimated from the activity level and cycle length during the Maunder minimum. Collectively the evidence makes it clear that the changing solar luminosity must be studied in greater detail with adequate instrumentation to establish the precise long-term variation and to understand the physics of the variation. The solar variability presents a challenge not only to the physics community but to the atmospheric sciences as well, because it is now imperative to understand the response of terrestrial climate to small changes in solar irradiance. Only then will it be possible to sort out the changes in climate attributable to the accumulating anthropogenic greenhouse gases, increasing monotonically at the present time, and the warming effect of the increasingly luminous Sun, which waxes and wanes according to its own unknown schedule.

In summary, the microscopic world of magnetic fibrils, their motions, and their internal structure is a fundamental scientific frontier. That microscopic world drives the large-scale effects with which we are familiar, including the terrestrial climate. There have been specific proposals for two decades to look into this new microscopic world with an orbiting diffraction limited telescope, and more recently through a balloon-borne telescope. The intellectual inertia of the scientific community and the hidden agendas of the funding agencies are sufficient, however, that no forward progress is visible at the time of writing. Fortunately there is the real prospect that the investigation of the microscopic solar activity will be carried out elsewhere, so that our basic knowledge of the world around us may yet include the cause of X-ray emission from a solitary star and the implications of the varying emissions from the Sun.

Now there have been relatively few remarks in this concluding chapter on the spontaneous formation of tangential discontinuities in the important circumstance of laboratory plasma confinement devices. This general omission is not for lack of interest, for we may hope that laboratory studies of plasma turbulence (Similon and Sudan, 1990), anomalous resistivity and rapid reconnection, mentioned briefly in §§2.4.3 and 10.4, will continue to push forward our present limited understanding of this very complex phenomenon. To date it is the complexity that is universal, and one might hope that at some time in the future a simplifying generality in the limit of large Lundquist number may come to light. On the other hand, it is not obvious that such a universal criterion exists, which is why observational studies are the only present means of determining the magnitude of the individual nanoflare and the detailed physics of the cooperative stimulated behavior of a sea of nanoflares in a web of surfaces of tangential discontinuity.

Laboratory plasma confinement in magnetic fields turns up a variety of suprathermal effects, including the production of current sheets and fast particles. The reader is referred to the discussion and references in Chapters 6 and 10, wherein specific measurements of the plasma dynamics under controlled conditions are described. Both laboratory measurements and numerical experiments are available to treat the formation of currents or tangential discontinuities, as a result of dynamical instabilities (Rosenbluth, Dagazian, and Rutherford, 1973; White, et al. 1977; Strauss and Otani, 1988; Finn, Guzdar, and Usikov, 1992). Unfortunately the limited scale of the terrestrial laboratory does not permit an experiment at the huge Lundquist numbers ($\gtrsim 10^{10}$) necessary to demonstrate the spontaneous formation of tangential discontinuities in the asymptotic relaxation of a stable interlaced field topology toward static equilibrium.

REFERENCES

Chapter 1

Arnold, V. 1965, *C.R. Acad. Sci. Paris* **261**, 17.

Arnold, V. 1966, *Ann. Inst. Fourier Grenoble* **16**, 361.

Arnold, V. 1974, *Proc. Summer School in Differential Equations* (Erevan: Armenian SSR Acad. Sci.).

Athay, R.G. 1981, *Astrophys. J.* **249**, 340.

Athay, R.G. and Illing, R.M.E. 1986, *J. Geophys. Res.* **91**, 10961.

Athay, R.G., Low, B.C. and Rompolt, B. 1987, *Solar Phys.* **110**, 359.

Babcock, H.D. 1959, *Astrophys. J.* **130**, 364.

Babcock, H.W. and Babcock, H.D. 1955, *Astrophys. J.* **121**, 349.

Batchelor, G.K. 1947, *Proc. Cambridge Phil. Soc.* **43**, 533.

Beckers, J. and Schröter, E.H. 1968, *Solar Phys.* **4**, 142.

Biskamp, D. 1984, *Phys. Letters A* **105**, 124.

Biskamp, D. 1986, *Phys. Fluids* **29**, 1520.

Biskamp, D. and Welter, H. 1980, *Phys. Rev. Letters* **44**, 1069.

Boozer, A.W. 1983, *Phys. Fluids* **26**, 1288.

Brueckner, G.E. and Bartoe, J.D.F. 1983, *Astrophys. J.* **272**, 329.

Brueckner, G.E., Bartoe, J.D.F., Cook, J.W., Dere, K.P. and Socker, D.G. 1986, *Adv. Space Res.* (8), 263.

Cary, J.W. and Littlejohn, R.G. 1983, *Ann. Phys. (New York Acad. Sci.)* **151**, 1.

Chandrasekhar, S. 1961, *Hydrodynamic and Hydromagnetic Stability*, (Oxford: Clarendon Press), pp. 83–85.

Chapman, G. 1973, *Astrophys. J.* **191**, 255.

Chapman, E.D. and Kendall, P.C. 1963, *Proc. Royal Soc. London A* **271**, 435.

Cowling, T.G. 1953, in *The Sun*, ed. G.P. Kuiper (Chicago: University of Chicago Press), pp. 583–590.

Cowling, T.G. 1958, in *Electromagnetic Phenomena in Cosmical Physics*, ed. B. Lehnert (Cambridge: Cambridge University Press), p. 105.

Craig, I.J.D. and Clymont, A.N. 1991, *Astrophys. J.* **371**, L41.

Dahlburg, J.P., Montgomery, D., Doolen, G.D. and Matthaeuss, W.H. 1986, *J. Plasma Phys.* **35**, 1.

Dere, K.P., Bartoe, J.D.F. and Brueckner, G.E. 1981, *J. Geophys. Res.* **96**, 9399.

Dungey, J.W. 1953, *Phil. Mag.* **44**, 725.

Dungey, J.W. 1958a, in *Cosmic Electrodynamics* (Cambridge: Cambridge University Press), pp. 98–102.

Dungey, J.W. 1958b, in *Electromagnetic Phenomena in Cosmical Physics*, ed. B. Lehnert (Cambridge: Cambridge University Press), p. 135.

Dunn, R.B. and Zirker, J.B. 1973, *Solar Phys.* **33**, 281.

Feldman, U. 1992, *Astrophys. J.* **385**, 758.

Feldman, U., Laming, J.M., Mandelbaum, P., Goldstein, W.H. and Osterheld, A. 1992, *Astrophys. J.* **398**, 692.

Finn, J.M. and Kaw, P.K. 1977, *Phys. Fluids* **22**, 2140.

Foukal, P.V. 1990, *Solar Astrophysics* (New York: John Wiley).
Frazier, E.N. and Stenflo, J.O. 1972, *Solar Phys.* **27**, 330.
Furth, H.P., Killeen, J. and Rosenbluth, M.N. 1963, *Phys. Fluids* **6**, 459.
Gabriel, A.H. 1976, *Phil. Trans. Royal Soc. London A* **281**, 339.
Giovanelli, R.G. 1947, *Mon. Not. Royal Astron. Soc.* **107**, 338.
Glencross, W.M. 1975, *Astrophys. J. Letters* **199**, L53.
Glencross, W.M. 1980, *Astron. Astrophys.* **83**, 65.
Gold, T. 1964, in *The Physics of Solar Flares* (AAS–NASA Symposium, NASA SP-50), ed. W.N. Hess (Maryland: Greenbelt), p. 389.
Golub, L., Herant, M., Kalata, K., Lovar, I., Nystrom, G., Pardo, F., Spillar, E. and Wilczynski, J. 1990, *Nature* **344**, 842.
Grad, H. 1967, *Phys. Fluids* **10**, 137.
Grad, H. 1984, in *Proc. Workshop on Mathematical Aspects of Fluid and Plasma Dynamics* (Trieste: International Center for Astrophysics).
Hahm, T.S. and Kulsrud, R.M. 1985, *Phys. Fluids* **28**, 2412.
Hale, G.E. 1908a, *Pub. Astron. Soc. Pacific* **20**, 220.
Hale, G.E. 1908b, *Pub. Astron. Soc. Pacific* **20**, 287.
Hale, G.E. 1908c, *Astrophys. J.* **28**, 100.
Hale, G.E. 1908d, *Astrophys. J.* **28**, 315.
Hale, G.E. 1913, *Astrophys. J.* **38**, 27.
Hale, G.E. and Nicholson, S.B. 1938, *Magnetic Observations of Sunspots*, 1917–1924 (Part I., Pub. Carnegie Inst. No. 498) (New York: Carnegie Institute).
Hassam, A.B. 1991, *Physics Paper No. 91–282*, (University of Maryland: Laboratory for Plasma Research).
Hones, E.W. 1984, *Magnetic Reconnection in Space and Laboratory Plasmas* (Geophysical Monograph No. 30) (Washington DC: American Geophysical Union).
Howard, R. 1959, *Astrophys. J.* **130**, 193.
Howard, R. and Stenflo, J.O. 1972, *Solar Phys.* **22**, 402.
Hundhausen, A.J. 1990, private communication.
Illing, R.M.E. and Hundhausen, A.J. 1986, *J. Geophys. Res.* **91**, 10951.
Jensen, T.H. 1989, *Astrophys. J.* **343**, 507.
Jokipii, J.R. and Parker, E.N. 1968, *Phys. Rev. Letters* **21**, 44.
Kerst, D.W. 1962, *J. Nucl. Energy* **CA4**, 253.
Kiepenheuer, K.O. 1953, in *The Sun*, ed. G.P. Kuiper (Chicago, University of Chicago Press), pp. 322–465.
Kopp, R.A. and Kuperus, M. 1968, *Solar Phys.* **4**, 212.
Kulsrud, R.M. and Hahm, T.S. 1982, Physica Scripta T2/2, International Conference of Plasma Physics, Göteborg, Sweden.
Laming, J.M. and Feldman, U. 1992, *Astrophys. J.* **386**, 384.
Leighton, R.B. 1959, *Astrophys. J.* **130**, 366.
Lichtenberg, A.J. 1984, *Nucl. Fusion* **24**, 1277.
Livingston, W. and Harvey, J. 1969, *Solar Phys.* **10**, 294.
Livingston, W. and Harvey, J. 1971, in *Solar Magnetic Fields* (IAU Symposium No. 43) (Dordrecht: Reidel), p. 51.
Low, B.C. 1987, *Astrophys J.* **323**, 358.
Low, B.C. 1989, *Astrophys J.* **340**, 558.
Low, B.C. 1991, *Astrophys J.* **381**, 295.
Low, B.C. and Wolfson, R. 1987, *Astrophys J.* **323**, 574.
Lundquist, S. 1950, *Arkiv Fysik* **2**, 35.
Machado, M.E., Moore, R.L., Hernandez, A.M., Rovira, M.G., Hagyard, M.J. and Smith, J.B. 1988, *Astrophys J.* **326**, 425.

Martin, S.F. 1984, in *Small-Scale Dynamical Processes in Quiet Stellar Atmospheres*, ed. S.L. Keil (Sacramento Peak, Sunspot, New Mexico: National Solar Observatory), p. 30.
Martin, S.F. 1988, *Solar Phys.* **117**, 243.
Martin, S.F. 1990, in *Solar Photosphere: Structure, Convection and Magnetic Fields*, IAU Symposium No. 138, ed. J.O. Stenflo (Dordrecht: Reidel), p. 129.
Matthaeus, W.H. and Lamkin, S.L. 1985, *Phys. Fluids* **28**, 303.
Matthaeus, W.H. and Lamkin, S.L. 1986, *Phys Fluids* **29**, 2513.
Mikic, Z., Schnack, D.D. and Van Hoven, G. 1989, *Astrophys. J.* **338**, 1148.
Moffatt, H.K. 1985, *J. Fluid Mech.* **159**, 359.
Moffatt, H.K. 1986, *J. Fluid Mech.* **166**, 359.
Moffatt, H.K. 1990, *Phil. Trans. Royal Soc. London* **333**, 321.
Montgomery, D. 1982, *Phys. Scripta* **T2**, (1), 83.
Moser, J. 1962, *Nachr. Akad. Wiss. Göttingen II, Math–Physik* Kℓ. 1.
Parker, E.N. 1957a, *Phys. Rev.* **107**, 830.
Parker, E.N. 1957b, *J. Geophys. Res.* **62**, 509.
Parker, E.N. 1963a, *Astrophys. J. Supplement* **8**, 177.
Parker, E.N. 1963b, *Interplanetary Dynamical Processes* (New York: John Wiley), pp. 47–48.
Parker, E.N. 1965, *Astrophys. J.* **142**, 584.
Parker, E.N. 1968, in *Stars and Stellar Systems* (Vol. VII: Nebulae and Interstellar Matter), ed. B.M. Middlehurst and L.H. Aller (Chicago: University of Chicago Press), pp. 707–754.
Parker, E.N. 1969, *Space Sci. Rev.* **9**, 651.
Parker, E.N. 1972, *Astrophys. J.* **174**, 499.
Parker, E.N. 1973, *Astrophys. J.* **180**, 247.
Parker, E.N. 1975, *Astrophys. J.* **201**, 494.
Parker, E.N. 1979, *Cosmical Magnetic Fields*, (Oxford: Clarendon Press).
Parker, E.N. 1981, *Astrophys. J.* **244**, 631, 644.
Parker, E.N. 1982, *Geophys. Astrophys. Fluid Dyn.* **22**, 195.
Parker, E.N. 1983a, *Astrophys. J.* **262**, 642.
Parker, E.N. 1983b, *Geophys. Astrophys. Fluid Dyn.* **23**, 85.
Parker, E.N. 1986a, *Geophys. Astrophys. Fluid Dyn.* **34**, 243.
Parker, E.N. 1986b, *Geophys. Astrophys. Fluid Dyn.* **35** ,277.
Parker, E.N. 1986c, in *Cool Stars, Stellar Systems, and the Sun*, ed. M. Zeilik and D.M. Gibson (New York: Springer-Verlag).
Parker, E.N. 1987, *Solar Phys.* **111**, 297.
Parker, E.N. 1988, *Astrophys.* **330**, 474.
Parker, E.N. 1989a, *Geophys. Astrophys. Fluid Dyn.* **45**, 159.
Parker, E.N. 1989b, *Geophys. Astrophys. Fluid Dyn.* **45**, 169.
Parker, E.N. 1989c, *Geophys. Astrophys. Fluid Dyn.* **46**, 105.
Parker, E.N. 1990a, in *Galactic and Intergalactic Magnetic Fields*, ed. R. Beck and P.P. Kronberg (Dordrecht: Reidel), pp. 169–175.
Parker, E.N. 1990b, *Geophys. Astrophys. Fluid Dyn.* **53**, 43.
Parker, E.N. 1990c, *Geophys. Astrophys. Fluid Dyn.* **52**, 183.
Parker, E.N. 1990d, *Geophys. Astrophys. Fluid Dyn.* **55**, 161.
Parker, E.N. 1991a, *Astrophys. J.* **372**, 719.
Parker, E.N. 1991b, *Phys. Fluids B* **3**, 2652.
Parker, E.N. 1992, *Astrophys. J.* **401**, 137.
Petschek, H.E. 1964, in *The Physics of Solar Flares* (AAS–NASA Symposium, NASA SP-50), ed. W.N. Hess (Greenbelt, Maryland), p. 425.
Petschek, H.E. and Thorne, R.M. 1967, *Astrophys. J.* **147**, 1157.
Porter, J.G. and Dere, K.P. 1991, *Astrophys. J.* **370**, 775.

Porter, J.G. and Moore, R.L. 1988, in *Proc. Ninth Sacramento Peak Summer Symposium*, ed. R.C. Altrock (Sacramento Peak, Sunspot, New Mexico: National Solar Observatory), p. 125.

Porter, J.G., Moore, R.L., Reichmann, E.J., Engvold, O. and Harvey, K.L. 1987, *Astrophys. J.* **323**, 380.

Priest, E.R. 1981, ed. *Solar Flare Magnetohydrodynamics* (New York: Gordon and Breach).

Priest, E.R. 1982, *Solar Magnetohydrodynamics* (Dordrecht: Reidel).

Priest, E.R. and Forbes, T.G. 1986, *J. Geophys. Res.* **91**, 5579.

Proudman, J. 1916, *Proc. Royal Soc. London A* **92**, 408.

Rosner, R. and Knobloch, E. 1982, *Astrophys. J.* **262**, 349.

Sakai, J. 1983a, *Solar Phys.* **84**, 109.

Sakai, J. 1983b, *J. Plasma Phys.* **30**, 109.

Simon, G.W. and Noyes, R.W. 1971, in *Solar Magnetic Fields* (IAU Symp. No. 43), ed. R. Howard (Dordrecht, Reidel), p. 663.

Sonnerup, B.U.O. 1970, *J. Plasma Phys.* **4**, 161.

Sonnerup, B.U.O. 1971, *J. Geophys. Res.* **76**, 8211.

Spicer, D.S. 1976, *Formal Report 8036* (Washington, DC: Naval Research Laboratory).

Spicer, D.S. 1977, *Solar Phys.* **53**, 305.

Spicer, D.S. 1982, *Space Sci. Rev.* **31**, 351.

Stenflo, J.O. 1973, *Solar Phys.* **32**, 41.

Sturrock, P.A., Dixon, W.W., Klimchuk, J.A. and Antiochos, S.K. 1990, *Astrophys. J. Letters* **356**, L31.

Svestka, Z. 1976, *Solar Flares* (Dordrecht: Reidel).

Sweet, P.A. 1958, in *Electromagnetic Phenomena in Cosmical Physics*, ed. B. Lehnert (Cambridge: Cambridge University Press), p. 123.

Syrovatskii, S.I. 1971, *Soviet Phys. JETP* **33**, 933.

Syrovatskii, S.I. 1978, *Solar Phys.* **58**, 89.

Syrovatskii, S.I. 1981, *Ann. Rev. Astron. Astrophys.* **19**, 163.

Tajima, T. and Sakai, J. 1986, *IEEE Trans. Plasma Phys.* **PS–14**, 929.

Taylor, G.I. 1917, *Proc. Royal Soc. London A* **93**, 99.

Title, A.M., Tarbell, T.D., Topka, K.P., Ferguson, S.H., Shine, R.A. and the SOUP Team, 1989, *Astrophys. J.* **336**, 475.

Tsinganos, K.C. 1981, *Astrophys. J.* **245**, 764.

Tsinganos, K.C. 1982a, *Astrophys. J.* **252**, 775.

Tsinganos, K.C. 1982b,*Astrophys J.* **259**, 820.

Tsinganos, K.C. 1982c, *Astrophys. J.* **259**, 832.

Tsinganos, K.C., Distler, J. and Rosner, R. 1984, *Astrophys. J.* **278**, 409.

Vainshtein, S.I. and Parker, E.N. 1986, *Astrophys. J.* **304**, 821.

Van Ballegooijen, A.A. 1985, *Astrrophys. J.* **298**, 421.

Van Ballegooijen, A.A. 1986, *Astrophys. J.* **311**, 1001.

Van Hoven, G. 1976, *Solar Phys.* **49**, 95.

Van Hoven, G. 1979, *Astrophys. J.* **232**, 572.

Van Hoven, G. 1981, in *Solar Flare Magnetohydrodynamics*, ed. E.R. Priest (New York: Gordon and Breach), p. 217.

Vasyliunas, V.M. 1975, *Rev. Geophys. Space Phys.* **13**, 303.

Walker Jr, A.B.C., Barbee Jr, T.W., Hoover, R.B. and Lindblom, J.F. 1988, *Science* **241**, 1781.

Webb, D.F. and Hundhausen, A.J. 1987, *Solar Phys.* **108**, 383.

Weiss, N.O. 1983, in *Stellar and Planetary Magnetism* (New York: Gordon and Breach), p. 115.

Withbroe, G.L. 1988, *Astrophys. J.* **325**, 442.

Withbroe, G.L. and Noyes, R.W. 1977, *Ann. Rev. Astron. Astrophys.* **15**, 363.

Yu, G. 1973, *Astrophys. J.* **181**, 1003.
Zweibel, E.G. and Li, H.S. 1987, *Astrophys. J.* **312**, 423.

Chapter 2

Alfven, H. and Carlquist, P. 1967, *Solar Phys.* **1**, 220.
Berman, R.H., Tetreault, D.J. and Dupree, T.H. 1983, *Phys. Fluids* **26**, 2437.
Berman, R.H., Tetreault, D.J. and Dupree, T.H. 1985, *Phys. Fluids* **28**, 155.
Bernstein, I., Greene, J. and Kruskal, M. 1957, *Phys. Rev.* **108**, 546.
Biermann, L. 1950, *Z. Naturforsch* **52**, 65.
Bittencourt, J.A. 1986, *Fundamentals of Plasma Physics* (Oxford: Pergamon Press), pp. 290–300.
Block, L.P. 1972, *Cosmic Electrodyn.* **3**, 349.
Block, L.P. 1978, *Astrophys. Space Sci.* **55**, 59.
Browne, P.F. 1968, *Astrophys. J.* **2**, 217.
Browne, P.F. 1985, *Astron. Astrophys.* **144**, 298.
Browne, P.F. 1988, *Astron. Astrophys.* **193**, 334.
Brueckner, K.A. and Watson, K.M. 1956, *Phys. Rev.* **102**, 19.
Buneman, O. 1958, *Phys. Rev. Letters* **1**, 8.
Buneman, O. 1959, *Phys. Rev.* **115**, 503.
Cattani, D. 1967, *Nuovo Cim. B* **152B**, 574.
Cattani, D. and Sacchi, C. 1966, *Nuovo Cim.* **46**, 258.
Chan, C., Cho, M.H., Herschkowitz, N. and Intrator, T. 1986, *Phys. Rev. Letters* **57**, 3050.
Chapman, S. 1954, *Astrophys. J.* **120**, 151.
Chapman, S. and Cowling, T.G. 1958, *The Mathematical Theory of Nonuniform Gases* (Cambridge: Cambridge University Press).
Chew, G.F., Goldberger, M.L. and Low, F.E. 1956, *Proc. Royal Soc. London* **236**, 112.
Coppi, B. and Friedland, A.B. 1971, *Astrophys. J.* **169**, 379.
Coroniti, F.V. and Eviatar, A. 1977, *Astrophys. J. Supplement* **33**, 189.
Cowling, T.G. 1953, in *The Sun*, ed. G.P. Kuiper (Chicago University of Chicago Press), pp. 532-591.
Cowling, T.G. 1957, *Magnetohydrodynamics* (New York: John Wiley).
Diamond, D., Hazeltine, R.D., An, Z.G., Carreras, B.A. and Hicks, H.R. 1984, *Phys. Fluids* **27**, 1449.
Dungey, J.W. 1953, *Phil. Mag.* **44**, 725.
Dungey, J.W. 1958, *Cosmic Electrodynamics* (Cambridge: Cambridge University Press), pp. 98–102.
Dupree, T.H. 1972, *Phys. Fluids* **15**, 334.
Dupree, T.H. 1982, *Phys. Fluids* **25**, 277.
Dupree, T.H. 1983, *Phys. Fluids* **26**, 2460.
Eddington, A.S. 1929, *Mon. Not. Royal Astron. Soc.* **90**, 54.
Eddington, A.S. 1959, *The Internal Constitution of Stars* (New York: Dover Publications), pp. 282-288.
Elsasser, W.M. 1954, *Phys. Rev.* **95**, 1.
Ferraro, V.C.A. and Plumpton, C. 1966, *An Introduction to Magneto-Fluid Mechanics* (Oxford: Clarendon Press).
Foukal, P. and Hinata, S. 1991, *Solar Phys.* **132**, 307.
Friedman, M. and Hamburger, S.M. 1969, *Solar Phys.* **8**, 104.
Gekelman, W., Stenzel, R.L. and Wild, N. 1982, *J. Geophys. Res.* **87**, 101.
Giovanelli, R.G. 1947, *Mon. Not. Royal Astron. Soc.* **107**, 338.
Goedbloed, J.P. 1973, *Phys. Fluids* **16**, 1927.
Haerendel, G. 1977, *J. Atmos. Terr. Phys.* **40**, 343.

Haerendel, G. 1990, in *Physics of Magnetic Flux Ropes* (Geophysical Monograph 58), ed. C.T. Russell, E.R. Priest and L.C. Lee(Washington DC: American Geophysical Union), pp. 539-553.
Hamburger, S.M. and Friedman, M. 1968, *Phys. Rev. Letters* **21**, 674.
Hershkowitz, N. 1985, *Space Sci. Rev.* **41**, 351.
Heyvaerts, J. 1981, in *Solar Flare Magnetohydrodynamics*, ed. E.R. Priest, (New York: Gordon and Breach), pp. 496–506.
Hollenstein, Ch., Guyot, M. and Weibel, E.S. 1980, *Phys. Rev. Leters* **45**, 2100.
Huba, J.D., Gladd, N.T. and Papadopoulus, K. 1977, *Geophys. Res. Letters* **4**, 125.
Jeffrey, A. 1966, *Magnetohydrodynamics* (London: Oliver and Boyd).
Kadomtsev, B.B. 1965, *Plasma Turbulence* (New York: Academic Press).
Kalinin, Y.G., Kingsep, A.S., Lin, D.N., Ryutov, V.D. and Skoryupin, V.A. 1970, *Soviet Phys. JETP* **31**, 38.
Liewer, P.C. and Krall, N.A. 1973, *Phys. Fluids* **16**, 1953.
Lundquist, S. 1952, *Arkiv. Fysik* **5**, 297.
Mestel, L. and Moss, D.L. 1977, *Mon. Not. Royal Astron. Soc.* **178**, 27.
Mestel, L. and Roxburgh, I.W. 1961, *Astrophys. J.* **136**, 615.
Moffatt, H.K. 1978, *Magnetic Field Generation in Electrically Conducting Fluids* (Cambridge: Cambridge University Press).
Moss, D.L. 1977a, *Mon. Not. Royal Astron. Soc.* **178**, 51.
Moss, D.L. 1977b, *Mon. Not. Royal Astron. Soc.* **178**, 61.
Parker, E.N. 1957, *Phys. Rev.* **107**, 924.
Parker, E.N. 1979, *Cosmical Magnetic Fields* (Oxford: Clarendon Press).
Perkins, F.W. and Sun, Y.C. 1981, *Phys. Rev. Letters* **46**, 115.
Priest, E.R. 1982, *Solar Magnetohydrodynamics* (Dordrecht: Reidel).
Quon, B.H. and Wong, A.Y. 1976, *Phys. Rev. Letters* **37**, 1393.
Roberts, P.H. 1967, *An Introduction to Magnetohydrodynamics* (New York: Elsevier).
Roxburgh, I.W. 1966, *Mon. Not. Royal Astron. Soc.* **132**, 201.
Sagdeev, R.Z. 1967, *Proc. Symp. Appl. Math.* **18**, 281.
Schamel, H. 1983, *Z. Naturforsch* **39a**, 1170.
Schindler, K., Hesse, M., and Birn, J. 1991, *Astrophys. J.* **380**, 293.
Schlüter, A. 1950, *Z. Naturforsch* **5a**, 72.
Schlüter, A. 1952, *Ann. Physik.* **10**, 422.
Schlüter, A. 1958, in *Electromagnetic Phenomena in Cosmical Physics*, ed. B. Lehnert (Cambridge: Cambridge University Press), p. 71.
Scudder, J.D. 1992a, *Astrophys. J.* **398**, 299.
Scudder, J.D. 1992b, *Astrophys. J.* **398**, 319.
Smith, R.A. and Goertz, C.K. 1978, *J. Geophys. Res.* **83**, 2617.
Sommerfeld, A. 1964, *Thermodynamics and Statistical Mechanics* (New York: Academic Press), pp. 328–332.
Spicer, D.S. 1982, *Space Sci. Rev.* **31**, 351.
Spitzer, L. 1952, *Astrophys. J.* **116**, 299.
Spitzer, L. 1956, *Physics of Fully Ionized Gases* (New York: John Wiley), pp. 17, 62, 72, 81-87.
Stenzel, R.L., Gekelman, W. and Wild, N. 1982, *J. Geophys. Res. Letters* **9**, 680.
Stenzel, R.L., Gekelman, W. and Wild, N. 1983, *J. Geophys. Res.* **88**, 4793.
Stern, D. 1981, *J. Geophys. Res.* **86**, 5839.
Stringer, T.E. 1964, *J. Nucl. Energy C* **6**, 267.
Sweet, P.A. 1950, *Mon. Not. Royal Astron. Soc.* **110**, 548.
Tanaka, M. and Sato, T. 1981, *J. Geophys. Res.* **86**, 5541.
Tanenbaum, B.S. 1967, *Plasma Physics* (New York: McGraw–Hill), pp. 210–221.
Temerin, W., Cerny, K., Lotko, W. and Mozer, F.S. 1982, *Phys. Rev. Letters* **48**, 1175.
Tetreault, D.J. 1983, *Phys. Fluids* **26**, 3247.

Tetreault, D.J. 1988, *Geophys. Res. Letters* **15**, 164.
Tetreault, D.J. 1989, *Phys. Fluids* **B**, 1511.
Tetreault, D.J. 1990, *Phys. Fluids* **B**, 253.
Tetreault, D.J. 1991, *J. Geophys. Res.* **96**, 3549.
Thorne, S.K. 1967, *Astrophys. J.* **148**, 51.
Tidman, D.A. and Krall, N.A. 1971, *Shock Waves in Collisionless Plasmas* (New York: John Wiley).
Von Zeipel, H. 1924, *Mon. Not. Royal Astron. Soc.* **84**, 665.
Watson, K.M. 1956, *Phys. Rev.* **102**, 12.
Westcott, E.M., Stenback-Nielsen, H.C., Hallinan, T.J. and Davis, T.N. 1976, *J. Geophys. Res.* **81**, 4495.
Williams, A.C. 1986 *IEEE Trans. Plasma Sci.* **14**, 800.
Williams, A.C. Weisskopf, M.C., Elsner, R.F., Darbro, W. and Sutherland, P.G. 1986, *Astrophys. J.* **305**, 759.

Chapter 3

Abbott, M.B. 1966, *An Introduction to the Method of Characteristics* (New York: Elsevier).
Arnold, V. 1965, *C.R. Acad. Sci. Paris* **261**, 17.
Arnold, V. 1966, *Ann. Inst. Fourier Grenoble* **16**, 361.
Arnold, V.J. 1972, *Mathematical Methods of Classical Mechanics* (New York: Springer).
Arnold, V. 1974, *Proc. Summer School in Differential Equations* (Erevan: Armenian SSR Acad. Sci.).
Berger, M.A. 1986, *Geophys. Astrophys. Fluid Dyn.* **34**, 265.
Bogdan, T.J. and Low, B.C. 1986, *Astrophys. J.* **306**, 271.
Courant, R. and Hilbert, D. 1962, *Methods of Mathematical Physics* (Vol. II) (New York: John Wiley).
Dicke, R. H. 1970, *Astrophys. J.* **159**, 25.
Grad, H. 1967, *Phys. Fluids* **10**, 137.
Hu, Y.Q. and Low, B.C. 1982, *Solar Phys.* **81**, 107.
Jensen, T.H. 1989, *Astrophys. J.* **343**, 507.
Low, B.C. 1972, *Solar Phys.* **77**, 43.
Low, B.C. 1975a, *Astrophys. J.* **197**, 251.
Low, B.C. 1975b, *Astrophys. J.* **198**, 211.
Low, B.C. 1987, *Astrophys. J.* **323**, 358.
Low, B.C. 1989, *Astrophys. J.* **340**, 558.
Low, B.C. and Hu, Y.Q. 1983, *Solar Phys.* **84**, 83.
Low, B.C. and Hu, Y.Q. 1990, *Astrophys. J.* **352**, 343.
Low, B.C. and Wolfson, R. 1988, *Astrophys. J.* **324**, 574.
Moffatt, H.K. 1985, *J. Fluid Mech.* **159**, 359.
Moffatt, H.K. 1986, *J. Fluid Mech.* **166**, 359.
Nakagawa, Y. 1973, *Astron. Astrophys.* **27**, 95.
Nakagawa, Y., Billings, D.E. and McNamara. D. 1971, *Solar Phys.* **19**, 72.
Nakagawa, Y. and Raadu, M.A. 1972, *Solar Phys.* **25**, 127.
Parker E.N. 1972, *Astrophys. J.* **174**, 499.
Parker E.N. 1979, *Cosmical Magnetic Fields* (Oxford: Clarendon Press), p. 361.
Parker E.N. 1986, *Geophys. Astrophys. Fluid Dyn.* **34**, 243.
Parker E.N. 1989, *Geophys. Astrophys. Fluid Dyn.* **45**, 159.
Priest, E.R. and Raadu, M.A. 1975, *Solar Phys.* **43**, 177.
Raadu, M.A. and Nakagawa, Y. 1971, *Solar Phys.* **20**, 64.
Sweet, P.A. 1969, *Ann. Rev. Astron. Astrophys.* **7**, 149.

Syrovatskii, S.I. 1969, in *Solar Flares and Space Research*, ed. C. deJager and Z. Svestka (Amsterdam: North Holland Pub. Co.).
Taylor, J.B. 1974, *Phys. Fluids* **33**, 1139.
Taylor, J.B. 1986, *Rev. Mod. Phys.* **58**, 471.
Tsinganos, K.C. 1982, *Astrophys. J.* **259**, 832.

Chapter 4

Arnold, V.J. 1965, *C.R. Acad. Sci. Paris* **261**, 17.
Arnold, V.J. 1966, *Ann. Inst. Fourier Grenoble* **16**, 361.
Arnold, V.J. 1978, *Mathematical Methods of Classical Mechanics* (New York: Springer).
Arnold V.I. and Avez, A. 1968, *Ergodic Problems of Classical Mechanics* (New York: Benjamin).
Boozer, A.H. 1983, *Phys. Fluids* **26**, 1288.
Cary, J.R. and Littlejohn, R.G. 1983, *Annals Phys.* (NY), **151**, 1.
Chandrasekhar, S. 1956, *Proc. Nat. Acad. Sci.* **42**, 1.
Chandrasekhar, S. 1958, *Proc. Nat. Acad. Sci.* **44**, 843.
Chandrasekhar, S. and Kendall, P.C. 1957, *Astrophys. J.* **126**, 457.
Chandrasekhar, S. and Woltjer, S. 1958, *Proc. Nat. Acad. Sci.* **44**, 285.
Edenstrasser, J.W. 1980, *J. Plasma Phys.* **24**, 515.
Filonenko, N.N., Sagdeev, R.Z. and Zaslavsky, G.M. 1967, *Nucl. Fusion* **7**, 253.
Gjellestad, G. 1954, *Astrophys. J.* **119**, 14.
Grad, H. 1967, *Phys. Fluids* **10**, 137.
Hamada, S. 1972, *Nucl. Fusion* **12**, 523.
Henon, M. and Heiles, S. 1964, *Astron. J.* **69**, 73.
Ince, E.L. 1926, *Ordinary Differential Equations* (New York: Dover Publications), pp. 381–384.
Kerst, D.W. 1962, *Plasma Phys.* (*J. Nuclear Energy, Part C*), **4**, 253.
Kruskal, M.D. and Kulsrud, R.M. 1958, *Phys. Fluids* **1**, 265.
Lüst, R. and Schlüter, A. 1954, *Z. Astrophys.* **34**, 263, 365.
Morozov, A.I. and Solovev, L.S. 1966, *Rev. Plasma Phys.* **2**, 1.
Moser, J. 1966, *SIAM Rev.* **8**, 145.
Moser, J. 1973, *Stable and Random Motion in Dynamical Systems* (Princeton: Princeton University Press).
Parker, E.N. 1957, *J. Geophys. Res.* **62**, 509.
Parker, E.N. 1963, *Astrophys. J. Supplement* **8**, 177.
Parker, E.N. 1979, *Cosmical Magnetic Fields* (Oxford: Clarendon Press).
Petschek, H.E. 1964, *The Physics of Solar Flares* (AAS–NASA Symposium, NASA SP-50), ed. W.N. Hess (Maryland: Greenbelt), p. 425.
Petschek, H.E. and Thorne, R.M. 1967, *Astrophys. J.* **147**, 1157.
Prendergast, K.H. 1956, *Astrophys. J.* **123**, 498.
Prendergast, K.H. 1957, *Astrophys. J.* **128**, 361.
Priest, E.R. 1982, *Solar Magnetohydrodynamics* (Dordrecht: Reidel), pp. 345–357.
Priest, E.R. and Forbes, T.G. 1986, *J. Geophys. Res.* **91**, 5579.
Rosenbluth, M.N., Sagdeev, R.Z., Taylor, J.B. and Zaslavsky, G.M. 1966, *Nucl. Fusion* **6**, 297.
Rosner, R. and Knobloch, E. 1982, *Astrophys. J.* **262**, 349.
Sonnerup, B.U.O. 1970, *J. Plasma Phys.* **4**, 161.
Sonnerup, B.U.O. 1971, *J. Geophys. Res.* **76**, 8211.
Sonnerup, B.U.O. and Priest, E.R. 1975, *J. Plasma Phys.* **14**, 283.
Sweet, P.A. 1958a, *Proc. IAU Symposium* No. 6, ed. B. Lehnert (Cambridge: Cambridge University Press), p. 123.

Sweet, P.A. 1958b, *Nuovo Cim. Suppl.* **8**, 10.
Sweet, P.A. 1969, *Ann. Rev. Astron. Astrophys.* **7**, 149.
Tsinganos, K.C. 1981a, *Astrophys. J.* **245**, 764.
Tsinganos, K.C. 1981b, *Astrophys. J.* **252**, 775.
Tsinganos, K.C. 1982a, *Astrophys. J.* **252**, 790.
Tsinganos, K.C. 1982b, *Astrophys. J.* **259**, 820.
Tsinganos, K.C. 1982c, *Astrophys. J.* **259**, 832.
Tsinganos, K.C., Distler, J. and Rosner, R. 1984, *Astrophys. J.* **278**, 409.
Vainshtein, S.I. and Parker, E.N. 1986, *Astrophys. J.* **304**, 821.
Van Ballegooijen, A.A. 1985, *Astrophys. J.* **298**, 421.
Vasyliunas, V.M. 1975, *Rev. Geophys. Space Phys.* **13**, 303.

Chapter 5

Lundquist, S. 1952, *Arkiv. Fysik.* **5**, 297.
Park, W., Monticello, D.A. and White, R.B. 1983, *Phys. Fluids* **27**, 137.
Parker, E.N. 1979, *Cosmical Magnetic Fields* (Oxford: Clarendon Press), pp. 38–41.
Parker, E.N. 1986a, *Geophys. Astrophys. Fluid Dyn.* **34**, 243.
Parker, E.N. 1986b, *Geophys. Astrophys. Fluid Dyn.* **35**, 277.
Parker, E.N. 1990, *Geophys. Astrophys. Fluid Dyn.* **52**, 183.
Roberts, P.H. 1967, *An Introduction to Magnetohydrodynamics* (New York: Elsevier), p. 46.
Rosenbluth, M.N., Dagazian, R.Y. and Rutherford, P.H. 1973, *Phys. Fluids* **16**, 1894.
Strauss, H.R. and Otani, N.F. 1988, *Astrophys. J.* **326**, 418.

Chapter 6

Aly, J.J. and Amari, T. 1989, *Astron. Astrophys.* **221**, 287.
Amari, T. and Aly, J.J. 1990, *Astron. Astrophys.* **227**, 628.
Biskamp, D. and Welter, H. 1989, *Phys. Fluids B* **1**, 1964.
Bobrova, N.A. and Syrovatskii, S.I. 1979, *Solar Phys.* **61**, 379.
Cheng, A.F. 1980, *Astrophys. J.* **242**, 326.
Hahm, T.S. and Kulsrud, R.M. 1985, *Phys. Fluids* **28**, 2412.
Hayashi, T. and Sato, T. 1978, *J. Geophys. Res.* **83**, 217.
Hu, Y.Q. and Low, B.C. 1982, *Solar Phys.* **81**, 107.
Jensen, T.H. 1989, *Astrophys. J.* **343**, 507.
Kraichnan, R.H. and Montgomery, D. 1980, *Rep. Prog. Phys.* **43**, 547.
Kulsrud, R.M. and Hahm, T.S. 1982, Physica Scripta T2/2, International Conference of Plasma Physics, Göteborg, Sweden.
Linardatos, D. 1992, *J. Fluid Mech.* **246**, 569.
Low, B.C. 1986a, *Astrophys. J.* **310**, 953.
Low, B.C. 1986b, *Astrophys. J.* **307**, 205.
Low, B.C. 1987, *Astrophys. J.* **323**, 358.
Low, B.C. 1989, *Astrophys. J.* **340**, 558.
Low, B.C. and Hu, Y.Q. 1983, *Solar Phys.* **84**, 83.
Low, B.C. and Wolfson, R. 1988, *Astrophys. J.* **324**, 574.
Matthaeus, W.H. 1982, *Geophys. Res. Letters* **9**, 660.
Matthaeus, W.H. and Montgomery, D. 1980, *Annals NY Acad. Sci.* **367**, 203.
Matthaeus, W.H. and Montgomery, D. 1981, *J. Plasma Phys.* **25**, 11.
Mikic, Z., Schnack, D.D. and Van Hoven, G. 1989, *Astrophys. J.* **338**, 1148.
Moffatt, H.K. 1987, in *Advances in Turbulence*, ed. G. Comte-Bellot and J. Meathieu (New York: Springer), p. 228.

Park, W., Monticello, D.A. and White, R.B. 1984, *Phys. Fluids* **27**, 137.
Parker E.N. 1981, *Astrophys. J.* **244**, 631.
Parker E.N. 1982, *Geophys. Astrophys. Fluid Dyn.* **22**, 195.
Parker E.N. 1983, *Geophys. Astrophys. Fluid Dyn.* **23**, 85.
Parker E.N. 1987, *Astrophys. J.* **318**, 876.
Parker E.N. 1989a, *Geophys Astrophys. Fluid Dyn.* **45**, 169.
Parker E.N. 1989b, *Geophys Astrophys. Fluid Dyn.* **46**, 105.
Parker E.N. 1990, *Geophys. Astrophys. Fluid Dyn.* **53**, 43.
Parker E.N. 1991, *Phys. Fluids B* **3**, 2652.
Priest, E.R. and Raadu, M.A. 1975, *Solar Phys.* **43**, 177.
Riyopoulos, S., Bondeson, A. and Montgomery, D. 1982, *Phys. Fluids* **25**, 107.
Rosenbluth, M.N., Dagazian, R.Y. and Rutherford, P.H. 1973, *Phys. Fluids* **16**, 1894.
Sato, T. and Hayashi, T. 1979, *Phys. Fluids* **22**, 1189.
Sato, T., Hayashi, T., Tamao, T. and Hasegawa, A. 1978, *Phys. Rev. Leters* **41**, 1548.
Sheeley, N.R. 1991, private communication.
Strauss, H.R. 1976, *Phys. Fluids* **19**, 134.
Strauss, H.R. and Otani, N.F. 1988, *Astrophys. J.* **326**, 418.
Syrovatskii, S.I. 1971, *Soviet Phys. JETP* **33**, 933.
Syrovatskii, S.I. 1978, *Solar Phys.* **58**, 89.
Syrovatskii, S.I. 1981, *Ann. Rev. Astron. Astrophys.* **19**, 163.
Tajima, T., Brunel, F. and Sakai, J. 1982, *Astrophys. J. Letters* **258**, L45.
Titov, V.S. 1992, *Solar Phys.* **139**, 401.
Vainshtein,S.I. and Parker, E.N. 1986, *Astrophys. J.* **304**, 821.
Van Ballegooijen, A. 1986, *Astrophys. J.* **311**, 1001.
Van Ballegooijen, A. 1988, *Geophys. Astrophys. Fluid Dyn.* **41**, 181.

Chapter 7

Boozer, A.H. 1982, *Phys. Fluids* **26**, 1288.
Brand, L. 1947, *Vector and Tensor Analysis* (New York: John Wiley), pp. 225–226.
Byrd, F.B. and Friedman, M.D. 1954, *Handbook of Elliptic Integrals for Engineers and Physicists* (Berlin: Springer-Verlag), pp. 240.05, 341.53.
Cary, J.R. and Littlejohn, R.G. 1983, *Annals NY Acad. Sci.* **151**, 1.
Field, G.B. 1990, *Proc. IUTAM Symp. Topological Fluid Dynamics*, ed. H.K. Moffatt and A. Tsinober, (Cambridge: Cambridge University Press).
Kerst, D.W. 1962, *Plasma Phys.* **4**, 253.
Lüst, R. and Schlüter, A. 1954, *Z. Astrophys.* **34**, 263, 365.
Parker, E.N. 1968, in *Stars and Stellar Systems (Vol. VII. Nebulae and Interstellar Matter)* ed. B.M. Middlehurst and L.H. Aller (Chicago:University of Chicago Press), Chapter 14.
Parker, E.N. 1979, *Cosmical Magnetic Fields* (Oxford: Clarendon Press).,
Parker, E.N. 1981a, *Astrophys. J.* **244**, 631.
Parker, E.N. 1981b, *Astrophys. J.* **244**, 644.
Parker, E.N. 1983, *Astrophys J.* **264**, 635.
Parker, E.N. 1987, *Astrophys J.* **318**, 876.
Parker, E.N. 1989a, *Geophys. Astrophys. Fluid Dyn.* **45**, 169.
Parker, E.N. 1989b, *Geophys. Astrophys. Fluid Dyn.* **46**, 105.
Parker, E.N. 1989c, *Geophys. Astrophys. Fluid Dyn.* **50**, 229.
Parker, E.N. 1990, *Geophys. Astrophys. Fluid Dyn.* **53**, 43.
Parker, E.N. 1991, *Phys. Fluids B* **3**, 2652.
Syrovatskii, S.I. 1971, *Soviet Phys. JETP* **33**, 933.
Syrovatskii, S.I. 1978, *Solar Phys.* **58**, 89.

Syrovatskii, S.I. 1981, *Ann. Rev. Astron. Astrophys.* **19**, 163.
Tsinganos, K.C. 1982, *Astrophys. J.* **259**, 832.
Tsinganos, K.C., Distler, J. and Rosner, R. 1984, *Astrophys. J.* **278**, 409.
Van Ballegooijen, A.A. 1988, *Geophys. Astrophys. Fluid Dyn.* **41**, 181.

Chapter 8

Linardatos, D. 1993, *J. Fluid Mech.* **246**, 569.
Lüsl, R. and Schlüter, A. 1954, *Z. Astrophys.* **34**, 263, 365.
Parker E.N. 1979, *Cosmical Magnetic Fields* (Oxford, Clarendon Press).
Parker E.N. 1981a, *Astrophys. J.* **244**, 631.
Parker E.N. 1981b, *Astrophys. J.* **244**, 644.
Parker E.N. 1986a, *Geophys. Astrophys. Fluid Dyn.* **35**, 277.
Parker E.N. 1986b, *Geophys. Astrophys. Fluid Dyn.* **34**, 243.
Parker E.N. 1989, *Geophys. Astrophys. Fluid Dyn.* **45**, 169.
Parker E.N. 1990, *Geophys. Astrophys. Fluid Dyn.* **53**, 43.
Parker E.N. 1991, *Phys. Fluids B* **3**, 2652.
Strachan, N.R. and Priest, E.R. 1991, *Geophys. Astrophys. Fluid Dyn.* **61**, 199.
Tsinganos, K.C. 1981, *Astrophys. J.* **245**, 764.
Tsinganos, K.C. 1982a, *Astrophys. J.* **252**, 775.
Tsinganos, K.C. 1982b, *Astrophys. J.* **252**, 790.
Tsinganos, K.C. 1982c, *Astrophys. J.* **259**, 820.
Tsinganos, K.C. 1982d, *Astrophys. J.* **259**, 832.
Vainshtein, S.I. 1990, *Astron. Astrophys.* **230**, 238.
Vainshtein, S.I. and Parker, E.N. 1986, *Astrophys. J.* **304**, 821.
Vekstein, G.E. and Priest, E.R. 1992, *Astrophys. J.* **384,** 333.
Vekstein, G.E. and Priest, E.R. 1993, *Solar Phys.* **146**, 119.

Chapter 9

Chapman, S. 1954, *Astrophys. J.* **120**, 151.
Chapman, S. and Cowling, T.G. 1958, *The Mathematical Theory of Nonuniform Gas* (Cambridge: Cambridge University Press).
Cowling, T.G. 1953, in *The Sun*, ed. G.P. Kuiper (Chicago: University of Chicago Press), pp. 536–543, 554–556.
Cowling, T.G. 1957, *Magnetohydrodynamics* (New York: John Wiley).
Kahler, S. 1991, *Astrophys. J.* **378**, 398.
Parker, E.N. 1979, *Cosmical Magnetic Fields* (Oxford: Clarendon Press).
Parker, E.N. 1981a, *Astrophys. J.* **244**, 631.
Parker, E.N. 1981b, *Astrophys. J.* **244**, 644.
Parker, E.N. 1990, *Geophys Astrophys. Fluid Dyn.* **50**, 229.
Spitzer, L. 1956, *The Physics of Fully Ionized Gases* (New York: John Wiley).
Title, A.M., Tarbell, T.D. and Topka, K.P. 1987, *Astrophys. J.* **317**, 892.
Title, A.M., Tarbell, T.D. and Topka, K.P., Ferguson, S.H. and Shine, R.A. 1989, *Astrophys. J.* **336**, 475.
Title, A.M., Topka, K.P., Tarbell, T.D., Schmidt, W., Balke, C. and Scharmer, G. 1992, *Astrophys. J.* **393**, 782.

Chapter 10

Anderson, L.S. and Athay, R.G. 1989, *Astrophys. J.* **336**, 1089.
Athay, R.G. 1981, *Astrophys. J.* **249**, 340.

Bhattacharjee, A., Dewar, R.L. and Monticello, D.A. 1980, *Phys. Rev. Letters* **45**, 347; 45, 1217 (E).
Bhattacharjee, A. and Wang, X. 1991, *Astrophys. J.* **372**, 321.
Birn, J., Hesse, M., and Schindler, K. 1989, *J. Geophys. Res.* **94**, 241.
Biskamp, D. 1986, *Phys. Fluids* **29**, 1520.
Biskamp, D. and Welter, H. 1989, *Phys. Fluids B* **1**, 1964.
Bueno, J.T. 1991, in *Mechanisms of Chromospheric and Coronal Heating*, ed. P. Ulmschneider, E.R. Priest and R. Rosner (Berlin: Springer Verlag), p. 60.
Coppi, B. 1983, *Astrophys. J. Letters* **273**, L101.
Coppi, B. and Friedland, A.B. 1971, *Astrophys. J.* **169**, 379.
Craig, I.J.D. and McClymont, A.N. 1991, *Astrophys. J. Letters* **371**, L41.
Craig, I.J.D. and McClymont, A.N. 1993, *Astrophys. J.* **405**, 207.
Craig, I.J.D. and Watson, P.G. 1992, *Astrophys. J.* **393**, 385.
Dahlburg, R.B., Antiochos, S.K. and Zang, T.A. 1992, *Naval Research Laboratory Report* (NRL/MR/4440–92–6966).
Dahlburg, R.B., Montgomery, D., Doolen, G.D. and Matthaeus, W.H. 1986, *J. Plasma Phys.* **35**, 1.
DeLuca, E. and Craig, I.J.D. 1992, *Astrophys. J.* **390**, 679.
Dickman, D.O., Morse, R.L. and Nielson, C.W. 1969, *Phys. Fluids* **12**, 1708.
Drake, J.F. and Lee, Y.C. 1977, *Phys. Rev. Letters* **39**, A53.
Dungey, J.W. 1953, *Phil. Mag.*, 44, 725.
Dungey, J.W. 1958, *Cosmic Electrodynamics* (Cambridge: Cambridge University Press), pp. 98, 125.
Dunn, R.B. and Zirker, J.B. 1973, *Solar Phys.* **33**, 281.
Finn, J.M. and Kaw, P.K. 1977, *Phys. Fluids* **22**, 2140.
Forbes, T.G. 1991, *Geophys. Astrophys. Fluid Dyn.* **62**, 15.
Forbes, T.G. and Malherbe, J.M. 1991, *Solar Phys.* **135**, 361.
Forbes, T.G. and Priest, E.R. 1982, *Solar Phys.* **81**, 303.
Forbes, T.G. and Priest, E.R. 1987, *Rev. Geophys.* **25**, 1583.
Furth, H.P., Killeen, J. and Rosenbluth, M.N. 1963, *Phys. Fluids* **6**, 459.
Gabriel, A.H. 1976, *Phil. Trans. Royal Soc. London A* **281**, 339.
Giovanelli, R.G. 1947, *Mon. Not. Royal Astron. Soc.* **107**, 338.
Giovanelli, R.G. 1948, *Mon. Not. Royal Astron. Soc.* **108**, 163.
Grad, H. 1978, *MF-92, C00-3077-152* (New York: Courant Institute of Mathematical Sciences, New York University).
Grad, H., Hu, P.N. and Stevens, D.C. 1975, *Proc. Nat. Acad. Sci.* **72**, 3789.
Greene, J.M. 1988, *J. Geophys. Res.* **93**, 8583.
Guzdar, P.N., Finn, J.M., Whang, K.W. and Bondeson, A. 1985, *Phys. Fluids* **28**, 3154.
Hahm, T.S. and Kulsrud, R.M. 1985, *Phys. Fluids* **28**, 2412.
Hassam, A.B. 1990, *Bull. Amer. Astron. Soc.* **22**, 853.
Hassam, A.B. 1991, *Plasma Preprint* (UM LPR 91-046) (University of Maryland: Laboratory for Plasma Research).
Hassam, A.B. 1992, *Astrophys. J.* **399**, 159.
Hu, Y.Q. and Low, B.C. 1982, *Solar Phys.* **81**, 107.
Jin, S.P. and Ip., W.H. 1991, *Phys. Fluids.*
Jokipii, J.R. 1973, *Astrophys. J.* **183**, 1029.
Jokipii, J.R. 1975, *Astrophys. J.* **198**, 727.
Jokipii, J.R. and Parker, E.N. 1968a, *Phys. Rev. Letters* **21**, 44.
Jokipii, J.R. and Parker, E.N. 1968b, *J. Geophys. Res.* **73**, 6842.
Jokipii, J.R. and Parker, E.N. 1969a, *Astrophys. J.* **155**, 777.
Jokipii, J.R. and Parker, E.N. 1969b, *Astrophys. J.* **155**, 799.
Kadomtsev, B.B. 1975, *Soviet J. Plasma Phys.* **1**, 389.

Kadomtsev, B.B. 1984, *Plasma Phys. Contr. Fusion* **26**, 217.
Kopp, R.A. and Kuperus, M. 1968, *Solar Phys.* **4**, 212.
Lau, Y.T. and Finn, J.M. 1990, *Astrophys. J.* **350**, 672.
Lau, Y.T. and Finn, J.M. 1991, *Astrophys. J.* **366**, 577.
Lee, L.C. and Fu, Z.F. 1986a, *J. Geophys. Res.* **91**, 13, 373.
Lee, L.C. and Fu, Z.F. 1986b, *J. Geophys. Res.* **91**, 4551.
Lee, L.C. and Fu, Z.F. 1986c, *J. Geophys. Res.* **91**, 6807.
Lee, L.C., Fu, Z.F., and Akasofu, S.I. 1987, in *Magnetotail Physics*, ed. A.T.Y. Lui (Baltimore: Johns Hopkins Press), p. 415.
Lichtenberg, J.A. 1984, *Nucl. Fusion* **221**, 1277.
Longcope, D.W. and Sudan, R.N. 1992, *Astrophys. J.* **384**, 305.
Low, B.C. and Hu, Y.Q. 1983, *Solar Phys.* **84**, 83.
Matthaeus, W.H. and Lamkin, S.L. 1985, *Phys. Fluids* **28**, 303.
Matthaeus, W.H. and Montgomery, D. 1981, *J. Plasma Phys.* **25**, 11.
Montgomery, D. 1982, *Phys. Scripta* **12** (1), 83.
Otani, N.F. and Strauss, H.R. 1988, *Astrophys. J.* **325**, 468.
Otto, A. 1990, *Comput. Phys. Comm.* **59**, 185.
Otto, A. 1991, *Geophys. Astrophys. Fluid Dyn.* **62**, 69.
Park, W., Monticello, D.A. and White, R.B. 1983, *Phys. Fluids* **27**, 137.
Parker, E.N. 1957a, *J. Geophys. Res.* **62**, 509.
Parker, E.N. 1957b, *Phys. Rev.* **107**, 830.
Parker, E.N. 1963, *Astrophys. J. (Supplement)* **8**, 177.
Parker, E.N. 1972, *Astrophys. J.* **174**, 499.
Parker, E.N. 1977, *Astrophys. J.* **215**, 374.
Parker, E.N. 1979, *Cosmical Magnetic Fields* (Oxford: Clarendon Press).
Parker, E.N. 1981a, *Astrophys. J.* **244**, 631.
Parker, E.N. 1981b, *Astrophys. J.* **244**, 644.
Parker, E.N. 1983a, *Astrophys. J.* **264**, 635.
Parker, E.N. 1983b, *Astrophys. J.* **264**, 642.
Parker, E.N. 1983c, *Geophys. Astrophys. Fluid Dyn.* **24**, 245.
Parker, E.N. 1990, *Geophys. Astrophys. Fluid Dyn.* **55**, 161.
Parker, E.N. 1992, *J. Geophys. Res.* **97**, 4311.
Parker, E.N. 1993a, *Astrophys. J.* **407**, 342.
Parker, E.N. 1993b, *Astrophys. J.* **414**, 389.
Petschek, H.E. 1964, in *The Physics of Flares* (AAS–NASA Symposium, NASA SP-50) ed. W.N. Hess (Maryland: Greenbelt), p. 425.
Petschek, H.E. and Thorne, R.M. 1967, *Astrophys. J.* **147**, 1157.
Priest, E.R. 1981, Solar Flare Magnetohydrodynamics (New York, Gordon and Breach), Chapter 3.
Priest, E.R. 1985, *Rep. Prog. Phys.* **48**, 955.
Priest, E.R. and Cowley, S.W.H. 1975, *J. Plasma Phys.* **14**, 271.
Priest, E.R. and Forbes, T.G. 1986, *J. Geophys. Res.* **91**, 5579.
Priest, E.R. and Forbes, T.G. 1989, *Solar Phys.* **119**, 211.
Priest, E.R. and Raadu, M.A. 1975, *Solar Phys.* **43**, 177.
Roberts, B. and Priest, E.R. 1975, *J. Plasma Phbys.* **14**, 417.
Rosenbluth, M.N., Dagazian, R.Y., and Rutherford, P.H. 1973, *Phys. Fluids* **16**, 1894.
Rutherford, P.H. 1973, *Phys. Fluids* **16**, 1903.
Ryopoulos, S., Bondeson, A. and Montgomery, D. 1982, *Phys. Fluids* **25**, 107.
Sakai, J. 1983a, *Solar Phys.* **84**, 109.
Sakai, J. 1983b, *J. Plasma Phys.* **30**, 109.
Sakai, J. and de Jager, C. 1991, *Solar Phys.* **134**, 329.
Sakai, J. and Washimi, H. 1982, *Astrophys. J.* **258**, 823.

Sakai, J., Tajima, T. and Brunel, F. 1984a, *Solar Phys.* **91**, 103.
Sakai, J., Tajima, T. and Brunel, F. 1984b, *Solar Phys.* **95**, 141.
Scholer, M. 1989, *J. Geophys. Res.* **94**, 345.
Scholer, M. 1991, *Geophys. Astrophys. Fluid Dyn.* **62**, 51.
Similon, P.L. and Sudan, R.N. 1989, *Astrophys. J.* **336**, 442.
Simon, G.W., Title, A.M., Topka, K.P., Tarbell, T.D., Shine, R.A., Ferguson, S.A., Zirin, H. and the SOUP Team 1988, *Astrophys. J.* **327**, 964.
Sonnerup, B.U.O. 1970, *J. Plasma Phys.* **4**, 161.
Sonnerup, B.U.O. 1971, *J. Geophys. Res.* **76**, 8211.
Sonnerup, B.U.O. and Priest, E.R. 1975, *J. Plasma Phys.* **14**, 283.
Soward, A.M. and Priest, E.R. 1977, *Phil. Trans. Royal Soc. London A* **284**, 369.
Sparks, L. and Van Hoven, G. 1988, *Astrophys. J.* **333**, 953.
Spicer, D.S. 1977, *Solar Phys.* **53**, 305.
Spicer, D.S. 1982, *Space Sci. Rev.* **31**, 351.
Spicer, D.S. 1990, *Adv. Space Res.* **10** (9) 43.
Strauss, H.R. 1988, *Astrophys. J.* **326**, 412.
Strauss, H.R. 1991, *Astrophys. J.* **381**, 508.
Strauss, H.R. and Otani, N.F. 1988, *Astrophys. J.* **326**, 418.
Sturrock, P.A. and Uchida, Y. 1981, *Astrophys J.* **246**, 331.
Sweet, P.A. 1958a, *Electromagnetic Phenomena in Cosmical Physics*, ed. B. Lehnert, (Cambridge: Cambridge University Press), p. 123.
Sweet, P.A. 1958b, *Nuovo Cim. (Supplement)* **8** (10), 188.
Sweet, P.A. 1969, *Ann. Rev. Astron. Astrophys.* **7**, 149.
Syrovatsky, S.I. 1971, *Soviet Phys. JETP* **33**, 933.
Tajima, T. and Sakai, J. 1986, *IEEE Trans. Plasma Sci.* **PS–14**, 929.
Tajima, T. and Sakai, J. 1989a, *Soviet J. Plasma Phys.* **15**, 519.
Tajima, T. and Sakai, J. 1989b, *Soviet J. Plasma Phys.* **15**, 606.
Taylor, J.B. 1974, *Phys. Fluids* **33**, 1139.
Tetreault, D. 1983, *Phys. Fluids* **26**, 3247.
Tetreault, D. 1988a, *Geophys. Res. Letters* **15**, 164.
Tetreault, D. 1988b, *Phys. Fluids* **31**, 2122.
Tetreault, D. 1989, *Phys. Fluids B* **1**, 511.
Tetreault, D. 1990, *Phys. Fluids B* **2**, 53.
Tetreault, D. 1991, *J. Geophys. Res.* **96**, 3549.
Tetreault, D. 1992a, *J. Geophys. Res.* **97**, 8531.
Tetreault, D. 1992b, *J. Geophys. Res.* **97**, 8541.
Title, A.M., Tarbell, T.D., Topka, K.P., Ferguson, S.H., Shine, R.A. and the SOUP Team 1989, *Astrophys. J.* **336**, 475.
Ugai, M. 1985, *Plasma Phys. Contr. Fusion* **27**, 1183.
Ugai, M. and Tsuda, T. 1977, *J. Plasma Phys.* **17**, 337.
Ugai, M. and Tsuda, T. 1979a, *J. Plasma Phys.* **21**, 459.
Ugai, M. and Tsuda, T. 1979b, *J. Plasma Phys.* **22**, 1.
Van Hoven, G. 1976, *Solar Phys.* **49**, 95.
Van Hoven, G. 1979, *Astrophys. J.* **23**2, 572.
Van Hoven, G. 1981, in *Solar Flare Magnetohydrodynamics*, ed. E.R. Priest (New York: Gordon and Breach), p. 217.
Vasyliunas, V.M. 1975, *Rev. Geophys. Space Phys.* **13**, 303.
Waddell, B.V., Rosenbluth, M.N., Monticello, D.A. and White, R.B. 1976, *Nucl. Fusion* **16**, 528.
Waelbroeck, F.L. 1989, *Phys. Fluids B* **1**, 2372.
Wang, X. and B hattacharjee, A. 1992, *Phys. Fluids B* **4**, 1795.

White, R.B., Monticello, D.A., Rosenbluth, M.N. and Waddell, B.V. 1977, *Phys. Fluids* **20**, 800.
Yang, C.K. and Sonnerup, B.U.O. 1976, *Astrophys. J.* **206**, 570.
Yang, C.K. and Sonnerup, B.U.O. 1977,*J. Geophys. Res.* **82**, 699.
Yeh, T. and Axford, W.I. 1970, *J. Plasma Phys.* **4**, 207.
Yeh, T. and Dryer, M. 1973, *Astrophys. J.* **182**, 301.

Chapter 11

Alfven, H. 1947, *Mon. Not. Royal Astron. Soc.*, **107**, 211.
Amnuel, P.R., Guseinov, O.H. and Rakhamimov, Sh.Y. 1979, *Astrophys. J. (Supplement)* **41**, 327.
An, C.H., Musielak, Z.E., Moore, R.L. and Suess, S.T. 1989, *Astrophys. J.* **345**, 597.
Anderson, L.S. and Athay, R.G. 1989, *Astrophys. J.* **336**, 1089.
Athay, R.G. and White, O.R. 1978, *Astrophys. J.* **226**, 1135.
Beaufume, P., Coppi, B. and Golub, L. 1992, *Astrophys. J.* **393**, 396.
Beckers, J.M. 1976, *Astrophys. J.* **203**, 739.
Beckers, J.M. 1977, *Astrophys. J.* **213**, 900.
Beckers, J.M. and Schneeburger, T.J. 1977, *Astrophys. J.* **215**, 356.
Belian, R.D., Conner, J.P. and Evans, W.D. 1976, *Astrophys. J. Letters* **207**, L33.
Berger, M.A. 1994, *Astron. Astrophys.* (in press).
Bhattacharjee, A. and Wang, X. 1991, *Astrophys. J.* **372**, 321.
Biermann, L. 1946, *Naturwiss* **33**, 118.
Biermann, L. 1947, *Naturwiss* **34**, 87.
Biermann, L. 1948, *Z. Astrophys.* **25**, 161.
Billings, D.E. 1966, *A Guide to the Solar Corona* (New York: Academic Press).
Blake, R.L., Chubb, T.A., Friedman, H. and Unzicker, A.E. 1963, *Astrophys. J.* **137**, 3.
Blake, R.L., Chubb, T.A. Friedman, H. and Unzicker, A.E. 1965, *Astrophys. J.* **142**, 1.
Bradt, H.V.D. and McClintock, J.E. 1983, *Ann. Rev. Astron. Astrophys.* **21**, 13.
Brants, J.J. 1985a, *Solar Phys.* **95**, 15.
Brants, J.J. 1985b, *Solar Phys.* **98**, 197.
Brown, S.F. 1991, *Astron. Astrophys* **249**, 243.
Brown, S.F. and Durrant, C.J. 1991, in *Mechanisms of Chromospheric and Coronal Heating*, ed. P. Ulmschneider, E.R. Priest and R. Rosner (Berlin: Springer-Verlag), p. 132.
Browning, P.K. and Priest, E.R. 1984, *Astron. Astrophys.* **131**, 283.
Brueckner, G.E. and Bartoe, J.D.F. 1983, *Astrophys. J.* **272**, 329.
Bueno, J.T. 1991, in *Mechanisms of Chromospheric and Coronal Heating*, ed. P. Ulmschneider and R. Rosner (Berlin: Springer-Verlag), p. 60.
Burnight, T.R. 1949, *Phys. Rev.* **76**, 165.
Burnight, T.R. 1952, in *Physics and Medicine of the Upper Atmosphere*, ed. C.S. White and O.O. Benson (Albuquerque: Albuquerque University of New Mexico Press), Chapter 13.
Byram, E.T., Chubb, T.A., Friedman, H., Kupperian, J.E. and Kreplin, R.W. 1958, *Astrophys. J.* **128**, 738.
Chandra, S. and Prasad, L. 1991, *Solar Phys.* **134**, 99.
Chapman, S. 1954, *Astrophys. J.* **120**, 151.
Chapman, S. 1958, *Proc. Phys. Soc.* **72**, 353.
Chen, L. and Hasegawa, A. 1974, *J. Geophys. Res.* **79**, 1024.
Cheng, C.C. 1991, in *Mechanisms of Chromospheric and Coronal Heating*, ed. P. Ulmschneider and R. Rosner (Berlin: Springer-Verlag), p. 77.
Cheng, C.C., Doschek, G.A. and Feldman, U. 1976, *Astrophys. J.* **210**, 836.

Cheng, C.C., Doschek, G.A. and Feldman, U. 1979, *Astrophys. J.* **227**, 1037.

Chubb, T.A., Friedman, H. and Kreplin, R.W. 1960, *J. Geophys. Res.* **65**, 1831.

Ciaravella, A., Peres, G. and Serio, S. 1991, *Solar Phys.* **132**, 279.

Collins, W. 1989a, *Astrophys. J.* **337**, 548.

Collins, W. 1989b, *Astrophys. J.* **343**, 499.

Collins, W. 1992, *Astrophys. J.* **384**, 319.

Cook, J.W. 1991, in *Mechanisms of Chromospheric and Coronal Heating*, ed. P. Ulmschneider and R. Rosner (Berlin: Springer-Verlag), p. 83.

Coppi, B. 1975, *Astrophys. J.* **195**, 545.

Coppi, B. and Friedland, A.B. 1971, *Astrophys. J.* **169**, 379.

Craig, I.J.D., McClymont, A.N. and Underwood, J.H. 1978, *Astron. Astrophys.* **70**, 1.

Davila, J.M. 1987, *Astrophys. J.* **317**, 514.

Davis, J.M., Gerassimenko, M., Krieger, A.S. and Vaiana, G.S. 1975, *Solar Phys.* **45**, 393.

de Jager, C. 1955, *Ann. Geophys.* **11**, 330.

de Jager, C. and Kuperus, M. 1961, *Bull. Astron. Inst. Netherlands* **16**, 71.

Dere, K.P., Bartoe, J.D.F., Brueckner, G.E., Ewing, J. and Lund, P. 1991, *J. Geophys. Res.* **96**, 9399.

Detwiler, C.R., Garrett, D.L., Purcell, J.D. and Tousey, R. 1961, *Ann. Geophys.* **17**, 263.

Donnelly, I.J., Claney, B.E. and Cramer, N.F. 1985, *J. Plasma Phys.* **34**, 227.

Doschek, G.A. and Feldman, U. 1977, *Astropys. J. Letters* **212**, L143.

Dwivedi, B.N. and Gupta, A.K. 1991, *Solar Phys.* **135**, 415.

Eddy, J.A. 1976, *Science* **192**, 1189.

Eddy, J.A. 1977, in *The Solar Output and its Variation*, ed. O.R. White (Boulder: Colorado Associated University Press), pp. 51–71.

Eddy, J.A. 1983, in *Weather and Climate Responses to Solar Variations*, ed. B.M. McCormac (Boulder: Colorado Associated University Press), pp. 1–15.

Eddy, J.A. Gilliland, R.L. and Hoyt, D.V. 1982, *Nature* **300**, 689.

Edlen, B. 1942a, *Ark. Mat. Astro. Fys.* **28**, No. 1B.

Edlen, B. 1942b, *Z. Astrophys.* **22**, 30.

Einaudi, G. and Mok, Y. 1985, *J. Plasma Phys.* **34**, 259.

Einaudi, G. and Mok, Y. 1987, *Astrophys. J.* **319**, 520.

Elwert, G. 1952, *Z. Naturforsch* **7a**, 202.

Elwert, G. 1954, *Z. Naturforsch* **9a**, 637.

Esser, R. 1992, in *Solar Wind Seven* (COSPAR Colloquium Series), ed. E. Marsch and R. Schwenn, (Oxford: Pergamon Press), pp. 31–36.

Feldman, U. 1983, *Astrophys. J.* **275**, 367.

Feldman, U., Doschek, G.A. and Mariska, J.T. 1979, *Astrophys. J.* **229**, 369.

Feldman, U. and Laming, J.M. 1993, *Astrophys. J.* **404**, 799.

Feldman, U., Purcell, J.D. and Dohne, B.C. 1987, *An Atlas of Extreme Ultraviolet Spectroheliograms from 170 to 625A* (NRL report 90-4100 and 91-4100) (Naval Research Laboratory: Washington, DC).

Feldman, U., Laming, J.M., Mandelbaum, P., Goldstein, W.H. and Osterheld, A. 1992, *Astrophys. J.* **398**, 692.

Fla, T., Habbal, S.R., helzer, T.E. and Leer, E. 1984, *Astrophys. J.* **280**, 382.

Forman, W., Jones, C., Cominsky, L., Julien, P., Murray, S., Peters, G., Tananbaum, H. and Giacconi, R. 1978, *Astrophys. J. (Supplement)* **38**, 357.

Friedman, H., Lichtman, S.W. and Byram, E.T. 1951, *Phys. Rev.* **83**, 1025.

Gaizauskas, V., Harvey, K.L., Harvey, J.W. and Zwaan, C. 1983, *Astrophys. J.* **265**, 1056.

Giacconi, R., Bechtold, J., Branduardi, G., Forman, W., Henry, J.P., Jones, C., Kellogg, E., Van der Laan, H., Liller, W., Marshall, H., Murray, S.S., Pye, J., Schreier, E., Sargent, W.L.W., Seward, F. and Tannenbaum, H. 1979, *Astrophys. J. Letters* **234**, L1.

Giampapa, M.S. Golub, L., Peres, G., Serio, S., and Vaiana, G.S. 1985, *Astrophys. J.* **289**, 203.

Gioia, I.M., Maccacaro, T., Schild, R.E., Wolter, A., Stocke, J.T., Morris, S.L. and Henry, J.P. 1990, *Astrophys. J. (Supplement)* **72**, 567.

Gold, T. 1963, in *The Physics of Solar Flares*, ed. W.N. Hess (Washington, DC: NASA Scientific and Technical Information Division), p. 389.

Gold, T. and Hoyle, F. 1960, *Mon. Not. Royal Astron. Soc.* **120**, 89.

Goldreich, P. and Kumar, P. 1988, *Astrophys. J.* **326**, 462.

Goldreich, P., Murray, N., Willette, G. and Kumar, P. 1991, *Astrophys. J.* **370**, 752.

Golub, L. 1991, in *Mechanisms of Chromospheric and Coronal Heating*, ed. P. Ulmschneider, E.R. Priest and R. Rosner (Berlin: Springer-Verlag), p. 115.

Golub, L., Harnden, F.R., Pallavicini, R., Rosner, R. and Vaiana, G.S. 1982, *Astrophys. J.* **253**, 242.

Golub, L., Herant, M., Kalata, K., Lovar, I., Nystrom, G., Pardo, F., Spillar, E. and Wilczynski, J. 1990, *Nature* **344**, 842.

Golub, L., Krieger, A.S. and Vaiana, G.S. 1976a, *Solar Phys.* **49**, 79.

Golub, L., Krieger, A.S. and Vaiana, G.S. 1976b, *Solar Phys.* **50**, 311.

Golub, L., Krieger, A.S. Harvey, J.W. and Vaiana, G.S. 1977, *Solar Phys.* **53**, 111.

Golub, L., Maxson, C., Rosner, R., Serio, S. and Vaiana, G.S. 1980, *Astrophys. J.* **238**, 343.

Grindlay, J., Gursky, H., Schnopper, H., Parsignault, D.R., Heise, J., Brinkman, A.C. and Schrijver, J. 1976, *Astrophys. J. Letters* **205**, L127.

Grossman, V. and Smith, R.A. 1988, *Astrophys. J.* **332**, 476.

Grossman, W. and Tataronis, J.A. 1973, *Z. Phys.* **261**, 217.

Grotrian, W. 1931, *Z. Astrophys.* **3**, 199.

Grotrian, W. 1933, *Z. Astrophys.* **7**, 26.

Grotrian, W. 1939, *Naturwiss* **27**, 214.

Habbal, S.R. 1991, in *Mechanisms of Chromospheric and Coronal Heating*, ed. P. Ulmschneider, E.R. Priest and R. Rosner (Berlin: Springer-Verlag), p. 127.

Habbal, S.R. 1992, in *Solar Wind Seven*, ed. E. Marsch and R. Schwenn (New York: Pergamon Press), p. 41.

Habbal, S.R. and Grace, E. 1991, *Astrophys. J.* **382**, 667.

Habbal, S.R. and Withbroe, G.L. 1981, *Solar Phys.* **69**, 77.

Hahm, T.S. and Kulsrud, R.M. 1985, *Phys. Fluids* **28**, 2412.

Haisch, B.M., Linsky, J.L., Harnden, F.R., Rosner, R., Seward, F.D. and Vaiana, G.S. 1980, *Astrophys. J. Letters* **242**, L99.

Hammer, R. and Ulmschneider, P. 1991, in *Mechanisms of Chromospheric and Coronal Heating*, ed. P. Ulmschneider, E.R. Priest and R. Rosner (Berlin: Springer-Verlag), p. 344.

Hardie, I.S., Hood, A.W. and Allen, H.R. 1991, *Solar Phys.* **133**, 313.

Harvey, K.L. and Martin, S.F. 1973, *Solar Phys.* **32**, 389.

Hayvaerts, J. and Priest, E.R. 1984, *Astron. Astrophys.* **117**, 220.

Hildner, E. 1974, *Solar Phys.* **35**, 123.

Hill, E.R. 1951, *Aust. J. Sci. Res.* **4a**, 437.

Hinteregger, H.E. 1961, *J. Geophys. Res.* **66**, 2367.

Hinteregger, H.E. and Watanabe, K. 1962, *J. Geophys. Res.* **67**, 3373.

Hollweg, J.V. 1984, *Astrophys. J.* **277**, 392.

Hollweg, J.V. 1985, *J. Geophys. Res.* **90**, 7620.

Hollweg, J.V. 1986, *J. Geophys. Res.* **306**, 730.

Hollweg, J.V. 1987a, *Astrophys. J.* **312**, 880.

Hollweg, J.V. 1987b, *Astrophys. J.* **320**, 875.

Hollweg, J.V. and Sterling, A.C. 1984, *Astrophys. J. Letters* **282**, L31.

Hollweg, J.V., Yang, G., Cadez, V.M. and Gakovic, B. 1990, *Astrophys. J.* **349**, 335.

Huber, M.C.E., Foukal, P.V., Noyes, R.W., Reeves, E.M., Schmahl, E.J., Timothy, J.C., Vernazza, J.E. and Withbroe, G.L. 1974, *Astrophys. J. Letters* **194**, L118.

Jokipii, J.R. 1973, *Astrophys. J.* **183**, 1029.

Jokipii, J.R. and Parker, E.N. 1968, *Phys. Rev. Letters* **21**, 44.

Jokipii, J.R. and Parker, E.N. 1969a, *Astrophys. J.* **155**, 777.

Jokipii, J.R. and Parker, E.N. 1969b, *Astrophys. J.* **155**, 799.

Kahler, S. 1976, *Solar Phys.* **48**, 255.

Kappraff, J.M. and Tataronis, J.A. 1977, *J. Plasma Phys.* **18**, 209.

Kopecky, M. and Obridko, V. 1968, *Solar Phys.* **5**, 354.

Kreplin, R.W. 1961, *Ann. Geophys.* **17**, 151.

Krieger, A.S., Vaiana, G.S. and Van Speybroeck, L.P. 1971, in *Solar Magnetic Fields (IAU Symp. 43)*, ed. R. Howard (Dordrecht: Reidel), p. 43.

Krishan, V. 1985, *Solar Phys.* **97**, 183.

Krishan, V. 1991, *Solar Phys.* **134**, 109.

Kulsrud, R.M. 1955, *Astrophys. J.* **121**, 461.

Kuperus, M., Ionson, J.A. and Spicer, D.S. 1981, *Ann. Rev. Astron. Astrophys.* **19**, 7.

Laming, J.M. and Feldman, U. 1992, *Astrophys. J.* **386**, 364.

Landini, M., Monsignori-Fossi, B.C., Krieger, A.S. and Vaiana, G.S. 1975, *Solar Phys.* **44**, 69.

Lee, M. and Roberts, B. 1986, *Astrophys. J.* **301**, 430.

Leighton, R.B. 1966, *Astrophys. J.* **140**, 1547.

Levine, R.H. 1974, *Astrophys. J.* **190**, 457.

Lewin, W.H.G. 1977, *Mon. Not. Royal Astron. Soc.* **179**, 43.

Lewin, W.H.G., Hoffman, J.A., Doty, J., Hearn, D.R., Clark, G.W., Jernigan, J.G., Li, F.K., McClintock, J.E. and Richardson, J. 1976a, *Mon. Not. Royal Astron. Soc.* **177**, 83P.

Lewin, W.H.G., Li, F.K., Hoffman, J.A., Doty, J., Buff, J., Clark, G.W. and Rappaport, S. 1976b, *Mon. Not. Royal Astron. Soc.* **177**, 93P.

Lighthill, M. 1960, *Proc. Royal Soc. London A* **252**, 49.

Lin, R.P., Schwartz, R.A., Kane, S.R., Pelling, R.M. and Hurley, K.C. 1984, *Astrophys. J.* **283**, 421.

Lothian, R.M. and Hood, A.W. 1992, *Solar Phys.* **137**, 105.

Lu, E.T. and Hamilton, R.J. 1991, *Astrophys. J. Letters* **380**, L89.

Lu, E.T., Hamilton, R.J., McTiernan, J.M. and Bromund, K.R. 1993, *Astrophys. J.* **412**, 841.

Lyot, B. 1939, *Mon. Not. Royal Astron. Soc.* **99**, 580.

Mariska, J.T., Feldman, U. and Doschek, G.A. 1980, *Astrophys. J.* **240**, 300.

Mariska, J.T. and Hollweg, J.V. 1985, *Astropphys. J.* **296**, 746.

Markert, T.H., Canizares, C.R., Clark, G.W., Hearn, D.R., Li, F.K., Sprott, G.F. and Winkler, P.F. 1977, *Astrophys. J.* **218**, 801.

Markert, T.H., Winkler, P.F., Laird, F.N., Clark, G.W., Hearn, D.R., Sprott, G.F., Li, F.K., Bradt, V., Lewin, W.H.G. and Schnopper, H.W. 1979, *Astrophys. J. (Supplement)* **39**, 573.

Marshall, F.E., Boldt, E.A., Holt, S.S., Mushotsky, R.F., Pravdo, S.H., Rothschild, R.E. and Serlemitsos, P.J. 1979, *Astrophys. J. (Supplement)* **40**, 657.

Martens, P.C.H., Van Den Oord, G.H.J. and Hoyng, P. 1985, *Solar Phys.* **96**, 253.

Martin, S.F. 1984, in *Small-Scale Dynamical Processes in Quiet Stellar Atmospheres*, ed. S.L. Keil (Sacramento Peak, Sunspot, New Mexico: National Solar Observatory), p. 30.

Martin, S.F., Livi, S.H.B. and Wang, J. 1985, *Aus. J. Phys.* **85**, 929.

Matthaeus, W.H. and Lamkin, S.L. 195, *Phys. Fluids* **28**, 303.

Matthaeus, W.H. and Lamkin, S.L. 1986, *Phys. Fluids* **29**, 2513.

Menzel, D.H. 1941, *The Telescope* **8**, 65.

Miyamoto, S. 1949, *Publ. Astron. Soc. Japan* **1**, 10.

Mok, Y. and Einaudi, G. 1990, *Astrophys. J.* **351**, 296.

Mok, Y., Schnack, D.D. and Van Hoven, G. 1991, *Solar Phys.* **132**, 95.

Moore, D.W. and Spiegel, E.A. 1964, *Astrophys. J.* **139**, 48.
Moore, R.L., Tang, F., Bohlin, J.D. and Golub, L. 1977, *Astrophys. J.* **218**, 286.
Musielak, Z.E. and Rosner, R. 1987, *Astrophys. J.* **315**, 371.
Musielak, Z.E. and Rosner, R. 1988, *Astrophys. J.* **329**, 376.
Musielak, Z.E., Rosner, R. and Ulmschneider, P. 1989, *Astrophys. J.* **337**, 470.
Nolte, J.T., Solodyna, C.V. and Gerassimenko, M. 1979, *Solar Phys.* **63**, 113.
Osterbrock, D.E. 1961, *Astrophys. J.* **134**, 347.
Otani, N.F. and Strauss, H.R. 1988, *Astrophys. J.* **325**, 468.
Pallavicini, R. 1989, *Astron. Astrophys. Rev.* **1**, 177.
Pallavicini, R., Serio, S. and Vaiana, G.S. 1977, *Astrophys. J.* **216**, 108.
Parker, E.N. 1963, *Interplanetary Dynamical Processes* (New York: John Wiley).
Parker, E.N. 1972, *Astrophys. J.* **174**, 499.
Parker, E.N. 1974, *Astrophys. J.* **191**, 245.
Parker, E.N. 1975, *Astrophys. J.* **201**, 494.
Parker, E.N. 1976, *Astrophys. Space Sci.* **44**, 107.
Parker, E.N. 1979, *Cosmical Magnetic Fields* (Oxford: Clarendon Press).
Parker, E.N. 1981a, *Astrophys. J.* **244**, 631.
Parker, E.N. 1981b, *Astrophys. J.* **244**, 644.
Parker, E.N. 1982a, *Astrophys. J.* **256**, 292.
Parker, E.N. 1982b, *Astrophys. J.* **256**, 302.
Parker, E.N. 1982c, *Astrophys. J.* **256**, 736.
Parker, E.N. 1982d, *Astrophys. J.* **256**, 746.
Parker, E.N. 1983a, *Astrophys. J.* **264**, 635.
Parker, E.N. 1983b, *Astrophys. J.* **264**, 642.
Parker, E.N. 1984, *Astrophys. J.* **283**, 343.
Parker, E.N. 1985, *Astrophys. J.* **294**, 57.
Parker, E.N. 1987, *Solar Phys.* **111**, 297.
Parker, E.N. 1988, *Astrophys. J.* **330**, 474.
Parker, E.N. 1991, *Astrophys. J.* **372**, 719.
Parker, E.N. 1993a, *Astrophys. J.* **414**, 389.
Parker, E.N. 1993b, *Astrophys. J.* **408**, 707.
Peres, G., Rosner, R., Serio, S. and Vaiana, G.S. 1982, *Astrophys. J.* **252**, 791.
Piddington, J.H. and Davies, R.D. 1953, *Mon. Not. Royal Astron. Soc.* **113**, 582.
Piddington, J.H. and Minnett, H.C. 1951, *Aus. J. Phys.* **4**, 131.
Pneuman, G.W. 1973, *Solar Phys.* **28**, 247.
Porter, J.G. and Dere, K.P. 1991, *Astrophys. J.* **370**, 775.
Porter, J.G. and Moore, R.L. 1988, in *Proc. 9th Sacramento Peak Summer Symposium*, ed. R.C. Altrock (Sacramento Peak, Sunspot, New Mexico: National Solar Observatory), p. 30.
Porter, J.G., Moore, R.L., Reichmann, E.J., Engvold, O. and Harvey, K.L. 1987, *Astrophys. J.* **323**, 380.
Porter, J.G., Toomre, J. and Gebbie, K.B. 1984, *Astrophys. J.* **283**, 879.
Rabin, D. 1992, *Astrophys. J.* **391**, 832.
Rabin, D. and Dowdy, J.F. 1992, *Astrophys. J.* **398**, 665.
Raymond, J.C. 1990, *Astrophys. J.* **365**, 387.
Reeves, E.M., Timothy, J.G., Foukal, P.V., Huber, M.C.E., Noyes, R.W., Schmahl, E., Vernazza, J.E. and Withbroe, G.L. 1976, in *Progress in Astronautics and Aeronautics* (Vol. 48), ed. M. Kent, E. Stuhlinger, and S.T. Wu (New York: AIAA), p. 73.
Robertson, J.A., Hood, A.W. and Lothian, R.M. 1992, *Solar Phys.* **137**, 273.
Rosner, R., Golub, L., Coppi, B. and Vaiana, G.S. 1978, *Astrophys. J.* **222**, 317.
Rosner, R., Tucker, W.H. and Vaiana, G.S. 1978, rophys. J. **220**, 643.
Saba, J.L.R. and Strong, K.T. 1991, *Astrophys. J.* **375**, 789.

Sakai, J. 1983a, *Solar Phys.* **84**, 109.
Sakai, J. 1983b, *J. Plasma Phys.* **30**, 109.
Sakai, J., Tajima, T. and Brunel, F. 1984, *Solar Phys.* **91**, 103.
Sakai, J. and Washima, H. 1982, *Astrophys. J.* **258**, 823.
Sakurai, T. and Granik, A. 1984, *Astrophys. J.* **277**, 404.
Schatzman, E. 1949, *Ann. d'Ap* **12**, 203.
Schnopper, H.W., Delvaille, J.P., Epstein, A., Helmken, H., Murray, S.S., Clark, G., Jernigan, G. and Doxsey, R. 1976, *Astrophys. J. Letters* **210**, L75.
Schwartz, R.A. and Stein, R.F. 1975, *Astrophys. J.* **200**, 499.
Schwarzschild, M. 1948, *Astrophys. J.* **107**, 1.
Scudder, J.D. 1992a, *Astrophys. J.* **398**, 299.
Scudder, J.D. 1992b, *Astrophys. J.* **398**, 319.
Sheeley, N.R. and Golub, L. 1979, *Solar Phys.* **63**, 119.
Sheeley, N.R. Nash, A.G. and Wang, Y.M. 1987, *Astrophys. J.* **319**, 481.
Sheeley, N.R., Wang, Y.M. and DeVore, C.R. 1989, *Solar Phys.* **124**, 1.
Shklovski, I.S. 1951, *Solar Corona* (Moscow).
Similon, P. and Sudan, R. 1989, *Astrophys. J.* **336**, 442.
Spicer, D.S. 1976, *Report 8036* (Washington, DC: Naval Research Laboratory).
Spicer, D.C. 1977, *Solar Phys.* **53**, 305.
Spicer, D.S. 1982, *Space Sci. Rev.* **31**, 351.
Spitzer, L. 1956, *Physics of Fully Ionized Gases* (New York: John Wiley).
Steele, C.D.C. and Priest, E.R. 1990, *Solar Phys.* **127**, 65.
Steele, C.D.C. and Priest, E.R. 1991a, *Solar Phys.* **132**, 293.
Steele, C.D.C. and Priest, E.R. 1991b, *Solar Phys.* **134**, 73.
Stein, R.F. 1968, *Astrophys. J.* **154**, 297.
Stein, R.F. and Leibacher, J. 1974, *Ann. Rev. Astron. Astrophys.* **12**, 407.
Stein, R.F. and Schwartz, R.A. 1972, *Astrophys. J.* **177**, 807.
Stein, R.F. and Schwartz, R.A. 1973, *Astrophys. J.* **186**, 1083.
Sterling, A.C. and Hollweg, J.V. 1985, *Astrophys. J.* **285**, 843.
Strauss, H.R. 1988, *Astrophys. J.* **326**, 412.
Strauss, H.R. and Lawson, W.J. 1989, *Astrophys. J.* **346**, 1035.
Sturrock, P.A., Dixon, W.W., Klimchuk, J.A. and Antiochos, S.K. 1990, *Astrophys. J. Letters* **356**, L31.
Sturrock, P.A. and Uchida, Y. 1981, *Astrophys. J.* **246**, 331.
Sweet, P.A. 1958a, *Proc. IAU Symp. No. 6, Electromagnetic Phenomena in Cosmical Physics*, ed. B. Lehnert (Cambridge: Cambridge University Press), p. 123.
Sweet, P.A. 1958b, *Nuovo Cim. (Supplement)* **8** (10), 188.
Tajima, T., Brunel, F. and Sakai, S. 1982, *Astrophys. J. Letters* **258**, L45.
Takakura, T. 1990, *Solar Phys.* **127**, 95.
Takakura, T. 1991, *Solar Phys.* **136**, 303.
Tataronis, J.A. 1975, *J. Plasma Phys.* **13**, 87.
Taylor, J.B. 1974, *Phys. Fluids* **33**, 1139.
Taylor, J.B. 1986, *Rev. Mod. Phys.* **58**, 471.
Topka, K., Fabricant, D., Harnden, F.R., Gorenstein, P. and Rosner, R. 1979, *Astrophys. J.* **229**, 661.
Tousey, R., Bartoe, J.D.F., Bohlin, J.D., Brueckner, G.E., Purcell, J.D., Scherrer, V.E., Sheeley, N.R., Schumacher, R.J. and Vanhoosier, M.E. 1973, *Solar Phys.* **33**, 265.
Tucker, W.H. 1973, *Astrophys. J.* **186**, 285.
Uchida, Y. 1963, *Publ. Astron. Soc. Japan* **15**, 376.
Ulmschneider, P., Priest, E.R. and Rosner, R. 1991, *Mechanisms of Chromospheric and Coronal Heating* (Heidelberg: Springer-Verlag).

Underwood, J.H. et al. 1976, in *Progress in Astronautics and Aeronautics* (Vol. 48), ed. M. Kent, E. Stuhlinger and S.T. Wu (New York: Amer. Inst. Astron. Aeron.).
Vaiana, G.S., Davis, J.M., Giacconi, R., Krieger, A.S., Silk, J.K., Timothy, A.F. and Zombeck, M. 1973, *Astrophys. J. Letters* **185**, L47.
Vaiana, G.S., Krieger, A.S., Timothy, A.F. and Zombeck, M.V. 1976, *Astrophys. Space Sci.* **39**, 75.
Vaiana, G.S., Reidy, W.P., Zehnpfennig, T. and Giacconi, R. 1968, *Astrophys. J.* **151**, 333.
Vainshtein, S.I., Parker, E.N. and Rosner, R. 1993, *Astrophys. J.* **404**, 773.
Van der Linden, R.A.M. and Goossens, M. 1991, *Solar Phys.* **134**, 247.
Van Hoven, G. 1976, *Solar Phys.* **49**, 95.
Van Hoven, G. 1979, *Astrophys. J.* **232**, 572.
Van Hoven, G. 1981, in *Solar Flare Magnetohydrodynamics*, ed. E.R. Priest (New York: Gordon and Breach), pp. 217–276.
Van Speybroeck, L.P., Krieger, A.S. and Vaiana, G.S. 1970, *Nature* **227**, 818.
Vegard, L. 1944, *Geofys. Publ.* **16**, No. 1.
Vekstein, G.E., Priest, E.R. and Steele, C.D.C. 1991, *Solar Phys.* **131**, 297.
Vette, J.I. and Casal, F.G. 1961, *Physa. Rev. Letters* **6**, 334.
Walbroeck, F.L. 1989, *Phys. Fluids B* **1**, 2372.
Waldmeier, M. 1945, *Mitt der Aarg Natur. Ges.* **22**, 185.
Walker, A.B.C., Barbee, T.W., Hoover, R.B. and Lindblom, J.F. 1988, *Science* **241**, 1781.
Wang, X. and Bhattacharjee, A. 1992, *Astrophys. J.* **401**, 371.
Wang, Y.M., Nash, A.G. and Sheeley, N.R. 1989, *Astrophys. J.* **347**, 529.
Wang, Y.M. and Sheeley, N.R. 1990, *Astrophys. J.* **365**, 372.
Watanabe, K. and Hinteregger, H.E. 1962, *J. Geophys. Res.* **67**, 999.
Wentzel, D.G. 1979a, *Astrophys. J.* **227**, 319.
Wentzel, D.G. 1979b, *Astrophys. J.* **233**, 756.
Withbroe, G.L. 1975, *Solar Phys.* **45**, 301.
Withbroe, G.L. 1988, *Astrophys. J.* **325**, 442.
Withbroe, G.L. and Noyes, R.W. 1977, *Ann. Rev. Astron. Astrophys.* **15**, 363.
Woolley, R.R. and Allen, C.W. 1948, *Mon. Not. Royal Astron. Soc.* **108**, 292.
Zhang, Q., Soon, W.H., Baliunas, S.L., Lockwood, G.W., Skiff, B.A. and Radick, R.R. 1993, *Nature* (in press).
Zwaan, C. 1985, *Solar Phys.* **100**, 397.
Zwaan, C., Brants, J.J. and Cram, L.E. 1985, *Solar Phys.* **95**, 3.

Chapter 12

Akasofu, S.I. and Chapman, S. 1972, *Solar-Terrestrial Physics* (Oxford: Clarendon Press).
Axford, W.I. and Hines, C.O. 1961, *Can. J. Phys.* **39**, 1433.
Axford, W.I., Petschek, H.E. and Siscoe, G.L. 1965, *J. Geophys. Res.* **70**, 1231.
Bak, P., Tang, C. and Wiesenfeld, K. 1988, *Phys. Rev. A.* **38**, 364.
Baliunas, S., Soon, W. and Zhang, Q. 1994, *Science* (in press).
Baranov, V.B. 1990, *Space Sci. Rev.* **52**, 89.
Berchem, J. and Russell, C.T. 1984, *J. Geophys. Res.* **89**, 6689.
Berger, M.A. 1991a, *Astron. Astrophys.* **252**, 369.
Berger, M.A. 1991b, in *Mechanisms of Chromospheric and Coronal Heating*, ed. P. Ulmschneider, E.R. Priest, and R. Rosner (Berlin: Springer-Verlag), p. 570.
Bertaux, J.L. and Blamont, J.E. 1971, *Astron. Astrophys.* **11**, 200.
Bertaux, J.L., Lallemont, R., Kurt, V.G. and Mironova, E.N. 1985, *Astron. Astrophys.* **150**, 1.
Bregman, J.N. and Glasgold, A.E. 1982, *Astrophys. J.* **263**, 564.
Breitschwerdt, D., McKenzie, J.F. and Volk, H.J. 1991, *Astron. Astrophys.* **245**, 79.

Bromund, K., McTiernan, and Kane, S. 1993, *Astrophys. J.* (in press).
Cheng, C.C. and Pallavicini, R. 1991, *Astrophys. J.* **381**, 234.
Chiang, W.H. and Foukal, F.A. 1985, *Solar Phys.* **97**, 9.
Choudhuri, A.R. 1988, *Geophys. Astrophys. Fluid Dyn.* **40**, 261.
Chupp, E.L. 1990, *Astrophys. J. Supplement* **73**, 111.
Clerke, A.M. 1894, *Knowledge* **17**, 206.
Cliver, E. and Kahler, S. 1991, *Astrophys. J. Letters* **366**, L91.
Collins, W.A. 1992, *Astrophys. J.* **384**, 319.
Cowley, S.W.H. 1982, *Rev. Geophys. Space Phys.* **20**, 531.
Daly, P.W., Saunders, M.A., Rijnbeek, R.P., Sckopke, N. and Russell, C.T. 1984, *J. Geophys. Res.* **89**, 3843.
Davis, T.N. 1962a, *J. Geophys. Res.* **67**, 59.
Davis, T.N. 1962b, *J. Geophys. Res.* **67**, 75.
Davis, T.N. 1967, in *Aurora and Airglow*, ed. B.M. McCormac (New York: Reinhold), p. 51.
Debrunner, H., Lockwood, J.A. and Ryan, J.M. 1992, *Astrophys. J. Letters* **387**, L51.
Deinzer, W., Hensler, G., Schüssler, M. and Weishaar, E. 1984, *Astron. Astrophys.* **139**, 435.
Dennis, B.R. 1985, *Solar Phys.* **100**, 465.
Dennis, B.R. 1988, *Solar Phys.* **118**, 49.
Dennis, B.R. and Schwartz, R.A. 1989, *Solar Phys.* **121**, 75.
Dere, K.P., Bartoe, J.D.F., Brueckner, G.E., Ewing, J. and Lund, P. 1991, *J. Geophys. Res.* **96**, 9399.
Dessler, A.J. 1962, *J. Geophys. Res.* **67**, 4897.
Dessler, A.J. 1964, *J. Geophys. Res.* **69**, 3913.
Dessler, A.J. and Juday, R.D. 1965, *Planet. Space Sci.* **13**, 63.
Dungey, J.W. 1961, *Phys. Rev. Letters* **6**, 47.
Eddy, J.A. 1976, *Science* **192**, 1189.
Eddy, J.A. 1977, *Scientific American* **236**, (5) 80.
Eddy, J.A. 1980, in *The Ancient Sun: Fossil Record in the Earth, Moon, and Meteorites*, ed. R.O. Pepin, J.A. Eddy and R.B. Merrill (New York: Pergamon Press), p. 119.
Eddy, J.A. 1983, in *Weather and Climate Response to Solar Variations*, ed. B.M. McCormac (Boulder: Colorado Associated University Press), p. 1.
Eddy, J.A. 1988, in *Secular Solar and Geomagnetic Variations in the Last 10,000 Years*, ed. F.R. Stephenson and A.W. Wolfendale (Dordrecht: Kluwer), p. 1.
Eddy, J.A. and Boornazian, A.A. 1979, *Bull. Amer. Astron. Soc.* **11**, 437.
Eddy, J.A., Gilman, P.A. and Trotter, D.E. 1976, *Solar Phys.* **46**, 3.
Ehmert, A. 1948, *Z. Naturforsch* **3a**, 264.
Eviatar, A. and Wolf, R.A. 1968, *J. Geophys. Res.* **73**, 5561.
Fabbiano, G. 1988, *Astrophys. J.* **330**, 67.
Fabbiano, G. and Trinchieri, G. 1984, *Astrophys. J.* **286**, 491.
Fahr, H.J. 1974, *Space Sci. Rev.* **15**, 683.
Fahr, H.J. and Fichtner, A. 1991, *Space Sci. Rev.* **58**, 193.
Feldstein, Y.I. 1969, *Rev. Geophys.* **7**, 179.
Finn, J.M., Guzdar, P.N. and Usikov, D. 1992, *UMLPR 93-010* (Maryland: University of Maryland).
Fla, T., Habbal, S.R., Holzer, T.E. and Leer, E. 1984, *Astrophys. J.* **280**, 382.
Forbes, T.G. 1991, *Geophys. Astrophys. Fluid Dyn.* **62**, 15.
Forbes, T.G. and Malherbe, J.M. 1991, *Solar Phys.* **135**, 361.
Forbush, S.E. 1946, *Phys. Rev.* **70**, 771.
Forman, W., Jones, C. and Tucker, W. 1985, *Astrophys. J.* **293**, 102.
Foukal, P. and Lean, J. 1986, *Astrophys. J.* **302**, 826.
Garcia-Munoz, M., Mason, G.M. and Simpson, J.A. 1975a, *Astrophys. J. Letters* **201**, L141.
Garcia-Munoz, M., Mason, G.M. and Simpson, J.A. 1975b, *Astrophys. J. Letters* **201**, L145.

Garcia-Munoz, M., Mason, G.M. and Simpson, J.A. 1977, *Astrophys. J.* **217**, 859.
Gold, T. 1959, *J. Geophys. Res.* **64,** 1665.
Gold, T. and Hoyle, F. 1960, *Mon. Not. Royal Astron. Soc.* **120**, 89.
Goldreich, P., Murray, N., Willette, G. and Kumar, P. 1991, *Astrophys. J.* **370**, 752.
Habbal, S.R. 1992a, in *Solar Wind Seven*,ed. E. Marsch and R. Schwenn (Oxford: Pergamon Press), p. 41.
Habbal, S.R. 1992b, *Ann Geophysicae* **10**, 34.
Hagen, F.A., Fisher, A.J. and Ormes J.F. 1977, *Astrophys. J.* **212**, 262.
Haisch, B.M., Strong, K.T., Harrison, R.A. and Gary, G.A. 1989, *Astrophys. J. (Supplement)* **68**, 371.
Haisch, B.M., Strong, K.T. and Rodono, M. 1991, *Ann. Rev. Astron. Astrophys.* **29**, 275.
Hamilton, R.J. and Petrosian, V. 1992, *Astrophys. J.* **398**, 350.
Hartquist, T.W. and Morfill, G.E. 1986, *Astrophys. J.* **311**, 518.
Hasan, S.S. 1985, *Astron. Astrophys.* **143**, 39.
Hasan, S.S. 1988, *Astrophys. J.* **332**, 499.
Hasan, S.S. and Sobouti, Y. 1987, *Mon. Not. Royal Astron. Soc.* **228**, 427.
Hawley, S.L. and Fisher, G.H. 1992, *Astrophys. J. Supplement* **78**, 565.
Hawley, S.L. and Pettersen, B.R. 1991, *Astrophys. J. Letters* **378**, 725.
Heikkila, W.J. 1990, *Space Sci. Rev.* **53**, 1.
Henry, G.W. and Hall, D.S. 1991, *Astrophys. J. Letters* **373**, L9.
Hoffert, M.I., Frei, A. and Narayanan, V.K. 1988, *Clim. Change* **13**, 267.
Holman, G.D. and Benka, S.G. 1992, *Astrophys. J. Letters* **400**, L79.
Holzer, T.E. and Axford, W.I. 1970, *Ann. Rev. Astron. Astrophys.* **8**, 31.
Hudson, H.S. 1987, *Rev. Geophys.* **25**, 651.
Hudson, H.S. 1988, *Ann. Rev. Astron. Astrophys.* **26**, 473.
Hudson, H. 1991, *Solar Phys.* **133**, 357.
Hultquist, B. 1969, *Rev. Geophys.* **7**, 129.
Jakimec, J., Fludra, A., Lemen, J.R., Dennis, B.R. and Sylwester, J. 1986, *Adv. Space Res.* **6** (6), 191.
Johnson, F.S. 1960, *J. Geophys. Res.* **65**, 3049.
Jokipii, J.R., Lerche, I. and Schommer, R.A. 1969, *Astrophys. J. Letters* **157**, L119.
Jokipii, J.R. and Morfill, G.E. 1985, *Astrophys. J. Letters* **290**, L1.
Jokipii, J.R. and Morfill, G.E. 1987, *Astrophys. J.* **312**, 170.
Jokipii, J.R. and Morfill, G.E. 1990, *Astrophys. J.* **356**, 255.
Jokipii, J.R. and Parker, E.N. 1968, *Phys. Rev. Letters* **21**, 44.
Jokipii, J.R. and Parker, E.N. 1969a, *Astrophys. J.* **155**, 777.
Jokipii, J.R. and Parker, E.N. 1969b, *Astrophys. J.* **155**, 799.
Kadanoff, L., Nagel, S.R., Wu, L. and Zhou, S. 1989, *Phys. Rev. A* **39**, 6524.
Kahn, F.D. and Brett, L. 1993, *Mon. Not. Royal Astron. Soc.* **263**, 37.
Kallenrode, M.B. and Wibberenz, G. 1991, *Astrophys. J.* **376**, 787.
Kaplan, S.A. 1966, *Interstellar Gas Dynamics* (Oxford: Pergamon).
Klecker, M.B. and McGuire, R.E. 1992, *Astrophys. J.* **401**, 398.
Ko, C.M., Dougherty, M.K. and McKenzie, J.F. 1991, *Astron. Astrophys.* **241,** 62.
Kopriva, D.A. and Jokipii, J.R. 1983, *Astrophys. J.* **267**, 62.
La Belle-Hamer, A.L., Fu, Z.F. and Lee, L.C. 1988, *Geophys. Res. Letters* **15**, 152.
Lallement, R., Vidal-Madjar, A. and Fexlet, R. 1986, *Astron. Astrophys.* **168**, 225.
Lee, M.A. and Jokipii, J.R. 1976, *Astrophys. J.* **206**, 735.
Lerche, I. 1967, *Astrophys. J.* **72**, 5295.
Lerche, I. and Parker, E.N. 1966, *Astrophys. J.* **145**, 106.
Levy, R.H., Petschek, H.E. and Siscoe, G.L. 1964, *Amer. Inst. Aeron. Astron. J.* **2**, 2065.
Lu, E.T. and Hamilton, R.J. 1991, *Astrophys. J. Letters* **380**, L89.
Lu, E.T., Hamilton, R.J., McTiernan, J.M. and Bromund, K.R. 1993, *Astrophys. J.* **412**, 841.

Lundin, R. 1988, *Space Sci. Rev.* **48**, 263.
Lyons, L.R. 1992, *Rev. Geophys.* **30**, 93.
Machado, M.E., Moore, R.L., Hernandez, A.M., Rovira, M.G., Hagyard, M. and Smith, J.B. 1988a, *Astrophys. J.* **326**, 425.
Machado, M.E., Xiao, Y.C., Wu, S.T., Prokaki, Th. and Dialetis, D. 1988b, *Astrophys. J.* **326**, 451.
Martin, S.F. 1984, in *Small-Scale Dynamical Processes in Quiet Stellar Atmospheres*, ed. S.L. Keil (Sacramento Peak, Sunspot, New Mexico: National Solar Observatory), p. 30.
Martin, S.F. 1988, *Solar Phys.* **117**, 243.
Mauas, P.J.D. 1990, *Astrophys. J. Suppl.* **74**, 609.
Mauas, P.J.D., Machado, M.E. and Avrett, E.H. 1990, *Astrophys. J.* **360**, 715.
Maunder, E.W. 1890, *Mon. Not. Royal Astron. Soc.* **50**, 251.
Maunder, E.W. 1894, *Knowledge* **17**, 173.
Maunder, E.W. 1922, *J. Brit. Astron. Assoc.* **32**, 140.
McTiernan, J.M. and Petrosian, V. 1991, *Astrophys. J.* **379**, 381.
Meyer, P., Parker, E.N. and Simpson, J.A. 1956, *Phys. Rev.* **104**, 768.
Miller, J.A. 1991, *Astrophys. J.* **376**, 342.
Miller, J.A. and Steinacker, J. 1992, *Astrophys. J.* **399**, 284.
Moore, R.L., McKenzie, D.L., Svestka, Z., Widing, K.G., et al. 1980, in *Solar Flares, A Monograph from Skylab Solar Workshop II*, ed. P.A. Sturrock (Boulder: Colorado Associated University Press), p. 341.
Moore, R.L., Musielak, Z.E., Suess, S.T. and An, C.H. 1991, *Astrophys. J.* **378**, 347.
Morfill, G.E., Meyer, P. and Lüst, R. 1985, *Astrophys. J.* **296**, 670.
Mouschovias, T. Ch. 1974, *Astrophys. J.* **192**, 37.
Mullan, D.J., Herr, R.B. and Bhattacharjee, A. 1992, *Astrophys. J.* **391**, 265.
Nesme-Ribes, E., Ferreira, E.N. and Vince, I. 1993, *Astron. Astrophys.* **276**, 211.
Nesme-Ribes, E. and Mangeney, A 1992, *Radiocarbon* **34**, 263.
Ness, N.F. 1969, *Rev. Geophys.* **7**, 97.
O'Brien, K., Wu, L. and Nagel, S.R. 1991, *Phys. Rev. A* **43**, 2052.
Parker, E.N. 1957a, *Phys. Rev.* **107**, 830.
Parker, E.N. 1957b, *Phys. Fluids* **1**, 171.
Parker, E.N. 1958a, *Astrophys. J.* **128**, 664.
Parker, E.N. 1958b, *Phys. Rev.* **109**, 1874.
Parker, E.N. 1960, *Astrophys. J.* **132**, 821.
Parker, E.N. 1963, *Interplanetary Dynamical Processes* (New York: John Wiley), Chapter XV.
Parker, E.N. 1965a, *Astrophys. J.* **142**, 584.
Parker, E.N. 1965b, in *Proc. Ninth Int. Conf. Cosmic Rays, Accel.* (London: The Institute of Physics and the Physical Society), **9**, 126.
Parker, E.N. 1965c, in *Stars and Stellar Systems (Vol. XII: Nebulae and Interstellar Matter)*, ed. B.M. Middlehurst and L.H. Aller (Chicago: University of Chicago Press), Chapter 16, pp. 741–747.
Parker, E.N. 1966a, *Astrophys. J.* **144**, 916.
Parker, E.N. 1966b, *Astrophys. J.* **145**, 811.
Parker, E.N. 1967a, *J. Geophys. Res.* **72**, 2315.
Parker, E.N. 1967b, *J. Geophys. Res.* **72**, 4365.
Parker, E.N. 1967c, *Astrophys. J.* **149**, 535.
Parker, E.N. 1968a, *J. Geophys. Res.* **71**, 3607.
Parker, E.N. 1968b, *Physics of the Magnetosphere*, ed. R.L. Carovillano, J.F. McClay and H.R. Radoski (Dordrecht: Reidel), p. 3.
Parker, E.N. 1968c, in *Stars and Stellar Systems (Vol. XII: Nebulae and Interstellar Matter)*, ed. B.M. Middlehurst and L.H. Aller (Chicago: University of Chicago Press), Chapter 14.

Parker, E.N. 1968d, *Astrophys. J.* **154**, 875.

Parker, E.N. 1969, *Space Sci. Rev.* **9**, 651.

Parker, E.N. 1979, *Cosmical Magnetic Fields* (Oxford: Clarendon Press), pp. 141–146, 325–341.

Parker, E.N. 1982a, *Astrophys. J.* **256**, 292.

Parker, E.N. 1982b, *Astrophys. J.* **256**, 302.

Parker, E.N. 1983, *Geophys. Astrophys. Fluid Dyn.* **24**, 245.

Parker, E.N. 1984, *Astrophys. J.* **283**, 343.

Parker, E.N. 1985, *Astrophys. J.* **294**, 57.

Parker, E.N. 1987, *Solar Phys.* **111**, 297.

Parker, E.N. 1991, *Astrophys. J.* **372**, 719.

Parker, E.N. 1992a, in *Solar Wind Seven*, ed. E. Marsch and R. Schwenn (Oxford: Pergamon Press), p. 79.

Parker, E.N. 1992b, *Astrophys. J.* **401**, 137.

Parker, E.N. 1992c, in *Sunspots: Theory and Observations*, ed. J.H. Thomas and N.O. Weiss (Dordrecht: Kluwer), p. 413.

Poletto, G., Gary, G.A. and Machado, M.E. 1993, *Solar Phys.* **144**, 113.

Porter, J.G. and Moore, R.L. 1988, in *Proc. 9th Sacramento Peak Summer Symposium*, ed. R.C. Altrock (Sacramento Peak, Sunspot, New Mexico: National Solar Observatory), p. 30.

Reid, G.C. 1991, *J. Geophys. Res.* **96**, 2835.

Ribes, E., Beardsley, B., Brown, T.M., Delache, Ph., Laclare, F., Kuhn, J.R. and Leister, N.V. 1991, in *The Sun in Time*, ed. Sonnett, C.P., Giampapa, M.S. and Matthews, M.S. (Tucson: University of Arizona Press), p. 59.

Rijnbeek, R.P., Farrugia, C.J., Southwood, D.J., Dunlop, M.W., Mier-Jedizejowicz, W.A.C., Chaloner, C.P., Hall, D.S. and Smith, M.F. 1987, *Planet. Space Sci.* **35**, 871.

Rosenbluth, M.N., Dagazian, R.Y. and Rutherford, P.H. 1973, *Phys. Fluids* **16**, 1894.

Ryan, J.M. and Lee, M.A. 1991, *Astrophys. J.* **368**, 316.

Sakai, J.I. 1990a, *Astrophys. J. Supplement* **73**, 321.

Sakai, J.I. 1990b, *Astrophys. J.* **365**, 354.

Sakai, J.I. and de Jager, C. 1991, *Solar Phys.* **134**, 329.

Sakai, J.I. and Ohsawa, Y. 1987, *Space Sci. Rev.* **46**, 113.

Saunders, M.A., Russell, C.T. and Schopke, N. 1984, *Geophys. Res. Letters* **11**, 131.

Savage, B.D. 1990, in *The Evolution of the Interstellar Medium* (Astron. Soc. Pac. Conf. Ser. **12**), ed. L. Blitz, p 33.

Schaefer, B.E. 1991, *Astrophys. J. Letters* **366**, L39.

Schatten, K.H., Mayr, H.G., Omidvar, K. and Maier, E. 1986, *Astrophys. J.* **311**, 460.

Schindler, K. and Birn, J. 1986, *Space Sci. Rev.* **44**, 307.

Schmidt, M. 1957, *Bull. Astron. Inst. Netherlands* **13**, 247.

Scholer, M. 1988, *Geophys. Res. Letters* **15**, 291.

Shapiro, M.M. and Silberberg, R. 1970, *Ann. Rev. Nucl. Sci.* **20**, 323.

Shibata, K., Nozawa, S., Matsumoto, T., Sterling, A.C. and Tajima, T. 1990, *Astrophys. J. Letters* **351**, L25.

Shibata, K., Tajima, T. and Matsumoto, R. 1990, *Phys. Fluids B* **2**, 1989.

Shibata, K., Tajima, T., Matsumoto, R., Horiuchi, T., Hanawa, T., Rosner, R. and Uchida, Y. 1989a, *Astrophys. J.* **338**, 471.

Shibata, K., Tajima, T., Steinolfson, R.S. and Matsumoto, R. 1989b, *Astrophys. J.* **345**, 584.

Similon, P.L. and Sudan, R.N. 1989, *Astrophys. J.* **336**, 442.

Similon, P.L. and Sudan, R.N. 1990, *Ann. Rev. Fluid Mech.* **22**, 317.

Skinner, S.L. 1991, *Astrophys. J.* **368**, 272.

Sofia, S., Oster, L. and Schatten, K. 1982, *Solar Phys.* **80**, 87.

Spicer, D.S. 1976, *Naval Res. Lab Rept. 8036* (Washington, DC: Naval Res. Lab.).

Spicer, D.S. 1977, *Solar Phys.* **53**, 305.

Spicer, D.S. 1982, *Space Sci. Rev.* **31**, 351.

Spiegel, E.A. and Weiss, N.O. 1980, *Nature* **287**, 616.

Sporer, F.W.G. 1889. *Bull. Astron.* **6**, 60.

Spruit, H.C. 1979, *Solar Phys.* **61**, 363.

Spruit, H. 1991, in *The Sun in Time*, ed. Sonnett, C.P., Giampapa, M.S. and Matthews, M.S. (Tucson: University of Arizona Press), p. 118.

Spruit, H.C., Title, A.M. and Van Ballgooijen, A.A. 1987, *Solar Phys.* **110**, 115.

Stern, D.P. 1984, *Space Sci. Rev.* **39**, 193.

Strauss, H.R. and Otani, N.F. 1988, *Astrophys. J.* **326**, 418.

Svestka, Z. 1976, *Solar Flares* (Dordrecht: D. Reidel).

Svestka, Z. 1986, in *The Lower Atmosphere of Solar Flares*, ed. D.F. Neidig (Tucson: National Solar Observatory), p. 332.

Sweet, P.A. 1958, *Nuovo Cim.* **8** (10), 188.

Tajima, T., Sakai, J., Nakajima, H., Kosugi, T., Brunel, F. and Kundu, M.R. 1987, *Astrophys. J.* **321**, 1031.

Thomas, G.E. and Krassa, R.F. 1971, *Astron. Astrophys.* **11**, 218.

Thomsen, M.F. Stansberry, J.A., Bame, S.J., Fuselier, S.A. and Gosling, J.T. 1987, *J. Geophys. Res.* **92**, 12, 127.

Title, A.M., Tarbell, T.D., Topka, K.P., Ferguson, S.H. and Shine, R.A. and the SOUP Team 1989, *Astrophys. J.* **336**, 475.

Title, A.M., Topka, K.P., Tarbell, T.D., Schmidt, W., Balke, C. and Scharmer, G. 1992, *Astrophys. J.* **393**, 782.

Tsinganos, K.C. 1979, *Astrophys. J.* **231**, 260.

Vallerga, J.V., Vedder, P.W., Craig, N. and Welsh, B.Y. 1993, *Astrophys. J.* **411**, 729.

Van Hoven, G. 1976, *Solar Phys.* **49**, 95.

Van Hoven, G. 1979, *Astrophys. J.* **2325**, 572.

Van Hoven, G. 1981, in *Solar Flare Magnetohydrodynamics*, ed. E.R. Priest (New York: Gordon and Breach), pp. 248–271.

Walbridge, E. 1967, *J. Geophys. Res.* **72**, 5213.

Weller, C.S. and Meier, R.R. 1974, *Astrophys. J.* **193**, 471.

Weller, C.S. and Meier, R.R. 1979, *Astrophys. J.* **227**, 816.

Welsh, B.V., Vedder, P.W., Vallerga, J.V. and Craig, N. 1991, *Astrophys. J.* **381**, 462.

White, R.B., Monticello, D.A., Rosenbluth, M.N. and Waddell, B.V. 1977, *Astrophys. J.* **20**, 800.

Willson, R.C. and Hudson, H.S. 1991, *Nature* **351**, 42.

Winglee, R.M., Dulk, G.A., Bornman, P.L. and Brown, J.C. 1991, *Astrophys. J.* **375**, 382.

Withbroe, G.L. 1988, *Astrophys. J.* **325**, 442.

Yi, Z. and Engvold, O. 1993, *Solar Phys.* **144**, 1.

Yu, G. 1974, *Astrophys. J.* **194**, 187.

Zhang, Q., Soon, W.H., Baliunas, S.L., Lockwood, G.W., Skiff, B.A. and Radick, R.R. 1994, *Nature* (in press).

Zhu, X. 1994a, *J. Geophys. Res.* (in press).

Zhu, X. 1994b, *Geophys. Res. Letters* (submitted).

Zirker, J.B. and Cleveland, F.M. 1993, *Solar Phys.* **144**, 341

Index

Action at a distance, 69, 70
Ampere's law, 38, 40
Anomalous resistivity, 50–54
Aurora, 368, 379

Biermann battery, 43
Bifurcation of fields, 5, 173, 176–178, 186–194, 198, 199, 202–204
Biot–Savart law, 38
Boltzmann equation, 30
Braiding, 23

Characteristics, 4, 5, 21, 24, 61–66
Clusters of galaxies, 367
Collisionless plasma, 37, 41–44
Continuity equation, 31
Continuous magnetic fields, 72–75, 114–123, 166–170, 205–207
Coronal heating, 8, 330, 331
Coronal heating by quasi-static fields, 348–351
  dissipation, 353–356
  energy input to fields, 351–353
Coronal heating by waves, 342–348
Coronal hole, 335
Coronal mass ejection, 6
Cosmic rays, 385–388
Current sheets, 2, 11, 16, 18, 23, 24, 63, 64, 86, 87, 108–114, 154–156, 159–160
  internal structure, 286–289, 320
  topology, 225–232, 289–291
  with continuing shear, 289–291
Cyclotron radius, 45, 53, 266

Debye radius, 36
Diffusion coefficient for heavy ions, 334
Discontinuity surfaces, 49, 108–114, 125–126, 210–212, 214–224
Displaced flux bundles, 219–224, 238, 252–254, 281–285
Dynamical dissipation, 292–298
Dynamical nonequilibrium, 87, 306–308

Eddington–Sweet circulation, 43
Eikonal equation, 175
Electric double layers, 51–54
Electromagnetic energy, 34
Electromagnetic momentum, 33
Electron conduction velocity, 49, 50
Elliptic equations, 61
Energy equation, 33, 3
Euler equations (hydrodynamics), 16
Euler equations (variational principle), 175, 196

Fermat's principle, 175
Field lines, 26, 123, 124
  near neutral points, 233, 295, 296
  stochastic, 26, 75–76, 95–99
  topology, 101–104, 120–123, 198–201, 211–212
Field rotation effect, 182–184
Flares, 7, 14, 370–373
  internal structure, 370–373, 377, 378
  microflare, 331, 336–338, 357, 369, 370
  milliflare, 357
  nanoflare, 2, 11, 16, 356–361, 367
  size distribution, 369, 375–377
Fluid motions, 255
Fluid pressure, 16, 55, 64, 180
  around gaps, 234
  at tangential discontinuities, 227–229
Flux surface, 173
Force-free field, 19, 56, 99–114
Free energy, 131, 137–145

Galactic halo, 385–388
Galactic X-ray emission, 2, 388
Galaxies, 367
Gap in flux surface, 5, 173, 176–178, 186–194, 198, 199, 202, 210–212, 234, 237
Geomagnetic field, 368, 378
Geomagnetic substorm, 15
Geomagnetic tail, 381
Grad–Shafranov equation, 22

Hamiltonian formulation, 21, 26, 90–94
Helicity, 19
Heliosphere, 367
Heliospheric wake, 367, 383
High-speed sheets, 202–205, 268–281

Index of refraction, 175
  motion of, 202–205
Instabilities, 156–160
Integral constraints, 78–83
Interstellar gas clouds, 367
Invariant magnetic fields, 73–75
Inviscid fluid, 255–261

Ion holes, 51
Isobaric surfaces, 180

Kinematic viscosity, 266
Kolmogoroff–Arnold–Moser theorem, 94

Laboratory plasma, 391, 392
Laplace's equation, 61, 68
Lundquist number, 40

Magnetic discontinuities, 63, 64, 154
Magnetic fibrils, 7, 56
Magnetic fields, 29
  continuous, 19, 57, 58
  cusp, 208
  solar, 6
Magnetic Reynolds number, 40
Magnetohydrodynamic equations, 29, 35, 39, 41–43
  deviations from, 45–47
Magnetospheric convection, 378–380
Magnetostatic equation, 16, 55
Magnetostatic equilibrium, 55–58, 95–123
Magnetostatic theorem, 1, 9, 15, 18
Maxwell equations, 29
Maxwell stress tensor, 1, 2, 13, 34, 38, 88
Mean free path, 266
Momentum equation, 32, 33

Nanoflare, 2, 11, 16, 356–361, 367, 373
Network activity, 336–338
Neutral points, 125
  X-type, 8, 9, 16, 132, 134, 137, 145–153, 169, 173
  Y-type, 16, 134, 146–153, 169
Non-Euclidean space, 174–176, 194–197
Noninsulating fluid, 37
Non-invariant magnetic fields, 75
Numerical simulations, 154–156, 158–160, 172

O–B associations, 384
Observational tests, 363, 388, 389–391
Optical analogy, 5, 25, 173–224, 268–272

Parallel electric field, 47
Particle drifts, 42
Particle energy, 30
Particle momentum, 30
Perturbations, 77, 84, 85
Planetary magnetic fields, 367
Planetary magnetosphere, 12
Plasma turbulence, 50
Poisson equation, 96
Poynting's theorem, 33, 34
Poynting vector, 33, 34, 39, 40

Radiative cooling, 333
Rapid reconnection, 2, 8, 11–16, 88, 89
  rate, 13, 14, 299–303, 305
Refraction of field lines, 174–194, 198–201, 268–272
Resistive diffusion, 39, 286–327
Resistive dissipation, 10, 11, 131, 137–145
Resistive fluid, 38
Resistive instability, 11, 303–305
Resistivity, 265
  nonuniform, 309–320, 321–327

Shock fronts, 49
Solar chromosphere, 338
Solar corona, 328, 329, 331–340
  heat input, 339–341
  loops, 339, 361–363
  variability, 336–338
Solar irradiance variations, 390, 391
Supernovae, 384
Surface locus, 173

Tangential discontinuity, 2, 11, 16, 18, 23, 24, 63, 64, 86, 87
  active, 10, 11, 18
  displaced flux bundles, 238, 252–259
  internal structure, 225–227, 229
  intersection of, 240–252
  irregular arrays, 236
  resistive dissipation, 11
  topology, 225
Tearing instability, 303–305
Thermal conductivity, 332
Thermal diffusion, 334
Time-dependent fields, 170–172
Topological invariants, 66
Topology of current sheets, 225
  changes, 321–327
  topology of field lines, 101–104, 120–123, 198–201, 211–212
Torsion coefficient, 19
Total internal reflection, 176
Turbulence, 27
Twisted flux bundles, 161–166

Unlimited winding, 66

Viscosity, 266
Viscous battery, 43
Viscous fluid, 261–264
Vortex sheets, 27
Vorticity equation, 27

X-ray astronomy, 328–330
X-ray corona, 7, 11
X-rays, 2
  galactic, 7
  solar, 2
  stellar, 7